Mass Spectrometry in Drug Metabolism

Mass Spectrometry in Drug Metabolism

Edited by
Alberto Frigerio
"Mario Negri" Institute for Pharmacological Research
Milan, Italy

and
Emilio L. Ghisalberti
Department of Organic Chemistry
University of Western Australia
Nedlands, Western Australia

PLENUM PRESS • NEW YORK AND LONDON

Library of Congress Cataloging in Publication Data

International Symposium on Mass Spectrometry in Drug Metabolism, Mario Negri Institute, 1976.
Mass spectrometry in drug metabolism.

"Proceedings of an International Symposium on Mass Spectrometry in Drug Metabolism held at the 'Mario Negri' Institute for Pharmacological Research, June, 1976."

Includes index.
1. Drug metabolism—Congresses. 2. Drugs—Analysis—Congresses. 3. Mass spectrometry—Congresses. I. Frigerio, Alberto. II. Ghisalberti, Emilio L. III. Title. [DNLM: 1. Spectrum analysis, Mass—Congresses. 2. Drugs—Metabolism—Congresses. 3. Drugs—Analysis—Congresses. 4. Drug abuse—Congresses. QV38 I605m 1975]
RM301.I57 1976 615'.7 76-53013
ISBN-13: 978-1-4613-4153-6 e-ISBN-13:978-1-4613-4151-2
DOI: 10.1007/978-1-4613-4151-2

Proceedings of an International Symposium on Mass Spectrometry in Drug Metabolism held at the "Mario Negri" Institute for Pharmacological Research, June, 1976

Softcover reprint of the hardcover 1st edition 1977
A Division of Plenum Publishing Corporation
227 West 17th Street, New York, N.Y. 10011

Foreword

When a dose of drug is administered, three main phases of drug action may be distingushed. In the "pharmaceutical" phase the dosage form disintegrates, the active substance dissolves and becomes available for absorption. The second phase ("pharmacokinetic" phase) includes absorption, distribution, metabolism, and excretion. That fraction of the dose which finally reaches the circulation after absorption will be available for biological action in the third, or "pharmacodynamic" phase when the drug reaches the target tissues and a drug-receptor interaction takes place. The objectives in studies of drug metabolism are: (a) to identify the pathways by which drugs are transformed in the body; (b) to ascertain quantitatively the importance of each pathway and intermediate; (c) to identify and quantify endogenous constituents influenced by the drug or its metabolites which may interfere with common metabolic processes.

Since metabolites usually differ from their precursors by only a single chemical group, the resulting metabolic pathways generally consist of a series of closely related compounds. Mass spectrometry is uniquely suited for the analysis of drugs and metabolites for several reasons: only a minimal amount of sample preparation is needed, closely related compounds can be analyzed in a single step, structures can often be deduced directly from the mass spectra without the need for pure reference spectra, and constituents can be quantified with relative ease even when present in fractional nanogram quantity. Recent developments in mass spectrometric and computer instrumentation and analytical techniques elevated mass spectrometry to a unique status: high sensitivity and high selectivity are provided simultaneously and are combined with general applicability to most drugs currently in use.

This volume presents the papers given during a 3-day international symposium on the use of mass spectrometry in the study of drug metabolism held in Milan, Italy, in June 1976. Most papers are contributed scientific papers describing original research. Several papers were invited reviews covering selected areas of techniques and applications. Three major areas were covered: drug metabolism, developments in methodology, and drug abuse. In the field of drug metabolism, sessions were held on the identification of metabolites, on quantification studies, and on the reactivity of intermediates. Special sessions in methodology development were devoted to new ionization techniques, to the use of stable isotopes in drug metabolism, and to ancillary instrumentation such as capillary columns, pyrolyzers and, of course, computers. The last session was devoted to drug abuse.

The careful selection and mixing of review articles and original papers should provide the reader both with a comprehensive picture of the current status of mass spectrometry in drug metabolism research and a detailed insight into a variety of applications of current interest.

Thanks to the efforts of Drs. A. Frigerio and E. Ghisalberti, the organizers of the Symposium, and also the staff of the Mario Negri Institute, the host for the Symposium, the participants of this event had a unique chance to meet, listen to, and sometimes interrogate - in a pleasant and conducive environment - several leading experts in this field. It was the consensus of those who attended that mass spectrometry is indeed a uniquely suitable tool for the study of drug metabolism. We hope that readers of this text will agree.

John Roboz

Preface

The papers collected in this volume were presented at the International Symposium on Mass Spectrometry in Drug Metabolism held at the "Mario Negri" Institute for Pharmacological Research in June 1976.

One of the primary aims of the Symposium was to bring together researchers of diverse experience but with some common interests. It was felt that this would promote critical discussion and a beneficial exchange of ideas and experiences.

For a symposium to be of value to those who did not participate, the proceedings should be made available as soon as possible. The problem confronting the editors was to choose between two possibilities: to publish all the papers presented, which might have led to a significant delay in publication, or to aim for a rapid publication of most of the papers presented. We have chosen the second alternative, hoping in this way better to serve the interests of those for whom the book is intended.

The Editors

Milan,
January, 1977

Contents

A STUDY ON THE METABOLISM OF FEPRAZONE UTILISING MASS SPECTROMETRY

A. Donetti°, L.G. Dring°°, P.C. Hirom°° and R.T. Williams°°

°Istituto De Angeli, Milan. °°St.Mary's

Hospital Medical School, Paddington, London W2 1PG,U.K.

INTRODUCTION

Feprazone is a recently synthesised non-steroidal anti--inflammatory drug (1). It is structurally closely related to phenylbutazone, the difference lying in the side chain which is dimethylallyl in feprazone (see formula in Fig. 1) and butyl in phenylbutazone. For the purposes of the study the drug was labelled with ^{14}C (Fig. 1) starting with $[^{14}C]$-diethylmalonate which was condensed with dimethylallyl chloride to give diethyldimethylallyl $[2-^{14}C]$-malonate. This was further condensed with hydrazobenzene to yield $[^{14}C]$-feprazone (2).

RESULTS AND DISCUSSION

Rats were dosed orally with the drug and the ^{14}C was found to be excreted in the urine and faeces within three days (half life of urinary radioactivity ≃ 1 day, table 1). The urine was examined by thin layer chromatography in a number of systems and it is clear that very little unchanged drug was excreted by examination of the chromatogram in Fig. 2. Three main areas of radioactivity were apparent and steps were taken to identify these as follows.

Nine Wistar albino female rats were dosed with feprazone (60 mg/kg) and the 24h urine collected, pooled and subjected to the extraction scheme outlined in Fig. 3.

From previous work on phenylbutazone it was anticipated that metabolites involving oxidation of the aromatic ring and /or the side chain might be produced. A number of compounds which were

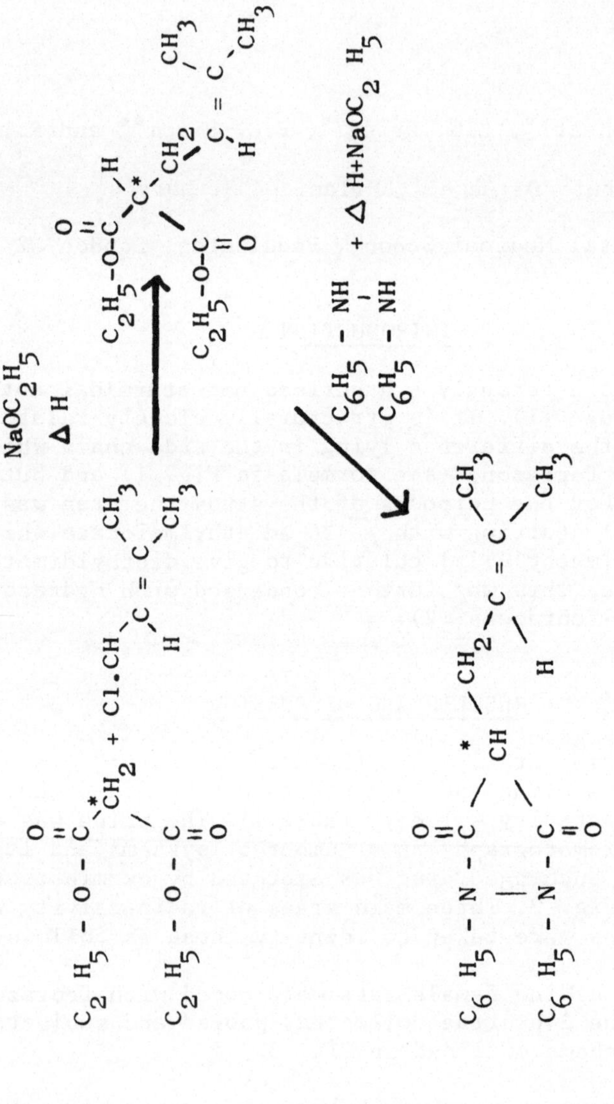

$$(C_2H_5-O-CO)_2\overset{*}{C}H_2 + Cl.CH_2-CH=C(CH_3)_2 \xrightarrow[\Delta H]{NaOC_2H_5} (C_2H_5-O-CO)_2\overset{*}{C}H-CH_2-CH=C(CH_3)_2$$

$$\xrightarrow{C_6H_5-NH-NH-C_6H_5\ +\ \Delta H + NaOC_2H_5} (C_6H_5-N-CO)_2\overset{*}{C}H-CH_2-CH=C(CH_3)_2$$

FIG. 1. Synthesis of $[^{14}C]$feprazone

TABLE 1

The excretion of ^{14}C by rats and human volunteers receiving / ^{14}C /-feprazone

Three female Wistar albino rats were dosed orally (60 mg/kg; 5.8μCi). Each man (total 3) took 200 mg orally (2.3μCi). Results are expressed as a % age of the dose and mean cumulative values are given.

	Rats ǂ		Man	
	Urine	Faeces	Urine	Faeces
Day 1	43.0	3.1	13.8	0.6
Day 2	50.2	11.9	17.2	5.3
Day 3	51.4	17.1	13.1	3.1
Day 4	51.7	17.7	8.4	2.3 °
Total	69.4		60.9†	

° One subject only.† Urine collection was continued in one subject for a further three days and another 8.7% of the activity was recovered. ǂ after Toft *et al.*(2).

possible metabolites were available, but their chromatographic properties did not match those of the metabolites; however, the mass spectra of a number of these were determined. It was noted that they always produced a characteristic cracking pattern which can be clearly seen on the examination of the spectra of feprazone and *p*-hydroxyfeprazone (Table 2). Thus feprazone which has an unsubstituted diphenylpyrazolidinedione nucleus gives major fragments of m/e 252, resulting from loss of the side chain, and m/e 183, 184 when the pyrazolidinedione nucleus is broken in half. The *p*-hydroxyfeprazone cleaves in the same way, the fragments produced however are 16 mass units higher, as would be expected (Fig. 4).

Surprisingly we were unable to identify any *p*-hydroxyfeprazone and accordingly we searched the literature for examples of compounds bearing a dimethylallyl side chain. A number of examples came to light, particularly with the terpenes such as geraniol and citral. More recently Pittman *et al.* (3) have studied the drug pentazocine and they found that either of the methyl groups could

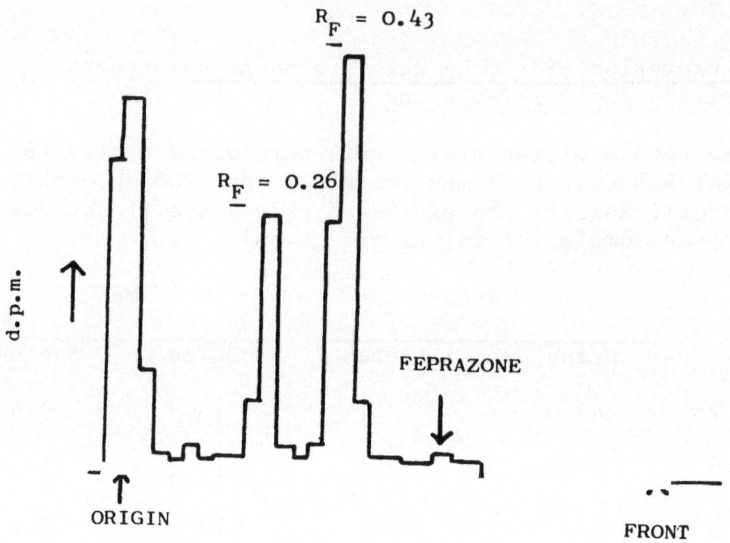

FIG. 2. Thin layer chromatography of 0-24 h rat urine Eluant $CHCl_3$/EtOH, 9 : 1 (v/v).

be oxidised to an alcohol and in the case of the trans isomer to the carboxylic acid too.

Of the two peaks eluted from the preparative t.l.c. plates only one (R_F 0.43) was amenable to GC-MS and this had a retention time of 8.5 min on an OV - 101 column at 240° with a helium flow rate of 40 ml/min. It was clear from the spectrum (Table 3) that the feprazone had been substituted with an oxygen (parent ion 336) but that the diphenylpyrazolidinedione nucleus itself had not been altered since m/e 183 and 252 were still present. It was concluded that the dimethylallyl side chain was the point of attack. Previous work on the metabolism of the dimethylallyl group suggested that a trans -alcohol was the most probable metabolite and accordingly this was synthesised and found to have identical chromatographic and mass spectral properties to the unknown.

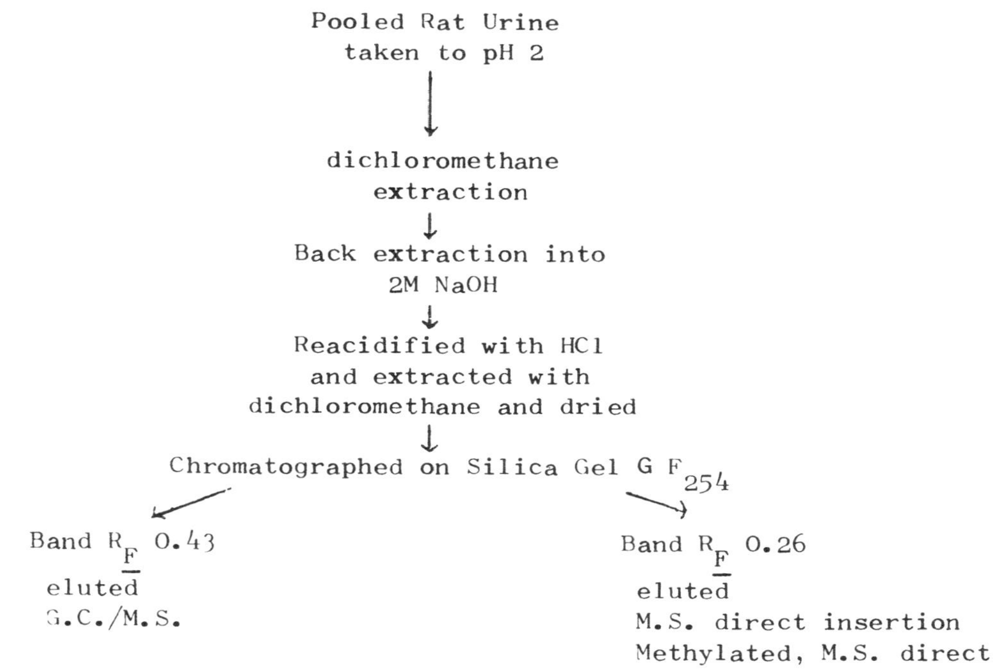

FIG. 3. Metabolite isolation scheme.

TABLE 2

Mass spectra of relevant references compounds

Only ions with intensities ≥ 10% of the base peak are given with the exception of the molecular ion when this is a lesser value. The figures represent m/e values with relative intensities in parentheses.

Feprazone (4-(3', 3'-dimethylallyl)-1,2-diphenyl-pyrazolidine-3,5-dione). 77 (100), 78 (10), 93 (11), 105 (13), 183 (95), 184 (48), 252 (52), 320 (21).
p-Hydroxyfeprazone (4-(3',3'-dimethylallyl)-1-(p--hydroxyphenyl)-2-phenylpyrazolidine-3, 5-dione). 77 (72), 79 (14), 81 (20), 83 (12), 85 (15), 93 (77), 94 (47), 95 (14), 104 (21), 105 (17), 109 (27), 119 (11), 120 (31), 121 (25), 132 (12), 133 (11), 135 (10), 199 (95), 200 (35), 268 (93), 269 (17), 336 (100), 337 (26).

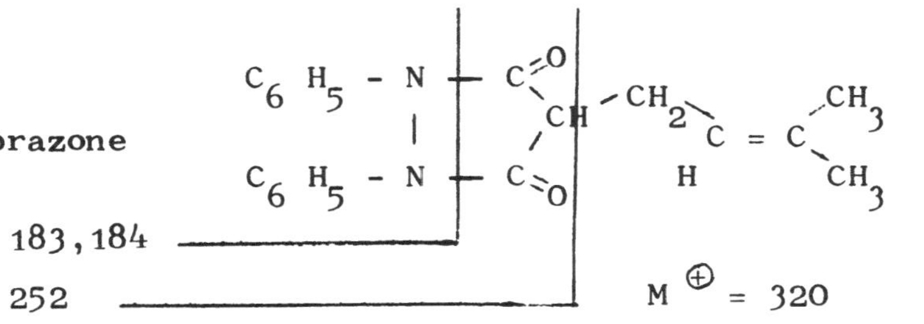

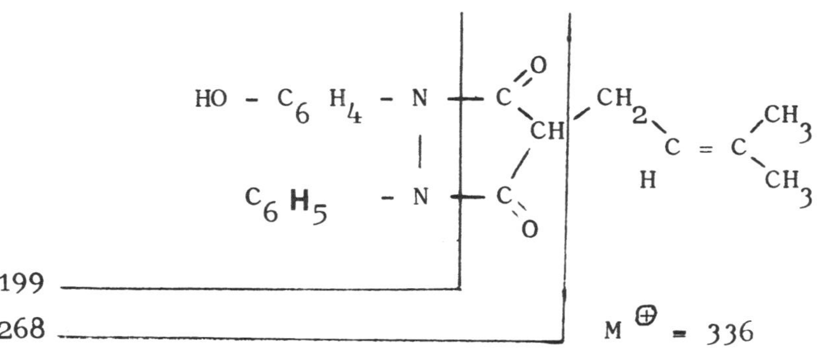

FIG. 4. Probable identity of the major fragments seen in the mass spectra of feprazone and p-OH-feprazone.

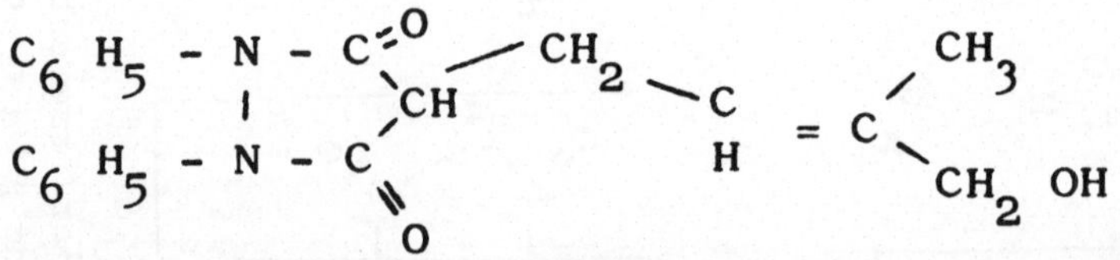

```
C6 H5 - N - C=O      CH2               CH3
        |     \    /     \            /
        |      CH          C    =   C
        |     /            H          \
C6 H5 - N - C                           CH2 OH
             \\
              O
```

trans-3'-hydroxy feprazone

```
HO.C6 H4 - N - C=O     CH2              CH3
           |     \   /     \           /
           |      CH         C   =   C
           |     /           H         \
   C6 H5 - N - C=O                       CH2 OH
```

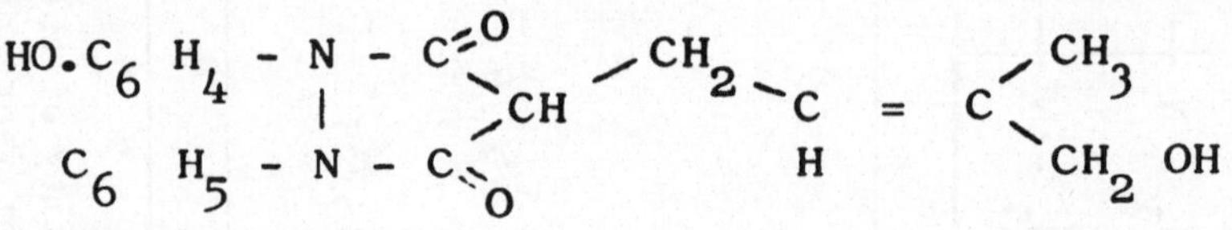

p-hydroxy-trans-3'-hydroxy feprazone

FIG. 5 Major metabolites found in rat urine.

TABLE 3

Mass Spectra of metabolites and reference compounds

In order to simplify the spectra only the main diagnostic peaks are given. The figures represent m/e values with relative intensities in parentheses (After Toft *et al.* (2)).

Metabolite R_F 0.43, T_R 8.5 min 77 (100), 93 (33), 105 (12), 183 (52), 184 (18), 252 (15), 253 (13), 336 (8).

3'-Hydroxyfeprazone (*trans*-4-(3'-hydroxymethylbut-2'-enyl)-1,2-diphenylpyrazolidine-3,5-dione) R_F 0.43, T_R 8.5 min. 77 (100), 93 (27), 105 (11), 183 (45), 184 (20), 252 (14), 253 (11), 336 (8).

Metabolite R_F 0.26 77 (23), 93 (100), 184 (3), 200 (4), 336 (3), 352 (1).

Methylated (diazomethane) metabolite R_F 0.26. 77 (100), 309 (32), 310 (40), 380 (19).

Methylated (diazomethane) *p*-methoxy-*trans*-3'-hydroxyfeprazone. 77(100), 309 (31), 310 (40), 380 (31).

Direct insertion mass spectrometry of the metabolite R_F 0.26 (Table 3) gave a molecular ion of 352 indicating the insertion of two atoms of oxygen into the feprazone nucleus. Also notable was a prominent ion at m/e 200 implying oxidation of the diphenylpyrazolidinedione nucleus, the other oxygen presumably being in the side chain. Bakke *et al.* (4) had identified a dihydroxy metabolite of phenylbutazone in the rat, therefore a dihydroxy metabolite of a diphenylpyrazolidinedione was not without precedent. With these facts in mind *p*-methoxy-3'-hydroxyfeprazone was synthesised and it and the metabolite methylated with diazomethane. The spectra of the methylated metabolite and the methylated reference compound were almost identical (Table 3). Thus the molecular ion in both spectra corresponds to the dimethylated *p*-hydroxy-3'-hydroxyfeprazone.

The two major metabolites of feprazone in rat urine appear to be *trans*-3'-hydroxyfeprazone and *p*-hydroxy-*trans*-3'-hydroxyfeprazone (Fig. 5).

Recently, human volunteers have taken the labelled drug (Table 1). The rate of excretion of ^{14}C after oral dosing was

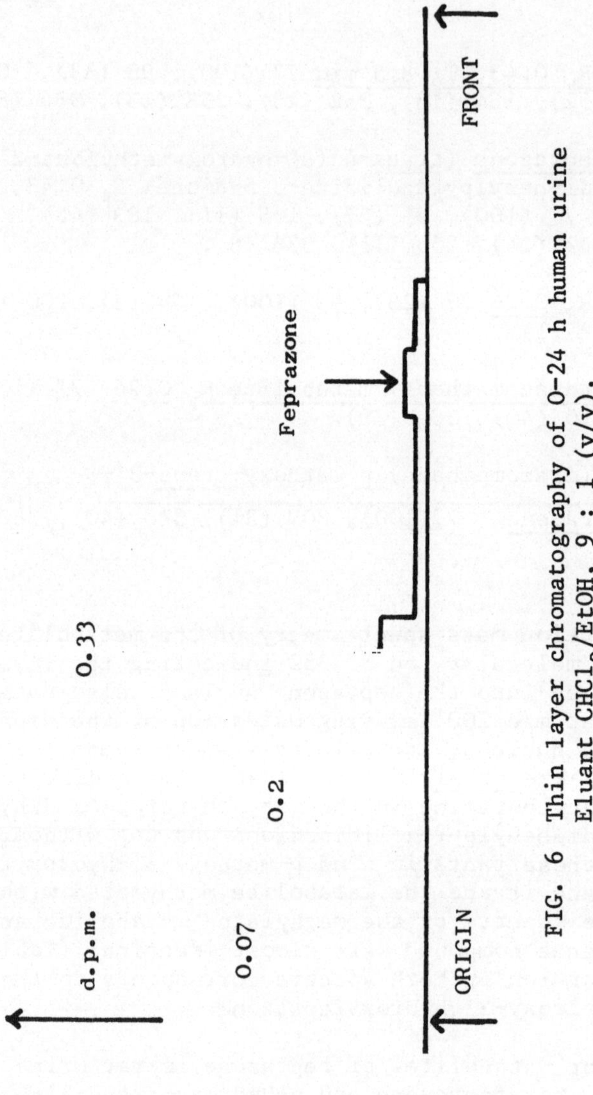

FIG. 6 Thin layer chromatography of 0-24 h human urine Eluant $CHCl_3$/EtOH, 9 : 1 (v/v).

significantly slower than in the rat and the half life of urinary ^{14}C was of the order of three days. Poor recoveries were obtained with the two subjects initially studied and in a third subject it was found that to obtain anywhere near a quantitative collection, the ^{14}C excreta had to be collected for at least one week after dosing.

Thin layer chromatographs of the first day's urine gave a pattern similar to, but not the same as the rat, since the three main peaks had different R_F values (Fig. 6).

Thus far the human metabolites have eluded characterisation largely because of the lower levels to be found in urine compared with the rat. What, however, is clear is that again very little unchanged feprazone was excreted.

REFERENCES

1) S. Casadio, G. Pala, E. Marazzi-Uberti, B. Lumachi, E. Crescenzi, A. Donetti, A. Mantegani, C. Bianchi, Arzneim. Forsch., 1972, 22, 171.
2) P. Toft, L.G. Dring, P.C. Hirom, R.T. Williams, A. Donetti, J.M. Midgley, Xenobiotica, 1975, 5, 729.
3) K.A. Pittman, D. Rosi, R. Cherniak, A.J. Merola, W.D. Conway, Biochem. Pharmacol., 1969, 18, 1673.
4) O.M. Bakke, G.H. Draffan, D.S. Davies, Xenobiotica, 1974, 4, 237.

IDENTIFICATION OF SOME URINARY METABOLITES OF IOPRONIC ACID IN HUMAN, RAT AND DOG

D. Pitrè, A. Frigerio ° and R. Maffei-Facino °°

Bracco Industria Chimica;°Istituto Ricerche Farmacologiche "Mario Negri";°°Ist.Chimica Farmaceutica,Milan;Italy

INTRODUCTION

The various techniques available for the radiological exploration of the biliary tract can be grouped under two essential methods, one direct and the other indirect (1). The former method involves filling the gallbladder and biliary passages with a radio-opaque solution.

This method is laborious and distressing to the patient and is rarely used.

The indirect method is based on the biliary excretion of a contrast medium following oral or intravenous administration. This method is particularly suitable for examining the gallbladder, diagnosing the presence of gallstones and detecting other biliary tract disturbances. Consequently this method is in general use with a preference for the oral administration of the contrast medium (1).

Biliary tract disturbances are so widespread as to represent a social issue.

In the United States alone, in 1968, examinations revealed that 15 million people had calculosis of the biliary system. Of these, some 300,000 were treated surgically with an expenditure of 500 million dollars.

The contrast media available for oral administration in biliary tract roentgenography are not exempt from undesirable side effects (nausea, vomiting, diarrhea, renal damage etc.). Consequently the development of a contrast agent which can readily be tolerated is of some importance.

Iopronic acid (Fig. 1), a compound recently developed in our

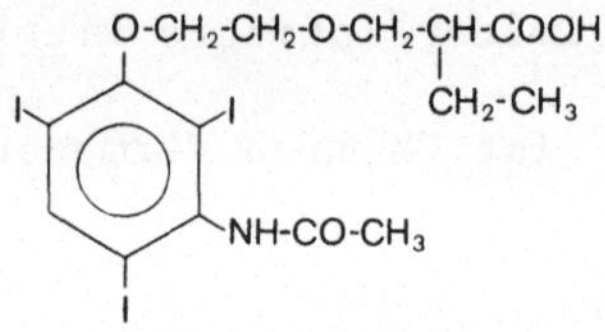

O-CH$_2$-CH$_2$-O-CH$_2$-CH-COOH

CH$_2$-CH$_3$

NH-CO-CH$_3$

M.W. 673,03 Iodine 56,57 %

B 11420

Acidum iopronicum (D.C.I., U.S.A.N.)

FIG. 1. Structure of iopronic acid.

research laboratories meets this requirement (2).

The chemical (3) and pharmacological (4) properties of iopronic acid have been described.

The purpose of this report is to present our studies on the metabolism of this new compound *in vitro*, with rat liver microsomes, and *in vivo* in the human, rat and dog.

RESULTS AND DISCUSSION

Preliminary *in vitro* experiments were carried out using rat liver homogenates, the 9000 g supernatant and microsomes. These experiments revealed that iopronic acid is metabolized by microsomal enzymes with consumption of NADPH and oxygen (5). Thin layer chromatography (TLC) and liquid chromatography of the incubation mixture revealed the presence of unchanged iopronic acid (M_3) and two metabolites (M_2 and M_4 respectively) (Fig. 2). On treatment

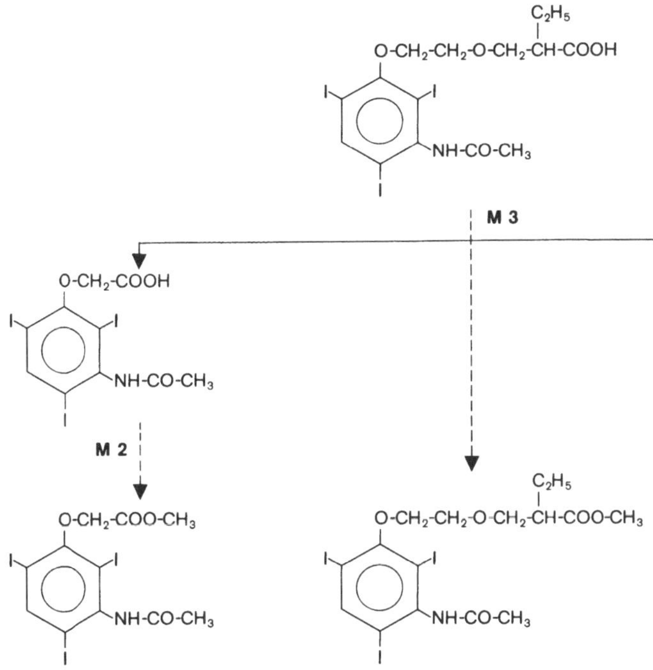

FIG. 2. Metabolism of iopronic acid with rat liver microsomes.

with diazomethane only M_2 was converted to a less polar derivative with the characteristics of a carboxylic acid ester. Preparative TLC afforded sufficient amounts of the metabolites to permit their identification.

Interpretation of the mass spectrum (M_5) of iopronic acid (Fig. 3) and its methyl ester provides a rationalization of a number of fragmentations as shown in Fig. 4. Comparison of the MS of M_2 (Fig. 5) its methyl ester (Fig. 6) and M_4 (Fig. 7) with these of iopronic acid and its methyl ester indicated that M_2 was 3-acetamido-2,4,6-triiodophenoxyacetic acid and M_4 was hydroxyethyl (3-acetamido-2,4,6-triiodophenyl)ether. Confirmation of the structures suggested for the two metabolites was obtained by

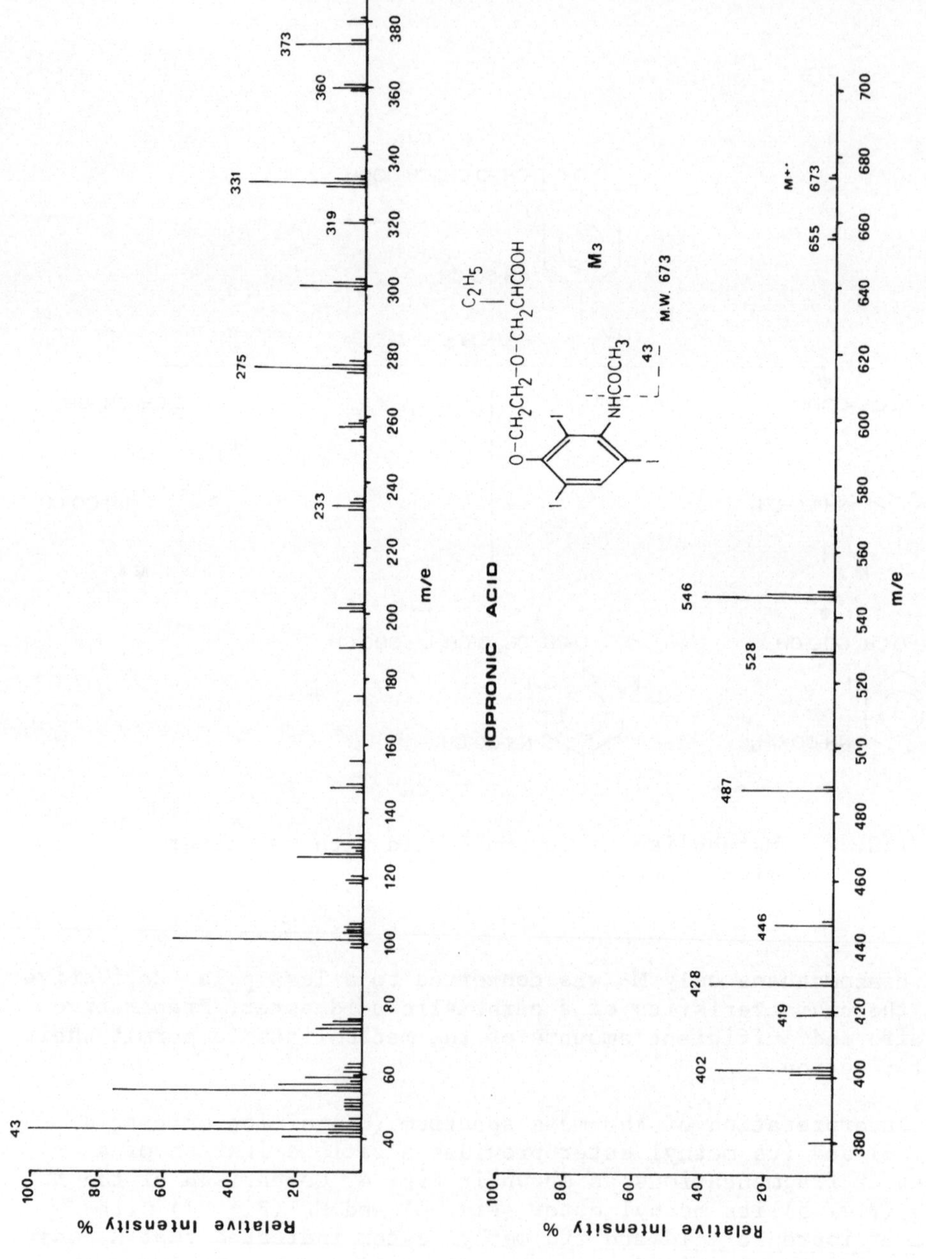

FIG. 3. Mass spectrum of iopronic acid.

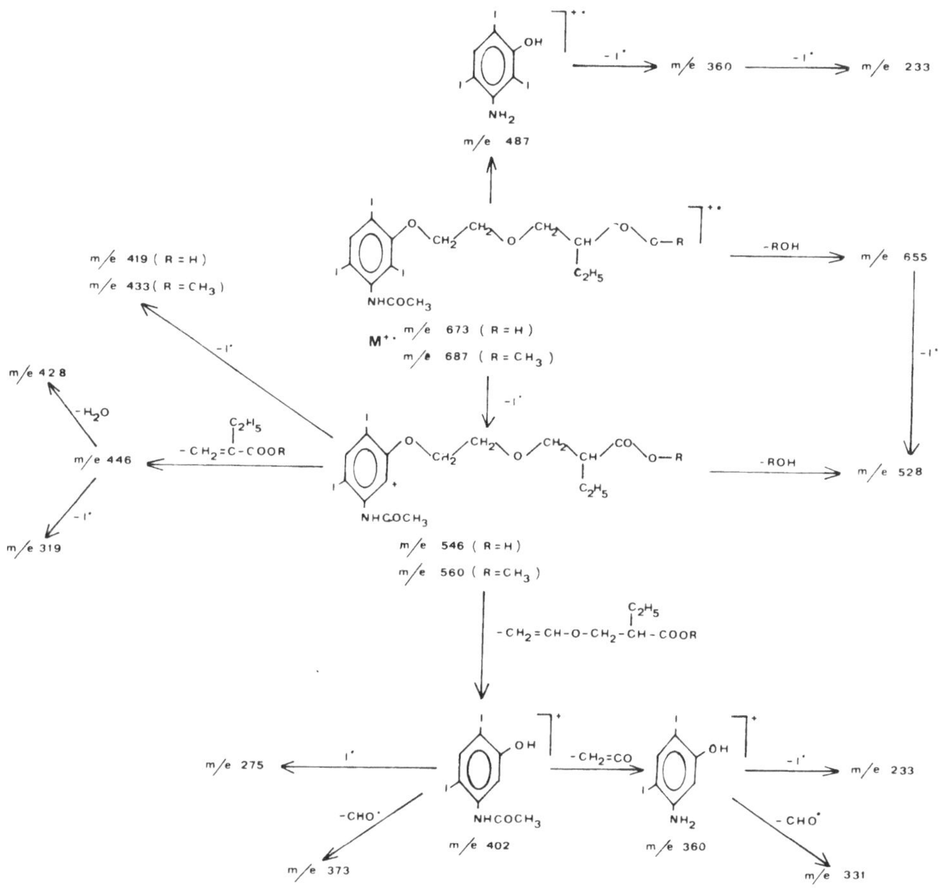

m/e 487 → -I˙ → m/e 360 → -I˙ → m/e 233

M⁺˙ m/e 673 (R = H), m/e 687 (R = CH$_3$) → -ROH → m/e 655 → -I˙ → m/e 528

m/e 546 (R = H), m/e 560 (R = CH$_3$) → -ROH → m/e 528

m/e 546/560 → -I˙ → m/e 419 (R = H), m/e 433 (R = CH$_3$)

m/e 546/560 → -CH$_2$=C(C$_2$H$_5$)-COOR → m/e 446 → -H$_2$O → m/e 428; m/e 446 → -I˙ → m/e 319

m/e 546/560 → -CH$_2$=CH-O-CH$_2$-CH(C$_2$H$_5$)-COOR → m/e 402

m/e 402 → I˙ → m/e 275; m/e 402 → -CHO˙ → m/e 373; m/e 402 → -CH$_2$=CO → m/e 360

m/e 360 → -I˙ → m/e 233; m/e 360 → -CHO˙ → m/e 331

FIG. 4. Mass spectral fragmentations of iopronic acid and its methyl ester.

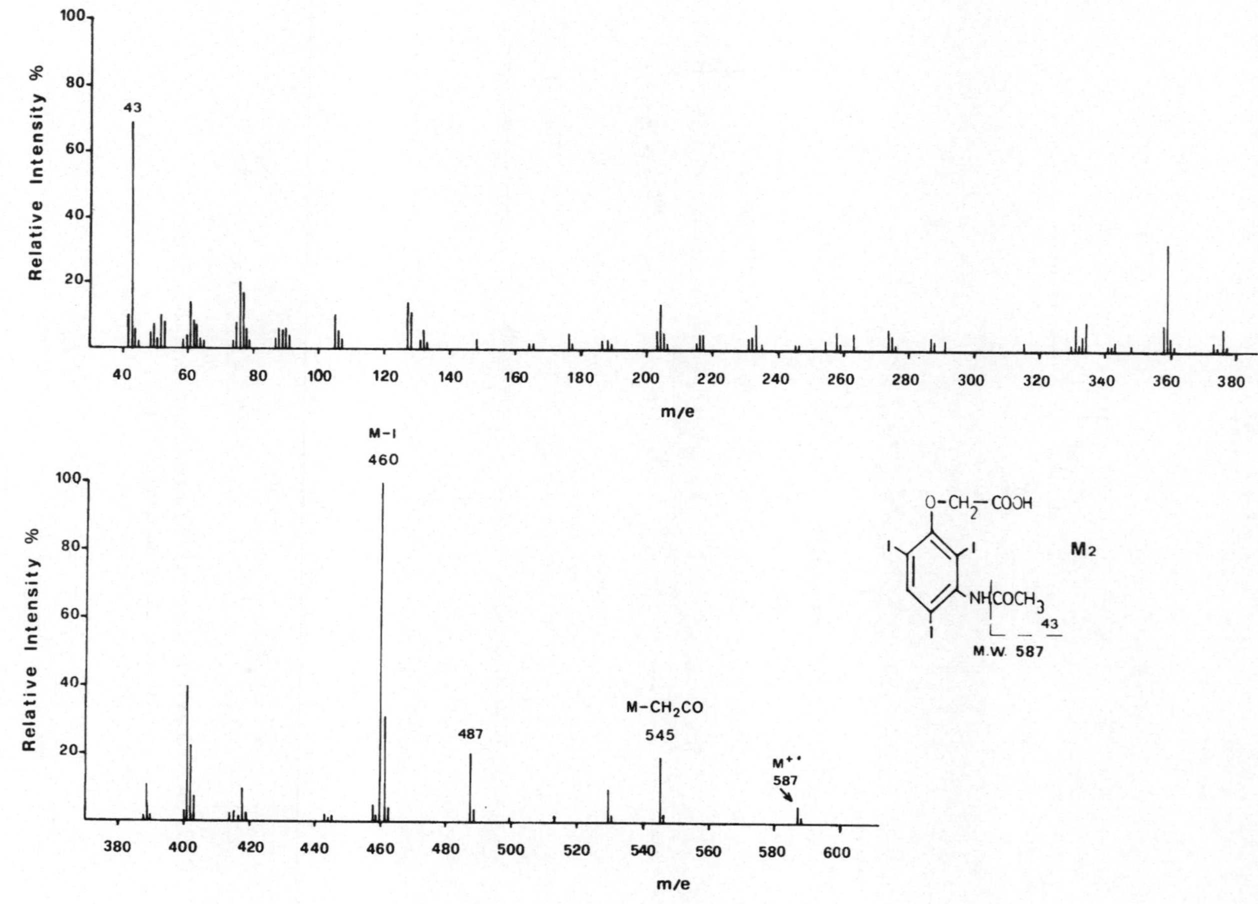

FIG. 5. Mass spectrum of 3-acetamido-2,4,6-triiodophenoxyacetic acid.

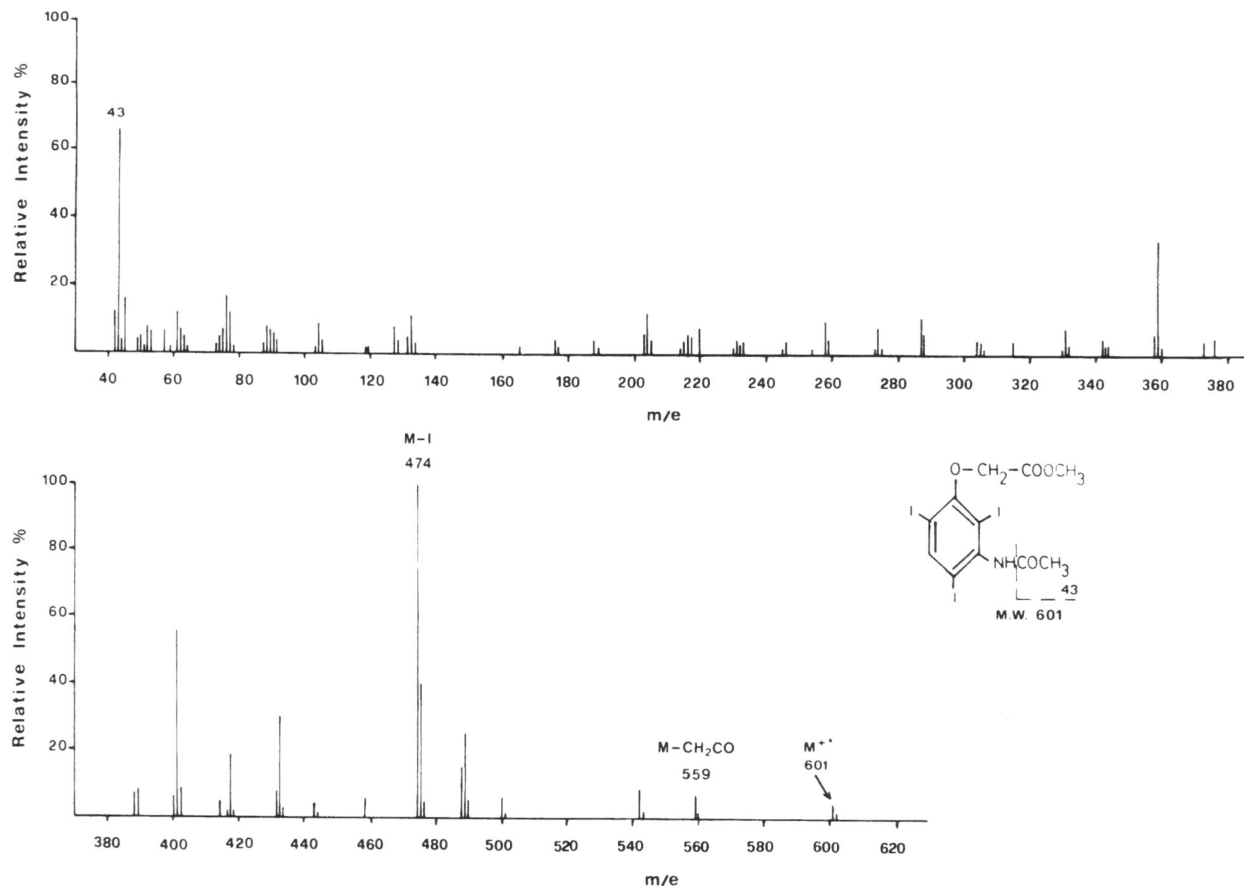

FIG. 6. Mass spectrum of the methyl ester of M_2.

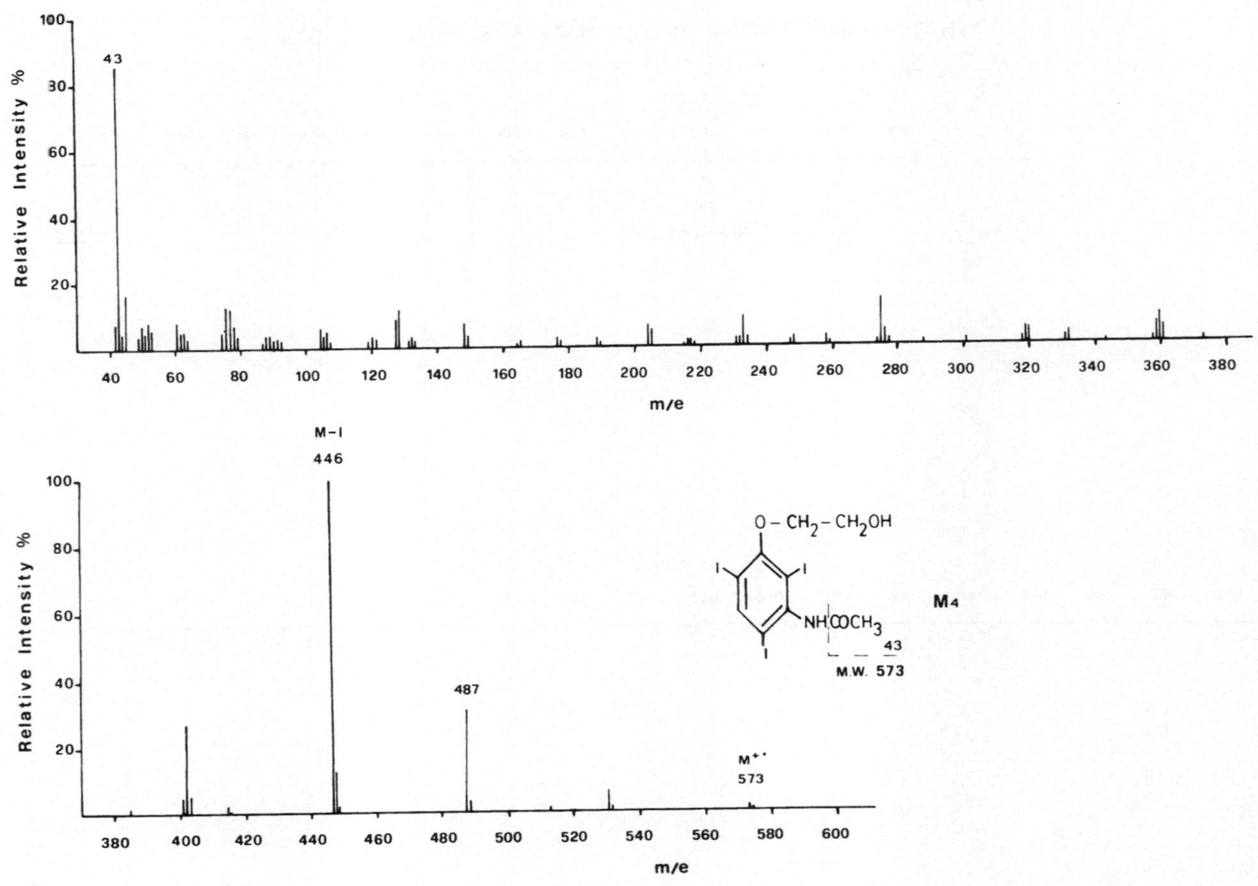

FIG. 7. Mass spectrum of hydroxyethyl(3-acetamido-2,4,6-triiodophenyl)ether.

comparison with synthetic stanaards. Both metabolites arise from cleavage of the bond of the ether in the side chain.

The phenol ether bond appears to resist enzymatic cleavage since 3-acetamido-2,4,6-triiodophenol was not detected as a metabolite. Also important is the observation that the metabolites do not include de-iodinization products. Oral administration of iopronic acid in amounts of 100 mg/kg to Wistar rats is followed by excretion of products containing iodine representing 13% (as iodine) in the urine and 61% in the faeces over the first 24 hrs (6). When the same dose is given intravenously to cannulated rats, 8% is excreted in the urine and 82% in the bile in the first 3 hours.

The urine of rats which had been administered iopronic acid was shown, by means of TLC and high pressure liquid chromatography (HPLC) to contain 4 iodinated metabolites together with the original compound. Two of these were identical with M_2 and M_4 mentioned previously.

The other two (M_1 and M_6, Fig. 8) when treated with naphtho-

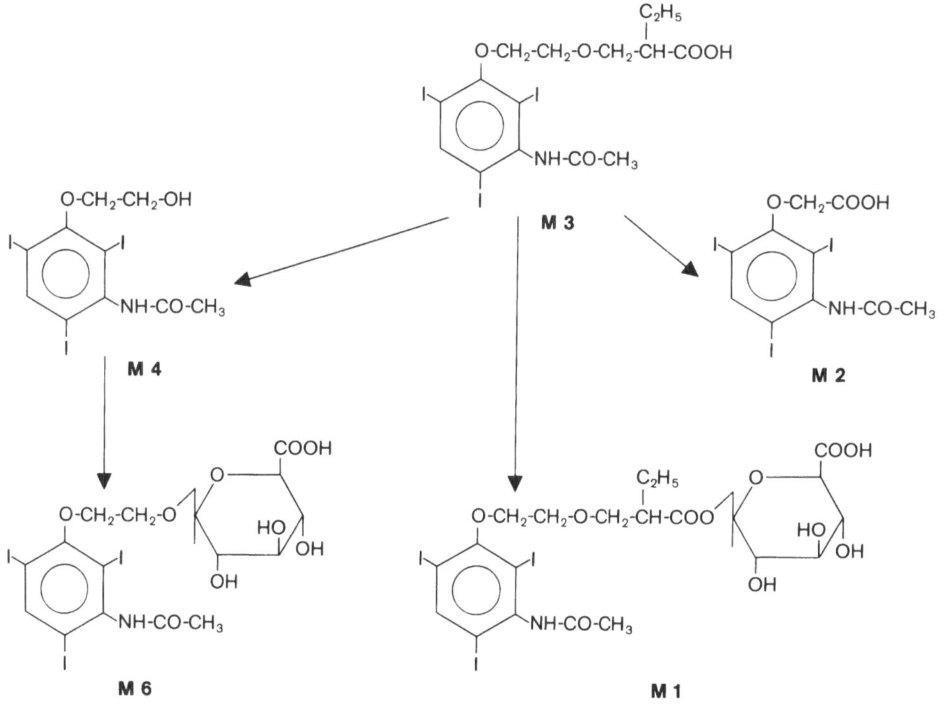

FIG. 8. Metabolism of iopronic acid in rat.

resorcinol gave a reaction characteristic of uronic acids. Fig. 9

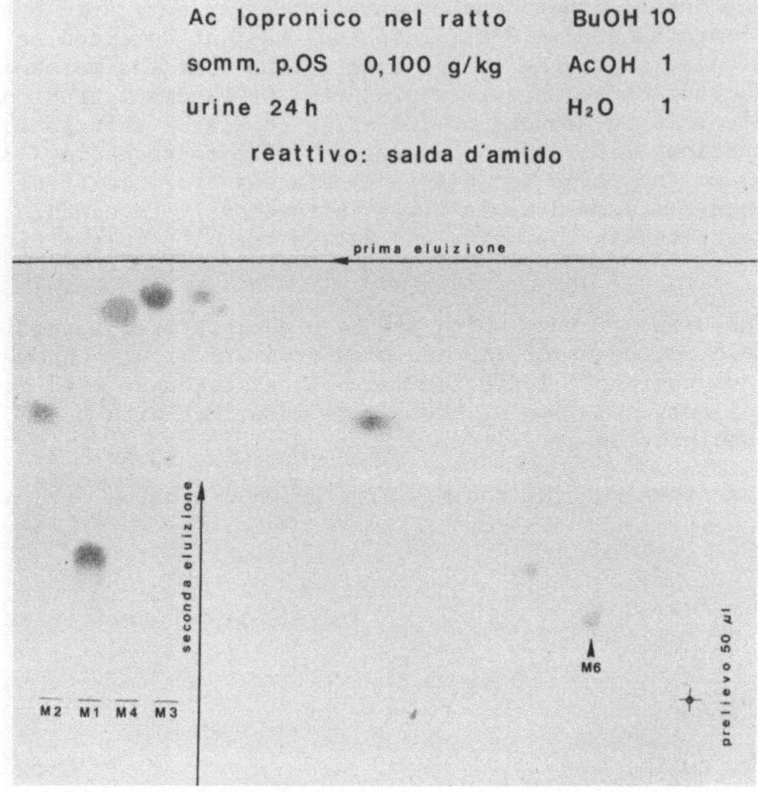

FIG. 9. Two-dimensional TLC of the metabolites of iopronic acid in rat.

shows a two-dimensional TLC of an urine extract with iopronic acid (M_3), 3-acetamido-2,4,6-triiodophenoxyacetic acid (M_2) and 2-hydroxyethyl(3-acetamido-2,4,6-triiodophenyl)ether (M_4) clearly visible. The two metabolites are the same as those obtained from the _in vitro_ incubation and appear typical of phase 1 of the "metabolic pathway" according to Williams (7).

The other two metabolites identified (M_1 and M_6) are glucuronic acid conjugates since hydrolysis by β-glucuronidase produces iopronic acid (M_3) and 2-hydroxylethyl(3-acetamido-2,4,6-triiodophenyl)ether (M_4) respectively. The structure of M_6 was confirmed by TLC, HPLC and by comparison with an authentic sample prepared as shown in Fig. 10. The mass spectrum of M_6 is shown in Fig. 11.

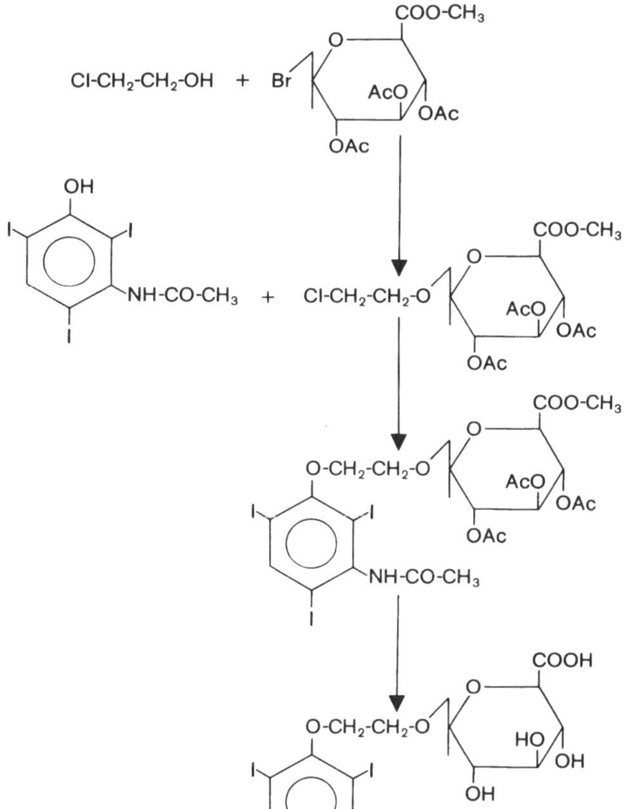

M 6

FIG. 10. Synthesis of M_6.

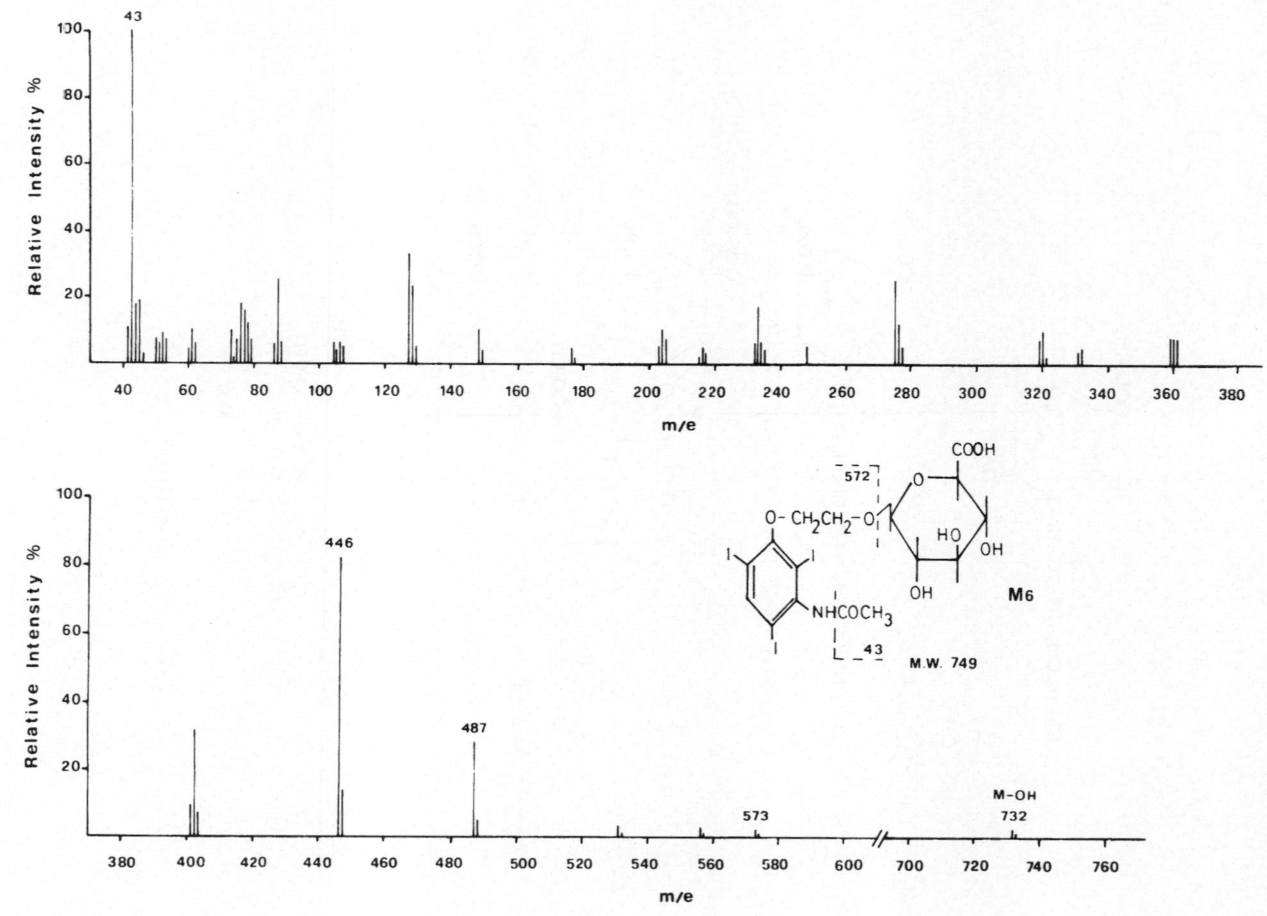

FIG. 11. Mass spectrum of M_6.

M_1 was found to be also formed as a metabolite of iopronic acid in dog (see below and Fig. 12).

Interestingly, the glucuronide of iopronic acid (M_1) was found to be more abundant than other metabolites in the normal rat whereas it was not detectable in the urine of cannulated rats. Oral doses of iopronic acid, 100 mg/kg, administered to dogs resulted in the excretion of products containing iodine representing 15% in the urine and 51% in the faeces over the first 24 hours. When the same dose was given intravenously 85% of the initial dose was found in the bile and 9% in the urine in the first 3 hours, confirming the marked biliary tropism of iopronic acid (8). The urine of dogs treated with iopronic acid contained three metabolites (M_1, M_2 and M_5) in addition to the parent compound. The absence of the alcohol, M_4, suggests the presence of a more efficient oxidizing enzyme system in the dog compared with the rat.

M_1 and M_5 are uronic acid conjugates which could be separated by countercurrent distribution.

Enzymatic hydrolysis with β-glucuronidase gave glucuronic acid and iopronic acid from either metabolite.

On the other hand alkaline hydrolysis resulted in the formation of glucuronic and iopronic acid from M_1 and iopronic acid and an unidentified saccharic acid from M_5. Elemental analysis IR, NMR and MS data are consistent with the structure proposed for M_1 (Fig. 13) although comparison with an authentic sample could not be carried out.

Oral administration of 3 g of iopronic acid to human subjects resulted in the urinary excretion of 71% (average of six measurements) of the administered dose in the first 48 hours (9). Two subjects were tested for faecal excretion until the faeces were free of iodine-containing compounds. The total excretion amounted to 7.3% in one case and 9% in the other. In addition to iopronic acid (M_3) the urine of the volunteers contained M_2, M_4 and M_1. Fig. 14 illustrates the metabolic fate of iopronic acid in man with the relative amount of each metabolite in the urine over 48 hours.

The only detectable compound in the faeces was iopronic acid. The small amount of iopronic acid recovered in the faeces reflects a remarkably thorough intestinal absorption of this compound. At the same time the roentgenograms obtained with iopronic acid demonstrate a marked biliary excretion of the compound, this conflicts with the limited faecal excretion observed unless one supposes a massive intestinal absorption followed by urinary excretion of the glucuronic acid conjugate of iopronic acid. A summary of the metabolism of iopronic acid in the three species studied is given in Fig. 15.

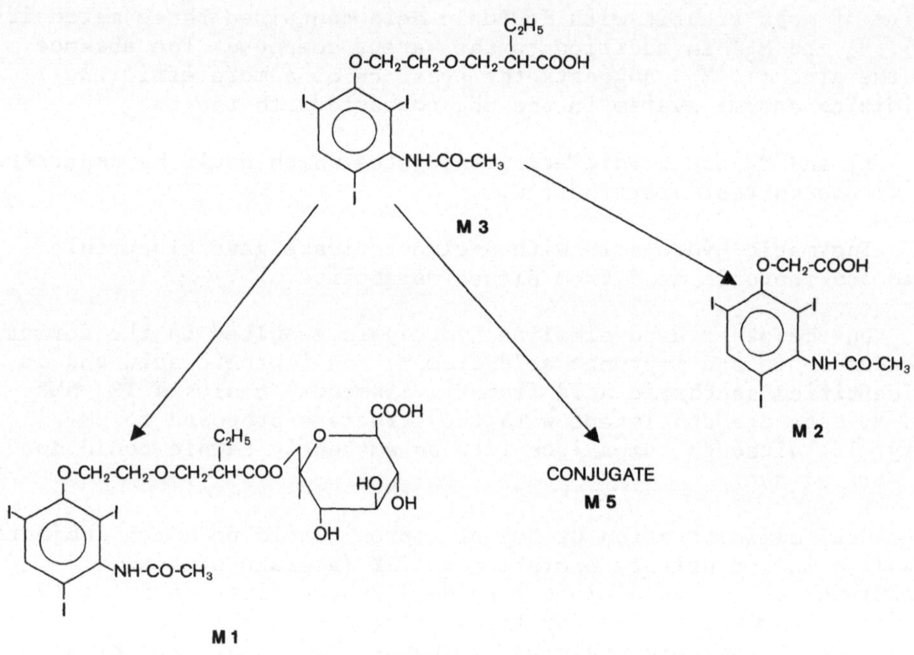

FIG. 12. Metabolism of iopronic acid in dog.

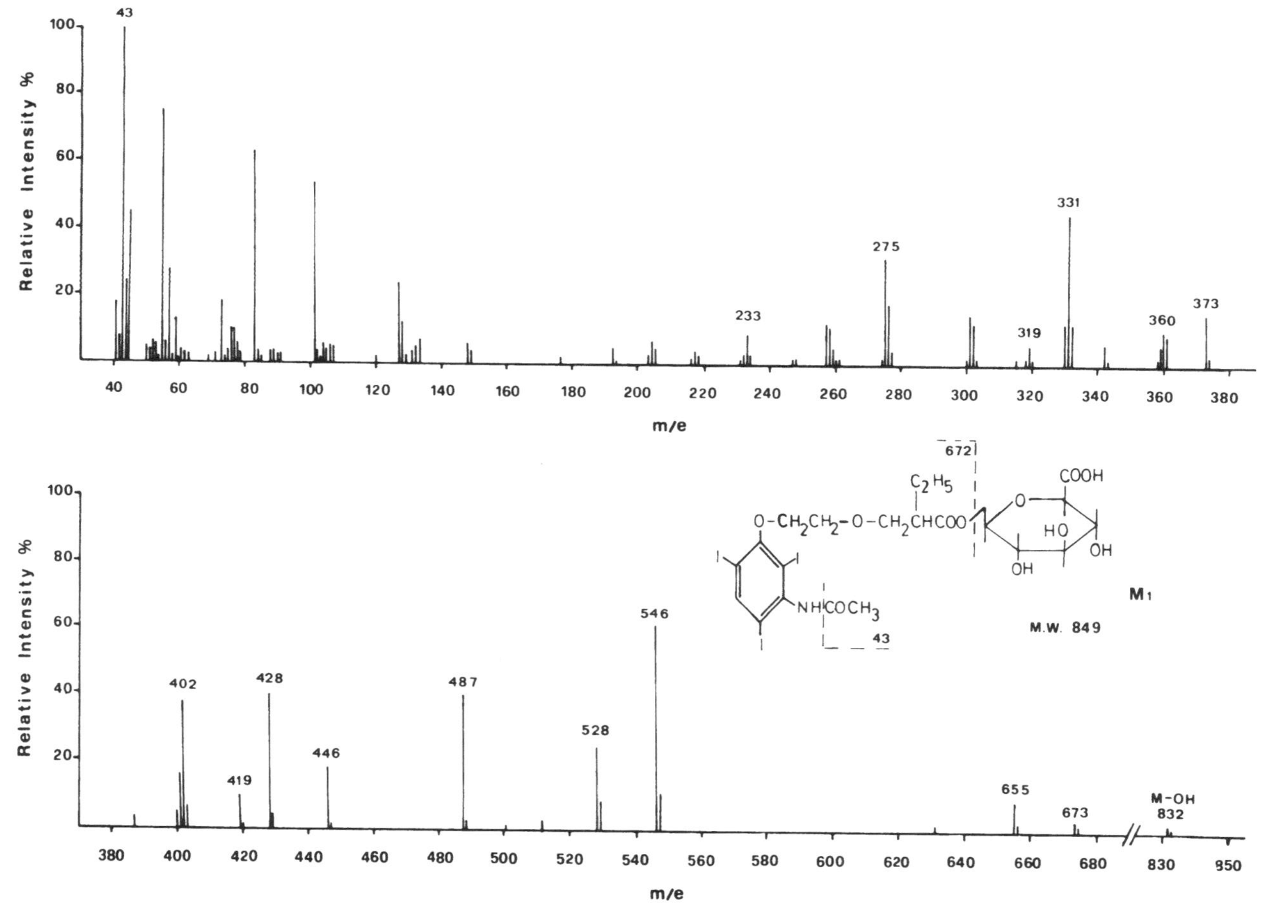

FIG. 13. Mass spectrum of M_1.

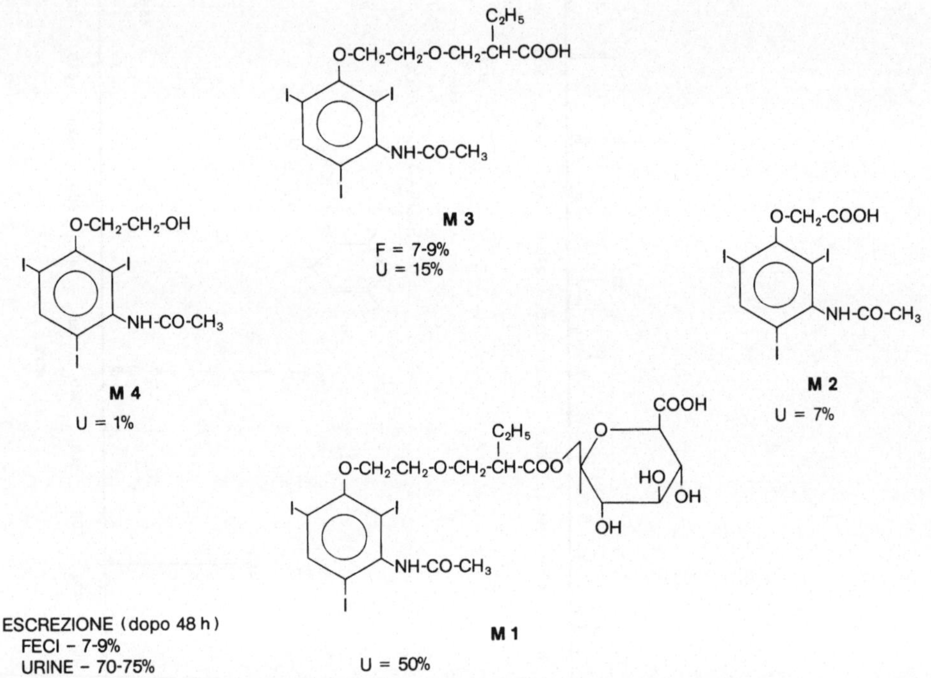

C_2H_5
O-CH_2-CH_2-O-CH_2-CH-COOH
I I
NH-CO-CH_3
I

M 3
F = 7-9%
U = 15%

O-CH_2-CH_2-OH
I I
NH-CO-CH_3
I

M 4
U = 1%

O-CH_2-COOH
I I
NH-CO-CH_3
I

M 2
U = 7%

COOH
C_2H_5 O
O-CH_2-CH_2-O-CH_2-CH-COO
HO
I I OH
OH
NH-CO-CH_3
I

M 1
U = 50%

ESCREZIONE (dopo 48 h)
FECI - 7-9%
URINE - 70-75%

FIG. 14. Distribution of metabolites of iopronic acid in man.

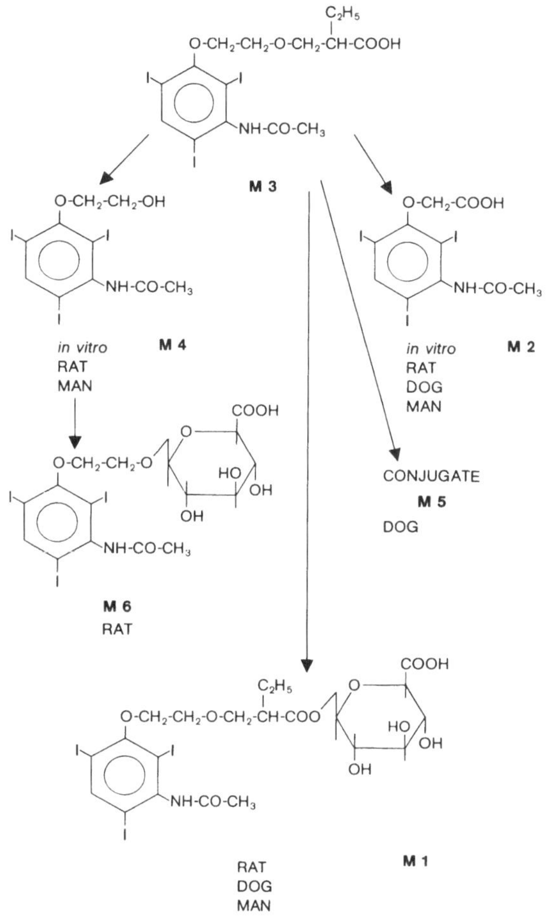

C_2H_5
O-CH_2-CH_2-O-CH_2-CH-COOH
I I
NH-CO-CH_3
I
M 3

O-CH_2-CH_2-OH
I I
NH-CO-CH_3
I
in vitro
RAT
MAN
M 4

O-CH_2-COOH
I I
NH-CO-CH_3
I
in vitro
RAT
DOG
MAN
M 2

COOH
O
O-CH_2-CH_2-O
HO
OH
OH
I I
NH-CO-CH_3
I
M 6
RAT

CONJUGATE
M 5
DOG

COOH
C_2H_5
O
O-CH_2-CH_2-O-CH_2-CH-COO
HO
OH
OH
I I
NH-CO-CH_3
I
RAT
DOG
MAN
M 1

FIG. 15. The metabolites of iopronic acid formed in man, rat and dog.

REFERENCES

1) E. Felder and C.B. Tanara, Farmaco Ed. Prat., 1976, 31, 283.
2) E. Felder, D. Pitrè, L. Fumagalli and E. Lorenzotti, Farmaco Ed. Sci., 1976, 31, 49; and E. Felder, D. Pitrè and M. Grandi, Farmaco Ed. Sci., 1976, 31, 426.
3) E. Felder, M. Zingales and U. Tiepolo, Farmaco Ed. Prat., 1976, 31, 383.
4) P. Tirone and G. Rosati, Farmaco Ed. Prat., 1976, 31, 437. and P. Tirone and G. Rosati, Farmaco Ed. Prat., 1976, 31, 397.
5) D. Pitrè and R. Maffei-Facino, Farmaco Ed. Sci., 1976, 31, 755.
6) D. Pitrè and L. Fumagalli, Farmaco Ed. Prat., 1976, 31, 529.
7) R.T. Williams, in "Fundamentals of Drug Metabolism and Drug Disposition" B.N. La Du, H.G. Mandel and E. Leong Way, Eds., Williams and Wilkins Co., Baltimore, 1971, p. 187.
8) D. Pitrè, Farmaco Ed. Prat., 1976, 31, 516.
9) D. Pitrè and E. Felder, Farmaco Ed. Prat., 1976, 31, 540.

METABOLITE IDENTIFICATION BY GC-MS: SPECIES DIFFERENCES IN THE METABOLIC PATTERNS OF ISOPROPYLBIPHENYL

H.R. Sullivan, R.E. McMahon, D.G. Hoffman and S. Ridolfo

Lilly Research Laboratories Indianapolis, Indiana 46206, U.S.A.

SUMMARY

Isopropylbiphenyl is an active anti-inflammatory agent in rats. The biological effects of the compound appeared to be mediated at least in part by metabolites. As a consequence an extensive study of the metabolism of this interesting hydrocarbon in several species, including man, was required. This objective was successfully accomplished in a relatively short time by extensive use of GC-MS technology. The complex metabolite sequences found proved to be different for each species studied (rat, dog, monkey, man). Metabolite patterns in each species were consistent with the pharmacological and toxicological properties in that species. Metabolite patterns were a consequence of sequential oxidation beginning with hydroxylation of the isopropyl group at the tertiary carbon in dog and man and at both the tertiary and primary carbon in rat and monkey.

INTRODUCTION

Today there are a number of nonsteroidal anti-inflammatory agents available to medical practice for the relief of the symptomatic pain of arthritis. These drugs are for the most part substituted phenylpropionic acids (Fig. 1). In the hope of improving the therapeutic ratio chemists at the Lilly Research Laboratories have investigated classes of compounds which are not acidic in character. Of the active compounds found the hydrocarbon, 4--isopropylbiphenyl (IPBP), was one of the more interesting.

This compound proved to be very effective when evaluated in

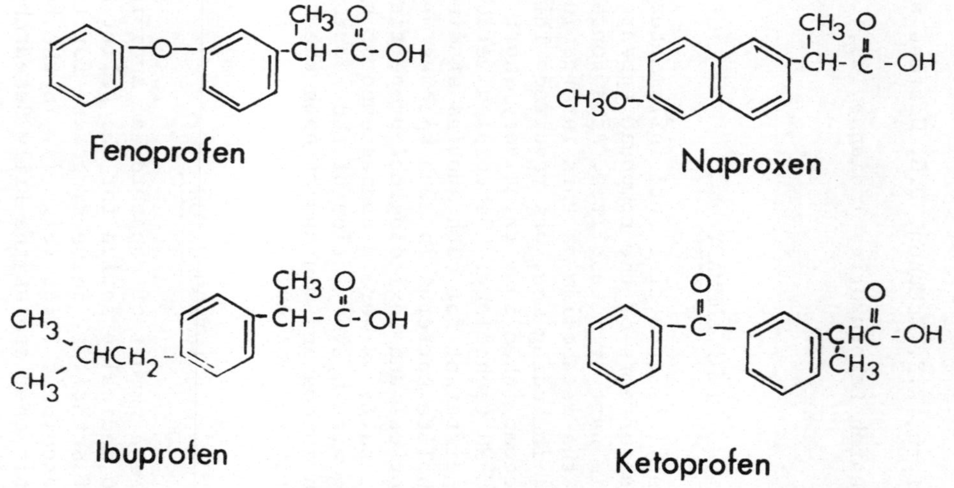

Fenoprofen

Naproxen

Ibuprofen

Ketoprofen

FIG. 1. Typical Nonsteroidal Anti-Inflammatory Agents

such assays as the erythema blocking test when administered orally to rats, guinea pigs and to a lesser extent dogs. The fact that the compound was less effective in the latter species suggested that metabolism might have a role in activating the compound. For this reason and because hydrocarbons do represent a unique class of drugs an extensive preclinical metabolism study was clearly indicated. Thus, an extensive toxicological evaluation as well as a thorough investigation of metabolism in a number of animal species were initiated. These studies were completed in a relatively short period of time through the extensive use of GC-MS for the separation and identification of metabolites. The results obtained from these studies are outlined in this paper.

METHODS

Combined gas chromatography - mass spectrometry (GC-MS). An LKB 9000 combined GC-MS fitted with an 2.4 m siliconized glass column (2.5 mm I.D.) packed with 1% W-98 silicone gum rubber on Gas Chrom Q was used for the separation and identification of urinary, biliary and circulating metabolites. Structures of all primary metabolites were confirmed by comparison of fragmentation patterns with those obtained from synthesized authentic compounds. The GC column was maintained isothermally at 150°, the GC flash heater and MS separator were maintained at 200°. The ionization potential and trap current were 70 eV and 60 μA respectively.

Animal studies. The species of laboratory animals in this metabolism study were mongrel dogs, Purdue Wistar rats and rhesus monkeys. IPBP as a suspension in 1% aqueous Methocel was administered orally to all animals. Biological fluids (urine, bile or plasma) collected from these animals were processed for GC-MS analysis in the following manner. Samples were adjusted to pH 7.0 with 1N NaOH and were extracted twice with equal volumes of dichloromethane. The combined organic phase was evaporated to dryness under N_2 and the residual material allowed to react with excess diazomethane. The biological sample was then acidified to pH 2.0 and again extracted twice with equal volumes of dichloromethane. The combined organic phase was evaporated to dryness under N_2 and the residual material allowed to react with excess ethereal diazomethane. The derivatized extracts were evaporated to dryness under N_2 and dissolved in 0.5 - 1.0 ml methanol for GC-MS analysis. The extracted fluid sample was then adjusted to pH 5.5 with 0.1N NaOH, innoculated with 0.05 ml of Glusulase (glucuronidase-sulfatase, Endo Products Inc.) and incubated at 37° for 24 hr. The hydrolyzed fluid samples were extracted at pH 7.0 and then at pH 2.0 as previously described. The extracted materials were derivatized by reaction with diazomethane prior to GC-MS analysis.

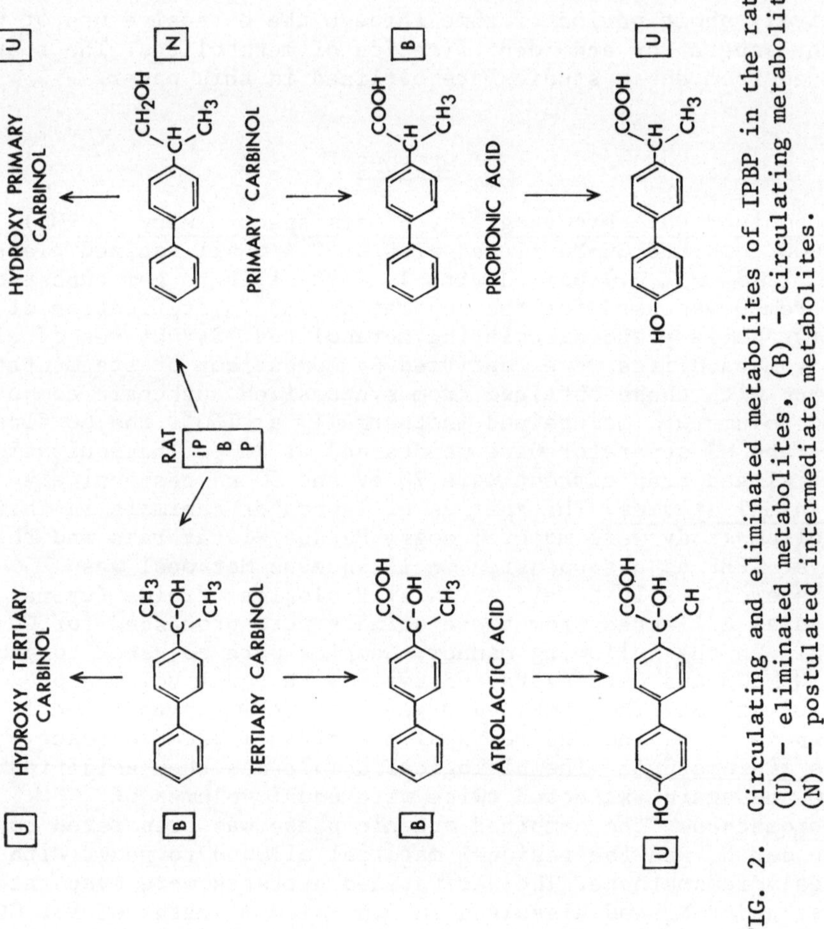

FIG. 2. Circulating and eliminated metabolites of IPBP in the rat. (U) - eliminated metabolites, (B) - circulating metabolites, (N) - postulated intermediate metabolites.

RESULTS AND DISCUSSION

Rat radiocarbon elimination studies. Initial radiocarbon elimination studies in rats showed 4-isopropylbiphenyl-methyl-^{14}C to be rapidly and quantitatively absorbed following oral administration. Biliary and urinary excretion were equally responsible for elimination of parent drug and/or metabolites. An appreciable portion, 15 to 20% of the administered radiocarbon, was found to be absorbed in epididymal fat and to be subsequently eliminated at a slower rate. No IPBP or metabolites were found to be eliminated *via* the lung.

Circulating metabolites. Since IPBP was rapidly and quantitatively absorbed in the GI tract, the identification of compounds circulating in the blood and available for distribution to sites of action was, therefore, of primary concern. GC-MS analysis of rat plasma extracts showed the presence of three metabolites of IPBP: 2-*p*-biphenylyl-2-propanol, *p*-phenylatrolactic acid and 2-*p*-biphenylylpropionic acid (Fig. 2). No detectable quantity of unchanged IPBP could be found in the plasma. Pharmacological testing showed the 2-propanol to possess pharmacological activity but less than that of IPBP and atrolactic acid to be devoid of anti-inflammatory activity. 2-*p*-biphenylyl-propionic acid was, however, found to be a potent anti-inflammatory agent possessing an ED_{50} in the erythema blockade in rats of 0.2 mg/kg, 20 times that of the parent IPBP.

GC-MS analysis of urine and bile extracts showed the major eliminated metabolites of IPBP to be ring hydroxylated derivatives of the tertiary carbinol and the propionic acid. Minor amounts of ring hydroxylated atrolactic acid and the primary carbinol were also found (Fig. 2). While the atrolactic acid was found to be a major circulating matabolite no detectable quantity of this acid could be found in the urine or bile. The failure to detect the primary carbinol in the plasma and the detection of only small eliminated quantities of its ring hydroxylated derivative implied its rapid metabolic conversion to the propionic acid. Other minor metabolites included 2-*p*-biphenylyl-1,2-propranediol a possible precursor for the atrolactic acid.

Plasma levels of tertiary carbinol, the atrolactic acid and the propionic acid were determined quantitatively by GC analysis after administration of 250 mg/kg of IPBP. Results obtained from these analyses, shown in Table 1, showed the major circulating metabolite of IPBP to be the propionic acid, the very active anti-inflammatory agent. The levels of the atrolactic acid were also quite high considering the fact that none was eliminated as such and only a small amount of its hydroxylated derivative was found to be eliminated.

Results obtained from these metabolism studies suggested that

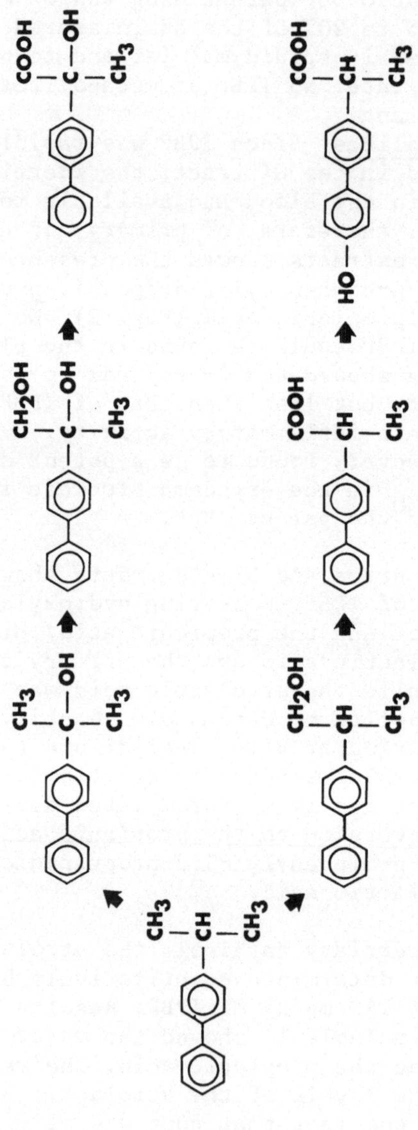

FIG. 3. Metabolic pathways involved in the biotransformation of IPBP in the rat.

TABLE 1. Plasma concentrations of parent drug and metabolites in the rat following oral administration of a single 250 mg/kg of IPBP.

	μg/ml Plasma at:				
Metabolite	0.5	1	2	4	8 hours
Propionic Acid	3.36	2.36	5.00	10.80	7.35
Atrolactic Acid	-	-	1.88	4.74	-
Tertiary Carbinol	0.6	0.6	0.7	1.4	0.8
Primary Carbinol	-	-	-	-	-
IPBP	1.00	0.80	0.97	0.77	0.22

two distinct pathways were involved in the biotransformation of IPBP in the rats (Fig. 3). Subsequent metabolism studies on the various intermediate metabolites confirmed this postulation. Both pathways involved aliphatic hydroxylation as the initial reaction. One, the propionic acid pathway, has as one of its products the pharmacologically active metabolite, 2-p-biphenylylpropionic acid. This pathway has as its initial reaction hydroxylation of a methyl group by oxygen insertion at a primary carbon atom to yield the primary carbinol. This primary carbinol was then an excellent substrate for enzymatic oxidation to the propionic acid. Aromatic hydroxylation was required for elimination. The second pathway, leading to the atrolactic acid, has as its initiating step hydroxylation by oxygen insertion on the secondary carbon to yield the tertiary carbinol which was then enzymatically oxidized to the atrolactic acid. None of the metabolites produced in this pathway possessed significant anti-inflammatory activity. All primary metabolites produced by this pathway required aromatic hydroxylation for elimination. One anomaly in this pathway was the presence of relatively high concentrations of the atrolactic acid in plasma and its near absence among the eliminated products. Was this acid being sequestered in the rat? At this time, however, it was learned from long term toxicity studies that chronic administration of relatively high doses of IPBP caused nephrotoxicity in rats. One feature of this toxicity was the presence of crystals in the tubules. The major component of these crystals was identified by GC-MS as conjugated atrolactic acid. It thus appeared that the rat was not capable of eliminating the atrolactic acid without aromatic hydroxylation and that it was subject to sequesterization in kidney tubules. These results suggested that one of the pathways

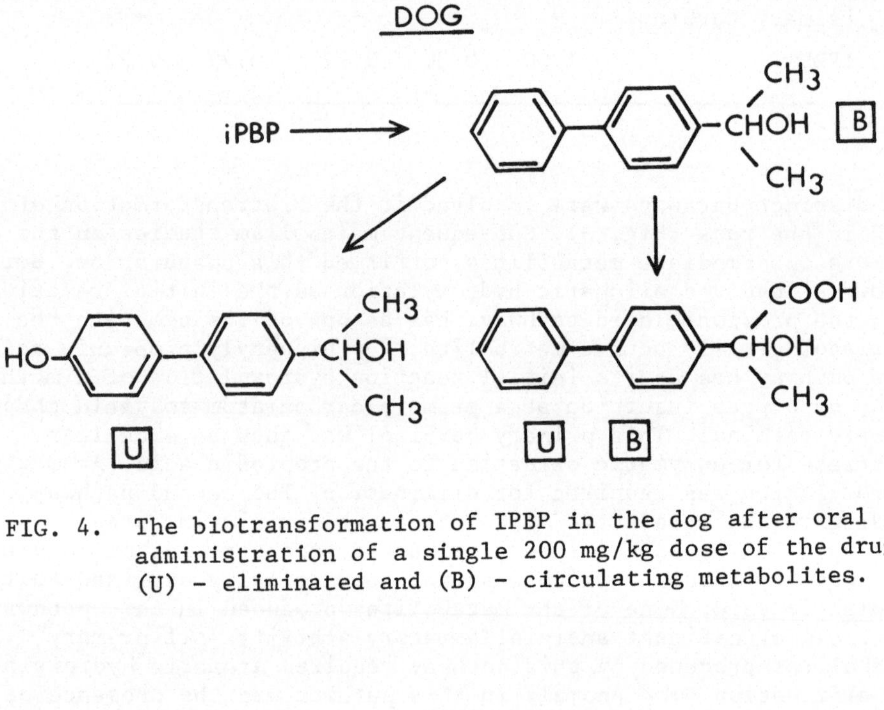

DOG

FIG. 4. The biotransformation of IPBP in the dog after oral administration of a single 200 mg/kg dose of the drug. (U) - eliminated and (B) - circulating metabolites.

involved in the biotransformation of IPBP in the rat, the propionic acid pathway, was responsible for a major part, if not all, of the anti-inflammatory activity. The second pathway, the atrolactic acid pathway, was responsible for much of the toxicity.

Dog. The atrolactic acid and the tertiary carbinol (Fig. 4) were the only circulating metabolites detected by GC-MS in the plasma of dogs following oral administration of IPBP. Similar analysis of urine extracts showed the eliminated metabolites to be the atrolactic acid and ring hydroxylated tertiary carbinol. These results implied that the biotransformation of IPBP in the dog proceeded *via* a single metabolic pathway, the atrolactic acid pathway. The 2-*p*-biphenylylpropionic acid the potent anti--inflammatory agent, was not a metabolite of IPBP in the dog. From these results it was also apparent that the dog, unlike the rat, was quite capable of elimination of the atrolactic acid without the requirement of aromatic hydroxylation and conjugation. From these results then it was possible to predict that IPBP would be less active pharmacologically but safer toxicologically in the dog. Chronic toxicity studies in the dog failed to demonstrate the development of nephrotoxicity.

Monkey. Attention was now focused on the metabolism of IPBP in yet a third species, the monkey. GC-MS analysis of plasma samples taken after oral administration of IPBP showed the presence of three circulating metabolites: 2-*p*-biphenylyl-1-propanol (primary carbinol), 2-*p*-biphenylyl-2-propanol (tertiary carbinol) and 2-*p*-biphenylylpropionic acid (Fig. 5). Quantitative plasma level determinations by GC showed the propionic acid to be the major drug related compound in the blood. The values listed represent plasma concentration of each component 1 and 3 hours after drug administration. The separation and identification of the primary carbinol and the β-hydroxypropionic acid marked the first instance of these compounds being identified as a metabolite in the monkey. The presence of the tertiary carbinol in the plasma indicated that the metabolism of IPBP in the monkey, as in the rat, involved the two pathways, the propionic acid and the atrolactic acid. Quantitatively, however, the propionic acid pathway appeared to be the major route of metabolism. Separation and identification of the urinary metabolites confirmed this observation. The major urinary metabolite of IPBP in the monkey was 2-*p*-biphenylylpropionic acid, the potent anti-inflammatory compound. The failure to detect the atrolactic acid metabolite indicated that the monkey was capable of eliminating the tertiary carbinol in either its free or conjugated form. These results were very promising in that they indicated IPBP to be a safe and effective drug in the monkey. Subsequent long term toxicological studies showed IPBP to be devoid of nephrotoxicity in the monkey.

Man. On the basis of the results obtained in the animal metabolism studies it was obvious that if IPBP was to be an effective anti-inflammatory in man it must be metabolized, at least in part, to 2-*p*-biphenylylpropionic acid. In addition, its

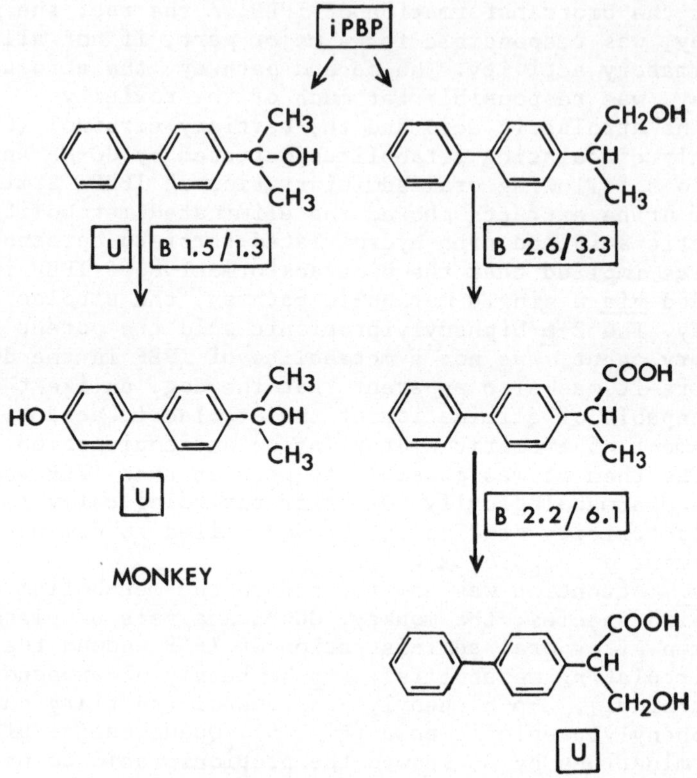

FIG. 5. Metabolic pathways involved in the metabolism of IPBP in the monkey.

failure to be metabolized, as in the monkey, to the atrolactic acid would attest to its safety. Following oral administration of a single 900 mg dose of IPBP trace quantities of the propionic acid were found in the plasma. The urinary metabolites (Fig. 6) included ring hydroxylated tertiary carbinol as the major metabolite, dihydroxylated tertiary carbinol, mono- and dihydroxylated olefins. The latter compounds are derived from the tertiary carbinol. Minor to trace amounts of the propionic acid and its monohydroxylated derivative were also found. No detectable quantity of the atrolactic acid was found. This latter finding was quite promising since it implied that IPBP would be safe in man. It

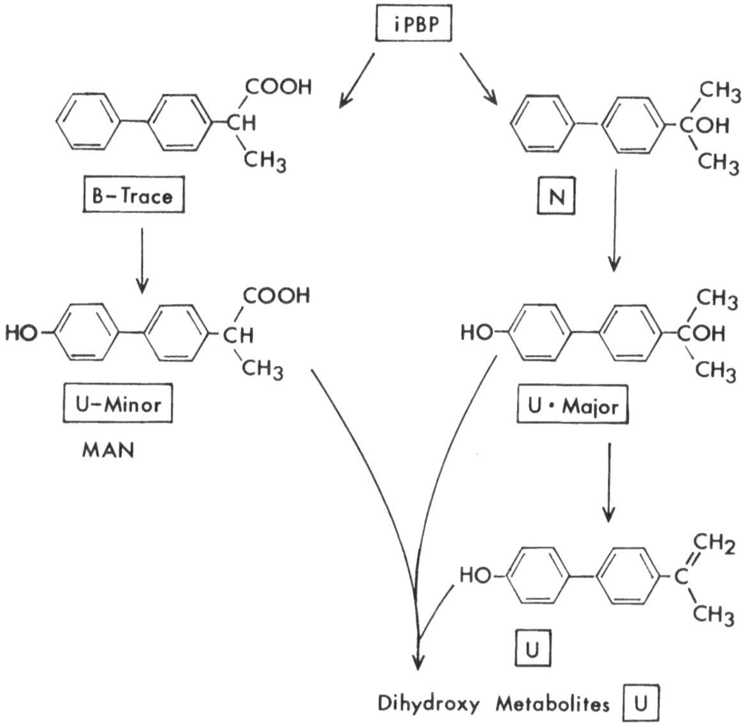

FIG. 6. Metabolic pathway involved in the metabolism of IPBP in man. (U) - eliminated, (B) - circulating, and (N) - postulated intermediate metabolites.

appeared, however, that the metabolism of IPBP in man, like the dog, involved principally the pathway initiated with the tertiary carbinol. Man was capable of eliminating the tertiary carbinol after aromatic hydroxylation and did not require conversion to the atrolactic acid for elimination.

Based on the proposition that 2-*p*-biphenylylpropionic acid was responsible for the anti-inflammatory activity of IPBP, these human metabolism results predicted that IPBP would be inactive in man as an anti-inflammatory agent. Subsequent clinical studies using the erythema blockade essentially confirmed the inactivity of IPBP in man. On the basis of these metabolism studies and the initial clinical studies, further clinical studies were felt to be unjustified and the project was terminated.

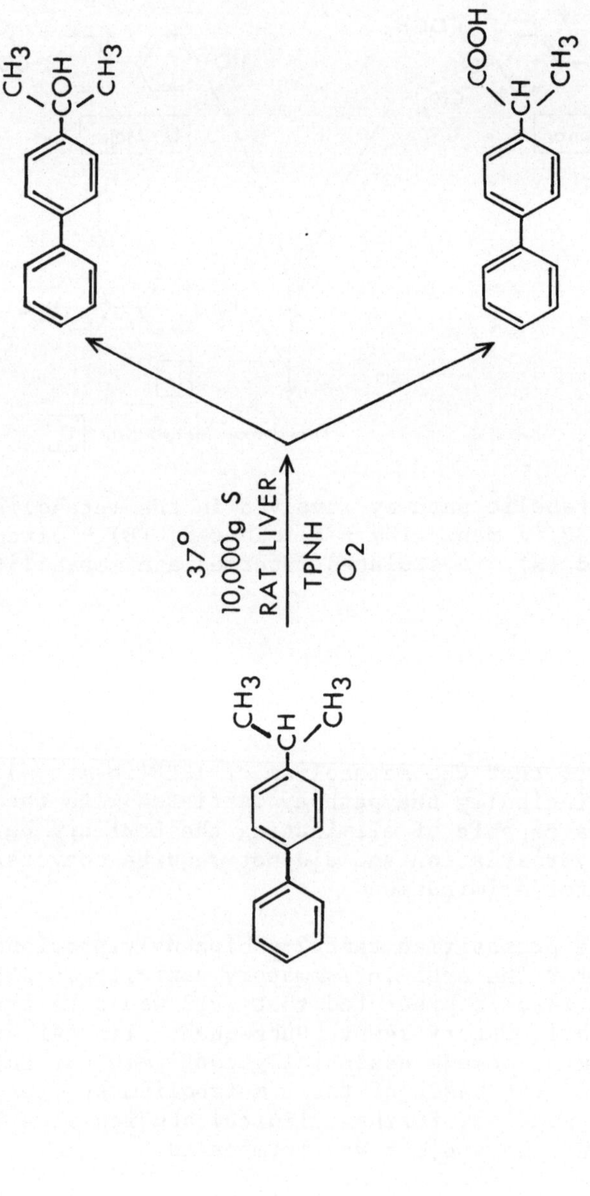

FIG. 7. In vitro metabolism of IPBP by the rat liver supernatant (15,000 x g) system.

COMMENTARY

These metabolism studies showed that two distinct metabolic pathways were involved in the biotransformation of IPBP in laboratory animals and in man. The first pathway, initiated by hydroxylation of the tertiary carbon, was involved to a major extent in two species, man and dog, and to a minor extent in two species, rat and monkey. The end product of this pathway, *p*-phenylatrolactic acid, was responsible for the nephrotoxicity observed in rats. The second pathway, that initiated by hydroxylation of the primary carbon, was the major route of metabolism of IPBP in the rat and monkey. The end product of this pathway was 2-*p*-biphenylylpropionic acid, a potent anti-inflammatory agent, which was undoubtedly responsible for a major portion of the pharmacological activity ascribed to IPBP.

In retrospect, the species differences observed in the metabolism of IPBP in this study raise an important question. Was there another method available by which these differences could have been more rapidly and accurately determined? Could an *in vitro* system such as a liver homogenate or a fraction thereof have been employed to predict the human metabolism of IPBP? IPBP was allowed to react *in vitro* with the rat liver supernatant and its appropriate co-factors (Fig. 7). The two products: 2-*p*-biphenyl-2-propanol, and 2-*p*-biphenylylpropionic acid, isolated from this reaction showed that, indeed, two pathways were involved in the metabolism of IPBP in the rat. Would a comparable *in vitro* reaction with human liver homogenates have implicated the involvement of only the atrolactic acid pathway in man?

STRUCTURAL ELUCIDATION OF AN S-OXIDIZED METABOLITE OF TRITHIOZINE IN RAT AND DOG

G. Pifferi, P. Ventura, C. Farina and A. Frigerio°

ISF-Italseber Research Laboratories,Trezzano s/N, Milan;

°Istituto "Mario Negri", Milan, Italy

INTRODUCTION

Trithiozine (ISF 2001), 4-(3,4,5-trimethoxythiobenzoyl) tetrahydro-1,4-oxazine (Fig. 1), is a new substance synthesized in our laboratories during a systematic research program on new alkoxythiobenzamides (1-4).

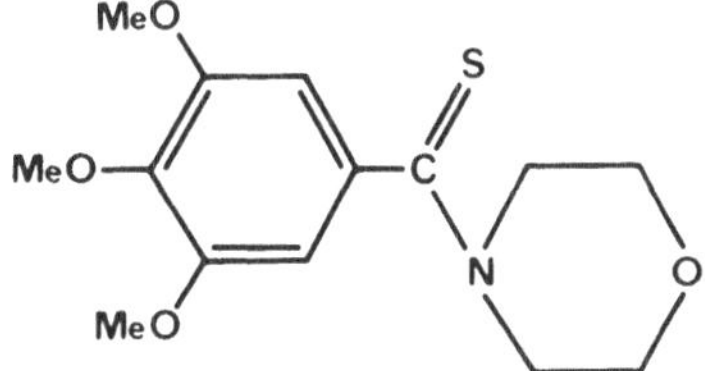

TRITHIOZINE (I.S.F. 2001)

4-(3,4,5,-Trimethoxythiobenzoyl)tetrahydro-1,4-oxazine

FIG. 1.

Pharmacological studies (5-7) indicate that Trithiozine (T) possesses considerable antisecretory and antiulcer activity, at the dose of 25-100 mg/kg per os, on different animals, while it is devoid of anticholinergic, antihistaminic, ganglioplegic and cardiovascular effects. The drug also has a mild tranquillizing action and a very low toxicity in various animal species.

The main pharmacological results have been confirmed in clinical trials (8-11).

Pharmacokinetic and metabolic studies in rats treated with 50 mg/kg ip and orally and dogs, 200 mg/kg ip and orally, showed that trithiozine (T) is quickly metabolized (12). Together with some unchanged T, two metabolites were isolated in the neutral toluene extracts from plasma and urine by TLC. One of them has been identified as the compound TBO by comparison with an authentic sample (Fig. 2) (12).

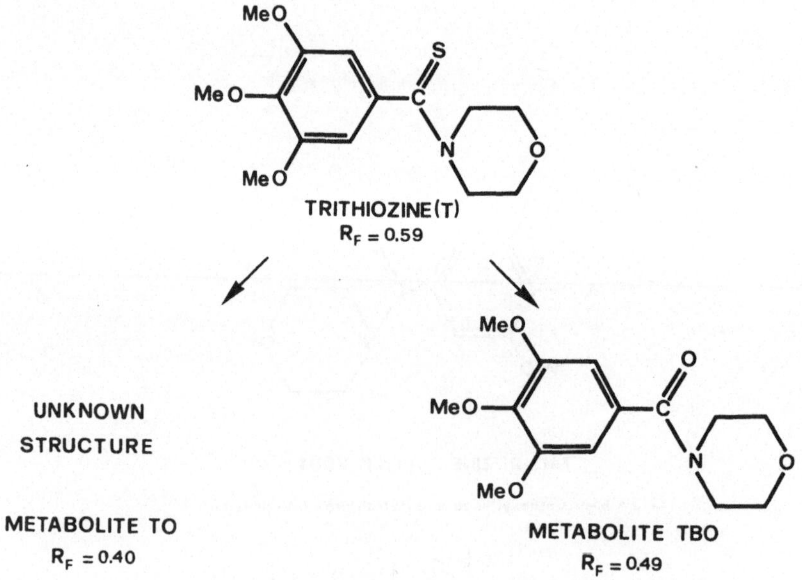

TRITHIOZINE (T)
$R_F = 0.59$

UNKNOWN STRUCTURE

METABOLITE TO
$R_F = 0.40$

METABOLITE TBO
$R_F = 0.49$

FIG. 2. Urinary excretion of Trithiozine: TLC of neutral extracts in rats and dogs.

This paper deals with the structural determination of the second metabolite by means of physico-chemical and synthetic methods. Moreover, a preliminary metabolic pathway of Trithiozine is proposed in relation to its pharmacological profile.

RESULTS AND DISCUSSION

Fig. 3 (top trace) shows the mass spectrum of Trithiozine (T)

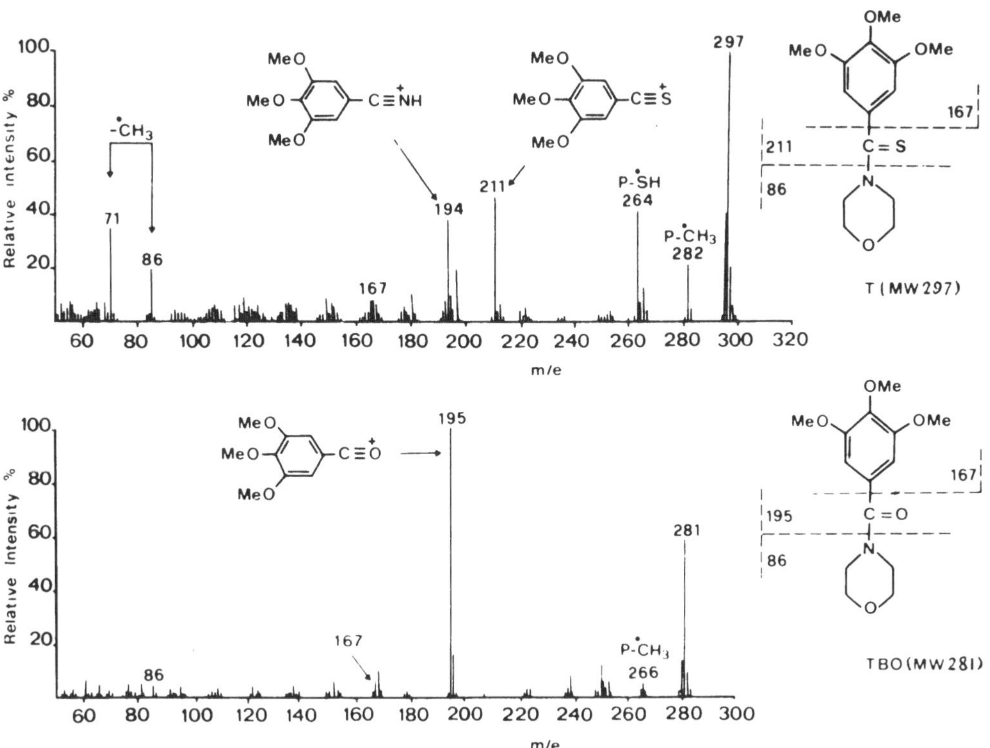

FIG. 3. Mass spectra of T and TBO.

recorded at 70 eV by direct inlet sampling. Characteristic features are: the molecular ion at m/e 297, which is also the base peak; the 3,4,5-trimethoxybenzonitrile at m/e 194 and the morpholine ion at m/e 86. The ion at 194 is rather unusual and its molecular composition has been confirmed by high resolution mass spectrometry.

The lower trace shows the mass spectrum of the first metabolite (TBO) with the molecular ion at m/e 281 and the base peak at m/e

195 corresponding to the 3,4,5-trimethoxybenzoyl ion. This compound on TLC (solvent system: toluene : EtOH: NH_3 - 90:20:1) appeared at R_f 0.49.

The lower trace of Fig. 4 shows the mass spectrum of the

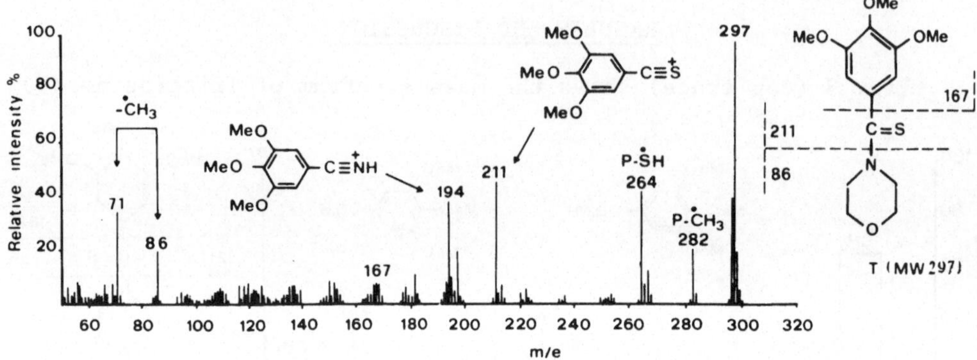

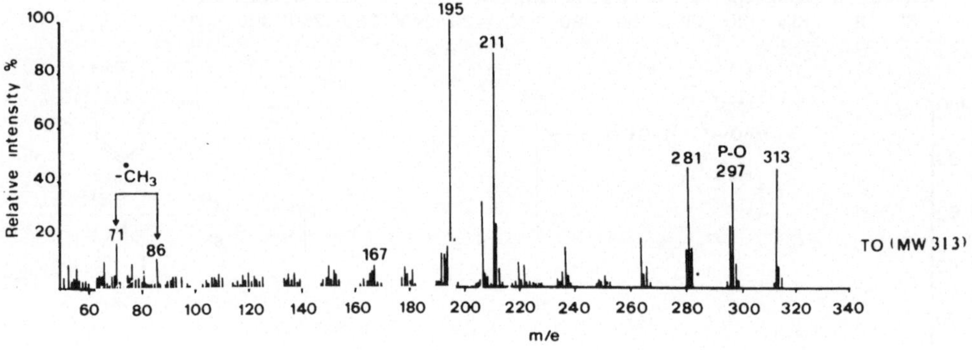

FIG. 4. Mass spectra of T and TO.

unknown metabolite TO (R_f 0.40). This is compared to the mass spectrum of Trithiozine (T). TO shows both a molecular ion 16 amu higher with respect to that of T and a peak at m/e P-16. Moreover this spectrum maintains the peaks at m/e 211 and m/e 86, characteristic of T.

These findings led us to assume that the compound might be an oxidized derivative of T. Therefore we chemically oxidized T with peroxides and obtained a product with an R_f and mass spectrum identical to those of the unknown metabolite TO.

On the basis of these data: loss of oxygen in the mass

spectrum, method of synthesis and extraction procedure (isolation of unconjugated metabolites (12)), we could reject a possible hydroxylated derivative of T. It was more likely that the additional oxygen atom was present in an N-oxide or an S-oxide function.

The infrared spectra of both Trithiozine and its metabolite TO (solvent CS_2; 2% w/v) differ only in the 1200-800 cm^{-1} region: the absence of hydroxylic groups in TO is therefore confirmed (no absorption in the 3600-3200 cm^{-1} region). A band at 1135 cm^{-1} is indicative of the presence of the 3,4,5-trimethoxyphenyl group and one at 1110 cm^{-1} is indicative of the morpholine ring, in both T and TO. On the other hand, a strong band at 1000 cm^{-1}, which is present only in the spectrum of TO, may be attributed to a nitrogen-oxygen or to a sulfur-oxygen stretching vibration (13).

The ultraviolet spectrum of T (solvent MeOH) shows bands at 281 nm (log ε = 4.12) and 369 nm (log ε = 2.65) characteristic of the thioamide chromophore (14). In comparison the ultraviolet spectrum of TO (solvent MeOH; absorption maxima at 328, 290, 276 and 245 nm with, respectively, log ε = 3.86, 3.89, 3.92 and 3.92) shows an intense band at 328 nm, with a small solvent effect, which may indicate a π-conjugation enhancement in this compound. This bathochromic effect is in agreement with an S-oxide structure, in which the π-electrons are delocalized throughout the thioamide S-oxide group.

In fact, the sulfur 3d orbitals, in contrast to those of nitrogen, are energetically available, thus allowing the formation of a $p_\pi \longrightarrow d_\pi$ retrodative bonding between oxygen and sulfur atoms, as depicted in Fig. 5.

From a comparison of the canonical forms of the thioamide function and the hypothetical canonical forms of the thioamide S-oxide and thioamide N-oxide functions it is apparent that the conjugation in the S-oxide structure is increased: the π electrons are delocalized throughout the thioamide S-oxide function, the π bond must be localized on the thione group. On this basis, the bathochromic shift in the UV spectrum of TO could be in agreement with an S-oxide structure.

The decreased double bond character of the carbon-nitrogen bond in both the S- and N-oxides and the consequent lower barrier to rotation around the carbon-nitrogen bond are also evident and supported by the PMR spectrum of TO, reported in Fig. 6.

The bottom trace shows the 5-3 ppm region of the PMR spectrum of T (30° C in $CDCl_3$). In this region we find the signals related to the proton of the aromatic methoxyl groups and of the morpholine ring. Since rotation around C-N bond is slow on the PMR time scale

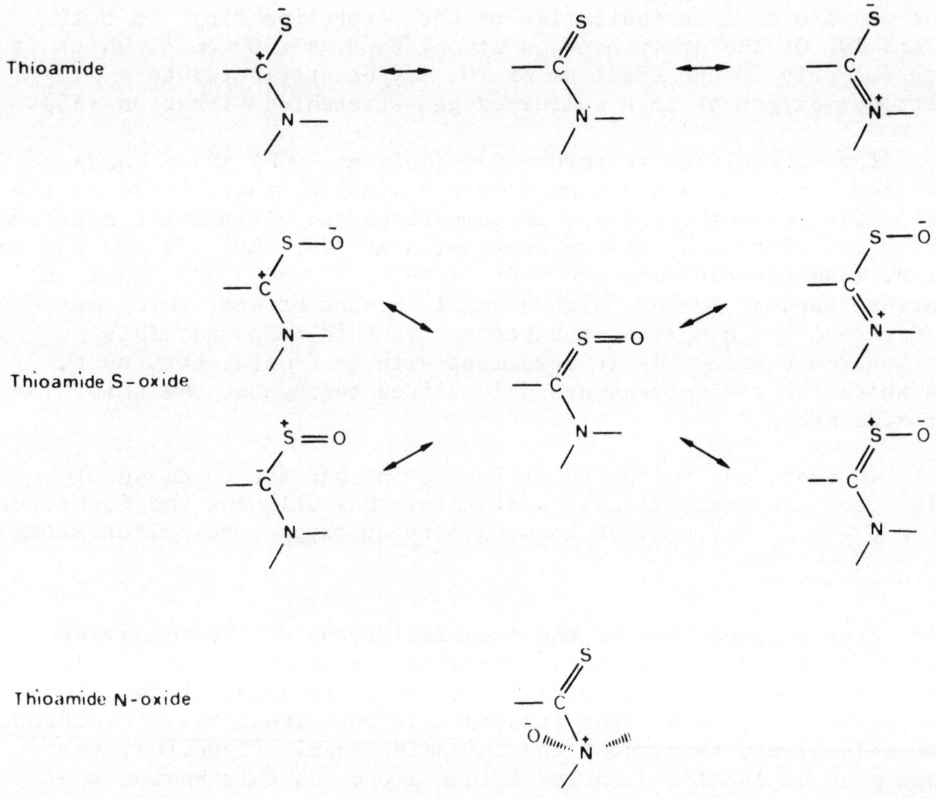

Thioamide

Thioamide S-oxide

Thioamide N-oxide

FIG. 5. Hypothetical canonical forms.

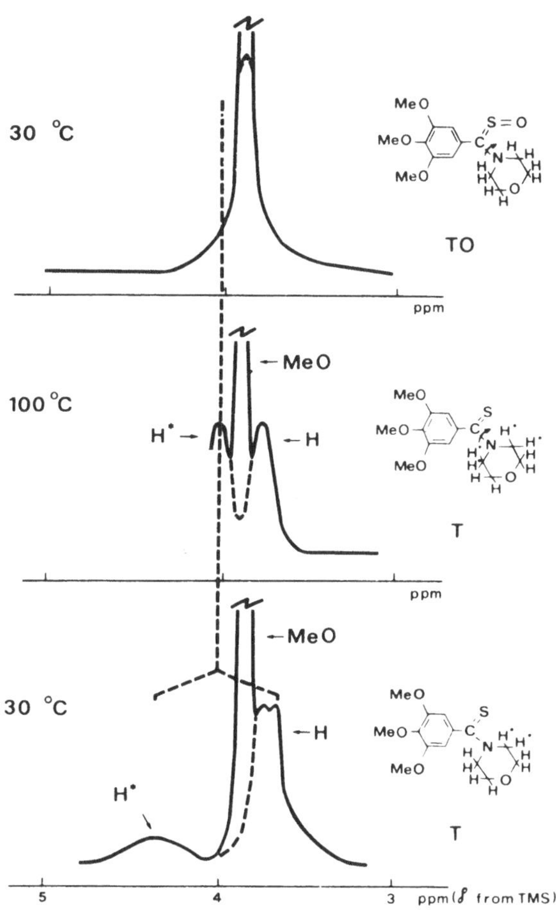

FIG. 6. Partial PMR spectra of T and TO.

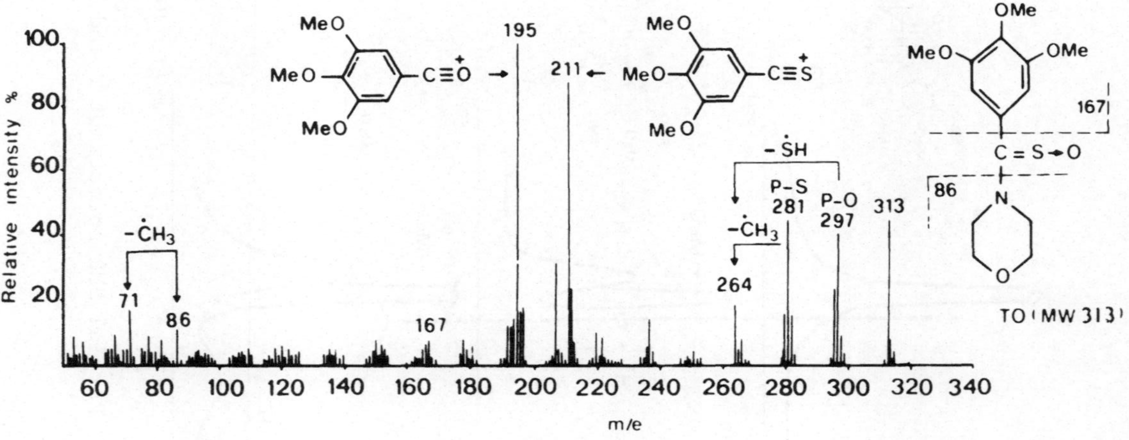

FIG. 7. Mass spectrum of TO.

at 30°C, separate signals are observed for the N-CH_2 groups situated *syn* and *anti* to the thiocarbonyl sulfur. By increasing the temperature to 100°C (solvent : hexachlorbutadiene), the signals related to these two groups collapse, as shown in the middle trace.

At the top of Fig. 6 the same region of the PMR spectrum of TO recorded at 30°C in $CDCl_3$ is shown. All the PMR signals have collapsed into a single signal, indicating a lower energy barrier to rotation around the C-N bond (in comparison with T).

This free rotation is in agreement with both S-oxide and N-oxide structures, but the chemical shifts of the protons of the morpholine ring in TO compared with the chemical shifts of T exclude the presence of a net positive charge on the nitrogen atom, because in such a situation the N-CH_2 signals should resonate at a lower field. Therefore the N-oxide structure can be excluded.

The mass spectrum of TO (Fig. 7; operating conditions: ion source temp. = 290°C, DIS temp. = 120°C), shows as mentioned before, a characteristic molecular ion at m/e 313, with a primary loss of 16 mass units, probably oxygen. This spectrum incorporates features present in the mass spectra of T and TBO. In fact, the ion at m/e 211 (the second major peak in the spectrum) is due to T, while that at m/e 195 (base peak) is due to TBO. Moreover the ion 86 is related to the morpholine moiety.

A tentative fragmentation of TO after electron impact is depicted in Fig. 8. It is reasonable to assume that this compound can undergo a thermal degradation reaction to give TBO, since this occurs also during the GLC analysis (12). At this point, TBO gives rise to the ions previously described. On the other hand, TO can undergo an electron impact reaction yielding the molecular ion at m/e 313, which, after loss of neutral oxygen, gives the ion at m/e 297. The alternative processes indicated with dotted lines also appear possible, although we have not yet obtained any direct evidence to support them.

The final proof of the S-oxide structure of the metabolite (TO) was obtained by chemical synthesis (Fig. 9). A sample of TO was obtained by oxidation of T with monoperphthalic acid in dichloromethane at -50°C and subsequent elution of the reaction mixture with chloroform on a dry column of basic alumina. This sample was tested for pharmacological activity.

The unambiguous synthesis of TO was accomplished starting from T. Treatment with iodomethane and then with hydrogen sulfide in pyridine, afforded the dithioester. This compound was oxidized with *m*-chloroperbenzoic acid in chloroform at 0°C, to give the thionoxide. This was reacted with morpholine in methanol and from the reaction

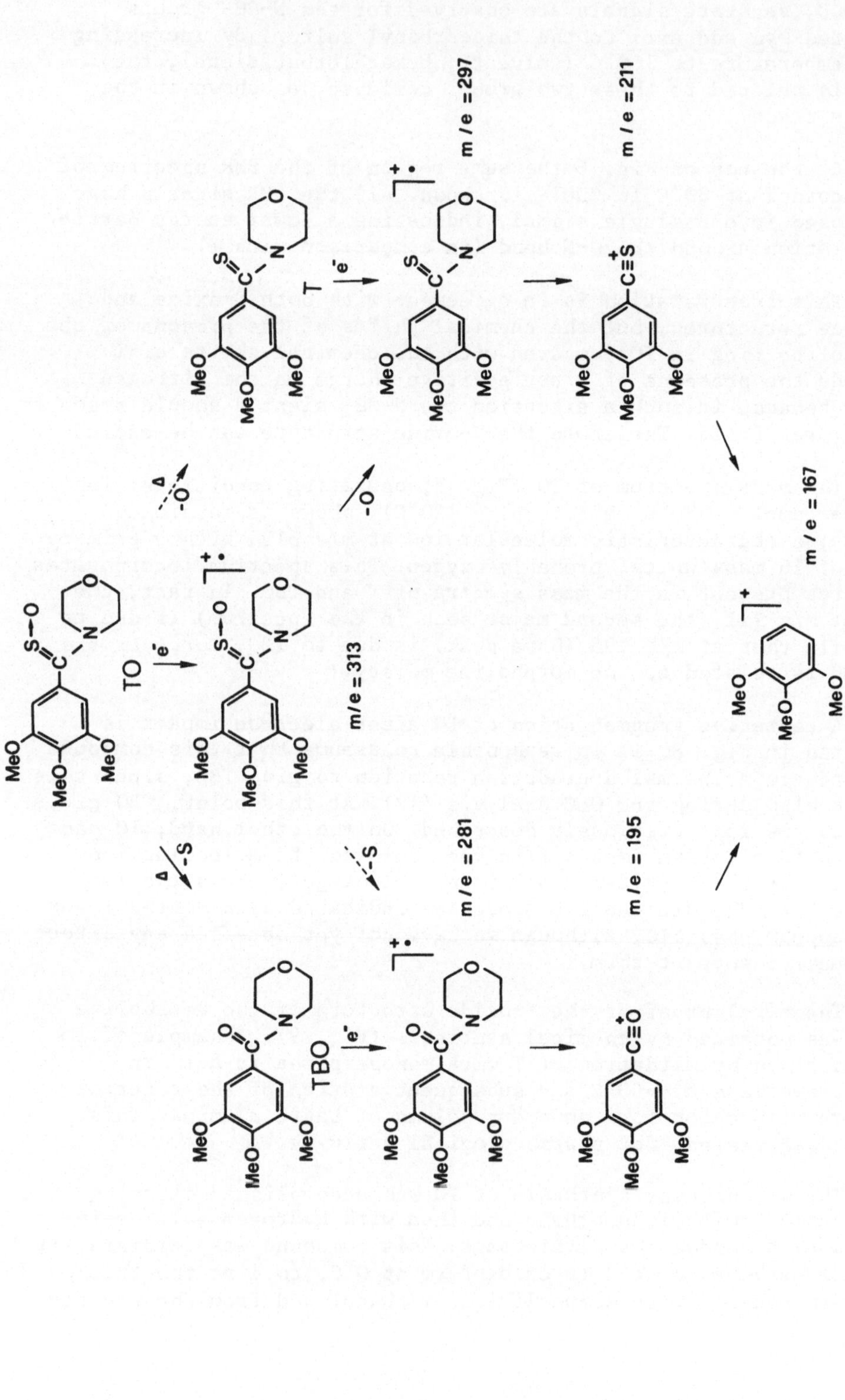

FIG. 8. Possible fragmentation pathway of TO.

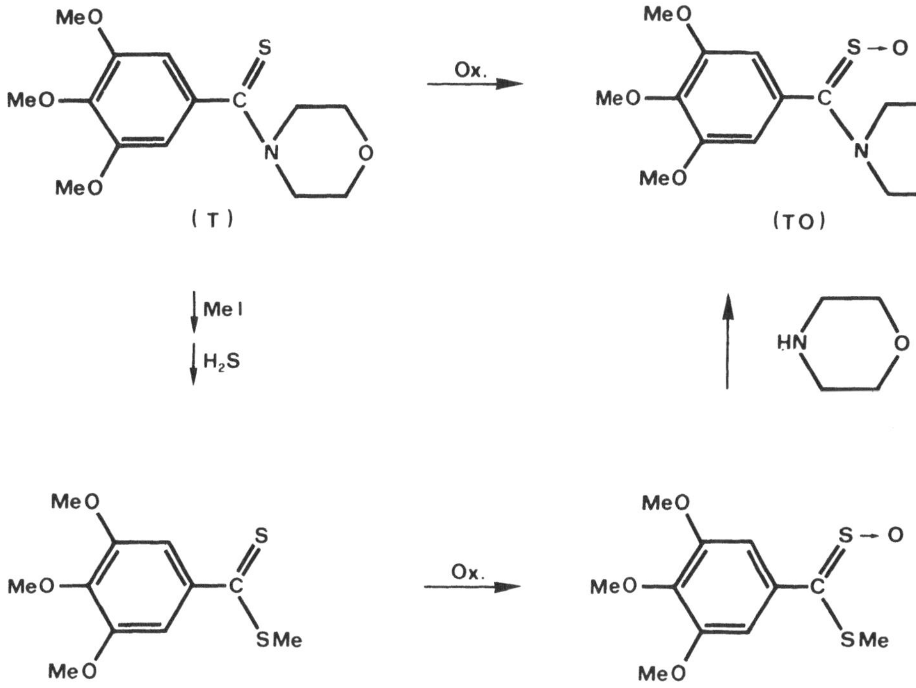

FIG. 9. Syntheses of Trithiozine S-oxide (TO).

mixture, containing TO, T and TBO, a pure sample of TO was obtained by thick layer cromatography on silica gel.

The IR, UV and mass spectra of the products obtained from the two different syntheses were identical.

To our knowledge only one example of a heterocyclic thioamide S-oxide has been reported previously (15). This is probably due to their low chemical stability.

We have therefore checked the stability of TO in aqueous solution at various temperatures, pH and concentrations (Fig. 10).

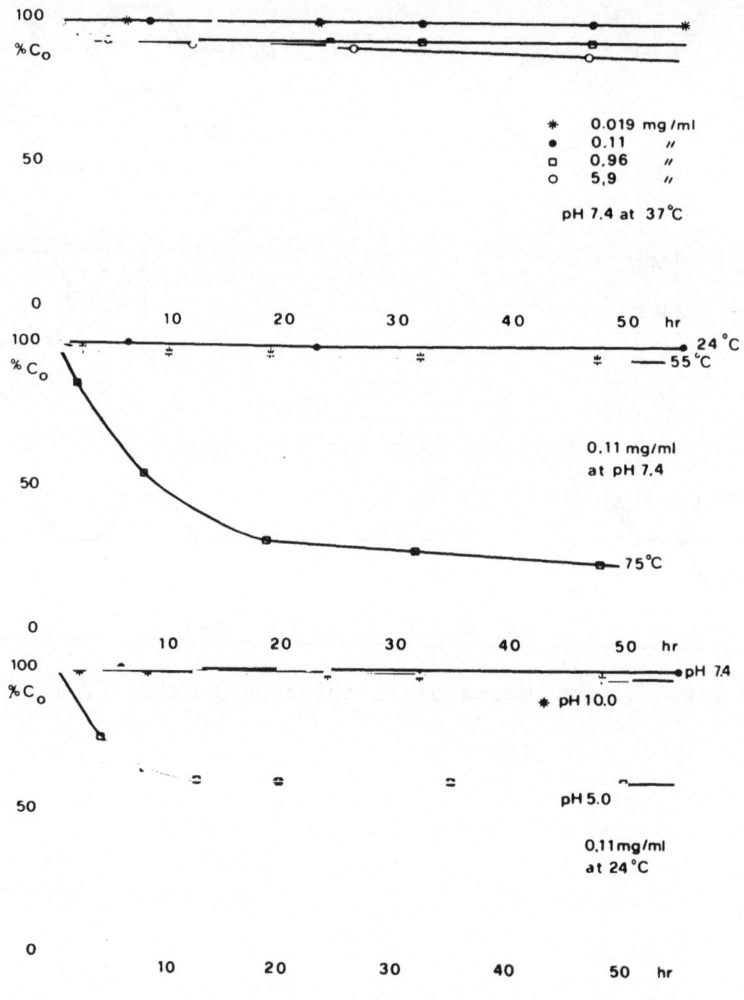

FIG. 10. Stability of TO in aqueous solution.

TO is unstable at acidic pH, at temperatures above 55°C and, to a lesser extent, at high concentrations. In all of these conditions, we always found the decomposition product to be TBO. It is interesting to observe that the decomposition of TO at acidic pH terminated after 20 min (bottom trace): further studies are in progress to explain these observations.

CONCLUSIONS

The identification of the metabolites TO and TBO in the neutral extract of rat and dog plasma and urine, the study of their chemical and pharmacological properties, allow us to gain a better understanding of the *in vivo* behaviour of Trithiozine (Fig. 11). In fact, T *in vivo* undergoes a rapid hepatic oxidation

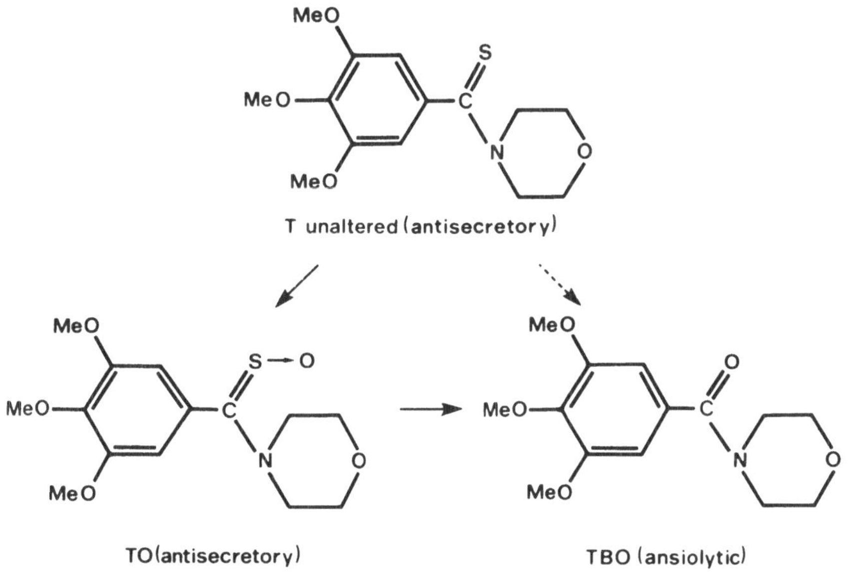

FIG. 11. Proposed metabolic pathway of Trithiozine.

on sulphur to give the unusual metabolite TO. In turn, this labile compound is transformed, at least partially, into the subsequent metabolite TBO (which is not an artifact). The alternative possibility of a direct metabolism of T to TBO (dotted line) by gastrointestinal bacterial flora seems less likely, since oral pretreatment with strong antibiotics did not alter the rate of formation of TBO. Moreover, intravenous administration of T, by--passing the GI tract, again did not have any effect.

Heteroaromatic alkoxybenzamides in general and TBO in particular, are pharmacologically active on conditioned and aggressive behaviour (3), while practically devoid of antisecretory properties. Thus, the mild tranquillizing activity of T seems to be due not to the parent drug but to TBO.

Finally, since T is quickly and uniformly absorbed and extensively and rapidly degraded (12), how can its longlasting antisecretory activity (5-6 h) be explained? The answer lies in that T gives rise _in vivo_ to other active metabolites, such as TO, possessing good or even better antisecretory and antiulcer action than T itself. Further studies to obtain confirmation of this are in progress.

REFERENCES

1) G. Pifferi, R. Monguzzi, S. Banfi and C. Carpi, Chim. Thér., 1973, 8, 462.
2) G. Pifferi and M. Pinza, in "XIème Renc. Intern. Chim. Thér., Lille 4-6 Sept.", 1974.
3) G. Pifferi and M. Pinza, in "Simposio Intern. Ricerca Scientifica nell'Industria Farmaceutica in Italia, Roma, 2-4 October", 1975.
4) G. Pifferi : U.S. pat. 3.755.317; U.S. pat. 3.862.138.
5) C. Carpi, S. Banfi and U. Cornelli, in "V° Congreso Mundial de Gastroenterologia - Ciudad de Mexico, 13-18 October, 1974.
6) C. Carpi, S. Banfi and U. Cornelli, in "6th Int. Congress of Pharmacology, Helsinki, July", 1975.
7) C. Carpi, S. Banfi and U. Cornelli, in "Simposio Intern. Ricerca Scientifica nell'Industria Farmaceutica in Italia, Roma 2-4 October", 1975.
8) L. Barbara, R. Corinaldes, M. Miglioli, M. Bortolotti, A. Luchetta, G. Busca and G. Labò, in "V° Congreso Mundial de Gastroenterologia, Ciudad de Mexico, 13-18 October", 1974.
9) L. Barbara, R. Corinaldes, M. Miglioli, R. Guidoboni, A. Luchetta, F. Lami and G. Labò, Minerva Gastroenterol., 1975, 21, 169.
10) G. Bianchi Porro, A. Ferrara, M. Petrillo and G. Abbondati, Minerva Gastroenterol., in press, 1976.
11) G. Bianchi Porro, A. Ferrara, M. Lazzaroni, M. Petrillo and R. Guidoboni, Minerva Gastroenterol., in press, 1976.
12) G.M. Pacifici, G. Bianchetti, A. Frigerio, R. Gomeni, P.L. Morselli, S. Garattini, G. Pifferi and C. Carpi, Eur. J. Drug Metab. Pharmacokinet., in press, 1976.
13) K. Nakanishi, "Infrared Absorption Spectroscopy, Practical", Holden-Day, S. Francisco, 1962, p. 51 and 54.

14) W. Walter and J. Voss, "Chemistry of Amides", Zabicky, Ed., Interscience, New York, 1970, p. 383.
15) W. Walter and K.D. Bode, Justus Liebigs Ann. Chem., 1966, 698, 131.

SOME METABOLITES OF 9-HYDROXY-19,20-BIS-NOR-PROSTANOIC ACID

R.Caponi°,R.Fumagalli°°,S.Innocenti°,P.Martelli°,U.Valcavi°

°Istituto Biochimico Italiano, Milan

°°Istituto di Farmacologia, Milan

INTRODUCTION

Some years ago (1,2) we began an investigation on the synthesis and the pharmacological properties (chiefly the hypolipemic activity and the activity against platelets aggregation) of very simple prostaglandin-like compounds having eighteen carbon atoms (instead of 20 carbon atoms of all known natural or synthetic prostaglandins).

Among many new prostaglandin-like compounds with eighteen carbon atoms prepared and pharmacologically tested, the 9-hydroxy-19,20-bis-nor-prostanoic acid (1, Fig. 1) shows very interesting properties as a hypolipemic agent. In fact it reduces the hypertriglyceridemia induced in different animal species by ethanol or by fructose administration and shows pharmacological activity after oral administration, whereas the prostaglandin compounds normally must be given by slow infusion. The 9-hydroxy-19,20-bis-nor-prostanoic acid differs from the prostaglandin $PGF_{1\alpha}$ (Fig. 1) in that it has eighteen carbon atoms instead of twenty carbon atoms of $PGF_{1\alpha}$, only three asymmetric centers ($PGF_{1\alpha}$ has five chiral centers) and lacks substituents at C_{11} and C_{15} and the double bond at C_{13}. It is a racemic mixture and can be easily produced in large amounts.

During the study of its pharmacological properties, we have considered the metabolic pathway of this compound in the rabbit.

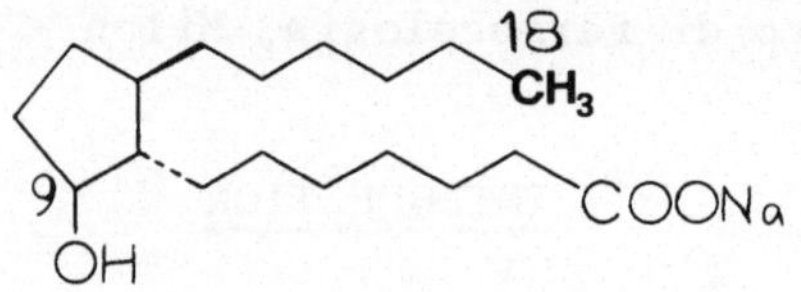

I

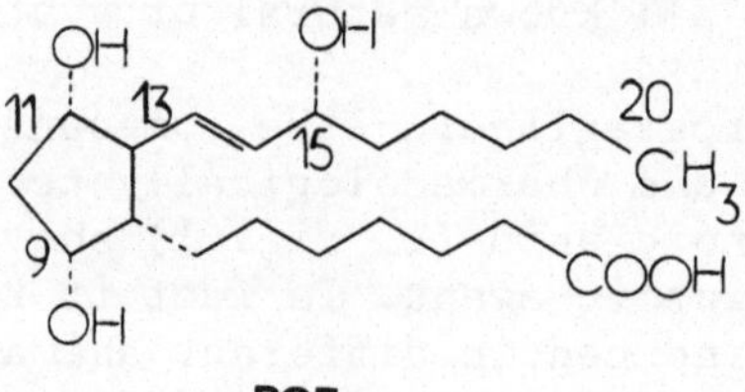

$PGF_{1\alpha}$

FIG. 1. Molecular structures of 9-hydroxy-19,20-bis-
-<u>nor</u>-prostanoic acid (sodium salt) and $PGF_{1\alpha}$.

RESULTS AND DISCUSSION

The 9-hydroxy-19,20-bis-nor-prostanoic acid was administered orally to adult male rabbits. The urine was collected and extracted at pH 8 to give a neutral fraction which by thin-layer chromatography (TLC) and gas-chromatography (GLC) did not contain metabolites.

Extraction of the urine at pH 1.5 gave an acidic fraction which by TLC and GLC, was shown to contain seven compounds not present in the urine of the untreated animals (Table 1).

The crude extract was esterified with an ethereal solution of diazomethane and separated into its components by column chromatography.

Purification was performed by preparative TLC. The compounds obtained were analyzed by Ultraviolet, Infrared and Nuclear Magnetic Resonance spectroscopy and by Gas Chromatography-Mass Spectrometry (GC-MS). The GLC analysis was carried out with a Perkin Elmer mod.990 equipped with a flame ionization detector, on a glass column OV 17 3%. NMR spectra were measured with a Varian 100 MHz for solutions in deuterated chloroform with tetramethylsilane as internal standard. IR spectra were performed on a Perkin Elmer mod. 157 G in chloroform solution. GLC-MS spectra were carried out with an LKB 9000 at 70 eV. and a glass column OV 17 at 235°C.

The starting compound was not detected in the urine extract, suggesting that it is rapidly metabolized. Furthermore, the total amount of the seven urinary metabolites isolated represents only about 5% of the administered compound. We have identified some of these seven metabolites isolated on the basis of the GC-MS data and their NMR and MS spectra (Fig. 2).

Metabolite A represents the principal urinary metabolite (1% of the administered compound) in the rabbit and derives from the starting compound by loss of 6 carbon atoms, perhaps by 3 consecutive β-oxidation steps, and is optically inactive.

The IR spectrum of the methylester derivative showed absorption for an hydroxyl group (3620 cm^{-1} in $CHCl_3$ solution) and an ester group at 1730 cm^{-1} and the NMR spectrum included a signal for the carbomethoxy group (δ 3.57). The mass spectrum shows (Fig. 3) the molecular ion at m/e 228 and the ions m/e 197 (M^+ - 31, loss of $-OCH_3$), 210 (M -18, loss of H_2O), 179 (loss of water and a methoxyl group), 151 (loss of water and carbomethoxy group), 125 (loss of $CH_2CH_2CH_2CH_2CH_2CH_3$ plus H_2O) and 112 (loss of side chain "a" plus OCH_3).

TABLE 1.

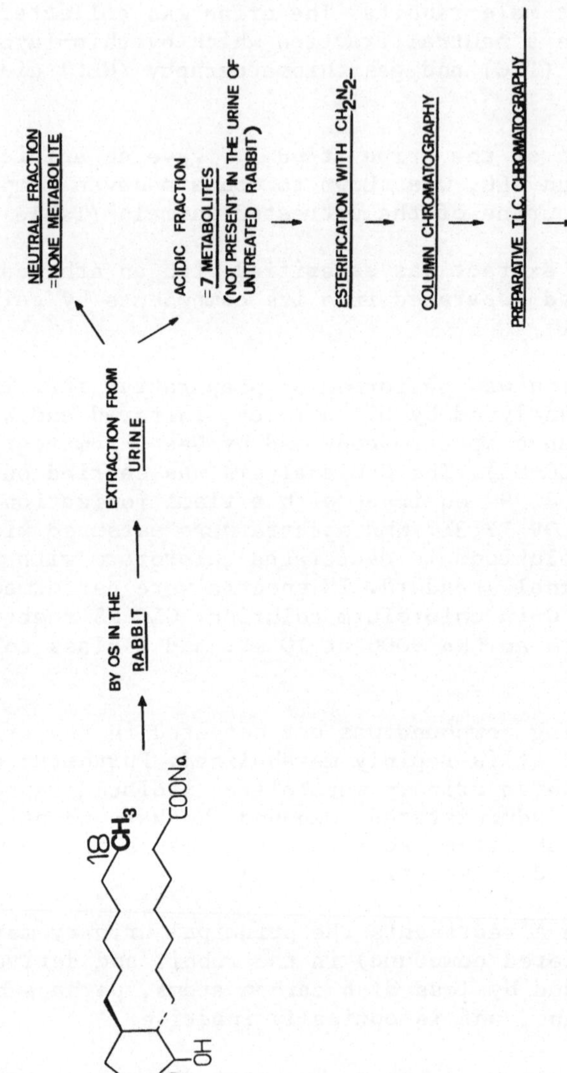
18
CH_3
COONa
9
OH
BY OS IN THE RABBIT
EXTRACTION FROM URINE
NEUTRAL FRACTION
=NONE METABOLITE
ACIDIC FRACTION
=7 METABOLITES
(NOT PRESENT IN THE URINE OF UNTREATED RABBIT)
ESTERIFICATION WITH CH_2N_2
COLUMN CHROMATOGRAPHY
PREPARATIVE T.L.C. CHROMATOGRAPHY
G.C.-MASS SPECTROMETRY

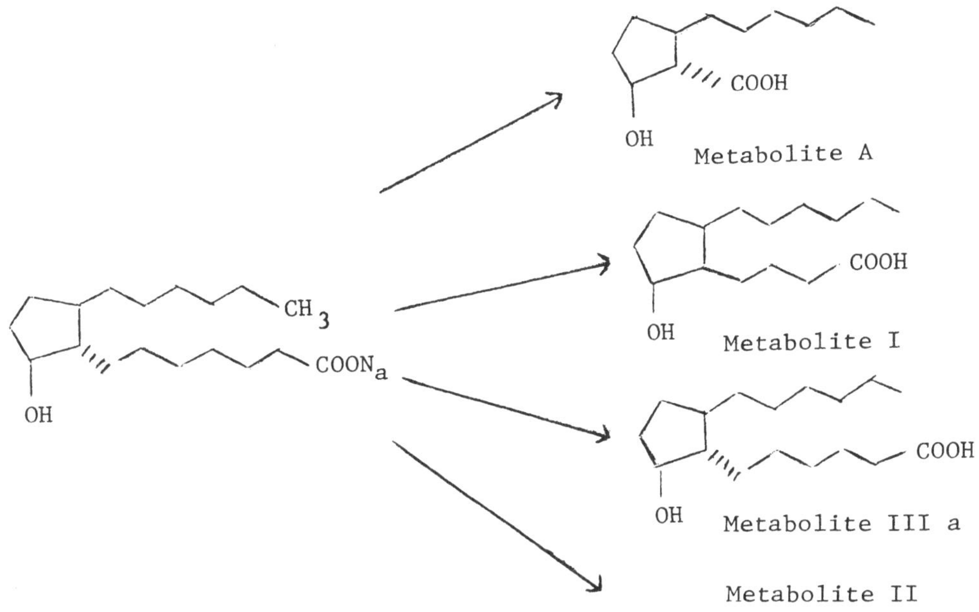

FIG. 2. Metabolic products of 9-hydroxy-19,20-bis-nor prostanoic acid after oral administration in rabbits.

Similar fragmentations were observed in the MS spectrum of the trimethylsilyl -derivative which also showed losses arising from the presence of one trimethylsilyl group. On this basis the compound A is assigned the structure shown in Fig. 2.

The structure of the second metabolite isolated, Metabolite I (0.6% of the administered compound) arises from the starting compound by loss of three carbon atoms, as confirmed by spectral data. In fact the IR spectrum showed the presence of an hydroxyl and a carbomethoxy group (3610 and 1725 cm^{-1} in a $CHCl_3$ solution) indicated also by the NMR signal at δ 1.95 which disappears after D_2O addition.

The molecular ion is not present in the GC-MS spectrum but ions are observed arising from loss of water (m/e 252) and the carbomethoxy group (m/e 239, M-OCH_3). The ion at m/e 185 arises from the loss of the six carbon atoms side chain and that at m/e 169 from the loss of $CH_2CH_2CH_2$-$COOCH_3$ group, as shown in Fig. 4. Further support for this structure came from the MS of its trimethyl-

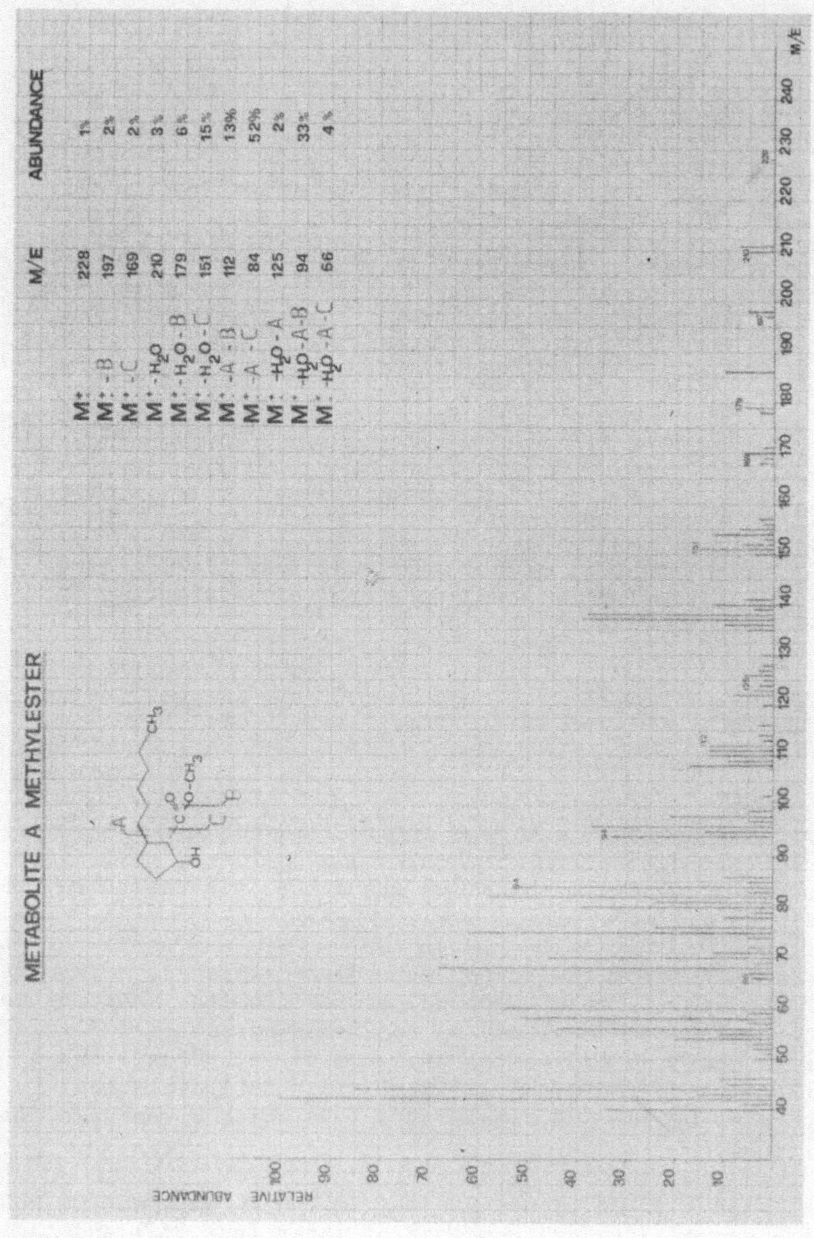

FIG. 3. GC -MS spectrum of Metabolite A methyl ester.

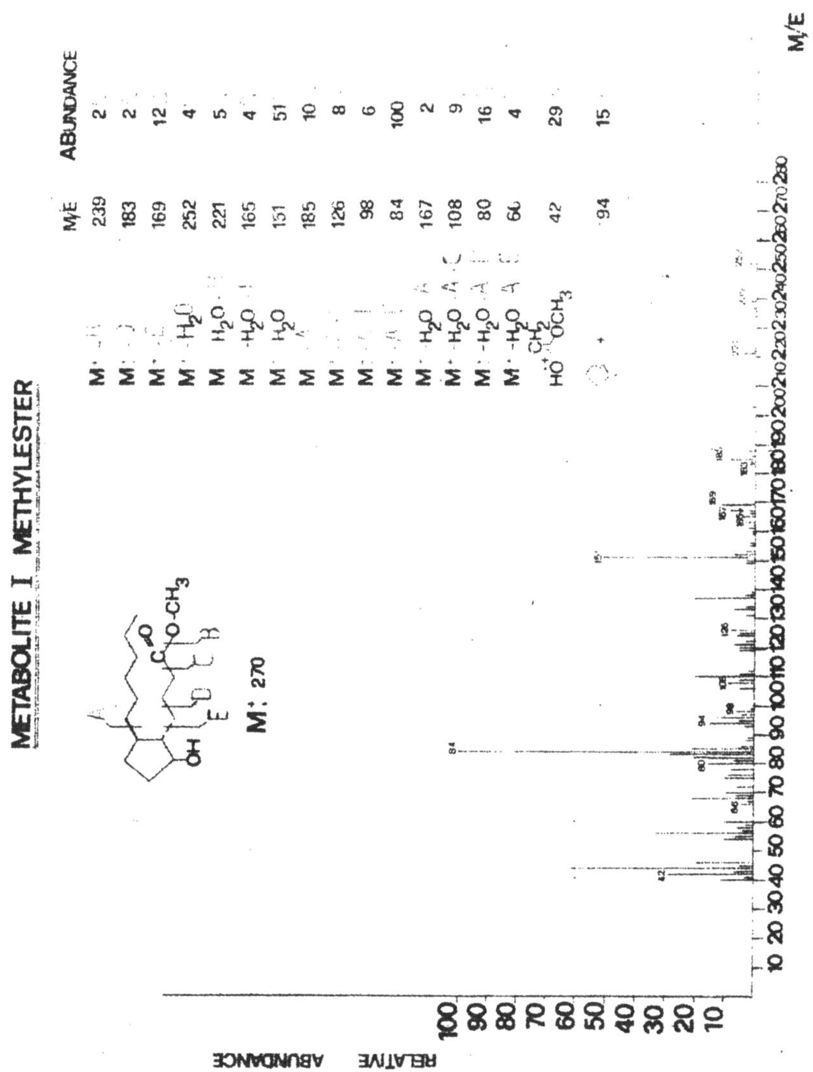

FIG. 4. GC -MS spectrum of Metabolite I methyl ester.

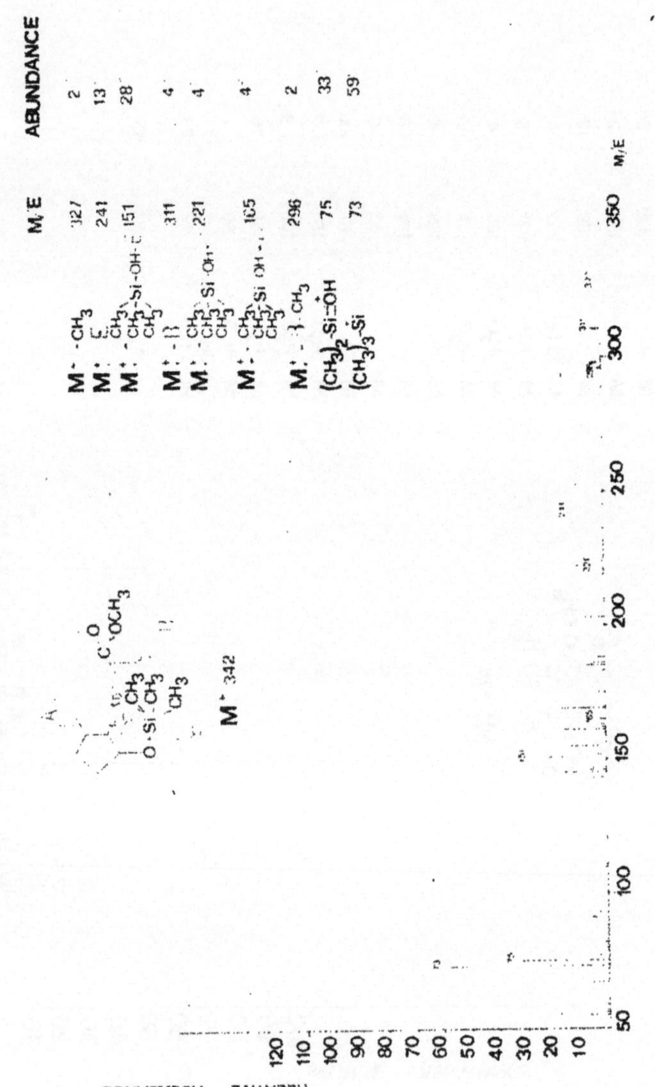

FIG. 5. GC-MS spectrum of Metabolite I methyl ester trimethyl silylether.

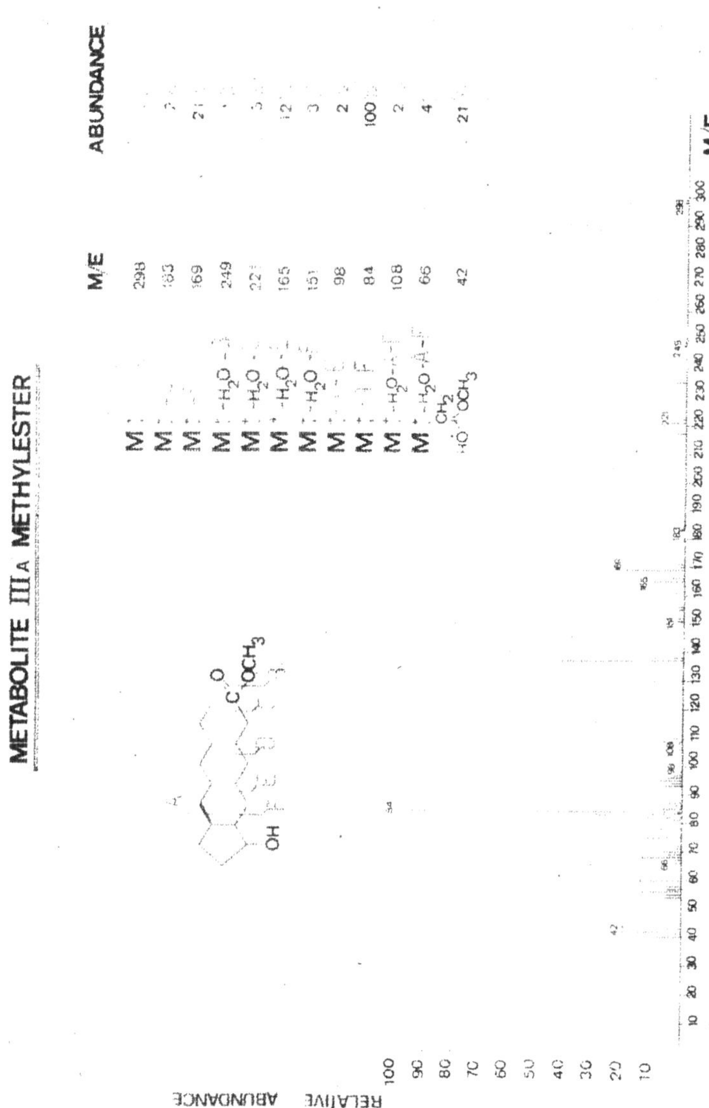

FIG. 6. GC -MS spectrum of metabolite IIIa methyl ester.

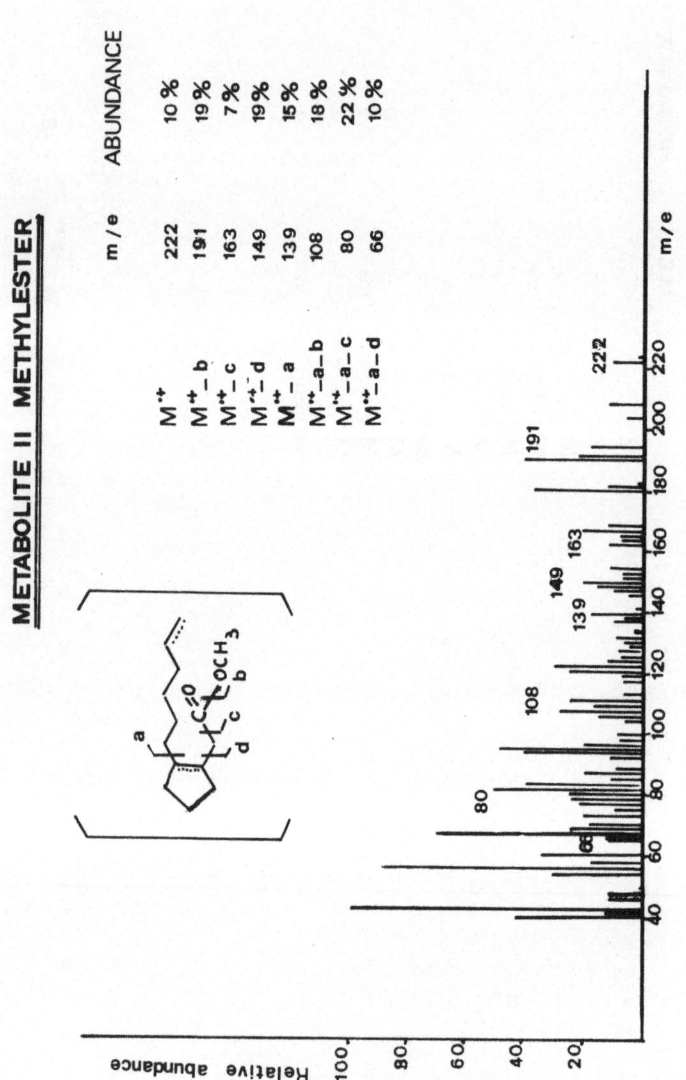

FIG. 7. GC -MS spectrum of metabolite II methyl ester.

silylester (Fig. 5) which contains ions compatible with the proposed structure: m/e 327 (M-15), 241 (M - CH_2CH_2-CH_2COOCH_2), 151 (M - CH_2CH_2-CH_2COOH - TMS:OH) as well as those arising from fragmentations observed in MS of the parent compound.

The structure of metabolite IIIa was assigned from the mass spectral data: molecular ion (m/e 298) and ions arising from loss of water plus methoxy group (m/e 249) or carbomethoxy group (m/e 221) and loss of the side chains were found as it is shown in Fig. 6, corresponding to a structure of the starting products minus a carbon atom. This metabolite represents about 0.2% of the administered compound.

Compound indicated as metabolite II, as shown from its IR spectrum, contains a carbomethoxy group (band at 1735 cm^{-1}) and no hydroxyl group (no bands in the 3600-3500 cm^{-1} region). In the mass spectrum the ion at highest mass (m/e 222) and fragment ions (Fig. 7) would seem to suggest the structure represented but the NMR spectrum shows a signal for only one hydrogen on the double bond. Thus, it has not yet been possible to determine the structure of this compound. All the metabolites of 9-hydroxy-19,20-bis-nor prostanoic acids are found as free acids, not as conjugated products.

The loss of carbon atoms from the side chain, seems the main metabolic pathway of these compounds, as found for the natural prostaglandins (3-13).

The loss of one carbon atom of the side chain is an uncommon degradation. Further studies with tritium labelled compound to determine the excretion and distribution of this compound in several animal species are now in progress, as are studies to fully determine the metabolic pathway of 9-hydroxy-19,20-bis-nor prostanoic acid and to confirm the structure proposed for all unknown metabolites.

For this purpose, we propose to synthesise all the identified metabolites and to determine their pharmacological activity in comparison with the parent drug.

REFERENCES

1) U. Valcavi, Farmaco, Ed. Sci., 1972, 270, 610.
2) U. Valcavi, S. Innocenti, G.B. Zabban and C. Pezzini, Farmaco, Ed. Sci., 1975, 30, 528.
3) E. Granström, J.Am.Chem.Soc., 1969, 3398.
4) K. Green, Acta Chem.Scand., 1969, 23, 1453.
5) H. Hambug, J.Biol.Chem., 1970, 245, 5107.
6) K. Green, Biochemistry, 1971, 10, 1072.
7) E. Granström, J.Biol.Chem., 1971, 246, 1470.

8) H. Hambug, Ann.N.Y.Acad.Sci., U.S.A., 1971, 180, 164.
9) H. Hambug, J.Biol.Chem., 1971, 246, 1073.
10) H. Hambug, J.Biol.Chem., 1971, 246, 6713.
11) E. Granström, Adv. Biosciences, 1972, 9, 39.
12) E. Granström, Prostaglandins, 1975, 9, 14.
13) H. Polet, J.Biol.Chem., 1975, 250, 35.

FORMATION OF A NEW AMINO ACID AS A METABOLIC INTERMEDIATE OF SODIUM DIPROPYLACETATE

I. Matsumoto, T. Kuhara, M. Yoshino and M. Tetsuo

Department of Biochemistry, Kurume University,

School of Medicine, Kurume, Fukuoka 830, Japan

SUMMARY

Using GC-MS a N-acetylated amino acid, 2-n-propyl-3-(N-acetyl-amino)pentanoic acid, was identified in the urine from rats treated with sodium dipropylacetate (DPA).

The urinary contents of β-alanine and methylated 4-amino-5--imidazole-carboxylate were found to be much more abundant in the urine of DPA-treated rats than the controls by the technique of GC-MF.

The N-acetylated amino acid and part of β-alanine may originate from DPA via the metabolic intermediate of β-oxidation, but the reason for the elevated excretion of methylated 4-amino-5-imidazole-carboxylate, degradation product of methylated purine bases, is not clear.

INTRODUCTION

Although sodium dipropylacetate (DPA), an anticonvulsant drug, has a relatively simple chemical structure, the metabolism *in vivo* seems to be considerably complicated due to chain branching. We have reported that DPA is metabolized *via* two metabolic pathways in addition to the glucuronide conjugation of DPA (1, 2). In one pathway DPA is metabolized to 2-*n*-propyl-5-hydroxypentanoic acid followed by 2-*n*-propylglutaric acid (3). 2-*n*-Propylglutaric acid, the end product of ω-oxidation of DPA, is not metabolized further and is excreted in urine without any changes.

In the other, DPA is metabolized to 2-*n*-propyl-3-hydroxypentanoic acid followed by 2-n-propyl-3-oxopentanoic acid (4). In order to know, in detail, the metabolism and the effect of DPA in rats, the urinary acids as well as the amino acids from DPA-treated rats were analyzed.

EXPERIMENTAL

One hundred mg per kg body weight of DPA was administered to male Wistar albino rats weighing about 250 g. Twenty-four hour urine specimens were collected, acidified to pH 1.0 with 6 N-HCl and extracted three times with ethyl acetate and twice with ethyl ether. The combined extracts were trimethylsilylated with bistrimethylsilylacetamide or methylated with dimethylformamide dimethylacetal and subjected to GLC and/or GC-MS as described previously (3). The remaining water-layer was applied to cation exchange resin, Amberlite IR-120 column. The neutral and acidic amino acids were eluted by 1M-pyrimidine and the basic amino acids by 2N-ammonium hydroxide. The eluates were concentrated separately under *vacuo* and analyzed by GLC and/or GC-MS after derivatization to the dimethylaminomethylene methyl ester with dimethylformamide dimethylacetal respectively (5).

The gas chromatograph used was JEOL type JGC-20K equipped with a hydrogen flame ionization detector. A glass column packed with 3% OV-17 on Gas chrom Q as a stationary phase was used.

The combined gas chromatography-mass spectrometric analyses were performed on the same gas chromatograph connected with JEOL type JMS-01SG-2 double focusing high resolution mass spectrometer.

RESULTS AND DISCUSSION

2-n-Propyl-3-(N-acetylamino)-pentanoic acid. A characteristic peak was detected at 20 minutes on the gas chromatogram with 3% OV-17 column of the derivatized urinary acids. As shown in Fig. 1, the molecular ion peak could not be detected, but ions corresponding to M-29 and M-31 were clearly detected. Moreover, abundant peaks at m/e 173 (M-42), m/e 156 (M-59), m/e 144 (M-71) and m/e 100 were detected. The relatively intense peak at m/e 173 may be formed by the elimination of ketene from the molecular ion and the ion at m/e 144 by loss of an ethyl from the ion at m/e 173. The base peak at m/e 100 had an exact mass of 100.0756, ($C_5H_{10}ON$). This ion probably corresponds to the N-acetylpropylamine moiety. If the compound were a glycine conjugate of DPA, a peak at m/e 88 corresponding to the methyl ester of the glycine moiety should be observed, but this was not the case. Although we could not prepare the authentic sample and/or isolate enough sample for full characterization, from the information given above the compound

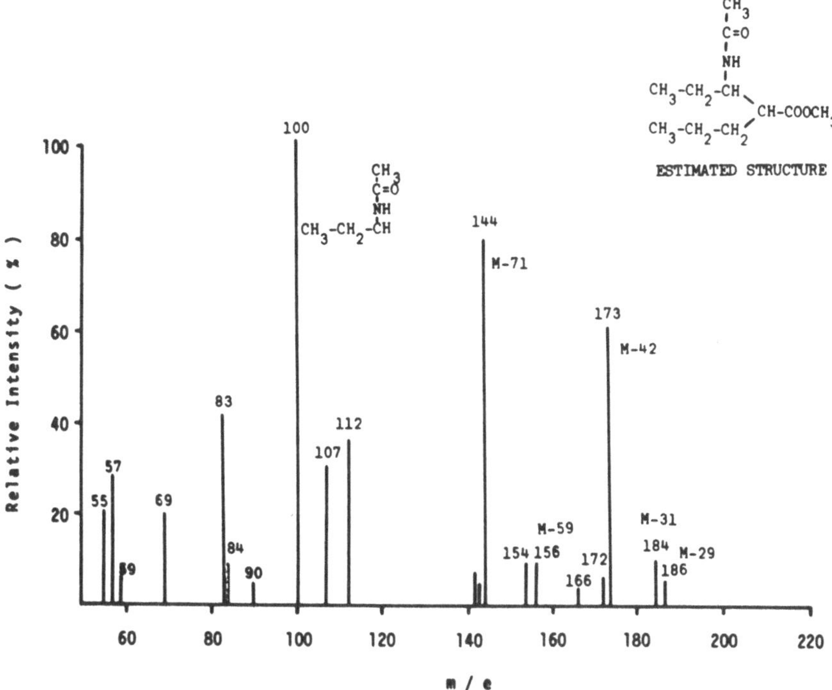

FIG. 1. Mass Spectrum of N-Acetylated Metabolite from Urine of DPA-treated rats.

appears to be the methyl ester of 2-n-propyl-3-(N-acetylamino)--pentanoic acid. This means that the compound in the urine was 2-n-propyl-3-(N-acetylamino)pentanoic acid.

Since DPA does not originally contain a nitrogen atom in its chemical structure, a transamination reaction must occur for the formation of this compound. As we have found that 2-n-propyl-3--oxopentanoic acid was formed as an intermediate of DPA by β--oxidation, transamination may take place between some amino acid and 2-n-propyl-3-oxopentanoic acid to form 2-n-propyl-3-amino-pentanoic acid. This amino acid is probably N-acetylated by acetyl--CoA (Fig. 2). The mechanism seems to be similar to that for Mephenesin which involves a preliminary oxidation followed by transamination and N-acetylation (6).

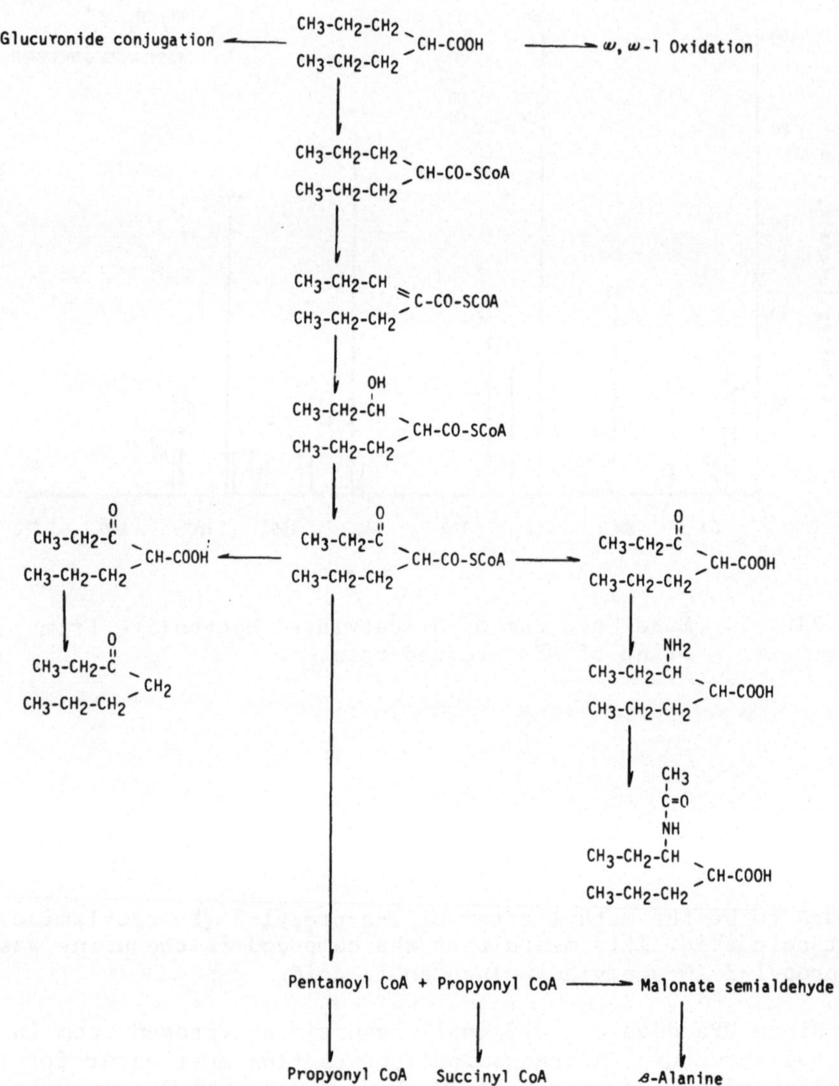

FIG. 2. Tentative metabolic fate of dipropylacetate.

β-Alanine. β-Alanine-N-dimethylaminomethylene methyl ester had a retention time of 8 minutes on the gas chromatogram with 3% OV-17. Its presence in the urine of DPA-treated rats is remarkable. The mass spectrum of the N-dimethylaminomethylene methyl ester of this compound is shown in Fig. 3a. Although the retention time of

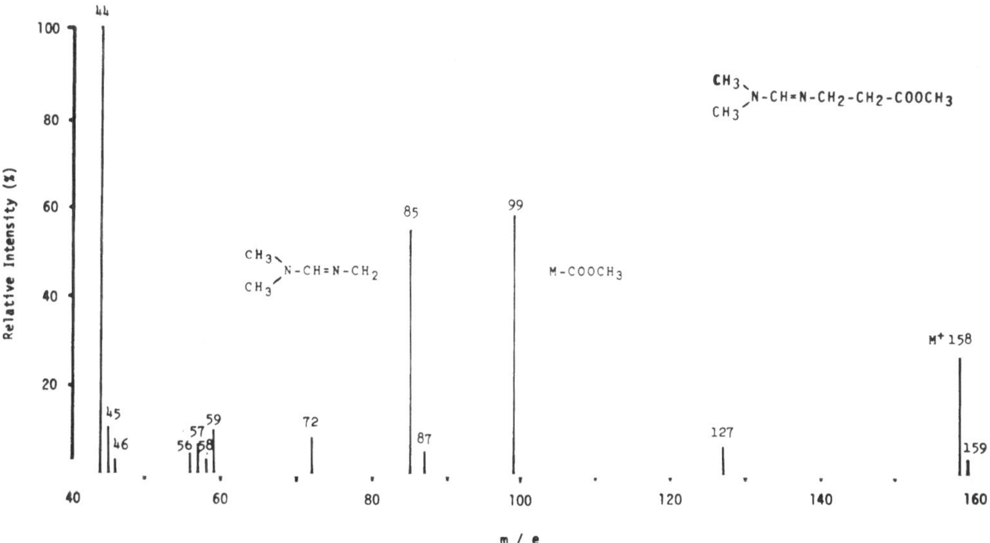

FIG. 3a. Mass spectrum of β-Alanine-N-dimethylaminomethylene methyl ester.

this compound is very close to that of α-alanine, as shown in Fig. 3b their mass spectra are quite different. The M^+ ion of β-alanine is more abundant than that of α-alanine and β-alanine has an ion at m/e 85 almost the same intensity as that at m/e 99. This amino acid which was present in greater quantity than in control rats was confirmed to be β-alanine by mass fragmentographs. β-Alanine known to be present in minor amount in the urine of control rats is derived via transamination of malonic semialdehyde which originates from many biological compounds. Because a large amount of propionic acid (2) as well as several intermediates of DPA via β-oxidation was found in the acidic extracts of the urine, a part of the β-alanine excreted in the urine of DPA-treated rats was suspected to originate from DPA through the proposed pathways shown in Fig. 2.

Methylated 4-amino-5-imidazolecarboxylate. Another peak characteristic of the urine of DPA-treated rats was also analyzed by GC-MS after the urinary amino acids were derivatized with

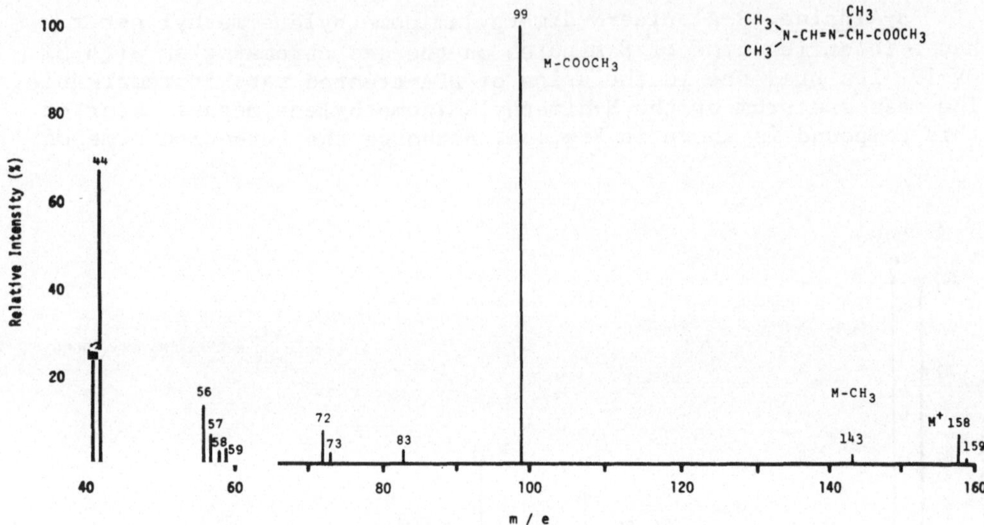

FIG. 3b. Mass spectrum of α-Alanine N-Dimethylamino-methylene methyl ester.

dimethylformamide dimethylacetal. The mass spectrum of the peak is shown in Fig. 4. The molecular ion was observed at m/e 210 and is the base peak in the spectrum. The ion at m/e 195 [M-15] was also intense. The high resolution mass spectrum of this compound is shown in Table 1. The exact mass of the molecular ion was 210.1116 and had an elemental composition of $C_9H_{14}N_4O_2$ and therefore the underivatized original molecule must contain three nitrogen atoms. From these and other data this compound appears to be methylated 4-amino-5-imidazolecarboxylic acid N-dimethylaminomethylene methyl ester. A possible precursor of the parent compound is methylated 4-ureido-5-imidazolecarboxylate derived from methylated purine nucleotides. By mass fragmentography, it was observed that the compound was also present in minor amount in the control rats. However, the reason for the elevated excretion of methylated 4-amino-5-imidazolecarboxylate into urine of DPA-treated rats is, at present, not clear.

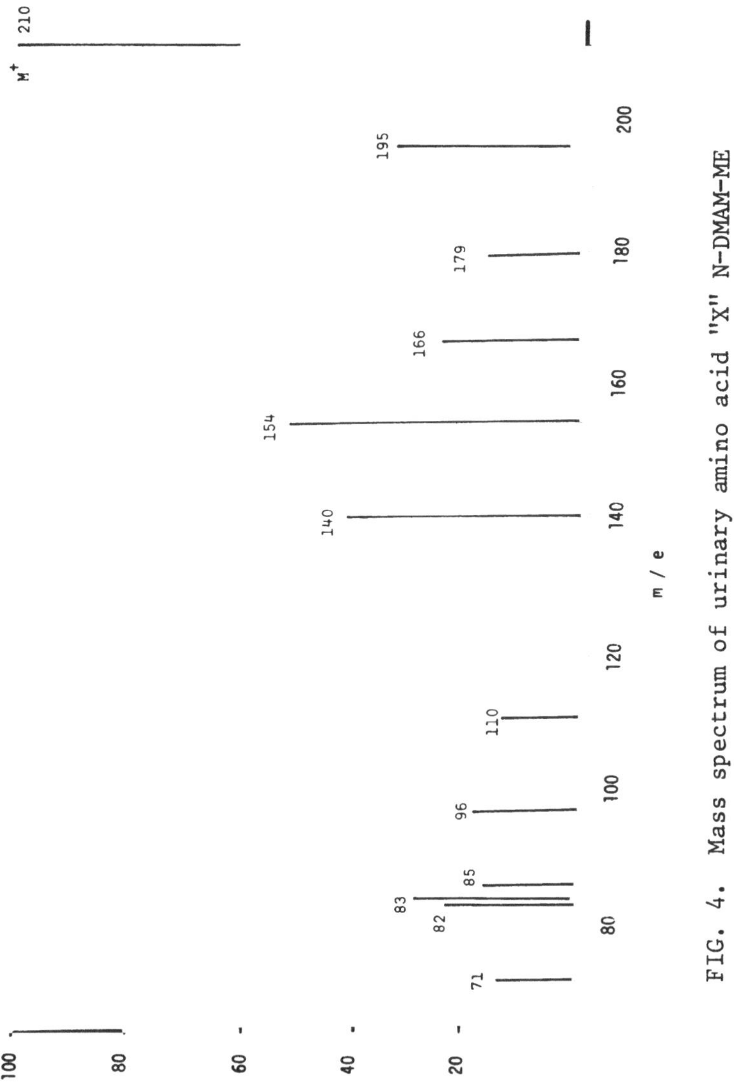

FIG. 4. Mass spectrum of urinary amino acid "X" N-DMAM-ME

TAB. 1.

HIGH RESOLUTION MASS SPECTRUM OF COMPOUND X

OBSD.	CALD.	ERROR*	C	H	N	O
210.1117	210.1116	0.1	9	14	4	2
195.0877	195.0881	0.4	8	11	4	2
179.0919	179.0932	-1.3	8	11	4	1
168.0782	168.0773	-1.0	7	10	3	2
166.0622	166.0616	0.5	7	8	3	2
154.0774	154.0742	3.2	7	10	2	2
140.0604	140.0585	1.9	6	8	2	2
110.0729	110.0718	1.0	5	8	3	0
96.0690	96.0687	0.2	5	8	2	0
83.0593	83.0609	-1.5	4	7	2	0
82.0517	82.0530	-1.3	4	6	2	0

* Error in mili-mass unit

REFERENCES

1) K. Kukino and I. Matsumoto, J. Kurume Med. Assoc., 1971, 34, 396.
2) I. Matsumoto, T. Kuhara and M. Yoshino, Biomed. Mass Spectrom., in press, 1976.
3) T. Kuhara and I. Matsumoto, Biomed. Mass Spectrom., 1974, 1, 291.
4) I. Matsumoto, T. Kuhara and M. Yoshino, in "Advances in Mass Spectrometry in Biochemistry and Medicine", Vol. 1, A. Frigerio and N. Castagnoli, Eds., Spectrum Publications, Inc., New York, 1976, p. 17.
5) J.P. Thenot, Anal. Lett., 1972, 5, 217.
6) T. Kuhara, M. Tetsuo, M. Yoshino and I. Matsumoto, in "Advances in Mass Spectrometry in Biochemistry and Medicine", Vol. 2, A. Frigerio, Ed., Spectrum Publications, Inc., New York, in press, 1976.

IDENTIFICATION OF THE N-ACETYL-β-ARYLOXYALANINES FROM RATS TREATED WITH 3-(*p*-TOLYLOXY)-1,2-PROPANEDIOL AND 3-PHENOXY-1,2-PROPANEDIOL

T. Kuhara, Y. Inoue and I. Matsumoto

Kurume University School of Medicine

Kurume, Fukuoka 830, Japan

SUMMARY

Using gas chromatography-mass spectrometry, N-acetyl-β--(*p*-tolyloxy)alanine was identified in the urine of rats treated with 3-(*p*-tolyloxy)-1,2-propanediol. Similary, N-acetyl-β-phenoxyalanine was identified in the urine of rats treated with 3-phenoxy-1,2-propanediol.

Thus the new metabolic conversion involving introduction of nitrogen which was found for the first time in the metabolism of Mephenesin, 3-(*o*-tolyloxy)-1,2-propanediol, was shown to occur also in the metabolism of the above two compounds.

INTRODUCTION

Mephenesin, 3-(*o*-tolyloxy)-1,2-propanediol, undergoes hydroxylation on the aromatic ring (1) and oxidation of the C_1-alcoholic hydroxyl group to form β-*o*-tolyloxylactic acid, which is the major metabolite of Mephenesin (2,3).

In the course of further study, a new metabolite which contained a nitrogen atom was detected in large amounts in the urine of rat and identified as N-acetyl-β-*o*-tolyloxyalanine by GC-MS, liquid chromatography and TLC. This represented the first case in which a nitrogen atom, not due to conjugation with amino acids, is introduced during the metabolism of a drug (4).

In order to investigate whether or not the new drug alteration was limited to this drug, we studied the metabolism of two

Mephenesin analogues, 3-(p-tolyloxy)-1,2-propanediol and 3-phenoxy-1,2-propanediol. The identification of the corresponding N-acetyl-amino acids are described in this paper.

MATERIALS AND METHODS

Chemicals. 3-(o-Tolyloxy)-1,2-propanediol and 3-phenoxy-1,2-propanediol were synthesized by acid-catalyzed hydrolysis of the corresponding epoxides according to the procedure of Ulbrich et al. (5). N-Acetylphenylalanine, hexamethyldisilazane, N,O-bistrimethylsilylacetamide (BSA) and tetramethylchlorosilane (TMCS) were purchased from Tokyo Kasei Kogyo Co. Ltd., Japan. Other chemicals were obtained from commercial sources.

Biological sampling and extraction procedure. Male Wistar albino rats weighing about 250 g were used. One group of rats received 150 mg/kg body weight of 3-(p-tolyloxy)-1,2-propanediol by stomach intubation and another group was given 150 mg/kg body weight of 3-phenoxy-1,2-propanediol. The animals were fasted and the twenty-four hour urine was collected. A portion of the urine was adjusted to pH 12.0 with 2N NaOH and extracted three times with ethyl ether. The remaining aqueous layer was acidified to pH 1.0 with 6N HCl and extracted in the same way. The combined acidic extracts were dried over $MgSO_4$ and evaporated to dryness under reduced pressure. The residues were trimethylsilylated in two ways, one withe BSA and TMCS, and the other with hexamethyldisilazane.

Gas chromatography and gas chromatography-mass spectrometry. A double focusing, high resolution JEOL JMS-01SG-2 mass spectrometer coupled with a gas chromatograph model JGC-20K was used. A glass column (2 m x 3 mm) packed with 3% OV-17 on Gas Chrom Q was used for separation. The electron energy was set at 75 eV, emission current at 200 µA and ion-accelating voltage at 8 kV. The temperatures of the ion source, the separator and the injection port were about 250 °C.

RESULTS

Identification of N-acetyl-β-(p-tolyloxy)alanine in the urine of rats treated with 3-(p-tolyloxy)-1,2-propanediol. The Mephenesin analogue, 3-(p-tolyloxy)-1,2-propanediol was administered to rats and urinary metabolites were identified by GC-MS. In the alkaline extracts, the unchanged substance was the major component. Fig. 1 (top) shows the gas chromatographic patterns of acidic extracts from the urine of rat treated with 3-(p-tolyloxy)-1,2-propanediol. For derivatization, hexamethyldisilazane was used as the trimethylsilylating agent. The column temperature was programmed from 200 °C at 3 °C/min. The large peak pointed by a small arrow was identified as β-(p-tolyloxy)-lactic acid.

The mass spectrum of the peak indicated by the large arrow is

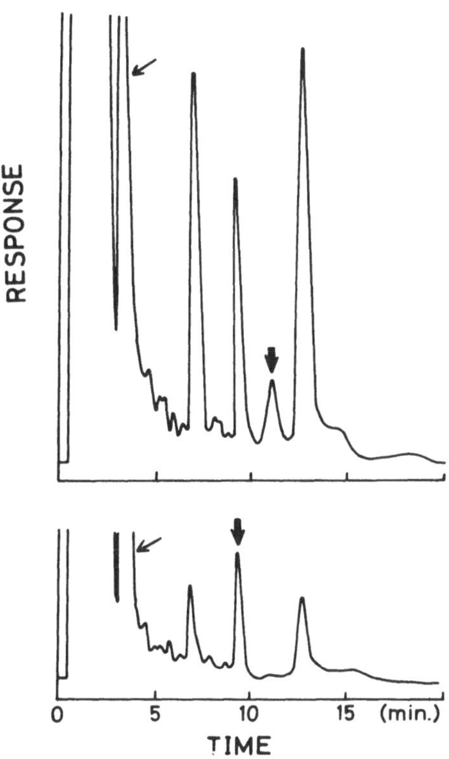

FIG. 1. The gas chromatograms of the urinary acids from 3-(p-tolyloxy)-1,2-propanediol-treated (top) and 3-phenoxy-1,2-propanediol-treated rats (bottom).

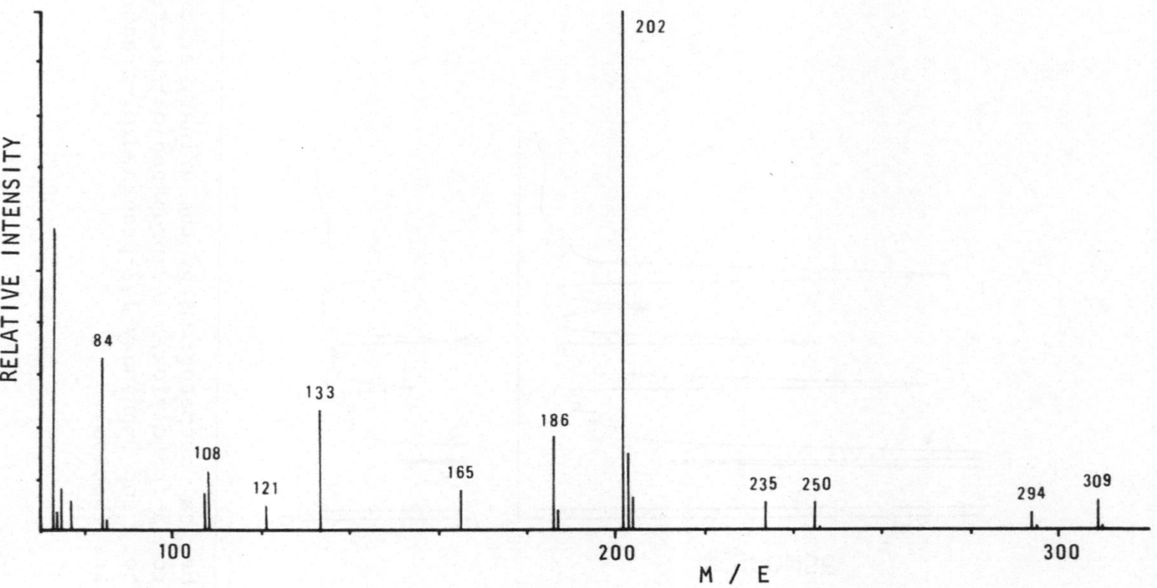

FIG. 2. The mass spectrum of the unknown metabolite from 3-(*p*-tolyloxy)-1,2-propanediol-treated rats. Trimethylsilylation was done with hexamethyldisilazane.

shown in Fig. 2. The molecular ion was seen at m/e 309. The odd number of the molecular ion indicated that the compound contained nitrogen. The ions corresponding to the tolyloxy group were observed at m/e 107, at m/e 108 with a hydrogen transfer, and at m/e 121. The base peak at m/e 202 (M-107) corresponded to the N-acetyl-alanine TMS ester moiety.

Ions due to M-59 and M-74 were observed in the mass spectrum of N-acetylphenylalanine TMS ester as shown in Fig. 3. The fragmentation pattern of the unknown compound was also similar to that of TMS ester of N-acetyl-β-(o-tolyloxy)alanine (Fig. 4). From these data the metabolite was estimated to be N-acetyl-β--(p-tolyloxy)alanine. In order to confirm the proposed structure, the acidic extracts from the urine were treated with BSA and TMCS. R.F. Coward et al. (6) reported that N-acetylphenylalanine after reaction with hexamethyldisilazane gave a peak with Methylene unit (M.U.) 20.7 on 3 % OV-17, and that a peak with M.U. 19.3 was obtained when BSA and TMCS were used. The former corresponds to monotrimethylsilylated and the latter to di-trimethylsilylated N-acetylphenylalanine. Their respective mass spectra are shown in Fig. 3 and Fig. 5. The di-TMS derivative of N-acetylphenylalanine gave the molecular ion at m/e 351. The ion at m/e 260 was formed by the release of the benzyl moiety, and the ion at m/e 234 was the elimination product of carbotrimethylsilyloxy radical. A characteristic fragmentation of the N-acetyl compounds was the loss of ketene, which gave ions at m/e 218 and at m/e 192. Fig. 6 illustrates the mass spectra obtained after the acidic extracts of the urine were reacted with BSA and TMCS. The molecular ion was seen at m/e 381 corresponding to the di-TMS derivative of N--acetyl-β-(p-tolyloxy)alanine. The ion at m/e 260 resulted from fission α to the nitrogen atom. The elimination of ketene from the fragment at m/e 260 produced the intense peak at m/e 218. The ion at m/e 274 was due to removal of the tolyloxy radical. From the two mass spectra the structure of the metabolite was confirmed to be N-acetyl-β-(p-tolyloxy)alanine.

Identification of N-acetyl-β-phenoxyalanine in the urine of rats treated with 3-phenoxy-1,2-propanediol. The other Mephenesin analogue, 3-phenoxy-1,2-propanediol was given to rats and the urine sample was analyzed by GC-MS. In the alkaline extracts, the main component was 3-phenoxy-1,2-propanediol. The gas chromatogram of the acidic extracts was obtained after trimethylsilylation (shown in Fig. 1, bottom). A major metabolite was β-phenoxylactic acid corresponding to the large peak indicated by the small arrow.

The gas chromatographic peak denoted by the large arrow was analyzed by mass spectrometry. Its mass spectra (Fig. 7) was quite similar to that of authentic N-acetylphenylalanine mono TMS derivative and also similar to those of TMS esters of N-acetyl-β--(o-tolyloxy)alanine and N-acetyl-β-(p-tolyloxy)alanine. The molecular ion was seen at m/e 295. The peak at m/e 202 was always

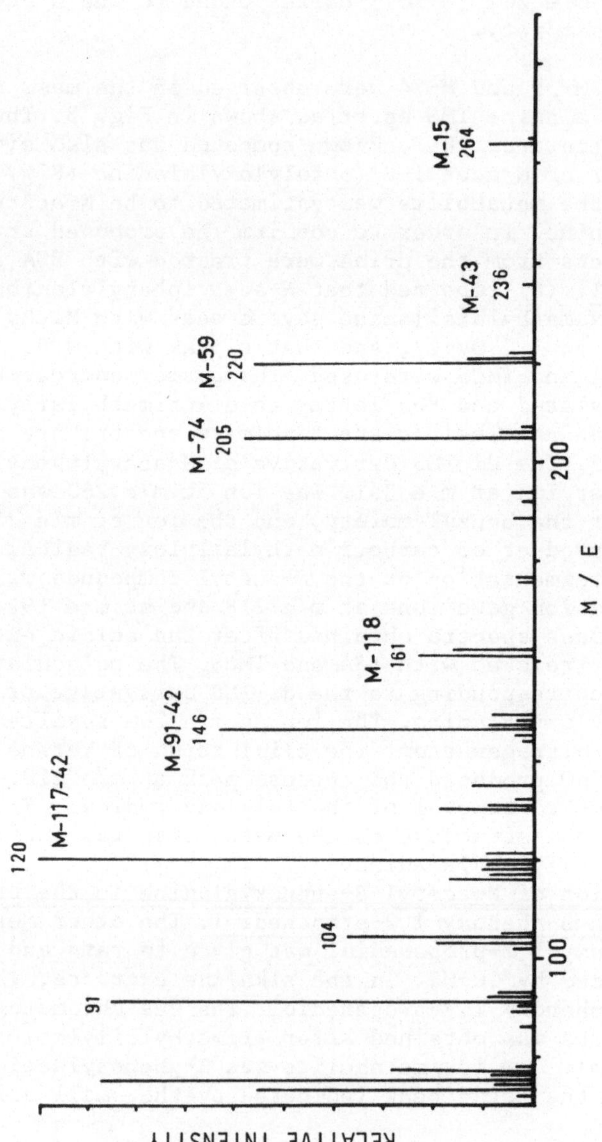

FIG. 3. The mass spectrum of trimethylsilyl ester of N-acetylphenylalanine.

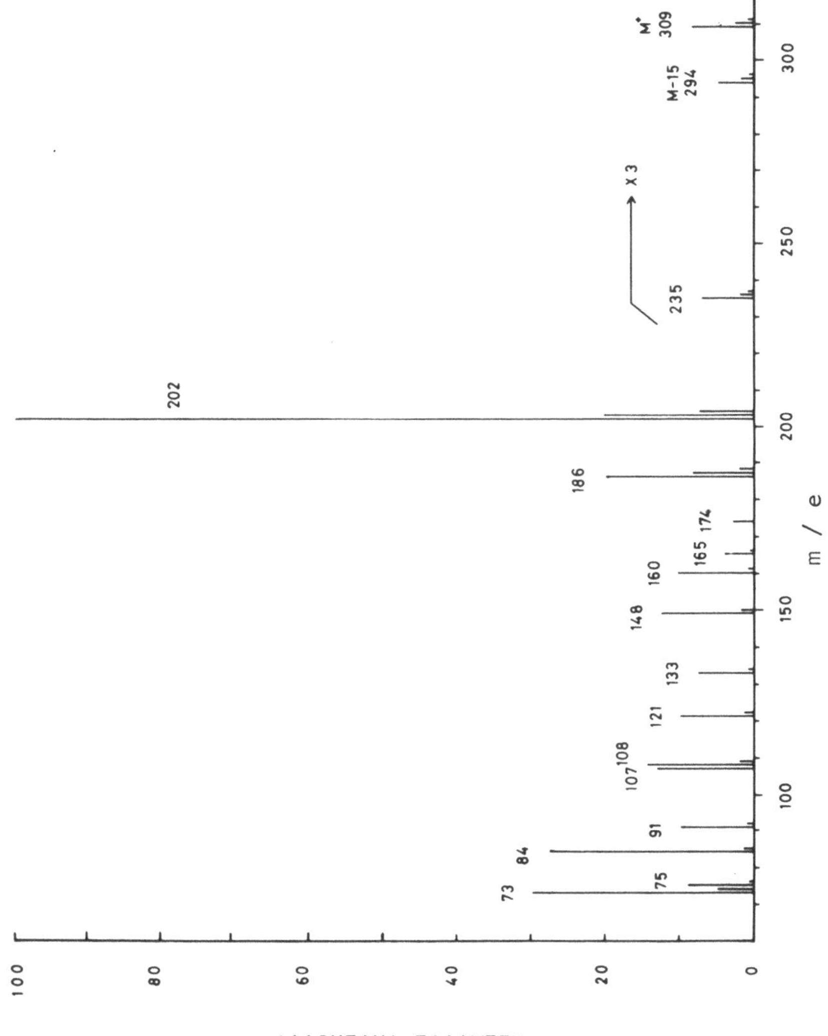

FIG. 4. The mass spectrum of trimethylsilyl ester of N-acetyl-β-(o-tolyloxy)alanine.

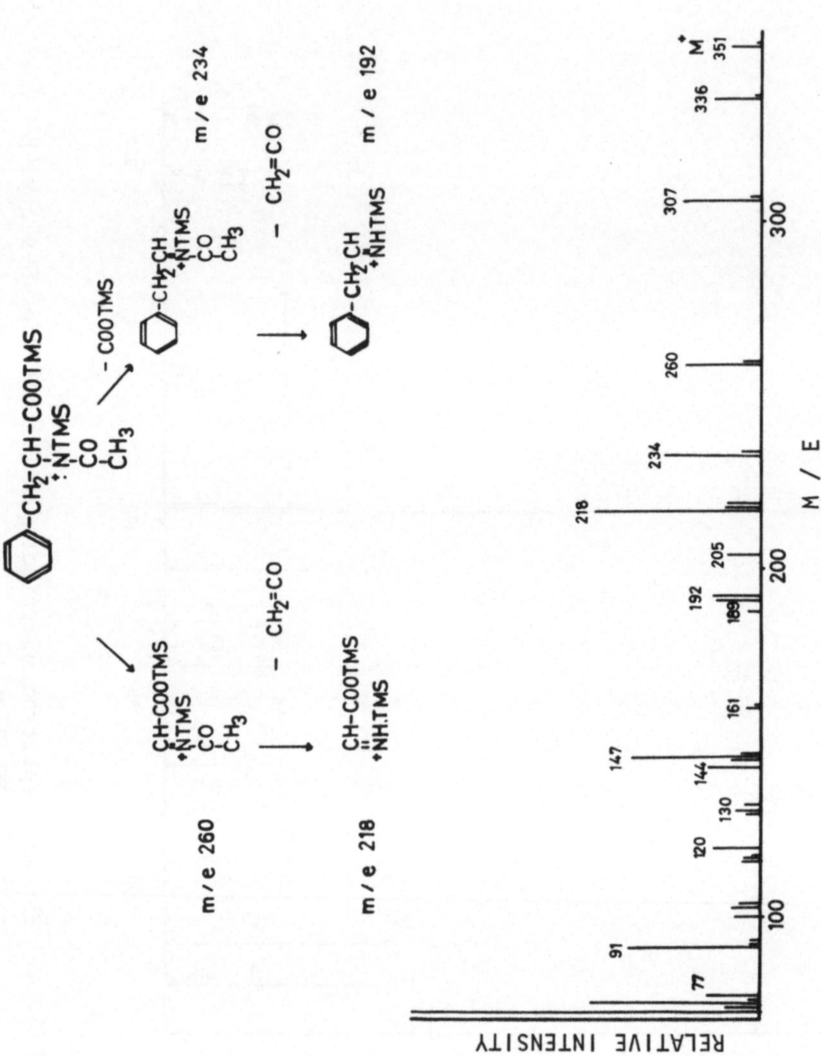

FIG. 5. The mass spectrum of N-acetylphenylalanine trimethylsilylated with BSA and TMCS.

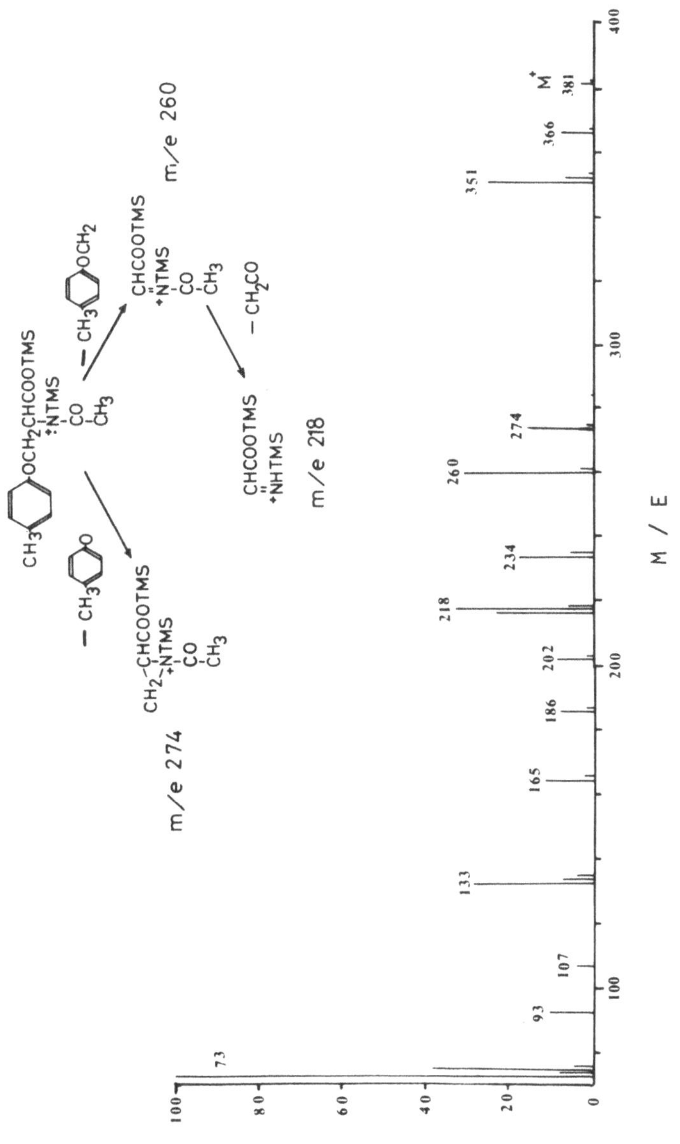

FIG. 6. The mass spectrum of the unknown metabolite from 3-(p-tolyloxy)-1,2-propanediol-treated rats. Trimethylsilylation was done with BSA and TMCS.

the base peak in the spectra of the N-acetyl-β-aryloxyalanine TMS esters. Ions assignable to the tropylium ion (m/e 91), phenol (m/e 94) and the phenoxymethyl fragment (m/e 107) were present. Accordingly, the unknown metabolite was estimated to be N-acetyl--β-phenoxyalanine. This was further supported by the mass spectra obtained after the urinary acids were reacted with BSA and TMCS (Fig. 8). As expected, the N-acetyl group underwent trimethylsilylation and the molecular ion shifted to m/e 367. The characteristic fragmentations observed supported the existence of the acetoamide fragment in the molecule. The ion at m/e 218 was formed by the sequential loss of phenoxymethyl radical and ketene from the molecular ion. A peak at m/e 274, is assigned to the ion resulting from the elimination of a phenoxy radical.

From these data, the structure of the unknown was assigned as N-acetyl-β-phenoxyalanine.

DISCUSSION

By gas chromatography-mass spectrometry N-acetyl-β-aryloxyalanine was identified as a metabolic product of 3-aryloxy-1,2-propanediol in rats. Although we have not yet obtained authentic N-acetyl-β-(p-tolyloxy)alanine and N-acetyl-β-phenoxyalanine, the proposed structure was supported by following data: 1. The gas chromatographic properties such as heat instability and retention time were very similar to that of N-acetyl-β-(o-tolyloxy)alanine TMS ester. 2. The mass spectra of mono-trimethylsilylated N-acetyl-β-(p-tolyloxy)alanine and N-acetyl-β-phenoxyalanine closely resembled that of N-acetyl-β-(o-tolyloxy)alanine, the structure of which was confirmed by high resolution mass measurement, acid hydrolysis and TLC. 3. These two metabolites showed very similar properties to N-acetylphenylalanine on trimethylsilylation and similar mass spectral fragmentations.

N-Acetyl-β-(o-tolyloxy)alanine as well as β-(o-tolyloxy)lactic acid are major metabolites of Mephenesin. In the rats treated with 3-(p-tolyloxy)-1,2-propanediol and 3-phenoxy-1,2-propanediol, the N-acetyl-β-aryloxyalanine was not as abundant as the β-aryloxylactic acid. However, the new metabolic conversion involving introduction of nitrogen was confirmed to occur also for the two Mephenesin analogues.

It is well known that foreign carboxylic acids undergo conjugation with glycine or other amino acids. In this way drugs without nitrogen are converted to the nitrogen containing compounds. Deamination by transamination was reported by Tanaka (7), in the metabolism of an unusual amino acid, Hypoglycin A. The drug biotransformation which involved introduction of nitrogen, not due to conjugation with amino acids, was demonstrated for the first time

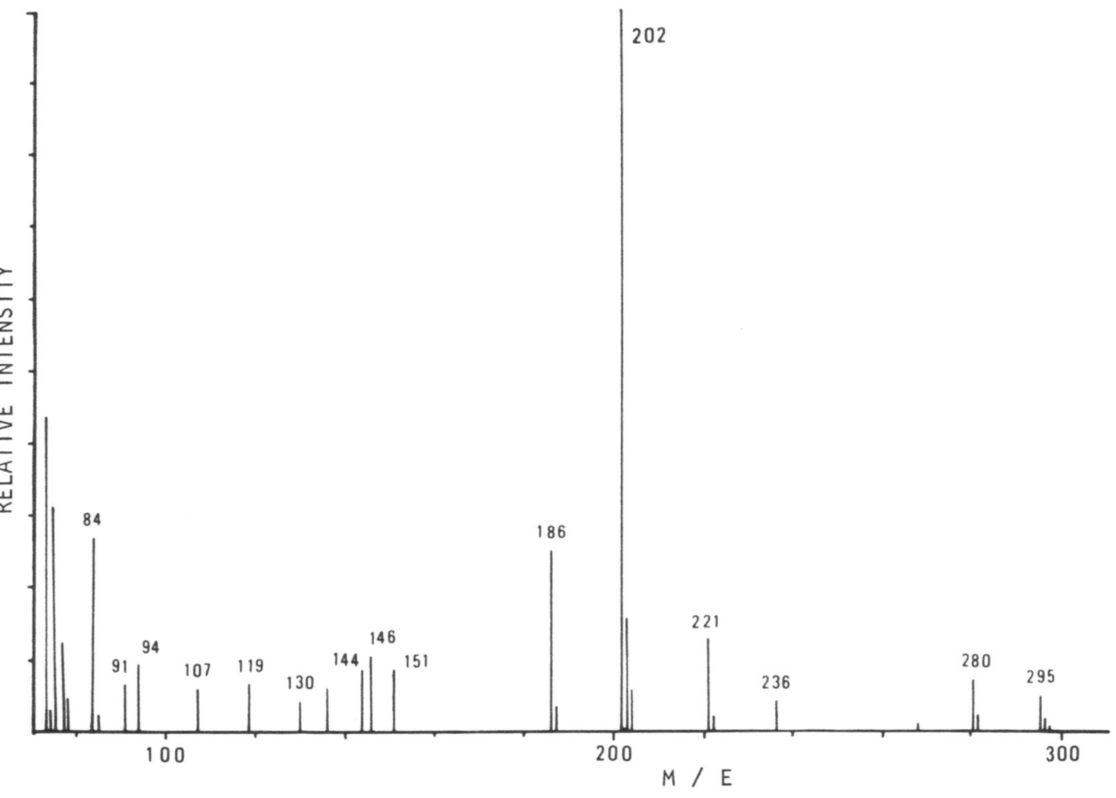

FIG. 7. The mass spectrum of the unknown metabolite from 3-phenoxy-1,2-propanediol-treated rats. Trimethylsilylation was done with hexamethyldisilazane.

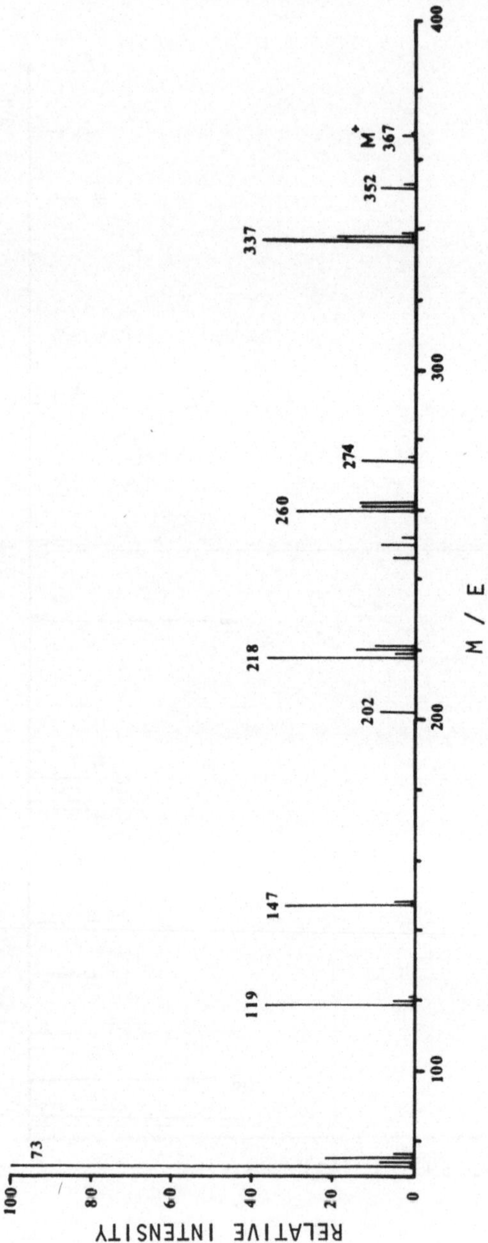

FIG. 8. The mass spectrum of the unknown metabolite from 3-phenoxy-1,2-propanediol-treated rats. Trimethylsilylation was done with BSA and TMCS.

in the case of Mephenesin (4).

Transamination is known to occur in the body on normal constituents of the cells and it appears that this is the only conceivable reaction responsible for the formation of the new metabolites. The substrate specificity of the transaminase and the localization of the enzyme are very interesting and remain to be understood. We cannot exclude the possibility of a non-enzymatic pyridoxamine-mediated reaction.

For the formation of N-acetyl-β-aryloxyalanine from 3-aryloxy--1,2-propanediol the following pathway was proposed. At first the administered substance undergoes oxidation at the C_1-hydroxyl group to form the β-aryloxylactic acid. Then the β-aryloxylactic acid is dehydrogenated to give β-aryloxypyruvic acid, a reaction which might be catalyzed by phenylpyruvic acid oxidoreductase. The α--oxo-acid thus formed undergoes transamination and is converted to β-aryloxyalanine. This unusual amino acid can be acetylated by acetyl-CoA transferase and excreted into the urine as N-acetyl--β-aryloxyalanine.

Oxidation, reduction, hydrolysis and conjugation are general reactions in drug metabolism. We propose a new metabolic pathway which consists of preliminary oxidations, transamination and N-acetylation.

REFERENCES

1) M. Tetsuo, T. Kuhara, M. Yoshino, J. Kunitomo, M. Ju-ichi and I. Matsumoto, in "Proceedings of the 2nd International Symposium on Mass Spectrometry in Biochemistry and Medicine", 1974.
2) E.L. Graves, T.J. Elliott and W. Bradley, Nature (London), 1948, 162,257.
3) R.F. Riley, J. Am. Chem. Soc., 1950, 72,5712.
4) T. Kuhara, M. Tetsuo, M. Yoshino and I. Matsumoto, in "Proceedings of the 3rd International Symposium on Mass Spectrometry in Biochemistry and Medicine", in press, 1975.
5) V. Ulbrich, J. Makes and M. Jurecek, Collect. Czech. Chem. Commun., 1964, 29,1466.
6) R.F. Coward and P. Smith, J. Chromatogr., 1969, 45,230.
7) K. Tanaka, J. Biol. Chem., 1972, 247,7465.

METABOLISM IN RAT, DOG AND MAN OF DIFENPIRAMIDE, A NEW ANTI-INFLAMMATORY DRUG

E. Grassi, G.L. Passetti, A. Trebbi and A. Frigerio°

Laboratori Zambeletti, Milan, Italy; °Istituto Ricerche Farmacologiche "Mario Negri", Milan, Italy

INTRODUCTION

Difenpiramide (α-pyridylbiphenyl acetamide) (1) synthesized in our research laboratory, is a promising non-steroidal anti--inflammatory drug which is well-tolerated and orally effective (LD_{50}= 1966 p.o., LD_{50}= 74000 in dog; ED = 19.9 mg/Kg in rat carrageneen oedema test). Its metabolism in rat, dog, man and rat tissue homogenate has been studied. The identification of the drug and its metabolites (biphenylylacetic acid, *p*-hydroxybiphenylylacetic acid and α-aminopyridine) was carried out using thin layer chromatography (TLC) and UV spectroscopy, gas chromatography-mass spectrometry (GC-MS) and high pressure liquid chromatography (HPLC).

MATERIALS AND METHODS

Reagents and standards. All the chemicals used were of analytical reagent grade and were tested for purity in blank runs. Ketodase was obtained from Warner-Chilcott, Morris Plains, N.J., U.S.A. Biphenylylacetic acid was synthesized from biphenyl and acetic chloride by a Friedel-Crafts reaction followed by a Willgerodt-Kindler reaction. *p*-Hydroxybiphenylylacetic acid was prepared in a similar way from *p*-methoxybiphenyl. The amino acid conjugates were obtained by reaction of biphenylylacetyl chloride and *p*-acetoxybiphenylylacetyl chloride with the appropriate amino acid.

Animals. Beagle dogs (10-12 Kg) and Wistar rats (200-250 g) bred in our animal house were used. Animals were maintained on a standard diet and fasted overnight before experiments. Difenpiramide suspended in 1% carboxymethylcellulose was administered orally to dogs (45 mg/Kg) and rats (50 mg/Kg). Human volunteers received a single oral dose of 250 mg in a gelatin capsule.

Isolation of metabolites. After hydrolysis, serum, urine, bile, stomach and intestine samples were adjusted to pH 2 and extracted with organic solvents. The aqueous layer was made alkaline (pH 9) and reextracted to isolate α-aminopyridine.

For the isolation of the amino acid conjugates the sample was hydrolysed (pH 2) at 60° for 1 hr. The solution was then extracted with methylisobutylketone and the extract purified by TLC. Drug and metabolites were purified by preparative TLC using plates coated with silica gel (Merck F 254; 2 mm). The plates were developed with cyclohexane:acetone:methanol (65:35:4; system "A") or chloroform :methanol:ammonium hydroxide (20:30:2; system "B").

The band with R_f corresponding to those of the standards (Table 1) were taken and eluted with methanol and the solvent was

TABLE 1.

Retention Times, R_f and $E^{1\%}_{1cm}$ Values of Difenpiramide and its Principal Metabolites

COMPOUND	RETENTION TIMES (min)		R_f VALUES TLC		$E^{1\%}_{1\,cm}$	λ_{max}
	HPLC	GLC	SYSTEM "A"	SYSTEM "B"		
DIFENPIRAMIDE	8.25	5*	0.53	0.93	810	248 nm
BIPHENYLYLACETIC ACID	3.92	2.7 (Methyl ester)	0.40	0.43	780	253 nm
pHYDROXY BIPHENYLYLACETIC ACID	0.88	7.5 (Methyl ester)	0.29	0.24	980	265 nm

SYSTEM "A": CYCLOHEXANE : ACETONE : METHANOL (65:35:4)

SYSTEM "B": CHLOROFORM : METHANOL : AMMONIUM HYDROXIDE (CONC.) (70:30:2)

then evaporated under N_2 at room temperature. The residues were then analysed by mass spectrometry using a direct inlet system (DIS). In all cases the mass spectra resulted identical with those of the authentic compounds.

Ultraviolet spectroscopy (UV). Quantitative analyses were carried out with a Beckman Model 25 Spectrophotometer. The $E^{1\%}_{1cm}$ and the R_f values for the three standards are shown in Table 1.

Gas liquid chromatography. Analyses were carried out with a Hewlett Packard Model 5750 gas chromatograph equipped with a Flame Ionization Detector. For halogenated derivatives a Perkin Elmer Model F 30 gas chromatograph equipped with a ^{63}Ni electron capture detector was used. The column consisted of a glass tube (90 cm x 6 mm ID) packed with 100-120 mesh Gas-chrom Q (Applied Science Lab., State College, Pennsylvania, U.S.A.) coated with 3% U.C.W.98.

The column was conditioned for 1 hr at 150° with a N_2 flow rate of 55 ml/min, for 4 hr at 300° without N_2 and then for 24 hr at 270° with a N_2 flow rate of 55 ml/min. The following operating conditions were used for the analysis of the compounds of interest and their derivatives; difenpiramide:column temperature, 270°; injection port and flame detector temperature, 300°; N_2 flow rate, 55 ml/min.

BPA and pHBA were analysed as the methyl esters: column temperature, 215°; injector port and flame detector temperature, 300°; N_2 flow rate; 55 ml/min. 2-Aminopyridine (as the perfluoropropionyl derivative) : column temperature, 130°; injector port temperature, 250°; electron capture detector temperature, 300°; N_2 flow rate, 30 ml/min.

Gas chromatography-mass spectrometry (GC-MS). A Finnigan Model 3100 quadrupole mass spectrometer equipped with a gas chromatograph was used. The GC conditions were those described above, helium being used as carried gas at a flow rate of 30 ml/min. The MS was set to the following conditions: molecular separator temperature, 250°; ion source temperature, 100°; ionization beam energy, 70 eV; ionization current, 200 A. For the direct inlet system (DIS) the temperature ranged from 180° to 200°.

High pressure liquid chromatography (HPLC). A Hewlett Packard Model 1084 liquid chromatograph equipped with an UV detector (254 nm) was used. Analyses were carried out at pressure of 30 atm using gradient elution with 5 to 70% of acetonitrile:water, at a flow of 1 ml/min.

RESULTS AND DISCUSSION

The metabolism of difenpiramide was first studied in rat. The acidic extract of urine of rats treated with difenpiramide on TLC contained three compounds not observed in the blank extract. Preparative TLC (system "A", Table 1) afforded three fractions of R_f 0.29, 0.40 and 0.53. The least polar compound was shown to be difenpiramide from its GC properties and mass spectrum (Fig. 1) which were identical with those of an authentic sample.

The mass spectrum of Difenpiramide is characterised by an abundant M^{+} at m/e 288 and the base peak at m/e 194 attributed to the ion arising from the loss of α-aminopyridine. The ion at m/e 167 may correspond to a phenyltropylium ion.

The mass spectrum of the fraction with R_f 0.40 is shown in Fig. 2. The major fragmentation corresponds to loss of 45 amu from the M^{+} suggesting loss of a $-CO_2H$ group to give the relatively stable phenyltropylium ion at m/e 167. On this basis the compound was assigned the structure of biphenylylacetic acid (BPA) and this was confirmed by comparison with an authentic sample of BPA.

The compound with R_f 0.29 gave the mass spectrum shown in

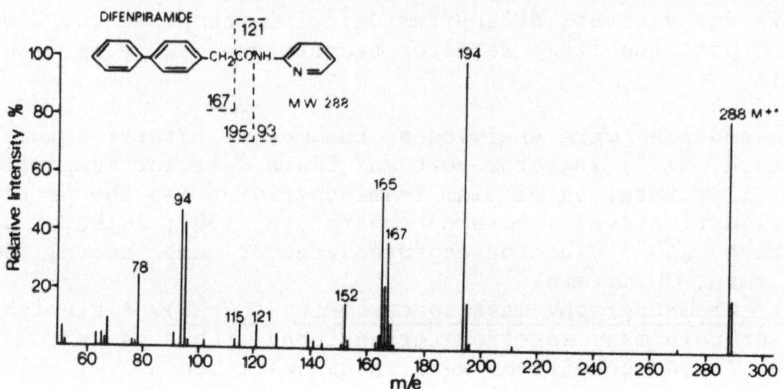

FIG. 1. Mass spectrum of difenpiramide.

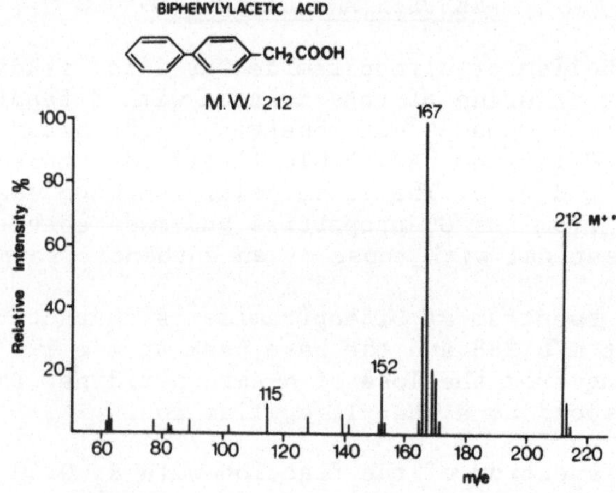

FIG. 2. Mass spectrum of biphenylylacetic acid.

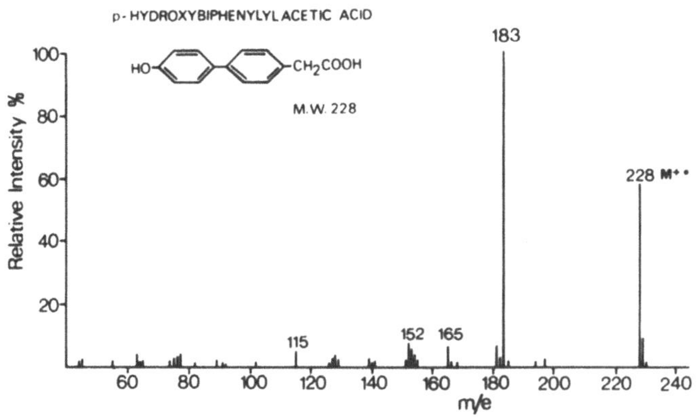

FIG. 3. Mass spectrum of p-hydroxybiphenylylacetic acid.

Fig. 3 and was assigned the structure of p-hydroxybiphenylylacetic acid. The mass spectrum shows the M^+ at m/e 228 and a base peak at m/e 183 both showing an increase of 16 amu when compared to the corresponding ions of BPA, suggesting the presence of aromatic hydroxyl group. The presence of this hydroxyl at the para - position was established by comparison with an authentic sample of p-hydroxybiphenylylacetic acid.

Extraction of urine or serum adjusted to pH 9 with chloroform afforded a compound which was analysed by GC-MS and shown to be α-aminopyridine.

The metabolic transformation of difenpiramide are shown in Fig. 4. The formation of the p-hydroxy metabolite can occur via an

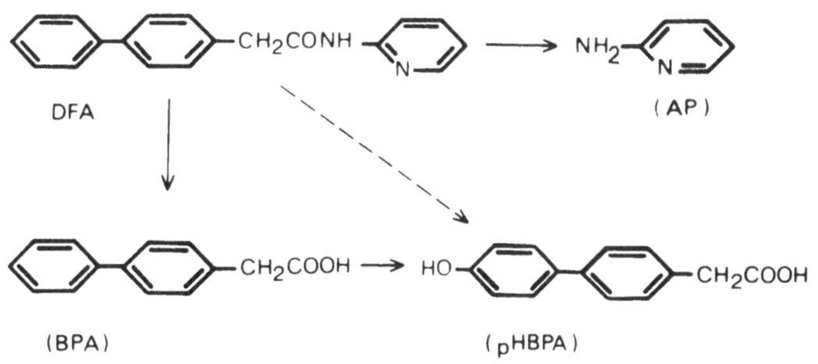

FIG. 4. Metabolic transformations of difenpiramide.

arene oxide intermediate or by direct hydroxylation (Fig. 5).

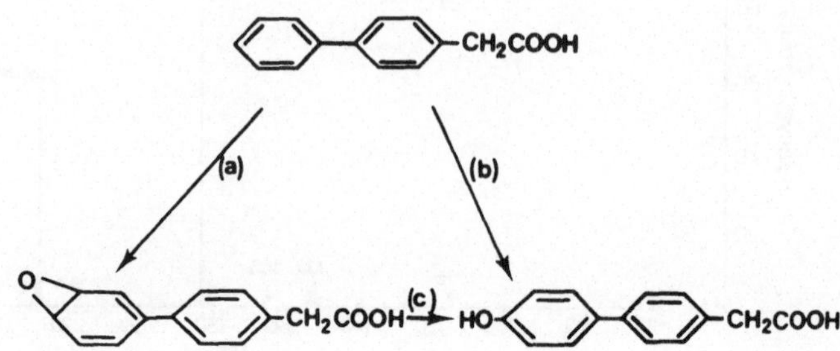

FIG. 5. Pathways for the formation of p-HBPA.

Given the instability of areneoxides it has not been possible to obtain evidence for the intermediacy of this compound.

Amino acid conjugates were found in the urine of humans treated with difenpiramide. Urine was hydrolyzed with HCl at pH 2 and the solution was extracted with methylisobutylketone. The components of the extract were separated by TLC (system "C" and "D"; Table 2). The two major compounds obtained were analyzed by

TABLE 2.
R_f Values of Amino Acid Coniugates of Biphenylyl Acetic Acid

COMPOUNDS	SYSTEM "C"	SYSTEM "D"
BIPHENYLYLACETYL TAURINE	0.54	0.51
BIPHENYLYLACETYL GLUTAMINE	0.58	0.15

SYSTEM "C": ACETIC ACID : WATER : n-BUTANOL (11:25:50)

SYSTEM "D": AMMONIUM HYDROXIDE (conc.) : METHANOL : CHLOROFORM (2:30:70)

MS and gave the spectra in Fig. 6 and 7. Both are characterized by a significant peak at m/e 194 representing the biphenylketene ion which also appears in the mass spectrum of difenpiramide. The fragments lost from the molecular ions correspond in the mass to taurine and glutamine respectively. This tentative assignment was confirmed by direct comparison with authentic compounds. No amino acid conjugates were found in the bile and urine of rat.

Quantitative studies. The quantitative determination of

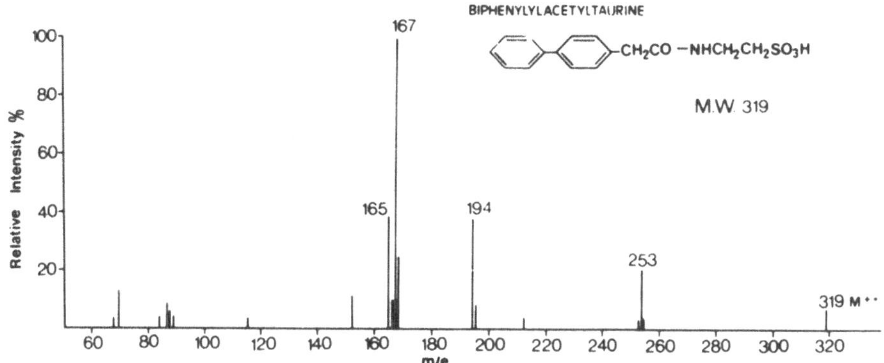

FIG. 6. Mass spectrum of biphenylylacetyltaurine.

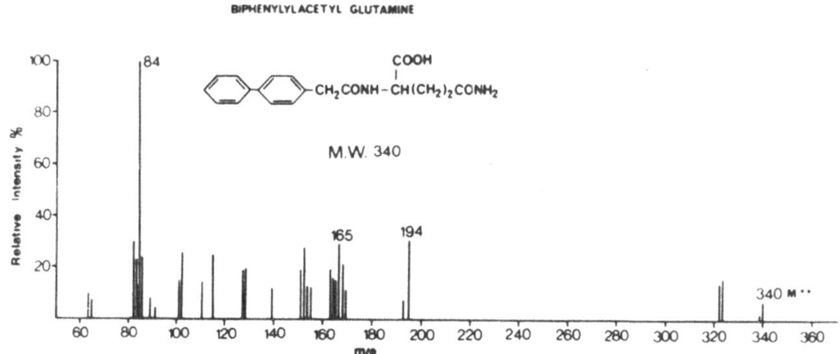

FIG. 7. Mass spectrum of biphenylylacetylglutamine.

difenpiramide and its metabolites was initially attempted by UV spectroscopy combined with TLC methods for the isolation of the compounds. This method is time consuming and not very sensitive. GC analysis has the disadvantage that the two acidic metabolites must be derivatized before analysis and even so the derivatives can be examined only under conditions differing from those required for difenpiramide (2). This method however was used for preliminary detection of the metabolites in dog serum and urine. Fig. 8 and 9

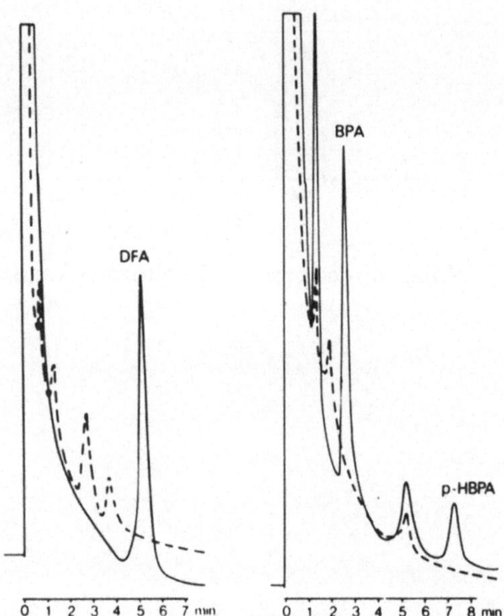

FIG. 8. Gas chromatograms of DFA and the methyl esters of BPA and p-HBPA in extracts of dog serum (broken line trace is the gas chromatogram of the control).

show the gas chromatograms of difenpiramide and of the two methylesters in dog serum and urine. Fig. 10 shows the GC calibration curves for difenpiramide and its three metabolites. α-Aminopyridine in serum, bile and urine was quantitated as its perfluoropropionyl derivative using an electron capture detector (Fig. 11).

For a more accurate determination of these compounds we resorted to HPLC which allowed the analysis of difenpiramide and its two acidic metabolites in one run as shown by the chromatograms for a mixture of three standard compounds (Fig. 12), for the extracts from human serum (Fig. 13) and for the extracts from human urine (Fig. 14). Of the three analytical methods tested HPLC provides a

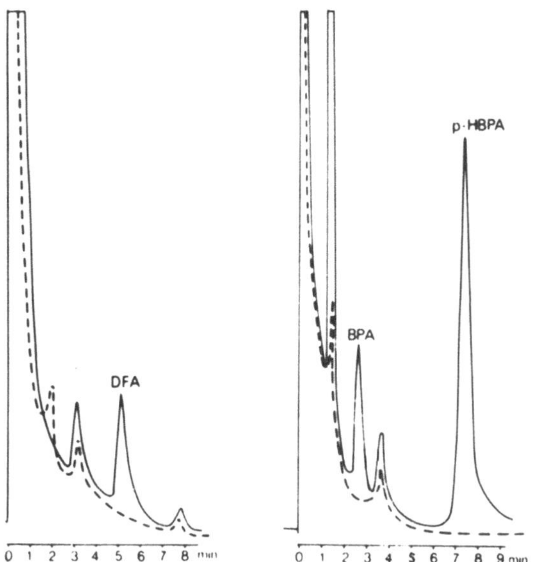

FIG. 9. Gas chromatograms of DFA and the methyl esters of BPA and p-HBPA in extracts of dog urine (broken line trace is the gas chromatogram of the control).

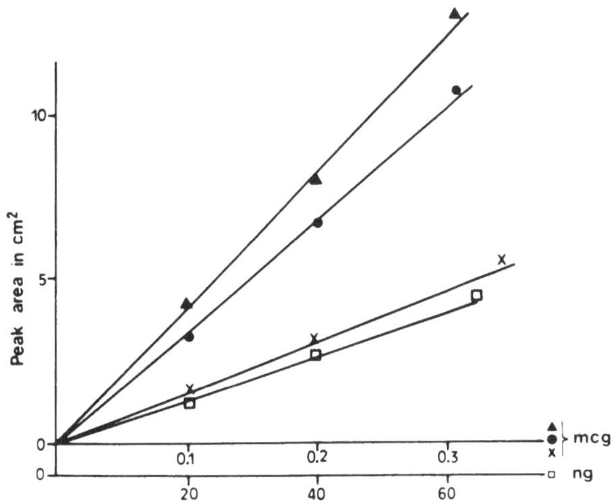

FIG. 10. GC calibration curve for difenpiramide (▲), BPA (□), p-HBPA (●) and α-aminopyridine (x).

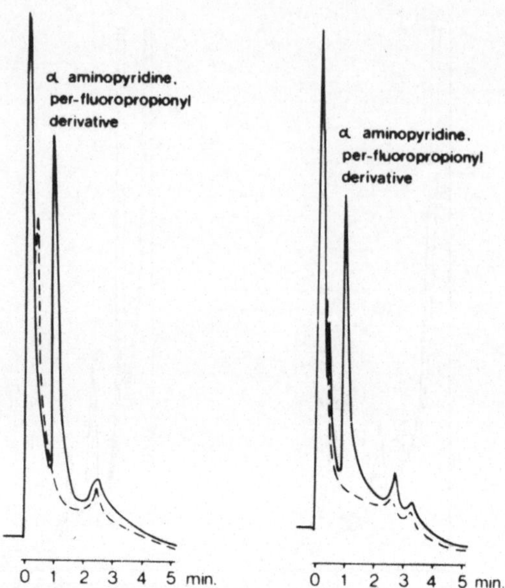

FIG. 11. Gas chromatograms of the perfluoropropionyl derivative of α-aminopyridine extracted from dog serum and urine (broken line traces are the gas chromatograms of the controls).

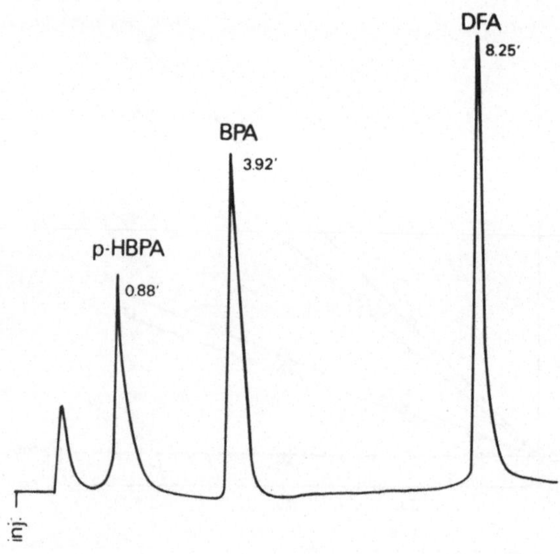

FIG. 12. HPLC profile of a mixture of the standard compounds.

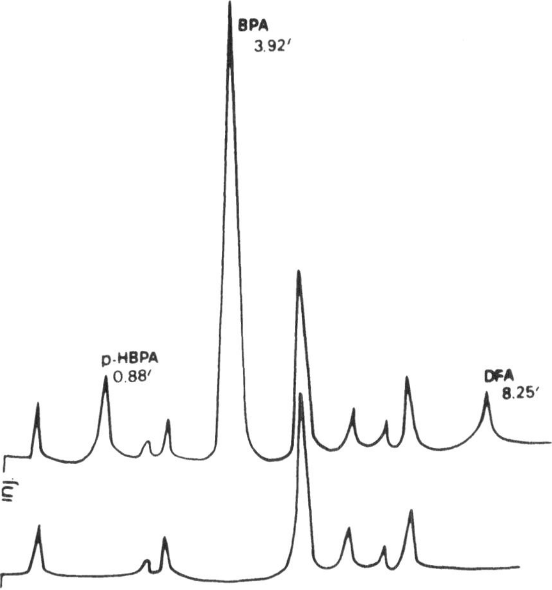

FIG. 13. Top trace, HPLC profile of extracts of human serum. Lower trace, profile of control.

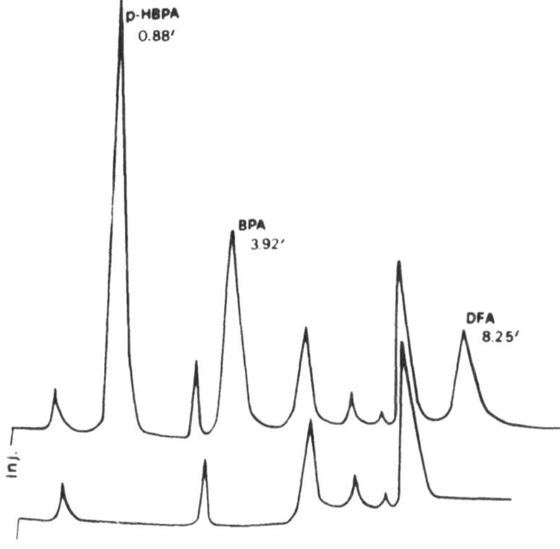

FIG. 14. Top trace, HPLC profile of extracts of human urine. Lower trace, profile of control.

greater sensitivity and convenience and this method was then used to study the pharmacokinetics of difenpiramide (Table 3). The

TABLE 3.
SENSITIVITY (mcg/ml)

COMPOUND	GLC		T.L.C.–U.V.	H.P.L.C.
	F.D.	E.C.D.		
DIFENPIRAMIDE	0.2		1	0.1
BPA			1	0.1
p-HBPA			1	0.1
BPA (METHYL ESTER)	0.1			
p-HBA (METHYL ESTER)	0.1			
α - AMINOPYRIDINE	0.5			
α - AMINOPYRIDINE PERFLUORO PROPIONYL DERIVATIVE		0.05		

percentage recovery of the drug BPA and p-HBPA from serum and urine is given in Table 4. The amount of free and conjugated BPA and

TABLE 4.
Recoveries of Difenpiramide and its Metabolites Added to Serum and Urine

Drug	Amount added (mg)	Recovery (%)
DFA	200	85
	150	82
	50	86
		mean ± S.D. 84.33 ± 1.20
BPA	200	83
	150	85
	50	86
		mean ± S.D. 84.66 ± 0.88
p–HBPA	200	81
	150	83
	50	85
		mean ± S.D. 83.00 ± 1.15

p-HBPA in the urine was also estimated in an experiment where the urine after extraction of the free compounds was hydrolyzed with ketodase at 37° for 24 hr followed by extraction with ethylacetate (Table 5). The levels of DFA and its metabolites in serum was monitored over 24 hr (Table 6). In rat maximum serum levels of BPA are reached after 3 hours and decrease slowly after 6 hours. In dog absorption is slower the maximum being reached after 6 hours,

TABLE 5.

Estimation of Free and Conjugated Forms of Biphenylylacetic and p-Hydroxybiphenylylacetic Acid in Animal Urine After Ingestion of Difenpiramide

Each result is the mean of six determinations

Animal	Initial dose per Kg. of body weight (mg)	% B P A		% p-HBPA	
		free	conjugated	free	conjugated
RAT	50	65	35	28	72
DOG	45	85	15	63	37
MAN	250 *	23	77	25	75

*total dose

TABLE 6 a)

Serum Levels (mcg/ml) Difenpiramide and its Metabolites After a Single Oral Administration

Each result (mean ± standard deviation) is the mean of six determinations

Species	Dose per Kg. of body weight mg	Time after difenpiramide administration								
		1 h			3 h			6 h		
		DFA	BPA	p-HBPA	DFA	BPA	p-HBPA	DFA	BPA	p-HBPA
RAT	50	2,47 ± 0,52	25,17 ± 3,49	6,7 ± 3,61	2,59 ± 0,36	30,25 ± 4.10	19,55 ± 0.13	2,09 ± 0,12	20,46 ± 3,16	31.03 ± 5.81
DOG	45	3.41 ± 0,34	8,4 ± 0,75	4,18 ± 0,83	2,7 ± 0,34	13,30 ± 2,13	2,98 ± 0,38	2,3 ± 0,46	22,12 ± 4,34	4.64 ± 0.77
MAN	250 *	0,253 ± 0,025	1,956 ± 0,211	0,428 ± 0,021	0,225 ± 0,022	1,018 ± 0,083	0,368 ± 0.039	0.16 ± 0,015	0,768 ± 0,08	0,303 ± 0,020

* total dose

TABLE 6 b)

9 h			12 h			24 h		
DFA	BPA	p-HBPA	DFA	BPA	p-HBPA	DFA	BPA	p-HBPA
2.45 ± 0.34	10.97 ± 0.96	28.39 ± 5.35	1.62 ± 0.14	9.8 ± 2.01	16.01 ± 2.03	1.96 ± 0.22	7.65 ± 0.9	13.31 ± 3.09
2.59 ± 0.27	15.75 ± 5.6	3.75 ± 1.01	2.01 ± 0.31	15.29 ± 3.65	4.06 ± 0.77	1.85 ± 0.38	13.15 ± 1.72	5.57 ± 1.54
			0.141 ± 0.01	0.624 ± 0.07	0.273 ± 0.027			

where as in man the levels decreased slowly after a maximum at 1 hour. The levels of difenpiramide and BPA in human serum after a single oral administration of DFA are plotted in Fig. 15.

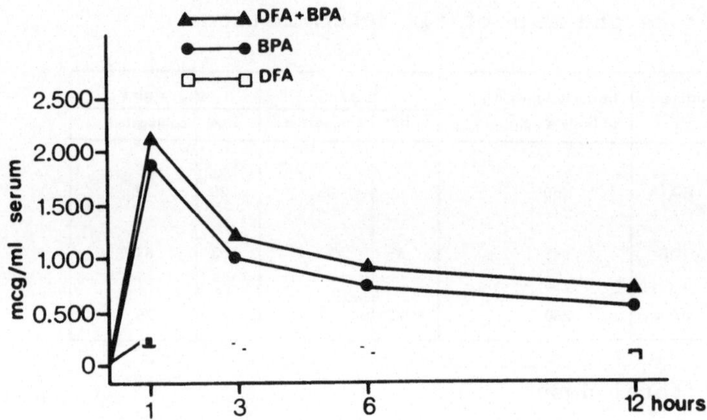

FIG. 15. Levels of DFA and BPA in human serum after a single oral administration.

TABLE 7.

Estimation of Difenpiramide and its Principal Metabolites in Urine After a Single Oral Dose of Difenpiramide

Each result is the mean of six determinations

SPECIES	DOSE PER KG. OF BODY WEIGHT	DIFENPIRAMIDE	BPA	p–HBPA
	mg.	mcg. tot.	mcg. tot.	mcg. tot.
RAT	50	36.8	182.34	728
DOG	45	37.2	468	667
MAN	250 *	39.3	826.17	2453.49

* total dose

N.B. –Total urine recovery of Difenpiramide and its metabolites after 24 hours single dose administration per os is: 1.3±0.05%

Excretion in the urine is moderate (Table 7). In the 72 hours after administration 9% of the drug and its metabolites is recovered in rat urine, 0.3% in dog. After 24 hours, 1.3% is recovered in human urine. The major metabolite in rat bile is p-hydroxyphenylylacetic acid (Table 8).

TABLE 8.

Estimation of Difenpiramide and its Principal Metabolites in Rat Bile After a Single Oral Administration (mcg)

Each result is the mean of six determinations

Dose per Kg. of body weight (mg)	Time after difenpiramide administrations					
	0–3 h			3 h – 6 h		
	DFA	BPA	p–HBPA	DFA	BPA	p–HBPA
50	12.15±2.91	62.01±17.35	135.3±27.69	11.55±32.07	39.44±9.04	124.95±38.78

TABLE 9.

Estimation of Difenpiramide and its Principal Metabolites in Rat Stomach and Intestine After a Single Oral Administration

Each result (Mean Standard Deviation) is the mean of six determinations

Dose per Kg of body weight	Time after difenpiramide administration											
	3 h (stomach and small intestine)				6 h (small and large intestine)				12 h (small and large intestine)			
mg	DFA mg	BPA mg	Quantity administered mg	%recovery	DFA mg	BPA mg	Quantity administered mg	%recovery	DFA mg	BPA mg	Quantity administered mg	%recovery
50	4.27± 1.11	0.783±0.15	13.05 ± 0.499	42.66	3.586±0.42	0.996±0.031	13.05±0.177	37.64	1.918±0.33	0.637±0.08	13.9 ± 0.40	19.94

In the stomach and in the intestine (Table 9) most of the difenpiramide is not metabolized. The recovery of the metabolites in the stomach is 42.66%: the amount of α-aminopyridine found is always a little less than the sum of the amount of biphenylylacetic acid and p-hydroxybiphenylylacetic acid. The total recovery of difenpiramide biphenylylacetic acid, p-hydroxybiphenylylacetic acid and α-aminopyridine, in urine, bile and stomach of rat is about 60%.

In conclusion difenpiramide is hydrolyzed in the body to biphenylylacetic acid and α-aminopyridine. Biphenylylacetic acid is metabolized to p-hydroxybiphenylylacetic acid by hydroxylation of the biphenyl group in the p-position. Biphenylylacetic acid is pharmacologically active, while p-hydroxybiphenylylacetic acid is inactive. An amount of these two acids made water soluble by conjugation (mainly with glucuronic acid, in part with amino acids) are excreted in the urine. Biphenylylacetic acid is found in greater quantity in serum, while p-hydroxybiphenylylacetic acid prevails in urine.

REFERENCES

1) Italian patent N. 24732 A/72, May 1972.
2) R.H. Greeley, J. Chromatogr., 1974, 88, 229.

POTENTIAL DRUGS FROM PROPOLIS

E.L. Ghisalberti, P.R. Jefferies and R. Lanteri

Department of Organic Chemistry, University of Western Australia, Nedlands, 6009, Western Australia

INTRODUCTION

Propolis or bee's glue is a resinous substance collected by bees from various plant sources for the purpose of sealing the top of the wax cells, filling cracks and crevices, cementing loose pieces of the hive and generally for the repair of any damage (1, 2). A mixture of resin, wax and pollen, it has been shown to possess antimicrobial activity (3- 6) and it has been suggested (3) that propolis may be responsible for the lower incidence of bacteria and moulds in the atmosphere within the hive relative to the outside. Propolis has been used as a folk medicine for several centuries, particularly for the cure of respiratory disorders (1, 7). In some countries it is used as a local anaesthetic in dental practice and it has been reported (8) that its anaesthetic action is comparable to that of procaine and cocaine. Propolis has also been shown to be beneficial in treating experimental stomach ulcers in the rat (9). A list of the actions attributed to propolis and some of its uses are presented in Table 1. Storage at either 0-4° or room temperature for three to four years does not reduce its antibacterial activity (10) although heating at 60° partially destroys its inhibitory action (6).

Limited work has been done on the isolation of the components of propolis. Until recently the compounds isolated and identified included a group of simple aromatic compounds (Fig. 1) and a large number of flavonoid compounds (Fig. 2; 7-18 plus the 7-methylether of 11). These compounds have been isolated from various samples of Russian and European propolis (11-13). Renewed interest in the composition of propolis has resulted in the identification

TABLE 1.

Reported actions and uses of propolis.

Action	Uses
Anaesthetic	Anaesthetic in dental practice
Antibacterial	Treatment of oral cavity diseases
Antifungal	Adjuvant for antibiotics
Antimicrobial	Additive in toothpaste, oral disinfectants, skin creams
Antiinflammatory	Treatment of dermatoses and burns Treatment of stomach ulcers
Antispasmodic	Treatment of vaginitis Treatment of respiratory disorders

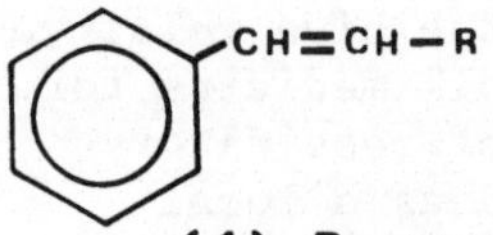

CH=CH—R

(1) R= CH_2OH

(2) R= CO_2H

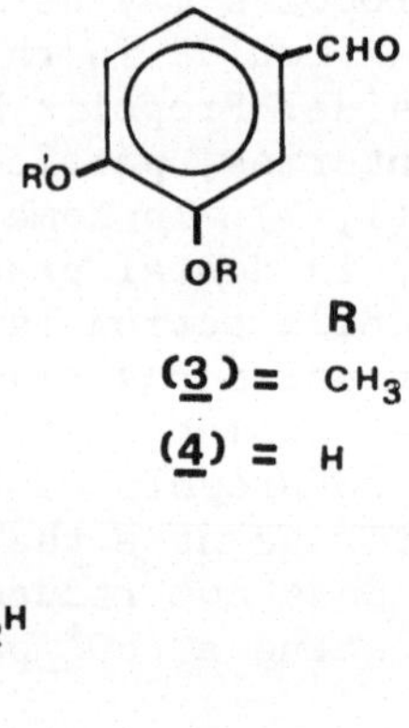

CHO, R'O, OR

	R	R'
(3) =	CH_3	H
(4) =	H	CH_3

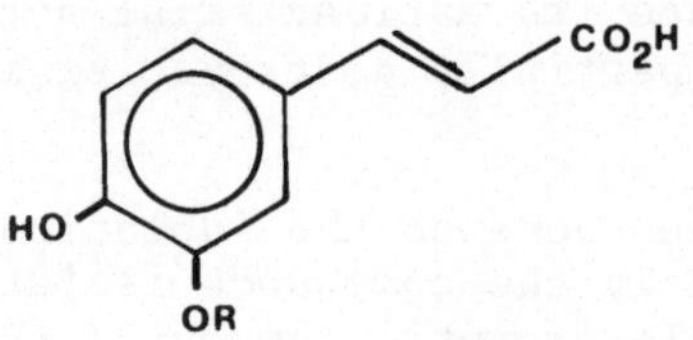

CO_2H, HO, OR

(5) R= H

(6) R= CH_3

FIG. 1. Aromatic compounds isolated from propolis.

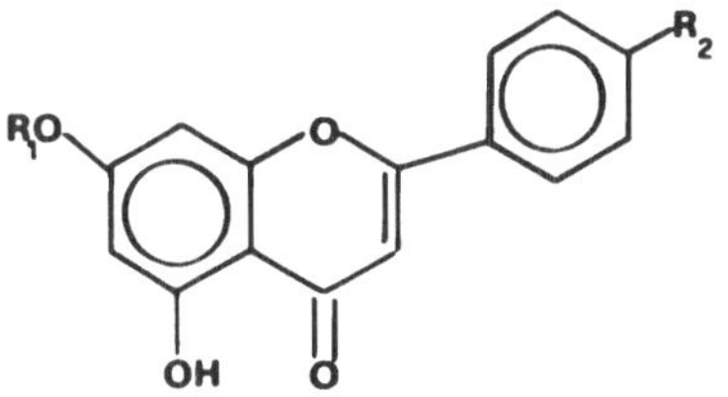

	R_1	R_2
(7)	H	H
(8)	CH_3	H
(9)	H	OCH_3
(10)	CH_3	OCH_3

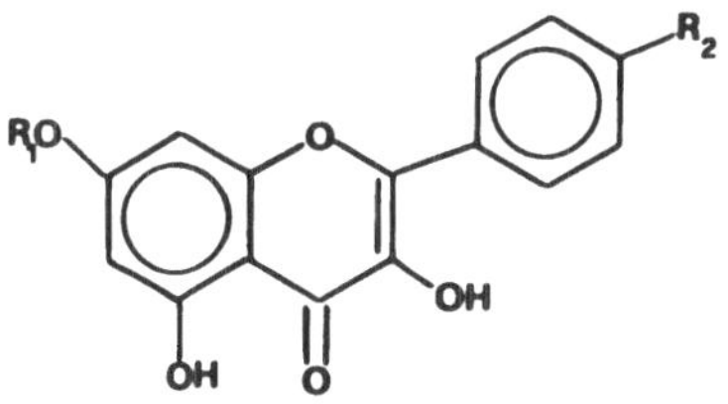

	R_1	R_2
(11)	H	H
(12)	H	OCH_3
(13)	CH_3	OH
(14)	CH_3	OCH_3

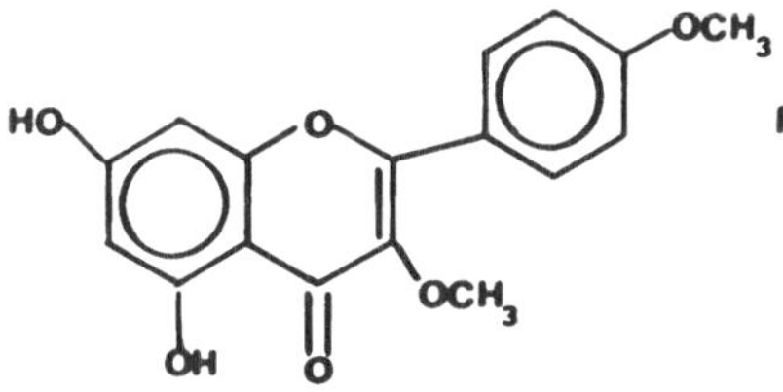

(15)

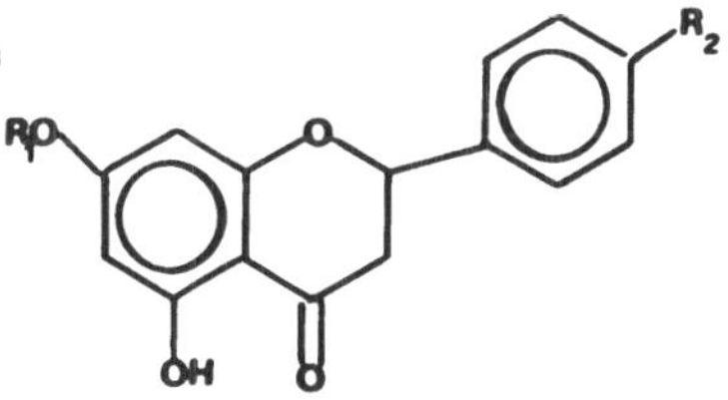

	R_1	R_2
(16)	H	H
(17)	CH_3	H
(18)	CH_3	OCH_3

FIG. 2. Flavonoid compounds isolated from propolis.

of a number of other compounds. Apart from the isolation of six flavones (7-9, 11, 13 and the 7-methyl ether of 11) and three flavanones previously identified in samples of propolis, Schneidewind *et al.*(14) further isolated and identified 5,7--dihydroxy-6,4'-dimethoxyflavone,5,7,4'-trihydroxy-3,3'-dimethoxyflavone, sakuranetin (19),3,5,7-trihydroxyflavanone and its 3-acetoxy derivative, *p*-coumaric acid benzyl ester and an ester of caffeic acid with an aromatic alcohol. In the steam volatile fraction of propolis Janes and Bumba (15) identified benzoic and sorbic acid, benzyl alcohol, phenyl vinyl ether, anisyl vinyl ether, cyclohexylbenzoate and vanillin. The presence of a large group of flavones affirms the plant origin of propolis. Of these compounds caffeic acid (5) has been shown to posses antibacterial, tuberculostatic and fungistatic activity towards selected organisms (11); ferulic acid (6),5,7-dihydroxyflavanone (pinocembrin, 16) and 3,5,7-trihydroxyflavone (galangin, 11) all have antibacterial activity (11, 12). A number of the compounds recently isolated by Schneidewind *et al.* appear to have antimycotic activity (14). We became interested in the constituents of propolis for the following reasons: (a) most of the pharmacological tests carried out only used crude extracts of propolis, (b) the wide spectrum of activity attributed to propolis did not appear to be justified by the type of compounds which had been isolated, (c) although no pharmacological tests had been done on Western Australian (W.A.) propolis one subjective opinion suggested that it had a greater anaesthetizing action than that of European propolis, consequently it could be taken only in smaller doses, and (d) even if one remained sceptical about its reported actions the fact remained that a number of people use it.

We thus decided to investigate the chemical constitutents of W.A. propolis with a view to isolating and identifying the compounds present, which could then be submitted for pharmacological screening.

RESULTS AND DISCUSSION

Propolis was extracted initially with 70% aqueous ethanol since this is the extract most often taken by people using propolis. The extract was then dissolved in ether and fractionated as shown in Scheme 1. The major part of the material (90%) was extracted with NaOH and the remainder was divided almost equally between the other three fractions. A preliminary examination indicated that the neutral fraction contained mainly fats and sterols and the $NaHCO_3$ fraction mostly flavones. These fractions were not investigated further. All the other fractions were examined in detail, most of the work involving the larger NaOH soluble fraction. In general, for the isolation and purification of the components column and preparative thin layer chromatography were used. Mass spectrometry

SCHEME 1.

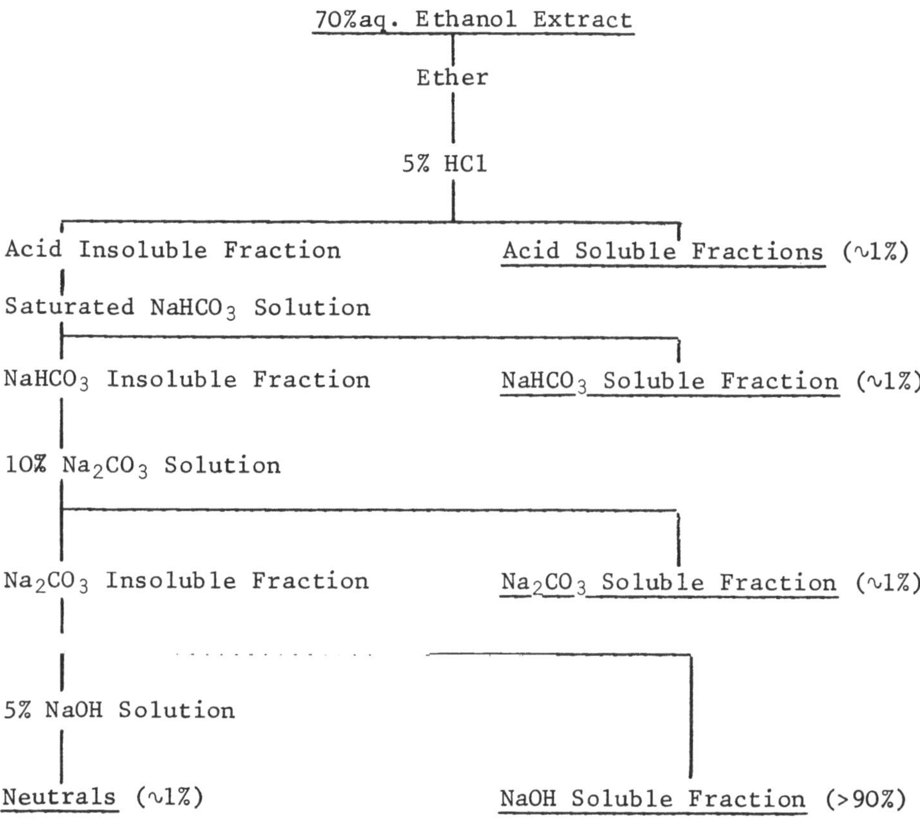

proved useful in monitoring the extent of separation achieved.

In all four flavanones were identified as components of the fractions examined. (S)-(-)-Pinostrobin (16, 17) (5-hydroxy-7-methoxyflavanone, 17) was isolated from all of the three fractions examined and was identified from its spectral characteristics and physicochemical properties (Fig. 3). The identification of flavanones by mass spectrometry (MS) is a relatively simple procedure. The major fragmentations of flavanones such as pinostrobin (Fig. 4) yield ions which are characteristic for the A and B rings of the molecule indicating the level of substitution on these rings (e.g. ions at m/e 166 and 104 respectively for pinostrobin) (18).

Two isomeric flavanones were also isolated, sakuranetin

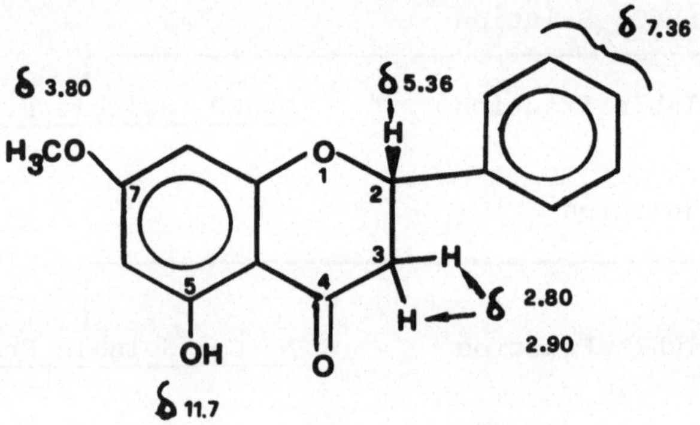

m.p. 110 — 111°, $[\alpha]_D$ −54°

(lit. 112−113°, $[\alpha]_D$ −56°)

FIG. 3. Structure, physico-chemical properties and NMR data of (S)-(−)-pinostrobin (5-hydroxy-7-methoxyflavanone, 17).

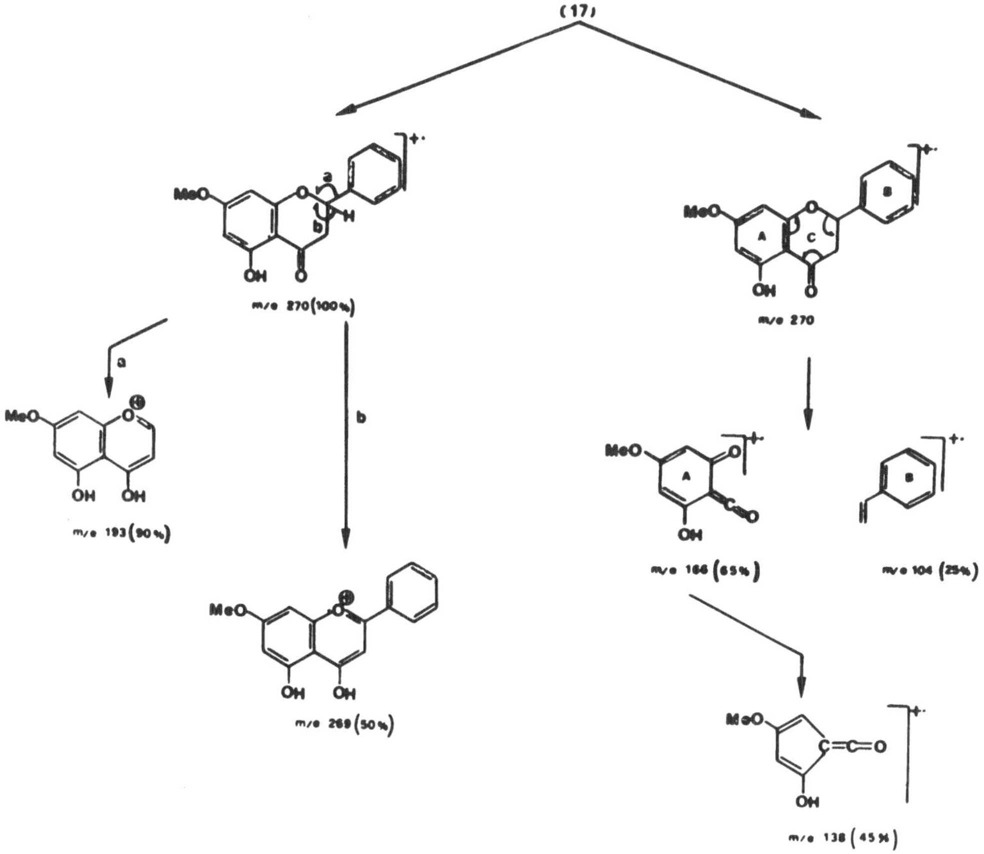

FIG. 4. Mass spectral fragmentations of pinostrobin (17).

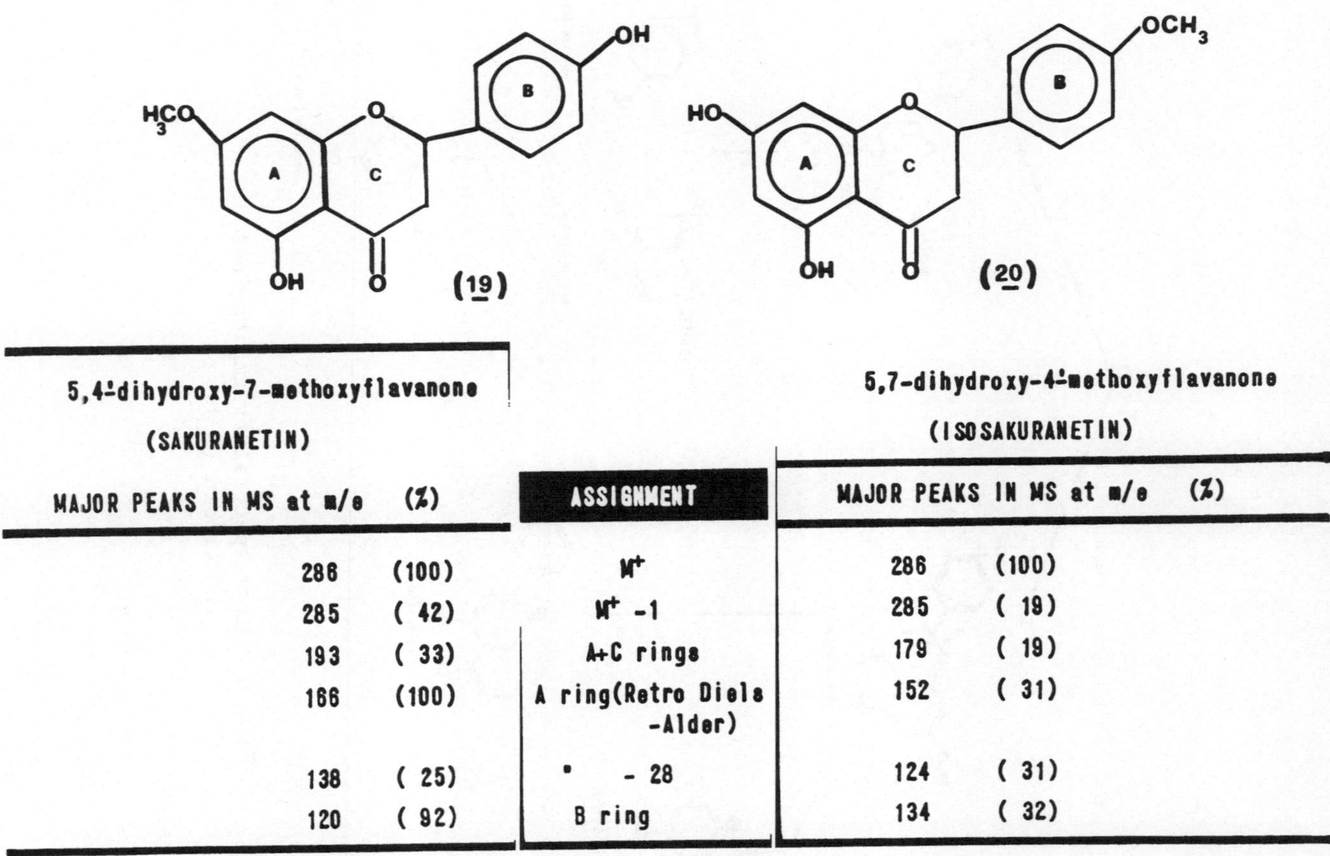

5,4'-dihydroxy-7-methoxyflavanone (SAKURANETIN)			5,7-dihydroxy-4'-methoxyflavanone (ISOSAKURANETIN)	
MAJOR PEAKS IN MS at m/e	(%)	ASSIGNMENT	MAJOR PEAKS IN MS at m/e	(%)
286	(100)	M^+	286	(100)
285	(42)	M^+ -1	285	(19)
193	(33)	A+C rings	179	(19)
166	(100)	A ring(Retro Diels -Alder)	152	(31)
138	(25)	" - 28	124	(31)
120	(92)	B ring	134	(32)

FIG. 5. Structures and major mass spectral fragmentations of sakuranetin (19) and isosakuranetin (20).

(5,4'-dihydroxy-7-methoxyflavanone,19) from the NaOH-soluble and acid-soluble fractions and isosakuranetin (5,7-dihydroxy-4'-methoxyflavanone,20) from the Na_2CO_3-soluble fraction. Isosakuranetin containing the p-hydroxyacetophenone moiety is more acidic than sakuranetin and can be selectively extracted into aqueous Na_2CO_3. Mass spectral analysis easily distinguishes between these two isomers (Fig. 5). The ion originating from the A-ring fragment appears at m/e 166 in sakuranetin and at m/e 152 in isosakuranetin. On the other hand the B-ring fragment appears at m/e 120 for sakuranetin and at m/e 134 for isosakuranetin, clearly indicating that the latter contains a methoxyl group in the B-ring and two free hydroxyls in the A-ring. The structure of sakuranetin was confirmed by comparison of its melting point and spectral properties with those reported in the literature (19) and that of isosakuranetin by direct comparison with an authentic sample. A fourth flavanone detected in the acid-soluble fraction and was assigned the structure of 5-hydroxy-4',7-dimethoxyflavanone (18) on the basis of its mass spectral fragmentations.

The next compound isolated was shown to be 4-acetyl-5-hydroxy-2-methyl-2H-3H-naphtho(1,8-b,c)pyran (xanthorrhoeol,21). The mass spectrum (Fig. 6) showed a molecular ion at m/e 242. The NMR spectrum (Fig. 7) showed a singlet (δ 2.57) attributed to an aromatic methyl ketone, and a doublet (δ 1.48) assigned to a secondary methyl adjacent to an oxygen. The presence of a third oxygen group in the molecule was indicated by the absorption in the IR spectrum due to an intramolecularly hydrogen bonded hydroxyl group. On this basis a molecular formula of $C_{15}H_{14}O_3$ was proposed.

The mass spectrum of 21 (Fig. 6) gave an indication of the type of structure involved. The abundant M^+ ion and the doubly-charged ion at m/e 113.5, arising from the ion at m/e 270, suggested a polycyclic aromatic compound which, from the molecular formula and the calculated degrees of unsaturation (9) must contain a naphthalene ring. The NMR spectrum was consistent with this and revealed the structure for this compound. A three-line pattern of an ABX system centred at δ 3.10 was indicative of two benzylic protons. The signal due to the X proton (δ 4.12) was also shown to be coupled to the doublet at δ 1.45 assigned to a secondary methyl group. The signals in the aromatic region gave an integral for four protons. Decoupling and INDOR (20) experiments revealed the presence of three doublets of doublets of a 3 proton spin system with coupling constants of J_{AB} 7.5, J_{AC} 1.5 and J_{BC} 8.0Hz expected for the hydrogens of a 1,2,3-trisubstituted benzene. The fourth aromatic proton (δ 6.95) showed no significant coupling to other protons and hence must be located on another aromatic ring. These data and careful consideration of the chemical shifts of the aromatic protons allowed the proposal of structure 21 for this compound. Confirmation was obtained by comparison of the NMR, MS, IR and UV spectra with

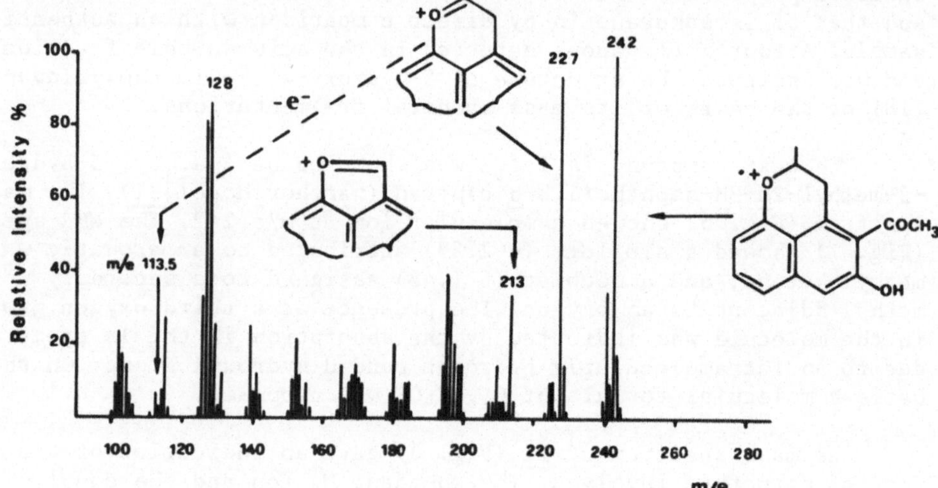

FIG. 6. Mass spectrum of xanthorrhoeol (21).

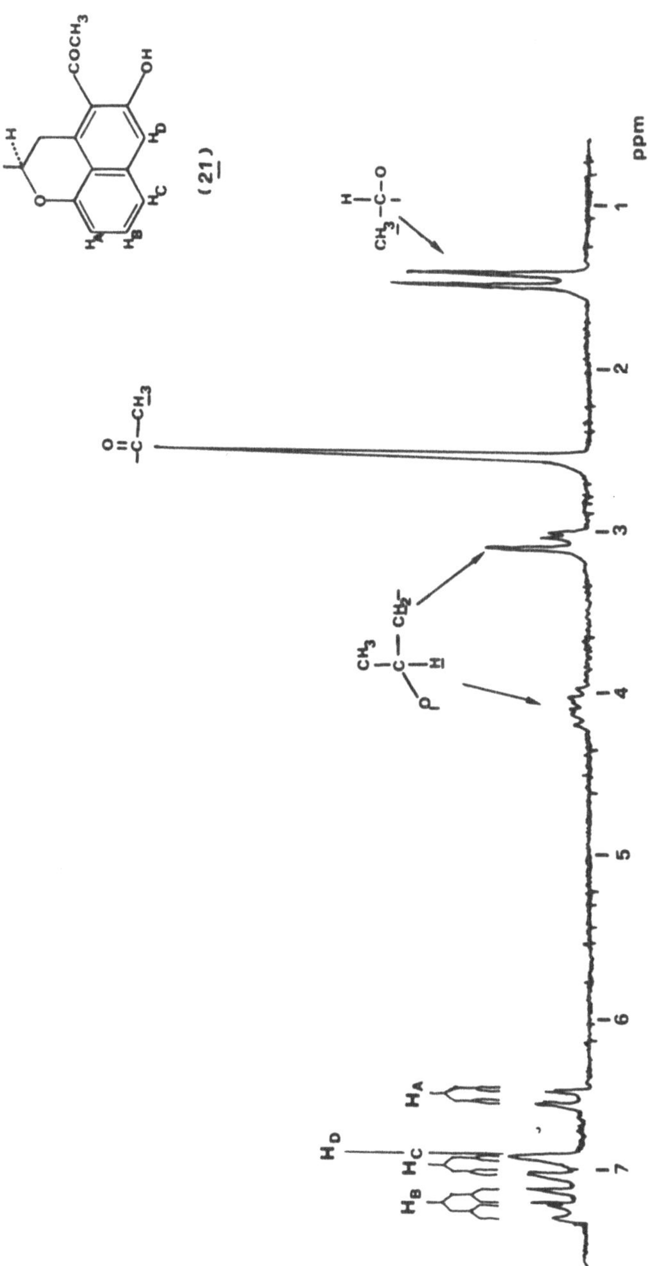

FIG. 7. NMR spectrum (90 MHz) of xanthorrhoeol (21).

those reported (21) for xanthorrhoeol and by mixed melting point with an authentic sample.

Another major component of the caustic-soluble fraction was assigned the structure of pterostilbene (22). The mass spectrum (Fig. 8) of this compound showed an M^+ at m/e 256 with an intense

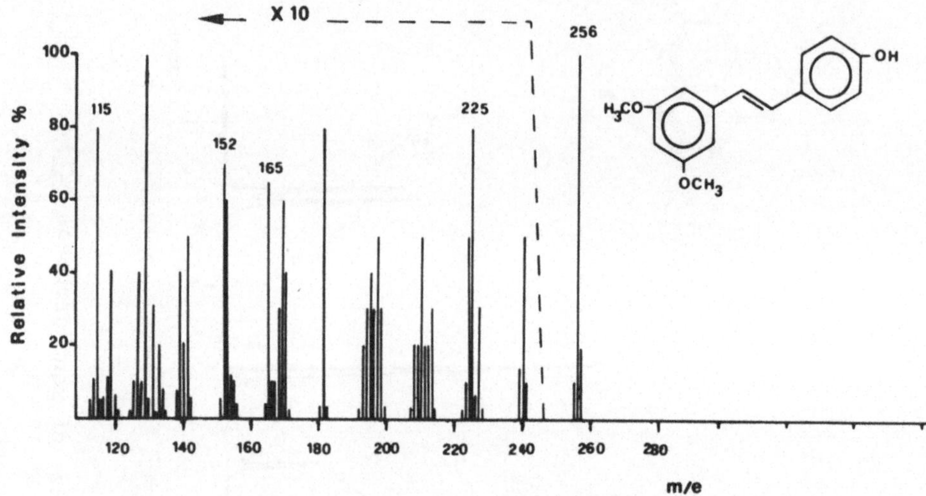

FIG. 8. Mass spectrum of pterostilbene (22).

M^{++} at 128 accompanied by an isotopic peak at m/e 128.5. The NMR spectrum (Fig. 9) showed a singlet at δ 3.81 for six protons attributed to two aromatic methoxyl groups. Furthermore the IR spectrum showed absorption due to free hydroxyl stretching at $3610cm^{-1}$. This implied that the compound contained three oxygens and suggested $C_{16}H_{16}O_3$ as the molecular formula for the structure which must have nine degrees of unsaturation. Decoupling experiments indicated that the triplet centred at δ 6.37 (J 2.5Hz) was coupled to a two-proton doublet at δ 6.63 suggesting a 1,3,5-trisubstituted benzene. The perturbed doublet centred at δ6.80 was shown to be coupled to a similar doublet at δ 7.37 with an apparent value of J 8.5Hz. Together these signals indicated an AA'BB' system of a 1,4-disubstituted benzene. INDOR experiments showed that the signals at δ 6.91 and 6.95 were part of an AB spin system with a coupling constant of 16Hz, suggesting the presence of a trans--(E)-disubstituted double bond. In view of this the only possible structure for this compound is that shown in 22. Confirmation was obtained by comparison of the melting point (80-81°) with that recorded (22) in the literature (84-86°). The alternative structure, 3-hydroxy-4',5-dimethoxystilbene has a melting point

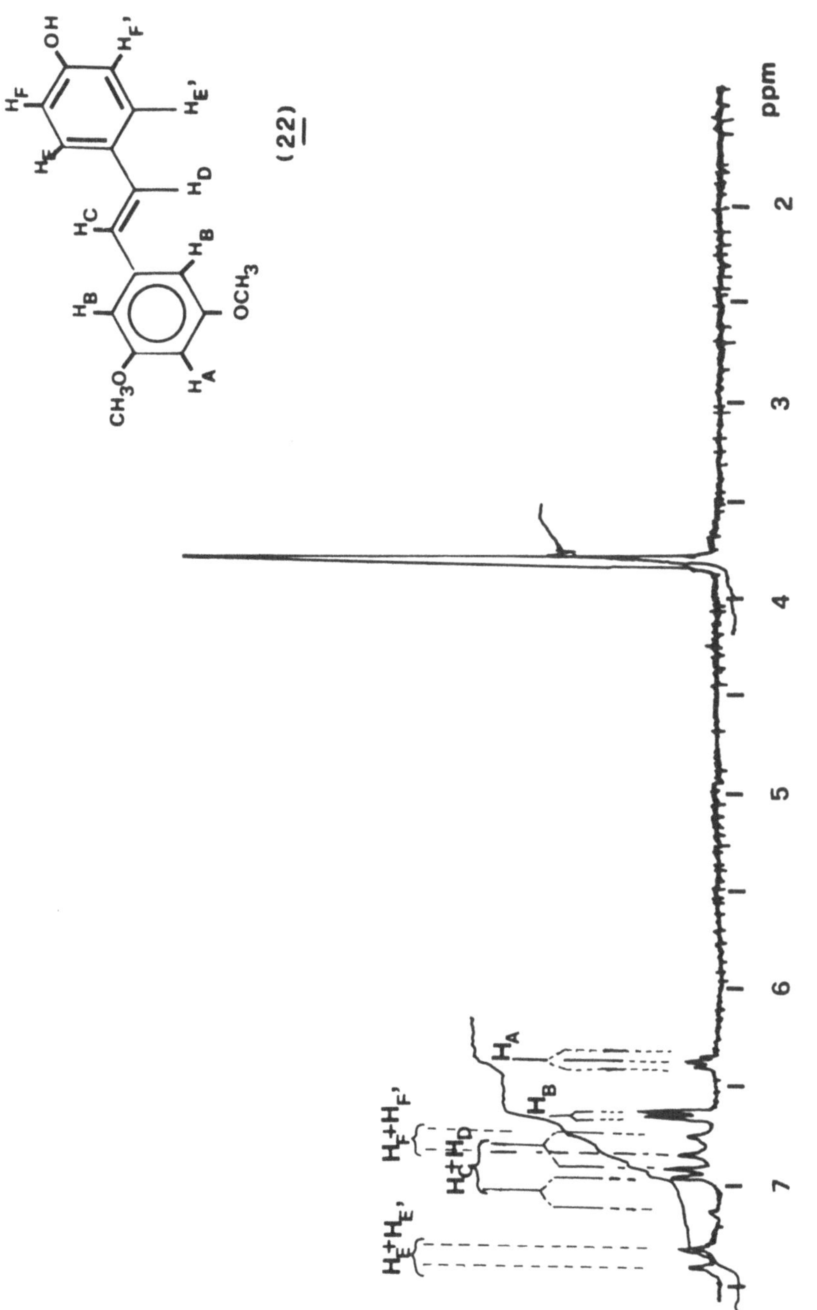

FIG. 9. NMR spectrum (90 MHz) of pterostilbene (22).

of 115-116° (23).

Investigation of the acid-soluble fraction yielded pinostrobin, sakuranetin, pterostilbene, 5-hydroxy-4',7-dimethoxyflavanone and a small amount of 3,5-dimethoxybenzylalcohol (<u>23</u>). Identification of this compound was achieved by interpretation of its spectral data (Fig. 10) and the structure was confirmed by comparison of its spectral data and GC retention time with those of an authentic sample.

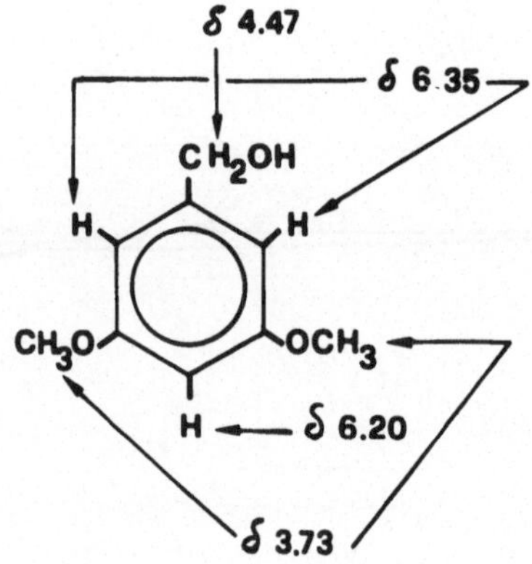

3, 5 – Dimethoxybenzyl alcohol

MS (m/e, %) **168 (M^+,100), 167 (11), 151 (12), 139 (54), 137 (14), 125 (40), 115.6 (m^* 167 → 139) 109 (26), 77 (24).**

FIG. 10. NMR and mass spectral characteristics of 3,5-dimethoxybenzyl alcohol.

The compounds obtained from the different fractions examined are listed in the experimental. Interestingly a number of compounds were isolated from the acid-soluble fraction. A possible explanation for this is that these compounds were initially present as water--soluble glycosides which were then hydrolyzed on contact with aqueous acid. Although alkaloidal compounds were not isolated from this fraction the aqueous extract gave a positive Mayer's test, suggesting interference from glycosides (24).

Given the present findings on the components of Western Australian propolis several points are worth mentioning. Pinostrobin (17) and 5-hydroxy-4',7-dimethoxyflavanone (18) also occur in European propolis. The presence of flavones in the $NaHCO_3$ extract was indicated and further examination of these may show similarities with the flavones of European propolis. This however may only reflect the ubiquitous nature of these flavanones and flavones in the plant kingdom. The possibility that 18 is an artifact arising from base-catalyzed cyclization of the corresponding chalcone during extraction cannot be excluded (25). Xanthorrhoeol (21), and (-)-sakuranetin (19) have also been found in *Xanthorrhoea presii* and isosakuranetin (20) and 18 in *X.australis* (26). These plants, commonly called "grass trees", are unique to Australia.

The origin of pterostilbene (22) is uncertain. Although *Eucalyptus* species are known to contain a number of hydroxy- and methoxy-stilbenes, pterostilbene has not yet been reported as a constituent (27). One explanation may be that bees actually transform one or more of the naturally-occurring *Eucalyptus* stilbenes by methylation or demethylation to give pterostilbene. Similarly intriguing is the presence of 3,5-dimethoxybenzylalcohol (23). Although alkyl resorcinols are commonly found in Australian *Grevillea* species (28), the substitution pattern indicates that it may be a degradation product of pterostilbene.

Although a variety of compounds have been isolated from different samples of propolis it is hard to understand which, if any, may be responsible for its pharmacological activity. Sakuranetin possesses antifungal activity (29) and isosakuranetin has been shown (30) to have papaverine-like spasmolytic activity. Pterostilbene and a number of other stilbenes may have antifungal activity (31) and pterostilbene has been reported to be useful in the treatment of diabetes (32). The pharmacological activity of xanthorrhoeol is at present unknown. Although propolis is generally regarded as non-toxic, at least in the short term, the presence of mainly aromatic compounds leaves room for some doubt.

Although no metabolic studies have been carried out for the specific compounds isolated from propolis, the metabolism of flavones, flavanones and *trans*-stilbene in several animal species has been studied.

The major metabolic pathways available to flavones can be illustrated for the case of 3,5,7-trihydroxy-4'-methoxyflavone (33).

Glucuronide formation is an important pathway for the

elimination of flavones. B-ring hydroxylation and demethylation can also occur. Biodegradation leads to the C_6-C_2 compounds, p- and m-hydroxyphenylacetic acid, arising from the B-ring.

The formation of the latter is thought to involve an hydroxylation-dehydroxylation reaction known to be mediated by intestinal bacteria (34). Flavanones, e.g. 4',5,7-trihydroxyflavanone undergo similar biotransformations (33) but C_6-C_3 degradation products are formed, p - and m-hydroxyphenylpropionic acid. The former can be dehydrogenated to coumaric acid which by β-oxidation gives rise to p-hydroxybenzoic acid.

The metabolism of stilbenes has been studied in the rat and *in vitro* with rat liver microsomes. Aromatic ring hydroxylation and reduction of the double bond are known to occur (35, 36). The following sequence has been established (37) for the *in vitro* metabolism of *trans*-stilbene: epoxidation of the internal double bond, hydration of the epoxide to the diol which can be oxidized to the benzoin and this in turn can be oxidatively cleaved to give the benzoic acid.

EXPERIMENTAL

Melting points were determined on a Kofler block and are uncorrected. Infrared (IR) spectra were recorded with a Perkin--Elmer 337 Grating Spectrophotometer or a Unicam SP. 200G Grating Spectrophotometer. Ultraviolet (UV) spectra were recorded with a Perkin-Elmer 137 Spectrophotometer using ethanol solutions (unless otherwise mentioned) in 1cm quartz cells. Mass Spectra (MS) were measured with a Varian MAT-CH7 mass spectrometer. Routine nuclear magnetic resonance (NMR) spectra were recorded with a 60 MHz Varian A-60A Spectrometer with tetramethyl silane (TMS) as internal standard. Signals are described as singlet (s), doublet (d), triplet (t), quartet (q), broad (br) or multiplet (m). Chemical shifts are quoted on the δ scale, relative to TMS at 0.00. Nuclear magnetic double resonance experiments were performed with a Bruker Spectrospin Spectrometer, operating at 90MHz. Optical rotations were determined with a Perkin-Elmer 141 Polarimeter for chloroform solutions in a 1 dm microcell. Analytical gas liquid chromatography (GLC) was performed using a Perkin-Elmer Model 880 flame ionisation Gas Chromatograph with nitrogen as the carrier gas and a 10ftx 1/8in. O.D. copper column of 5% Carbowax 20M 2½% F.F.A.P. on Chromosorb W.Analytical and preparative thin layer chromatography (TLC) were carried out with layers of silicic acid, 0.25 mm and 0.75 mm thickness respectively. The position of the compounds was determined by spraying with a 0.03M ceric sulphate solution followed by heating in an oven at 100° for 10 minutes in the case of analytical plates, and by viewing under ultraviolet light in the case of preparative plates. Solvent systems used were "a" chloroform, "b" methanol:chloroform (1 :19), "c" benzene-n--propanol (100:7), "d" acetone:isopropyl ether (1 :39) and "e" acetone:isopropyl ether (1 : 4).

Extraction of propolis. Crude propolis (300g), from the Mandurah area in Western Australia was broken up and extracted exhaustively with 70% aqueous ethanol solution. The extract was concentrated to half its original volume, diluted with water and extracted with ether. The ether solution was fractionated as shown in Scheme 1.

Analysis of NaOH soluble extracts. A quantity of the extract (10 g) was adsorbed on silicic acid (550 g) and eluted with $CHCl_3$. The 47 fraction obtained were analysed by TLC and similar fractions were combined. Preparative TLC was then used to isolate the compounds listed below.

S-(-)-Pinostrobin (5-hydroxy-7-methoxyflavanone,17). Fractions 4-6 (0.42g) were combined and recrystallized from $CHCl_3$-MeOH togive 17 as white needles, m.p. 110-111°, $[\alpha]_D$-54°(c,0.7)(lit. (16) 112-113°, $[\alpha]_D$-56°).
M.S. major peaks are given in Fig. 4. NMR(CCl_4,60MHz): δ 2.80 (d, J 5Hz, 3-H_{cis}),2.90(d, J11Hz,3-H_{trans}), 3.80(s, aromatic methoxyl),5.36(ABXq,$|J_{AX}+J_{BX}|$16Hz,2-H),5.94(s,6- and 8-H),7.36 (s,Ar-H) and 11.92(s,OH). IR(CCl_4):ν_{max} cm^{-1} 3275,3100,3072,3038, 3012,2972,2938,2855,1650. UV(MeOH):λ_{max} nm 227(sh,ε 33800),287 (ε 33800),325 (sh, ε 6800), after addition of 1 drop 1% NaOH, 240 (sh, ε 20300),287 (ε 32100), 335(ε 8500).

Xanthorrhoeol (21). Fractions 14-17 (0.17g) were combined and the major component isolated by preparative TLC (system c). The compound recovered was recrystallized from ether-light petroleum as pale yellow flakes of 21,m.p. 120-121°, $[\alpha]_D$ + 136° (c,0.1) (lit. (21) 121°, $[\alpha]_D$ + 143°). The melting point was undepressed on admixture with an authentic sample. MS: given in text (Fig. 6). NMR(CCl_4,90MHz) : δ 1.48(d, J6Hz,secondary methyl), 2.57(s, $COCH_3$), 3.00-3.22(m,3-H_2), 4.11(ABXq,$|J_{AX}+J_{BX}|$20Hz,2-H), 6.54(d of d,J7.5,1.5Hz, 9-H), 6.95(s,6-H), 7.00(d of d, J8.0, 1.5Hz,7-H), 7.32(d of d,J8.0,7.5Hz,8-H). IR(CCl_4) : ν_{max} cm^{-1} 3591,3055,2975,2931,2891,2867,1651. UV: λ_{max} nm 222(ε 12400),257 (ε 8000),290(ε 3900).

Pterostilbene(3,5-dimethoxy-4'-hydroxy-(E)-stilbene,22). The combined fractions 26-28 (1.04g) were purified by preparative TLC(double development,system c). The compound isolated was recrystallized from light petroleum as white needles of 22,m.p. 80-81°(lit. (22) 85-86°). MS : given in text (Fig. 8). NMR ($CDCl_3$,90MHz) : δ 3.81(s,aromatic methoxyl),6.37(t,J2.5Hz,4-H), 6.63(d,J2.5Hz,2- and 6-H),6.87 and 7.00(ABq,J16Hz,vinylic protons),

6.81 and 7.36(AA'BB' system,4 aromatic protons). IR ($CHCl_3$) : ν_{max} cm^{-1} 3610,2850. UV : λ_{max} nm 217(ε 20000),315(ε 22500).

Sakuranetin (19). Fractions 39-41(0.33g) were combined and the major compound isolated by preparative TLC (system d) was recrystallized from benzene as cream needles of 19,m.p. 152-154° (lit. (19) 152-154°). MS : major ions given in Fig. 5. NMR($CDCl_3$, 60MHz) : δ 2.70-3.13(m,3-H_2),3.76(s,aromatic methoxyl),5.30 (ABXq,$|J_{AX}+J_{BX}|$16Hz,2-H),6.00(s, 6- and 8-H),6.87 and 7.25 (AA'BB' system, apparent J 8.5Hz),11.95(s,OH). IR($CHCl_3$) : ν_{max} cm^{-1} 3610,3450,2950,1700,1640. UV :λ_{max} nm 227sh,287(ε 7200), (EtOH+1 drop of NaOH),240sh(ε 10700),287(ε 7200),424(ε 2900).

Analysis of the sodium carbonate fraction. Preparative TLC (system b) of the extract yielded, as the main components, pinostrobin (17), xanthorrhoeol (21), and isosakuranetin (20) which showed the following spectral properties. MS : major ions given in Fig. 5. NMR($CDCl_3$, 60MHz) : δ 2.75-3.17(m,3-H_2),3.8(s, aromatic methoxyl),5.34(m,2-H),6.11(s,6- and 8-H),6.91 and 7.30 (AA'BB' system , apparent J8.5Hz).

Analysis of the acid soluble fraction. The 5% HCl washings were made alkaline (pH 11) and extracted with ether. The extract was fractionated by preparative TLC (system b) to give small amounts of pinostrobin (17), pterostilbene (22), and 3,5-dimethoxybenzylalcohol (23), whose NMR, MS (see Fig. 10), IR data and GLC retention time were identical to those of an authentic sample. The aqueous layer was adjusted to pH 7 and extracted with ether. Preparative TLC (system b) of the extract gave sakuranetin (19) and a mixture of 19 and 5-hydroxy-4'7-dimethoxyflavanone. The MS of this mixture showed ions for both compounds. Subtraction of the peaks known to arise from 19 left peaks characteristic of the dimethoxyflavanone.

Acknowledgments. We wish to thank Dr. H. Duewell for kindly providing a sample of xanthorrhoeol and to Mr. J. Matisons for supplying samples of propolis and for his continued interest in this work.

REFERENCES

1) Thorpe's Dictionary of Applied Chemistry, J.F. Thorpe and M.A. Whiteley, Eds., Vol. X, Fourth Edition, Longmans, Green and Co., 1950, p. 227.
2) A.I. Root, in "The ABC of Bee Culture-A Cyclopedia of Everything Pertaining to the Care of the Honey Bee",The A.I. Root Co., Ohio, 1901, p. 254.
3) A. Derevici, A. Popescu and N. Popescu, Rev. Pathol. Comp.,

1965, 2, 21 (C.A., 1965, 62, 12183h).
4) L.A. Lindenfelser, Amer. Bee. J., 1967, 107, 90 and 130 (C.A., 1967, 67, 29958n).
5) A. Popescu, C. Brailenau and A. Ghiorghiu, Dermato-venerol, 1967, 12, 57 (C.A., 1967, 67, 62762s).
6) A. Derevici, A. Popescu and N. Popescu, Ann. Abeille, 1964, 7, 191 (C.A., 1965, 62, 12184a).
7) Dr. E. Rosenburg, private communication to J. Matisons.
8) V. Todorov, St. Drenovski and V. Vasilev, Farmatsiya (Sofia), 1968, 18, 23 (C.A., 1969, 70, 113666k).
9) Kh. A. Aripov, I.K. Kamilov and Kh. V. Aliev, Med. Zh. Uzb., 1968, 50 (C.A., 1969, 70, 2300s).
10) T.V. Vakhonina, L.G. Breeva, R.N. Brodova and E.S. Dushkova, 22nd Int. Beekeep. Congr. Summ., 1969, 185 (C.A., 1971, 74, 10750m).
11) J. Cizmarik and I. Matel, Experientia, 1970, 26, 713; and J. Apic. Res., 1973, 12, 52; and references therein.
12) V.R. Villaneueva, M. Barbier, M. Gonnet and P. Lavie, Ann. Inst. Pasteur, Paris, 1970, 118, 84 (C.A., 1970, 72, 107167y); and V.R. Villaneueva, D. Bogdanovsky, M. Barbier, M. Gonnet and P. Lavie, Ann. Inst. Pasteur, Paris, 1964, 106, 282 (C.A., 1964, 60, 14875b).
13) S.A. Propavko, A.I. Gurevich and M.N. Kolosov, Khim. Prir. Soedin., 1969, 5, 476 (C.A., 1970, 73, 85062b); and S.A. Propavko, 22nd Int. Beekeep Summ., 1969, 163 (C.A., 1971, 74, 10749t).
14) E.M. Schneidewind, H. Kala, B. Linzer and J. Metzner, Pharmazie, 1975, 30, 803 (C.A., 1976, 84, 86747t).
15) K. Janes and V. Bumba, Pharmazie, 1974, 29, 544 (C.A., 1976, 84, 148035v).
16) Dictionary of Organic Compounds, R. Stevens, Ed., Fifth and Cumulative Supplement, Eyre and Spottiswoods Ltd., London, 1969, p. 770.
17) J.W. Clark-Lewis, Rev. Pure and Appl. Chem. (Australia), 1962, 12, 96.
18) H. Audier, Bull. Soc. Chim. Fr., 1966, 2892.
19) Dictionary of Organic Compounds, J.R.A. Pollock and R. Stevens, Eds., Vol. 5, Fourth Edition, Eyre and Spottiswoods Ltd., London, 1965, p. 2880.
20) W. Von Philipsborn, Angew. Chem. Int. Ed. Engl., 1971, 10, 472.
21) H. Duewell, Aust. J. Chem., 1965, 18, 575.
22) E. Spath and J. Schlager, Chem. Ber., 1940, 73, 881.
23) D.E. Hathway and J.W.T. Seakins, Biochem. J., 1959, 72, 369.
24) Encyclopedia of Chemical Technology, R.E. Kirk and D.F. Othmer, Eds., Vol. 1, The Interscience Encyclopedia Inc., New York, 1947, p. 468.
25) B.A. Bohm, Phytochemistry, 1968, 7, 1687.
26) A.J. Birch and C.J. Dahl, Aust. J. Chem., 1974, 27, 331.
27) D.E. Hathway, Biochem. J., 1962, 83, 80; and W.E. Hillis and M. Hasegawa, Biochem. J., 1963, 83, 503.

28) J.R. Cannon, P.W. Chow, M.W. Fuller, B.H. Hamilton, B.W. Metcalf and A.J. Power, Aust. J. Chem., 1973, 26, 2857; and references therein.
29) R.M.V. Assumpcao, S.M. Koop and O.R. Gottlieb, An. Acad. Brasil Cienc., 1968, 40, 297 (C.A., 1969, 71, 46664p); and P. Rudman, Holzforschung, 1963, 17, 54 (C.A., 1963, 59, 5712c).
30) S. Shibata, M. Harada and W. Budidarno, Yakugaku Zasshi, 1960, 80, 620 (C.A., 1960, 54, 21488e).
31) H. Lyr, Enzymologia, 1961, 23, 231 (C.A., 1962, 57, 3870g); and A. Akisanya, C.W.L. Bevan and J. Hirst, J. Chem. Soc., 1959, 2679.
32) T.R. Seshadri, Phytochemistry, 1972, 11, 891.
33) F. De Eds, in Flavonoid Metabolism in"Metabolism of Cyclic Compounds",Comprehensive Biochemistry, M. Florkin and E.H. Stotz, Eds., Vol. 20, Elsevier Publishing Company, Amsterdam, 1968, p. 127; and J. Mosser, M. Trouilloud, F. Fauran and P. Cros, Fitoterapia, 1975, 46, 135.
34) R.R. Scheline, Pharmacol. Rev., 1973, 25, 451.
35) R.R. Scheline, Experientia, 1974, 30, 880.
36) J.E. Sinsheimer and R.V. Smith, Biochem. J., 1969, 111, 35.
37) T. Watanabe and K. Akamatsu, Biochem. Pharmacol., 1975, 24, 442.

MASS FRAGMENTOGRAPHY (SINGLE AND MULTIPLE ION MONITORING) IN DRUG RESEARCH

A. Frigerio ° and E.L. Ghisalberti °°

°Istituto "Mario Negri", 20157 Milan, Italy.

°°University of W.A., Nedlands, Western Australia

INTRODUCTION

One of the major problems in drug research is the accurate measurement of very low levels of drugs and their metabolites in biological tissues and fluids. The technique of selected ion monitoring (SIM)¨, or mass fragmentography, has been shown to be one of the most sensitive detection systems known. Simply defined, SIM is a technique in which the detection and recording systems of a mass spectrometer (MS) are dedicated to acquiring the ion current due to one or more selected ions. The sensitivity and selectivity of SIM have been amply demonstrated in the measurement of low levels of endogenous compounds, drugs and their metabolites in biological samples (1-7). Recently the technique has also been used for the detection and quantitation of pesticides and environmental pollutants (8). Since the first application of SIM for the identification of chlorpromazine and some of its metabolites in human blood (9), several reviews have appeared which discuss the principles and applications of SIM (1-7).

In this review some specific examples of the application of SIM to problems in drug research are presented. Some of the more important factors which can influence the sensitivity, specificity and reliability of SIM measurements are also discussed.

¨Over the last few years a number of names have been coined to describe this technique. The name more commonly used and adopted in this review is "selected ion monitoring" or SIM. A rationale for this choice has been advanced (1).

APPLICATIONS

Selected ion monitoring can be divided into two classes: (a) single ion and (b) multiple ion monitoring (Table 1), each

TABLE 1.

Selected ion monitoring (mass fragmentography)

A. Single ion monitoring
   1. Single compound analysis
   2. Functional group analysis

B. Multiple ion monitoring
   1. For detection of drugs and metabolites containing
      (a) Naturally occurring isotopes
      (b) Artificially generated isotope doublets
   2. For quantitation of drugs and metabolites using
      (a) Unlabelled internal standard
      (b) Labelled internal standard

C. Use of other ionization techniques
   e.g. Chemical ionization
   Field ionization
   Field desorption

presenting some advantages and limitations with respect to the other.

Single ion monitoring can be used for the detection and quantitation of a single compound by focusing on an ion unique to the compound of interest. Calibration curves must however be obtained to determine the losses in extraction, purification and derivatization on the one hand and losses in column absorption and variation in instrumental response on the other. This is time consuming and often unreliable.

Functional group analysis allows quantitation of more than one compound. In most cases the drug and the metabolites of interest have common aspects to their structure. If this is reflected in their mass spectra, by formation of a common fragment, then they can be separately detected by single ion monitoring if their retention time is different. Quantitation can be achieved by selecting an internal standard which also meets these requirements. This approach has been used (10) for the simultaneous quantitation of propranolol (1) and one of its metabolites 4-hydroxypropranolol

(2). The latter also appears to have equal adrenergic β-blocking activity. The trifluoroacetyl derivatives of the two compounds (3 and 4) and the internal standard chosen, oxprenolol (5), give an ion at m/e 308 as the base peak in their mass spectra (Fig. 1).

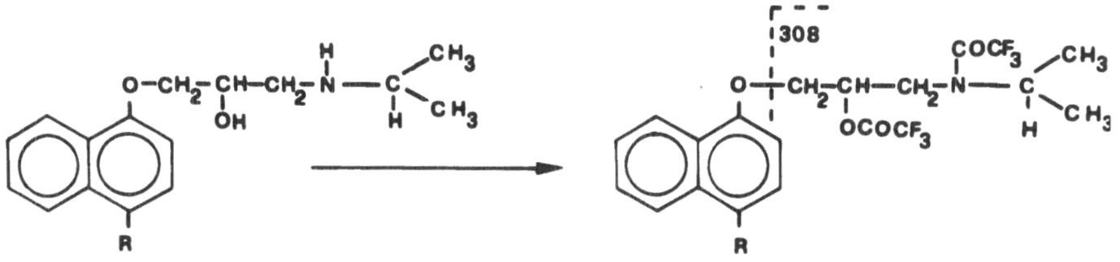

(1) R = H Propranolol

(2) R = OH 4-Hydroxy Propranolol

(3) R = H

(4) R = $OCOCF_3$

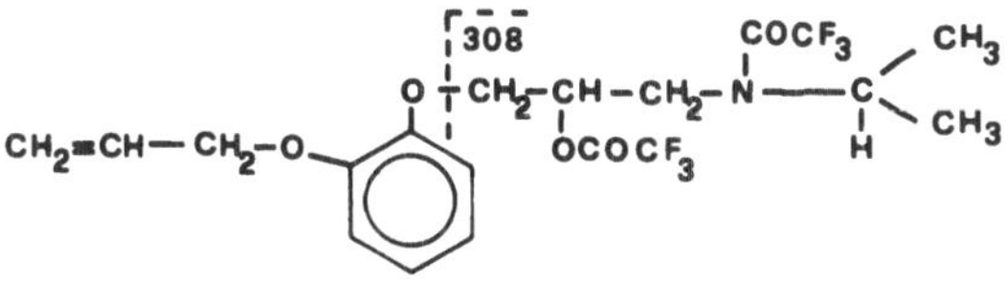

(5) Oxprenolol

FIG. 1. Structure of propranolol, 4-hydroxypropranolol their trifluoroacetyl derivatives and the ditrifluoracetyl derivative of oxprenolol.

The selected ion profiles of the extracts from plasma of human subjects treated with propranolol are shown in Fig. 2. Calibration curves were obtained for both compounds and are linear in the range 5-150 ng/ml. The minimum detectable concentrations for propranolol and 4-hydroxypropranolol were 1 ng and 5 ng/ml respectively. Interestingly, the standard curve obtained from plasma did not differ significantly from that obtained from water, indicating no interference from plasma endogenous compounds. A similar curve was obtained even when the extraction step was omitted.

Other examples of functional group analysis have been reported

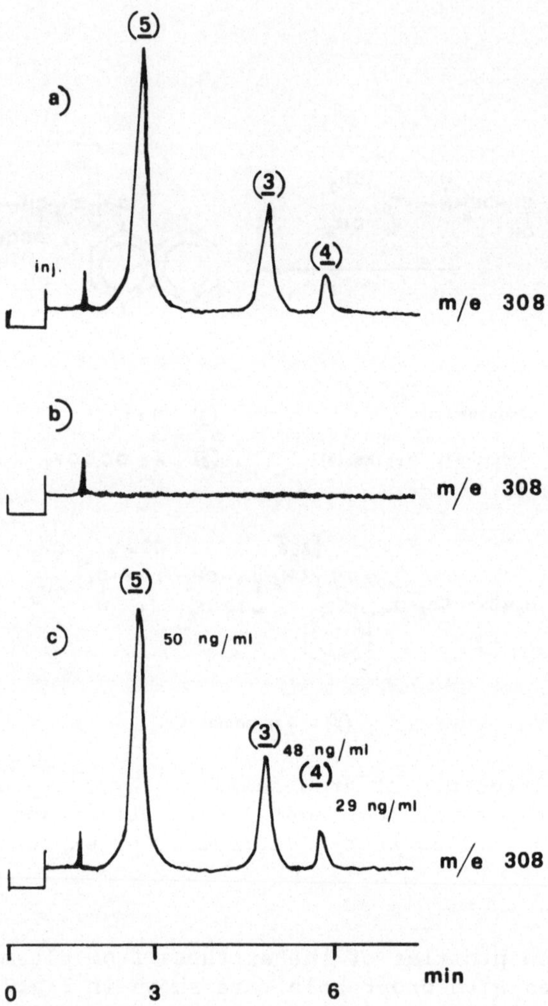

FIG. 2. Selected ion profiles of (a) extract of plasma from humans treated with propranolol (b) blank and (c) mixtures of standards.

(11). Quantitation of propranolol, 4-hydroxypropranolol and N--deisopropyl-propranolol was achieved by single ion monitoring. The heptafluorobutyrate derivatives of the three compounds showed an ion at m/e 252 in their mass spectra. Similarly, diphenylhydantoin and 5-(p-hydroxyphenyl)-5'-phenylhydantoin were quantitated by single ion monitoring of the ion at m/e 165 formed in the mass spectra of their trimethylsilyl derivatives.

Functional group analysis is potentially very sensitive since monitoring of a single ion allows the output from the electron multiplier to be integrated over a longer time, enhancing signal to noise ratio and reducing random noise. Furthermore wider slit settings can be used since resolution is not critical. It is expected that this technique will find wider applications.

Multiple ion monitoring can be applied in a number of ways (Table 1). It has been shown to be extremely useful for the detection of drugs and metabolites containing elements with more than one naturally-occurring stable isotope (e.g. Cl and Br) (12). For drugs which do not contain these elements isotope doublets can be artificially generated using equimolar mixtures of unlabelled and ^{15}N and/or deuterium-labelled drug (13, 14). The multiple ion monitoring capacity is used to detect the isotope peaks characteristic of the drug and metabolites. This elegant method has been used in the study of the metabolism of nortriptyline (13).

The advantage of this method is that it allows metabolic studies to be carried out in humans after very small, even single, doses of drugs containing non-toxic stable isotopes. However both of the approaches mentioned still require the addition of an internal standard before isolation of the extracts if quantitation is required. In this case they can both be considered under the general approaches listed below.

Multiple ion monitoring is a powerful method for the simultaneous quantitation of drugs and their metabolites and two approaches are available. The first relies on quantitation using an unlabelled standard. A representative application (15) is that of the determination of the plasma level of methsuximide (6), an anticonvulsant, and its metabolite N-desmethylmethsuximide (7) (Fig. 3). Surprisingly, it was shown that the plasma levels of the metabolite were approximately 700 times those of the parent drug. Since the effectiveness of these two compounds is equivalent this result strongly implicates the metabolite as responsible for the anti-convulsant activity.

Both drug and metabolite give a base peak at m/e 118 in their mass spectra whereas the internal standard used, phensuximide (8), has a base peak at m/e 104 (Fig. 3). These two ions were monitored as well as those representing the molecular weights of the drug

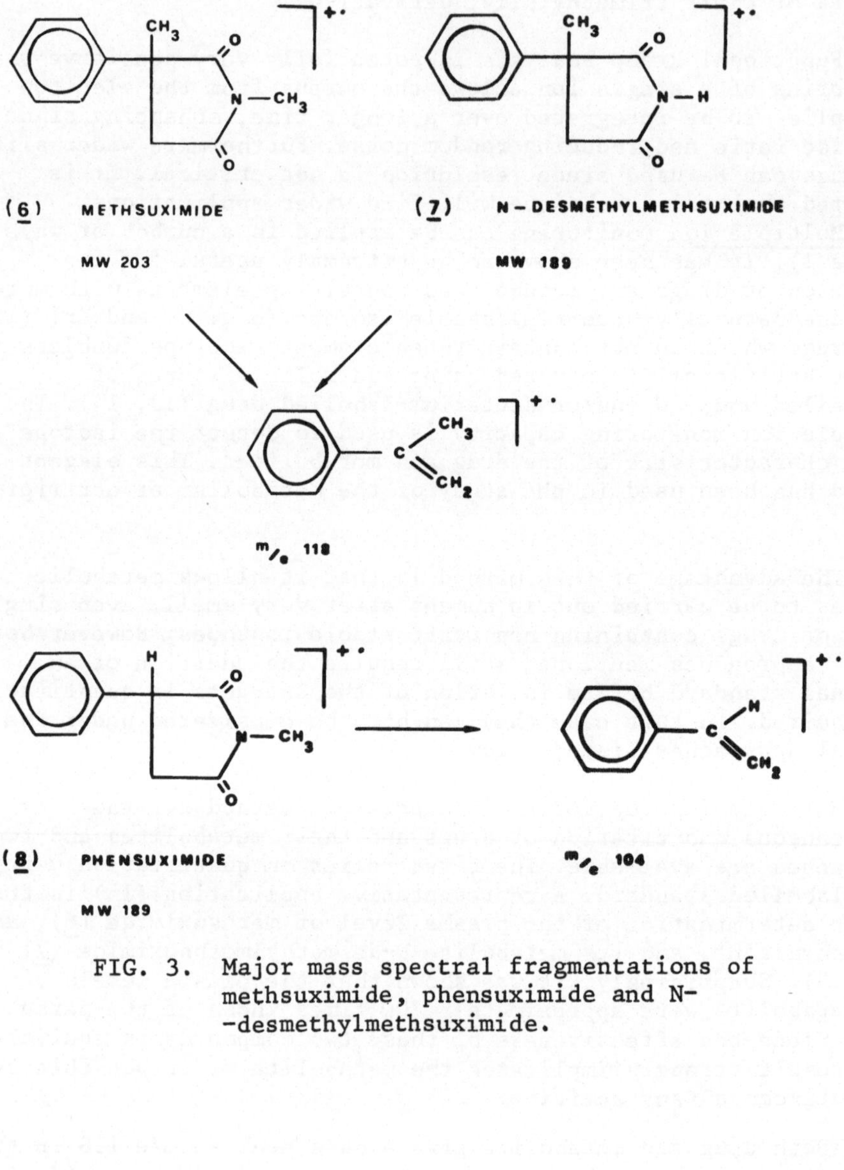

(6) METHSUXIMIDE

MW 203

(7) N-DESMETHYLMETHSUXIMIDE

MW 189

m/e 118

(8) PHENSUXIMIDE

MW 189

m/e 104

FIG. 3. Major mass spectral fragmentations of methsuximide, phensuximide and N-desmethylmethsuximide.

(m/e 203), the metabolite and the internal standard (m/e 189). Since the structure, and presumably the physical properties of the standard, are similar to those of the drug and its metabolite, the choice of this standard for the quantitation of both compounds appears justifiable. The selected ion profiles of the extracts of plasma from subjects treated with methsuximide are shown in Fig. 4. The relative levels of the two compounds are quite evident. Standard curves were prepared for methsuximide (10 ng - 1 µg/ml, RSD 6%) and N-desmethylmethsuximide (1 - 100 µg/ml, RSD 3%). It is noteworthy that in a single analysis widely different concentrations, in this case varying by three orders of magnitude, can be measured.

The use of a labelled internal standard offers a number of advantages from the point of view of convenience and reliability. Since the chemical and physical properties of such a standard are similar to those of the compound of interest, negligible differences in extraction, derivatization and chromatographic properties are observed. An added advantage is the use of an internal standard to minimize the uncontrollable but variable losses encountered in the analysis of low levels of drugs and their metabolites. Conversely, the lack of suitably-labelled standards, particularly for metabolites, presents a serious limitation. Recently, multiply-labelled internal standards have been increasingly used. This is justified by the need to separate the ions monitored so that the isotopic contribution of an ion will not interfere with another ion. However this approach can present problems which have not been widely recognized or investigated. Multiply-labelled standards can show GC or HPLC properties slightly different to those of the compound of interest. The presence of isotopes can also introduce bias, *via* secondary isotope effects, in either the derivatization step or in the ionization and fragmentation processes in the mass spectrometer. This last point may be quite significant since isotope effects in dissociation processes are much larger than those involved in ionization processes.

An example of multiple ion monitoring using a standard labelled with stable isotopes is the quantitation of morphine (**9**) and 6-acetylmorphine (**11**) in the blood of rabbits (16). The compounds were analyzed as the trifluoroacetyl derivatives (**10** and **11**) using the corresponding N-trideuteriomethyl derivatives (**13** and **14**) as the internal standards. The base peak in the mass spectra of each pair of compound appeared at m/e 364 and 367 (Fig. 5). As an internal check, two other ions at m/e 363 and 365 were also monitored. An example of the selected ion profiles for blank plasma and the plasma extracts of rabbits administered heroin are shown in Fig. 6. The sensitivity of detection is such that 500 pg can be detected although the pure compound can be detected in the range of 5 to 10 pg. In comparison, immunoassay methods for morphine

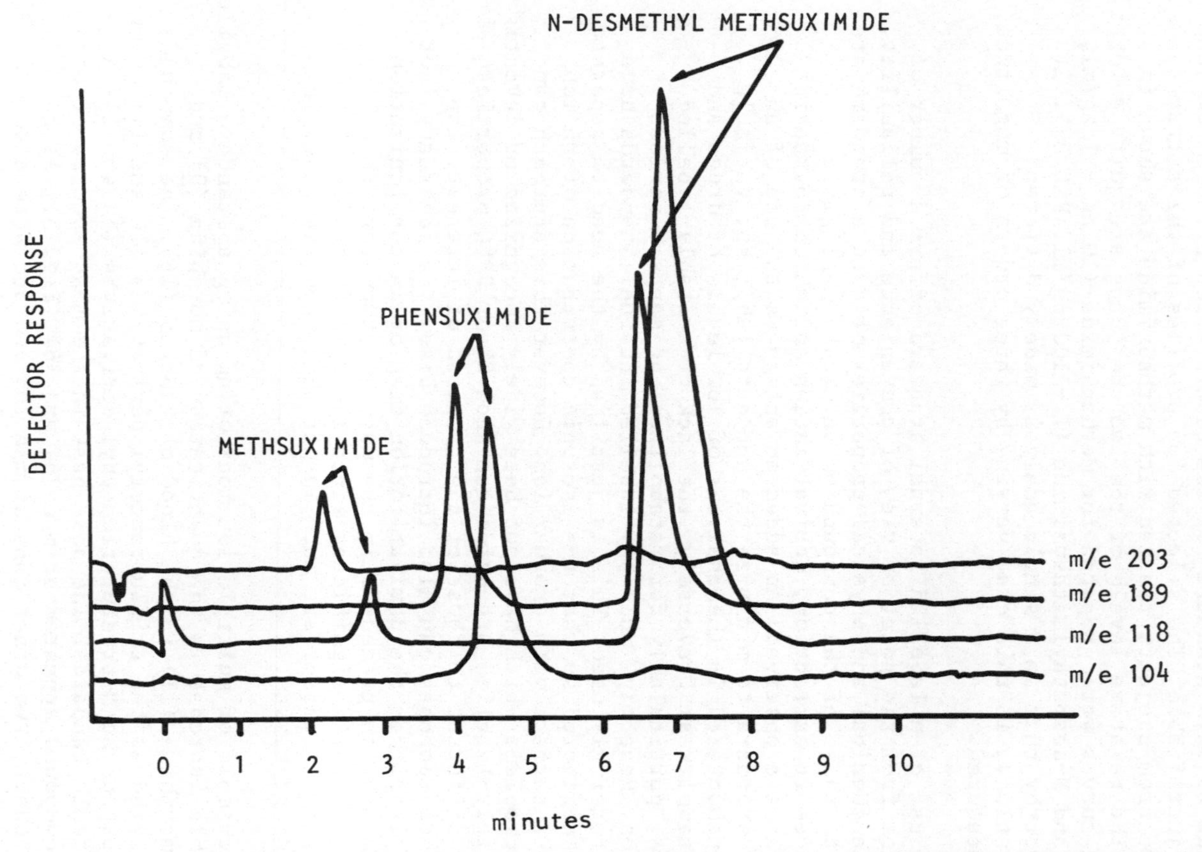

FIG. 4. Selected ion profiles of plasma extract of subjects treated with methsuximide (Lowest trace is aligned with time scale. After the elution of methsuximide the detector response was attenuated x 10).

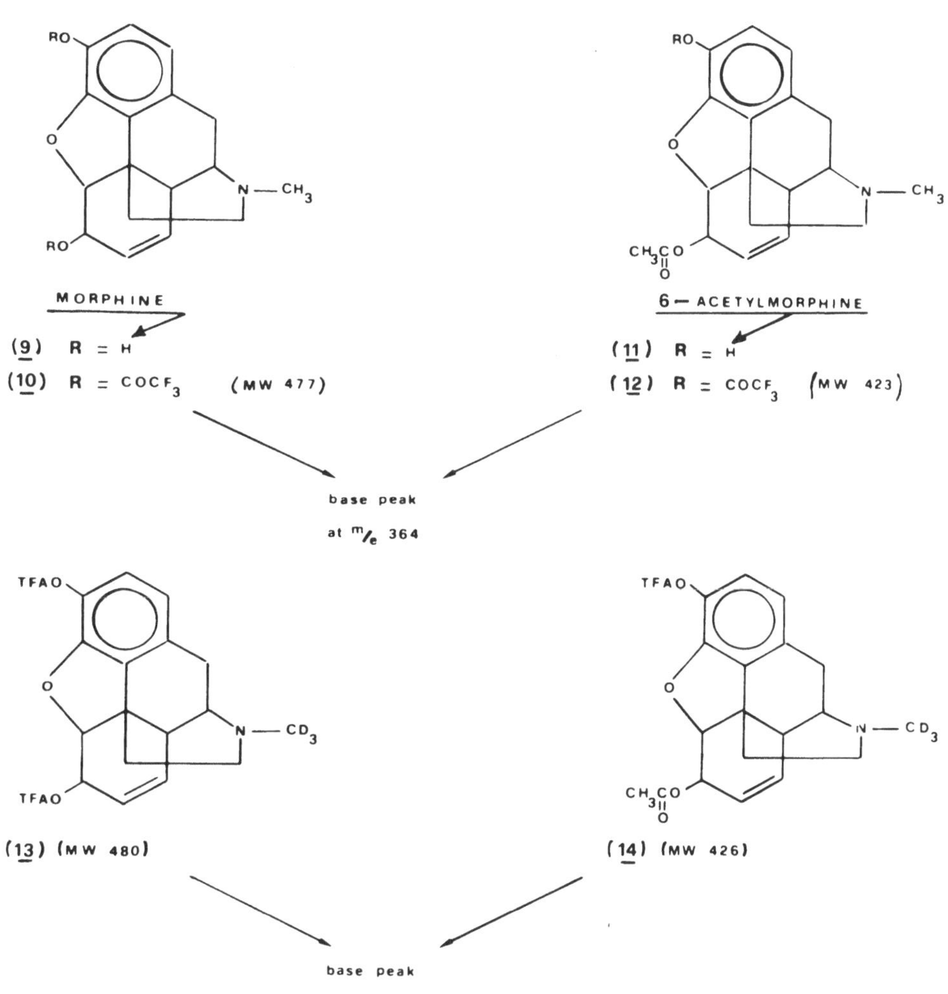

FIG. 5. Structure of morphine, 6-acetylmorphine and their derivatives.

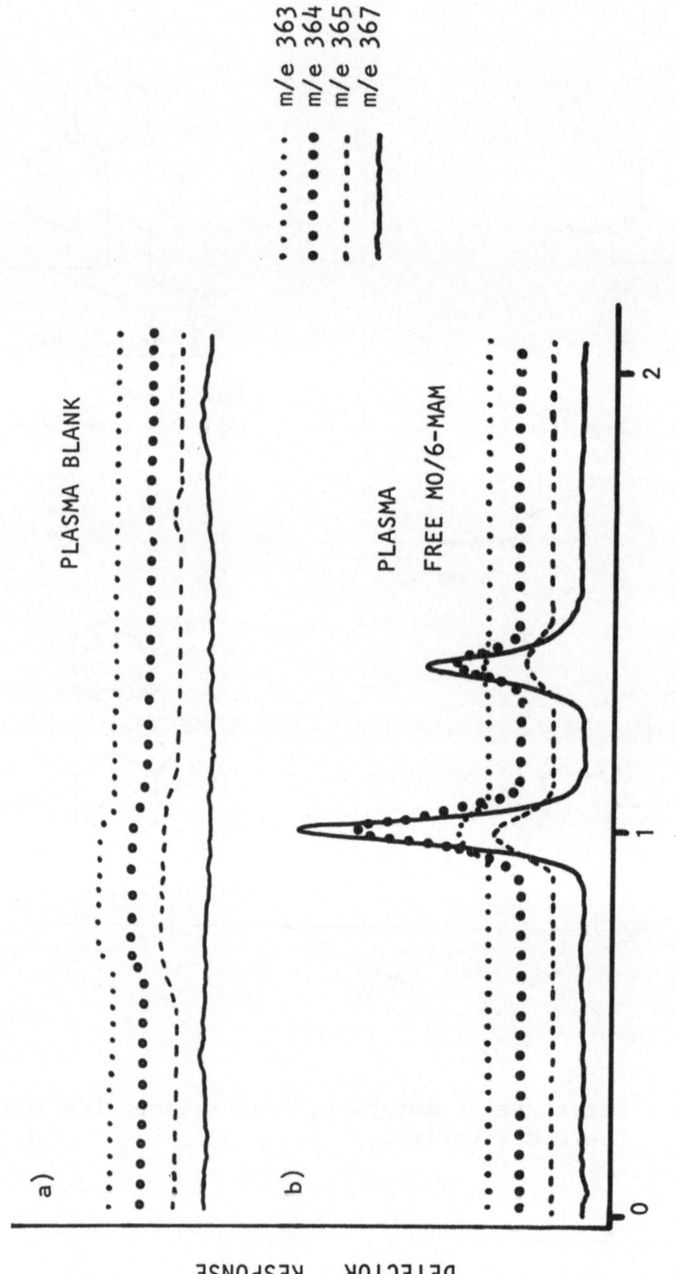

FIG. 6. Selected ion profiles of (a) plasma blank and (b) plasma extracts from rabbits administered heroin intraperitoneally.

are capable of measuring levels of morphine down to 50 - 100 pg. SIM analysis of morphine using CI has been reported to achieve comparable sensitivity with a specificity which is not equalled by the immunoassay or haemagglutination analyses (17).

In a series of publications Horning et al. have described the use of GC-MS-computer techniques for the investigation of a variety of problems in toxicology and perinatal pharmacology (18, 19). Using ^{13}C- and ^{2}H-labelled internal standards a number of anti-convulsant drugs and compounds of clinical interest have been quantitated in biological fluids by SIM. Because of its high sensitivity and specificity SIM analyses allow quantitation of nanogram or picogram amounts of several drugs and drug metabolites in a single run. This is particularly important in the study of drug metabolism in the human neonate and in monitoring drug levels in multiple drug therapy cases.

All the applications so far described used electron impact ionization for the MS analysis. Alternative methods of ionization, CI,FI and FD, most often provide simpler mass spectra dominated by a few highly intense ions. Furthermore these ionization techniques allow the analysis of non-volatile compounds to be achieved without recourse to derivatization. Although as yet only a few comparative experiments have been carried out, the indications are that these techniques, apart from increasing the range of drugs that can be analysed, are as sensitive as EI methods and offer increased specificity.

One of the first applications of CI-SIM to drug analysis involved the development of an analytical technique to quantitate selected anti-neoplastic agents (20). To test the method, an equimolar mixture of 5-fluoro-2'-deoxyuridine, 6-mercaptopurine ribonucleoside and arabinosylcytosine were added to mouse serum. Extraction and flash methylation yielded the methylated derivatives (15, 16 and 17 respectively; Fig. 7) which could be separated by GC. The selected ion profiles of this extract, obtained by monitoring the MH^+ of each derivative are shown in Fig. 8. This method has been used for the quantitation of arabinosylcytosine in mice serum (20).

CI has also been used for the quantitation of diphenoxylic acid (19) in human plasma following the administration of diphenoxylate (18), a drug used in the treatment of diarrhoea (21). The acid was analysed as the methyl ester derivative (20) using the d_4-analogue as the internal standard (22) (Fig. 8). In this case four ions were selected for monitoring. To compensate for instrument drift the ions at m/e 439.3, 439.4 for methyl diphenyloxalate and m/e 443.3 and 443.4 for the internal standard were chosen. Levels down to 20 ng/ml could be measured. The limit of detection of the gas chromatographic method, previously used

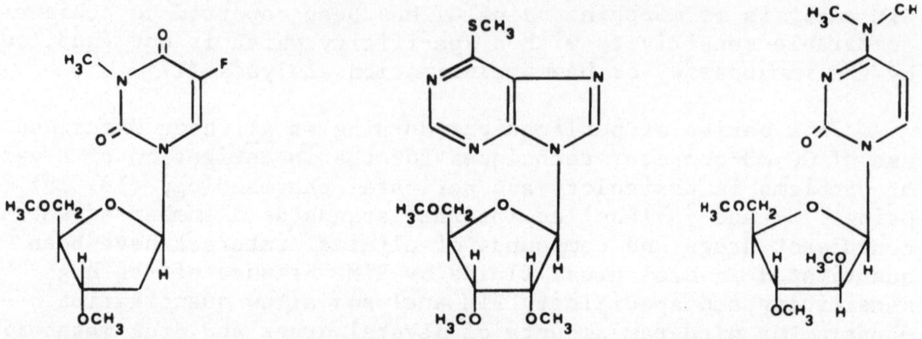

(15) 5-FLUORO-2'-DEOXYURIDINE-METHYL-DERIVATIVE

MW 288

(16) 6-MERCAPTO-PURINE-RIBONUCLEOSIDE-METHYL-DERIVATIVE

MW 340

(17) ARABINOSYLCYTOSINE-METHYL-DERIVATIVE

MW 313

FIG. 7. Structure of the derivatives of selected anti-neoplastic agents.

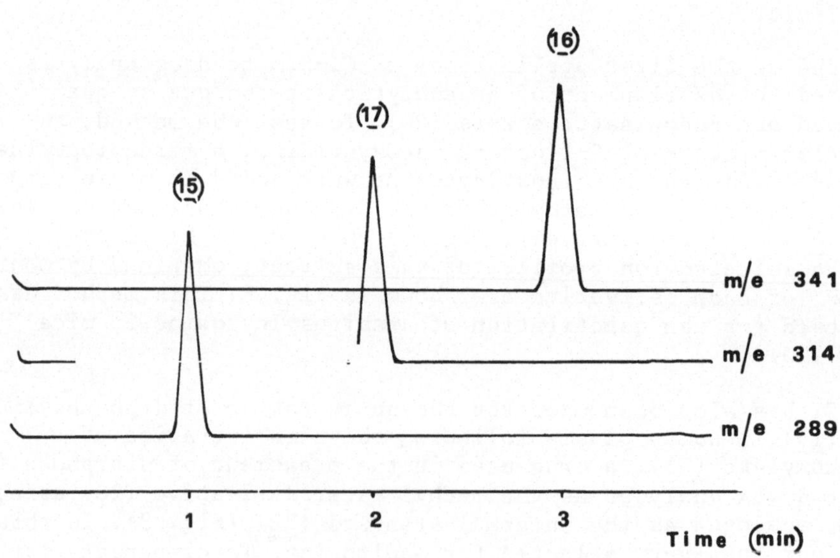

FIG. 8. Selected ion profiles of <u>15</u>, <u>16</u> and <u>17</u> extracted from mouse serum.

to quantitate diphenoxylic acid, was 200 ng/ml.

Chemical ionization is now more frequently used for the quantitation of drugs and drug metabolites. This technique has been applied to the quantitation of tolbutamide and its metabolites (22), quinidine, lidocaine and monoethylglycinexylidide (23), phenformin (24), and N,N-bis(2-chloroethyl)phosphoro-diamidic acid (25) in human plasma. Phencyclidine has also been quantitated in blood and urine (26). In some cases compounds which can not be derivatized easily or which show poor chromatographic properties can be quantitated using the direct insertion-SIM method (22, 23).

Field ionization (27) and field desorption techniques (28) have also been used for SIM in drug analysis. The examples given above were chosen to indicate the variety of applications of SIM. Comprehensive lists of references on SIM have been published for the period up to to the beginning of 1975 (29, 30).

The development of more potent drugs and the usefulness of monitoring lower levels of drugs during the treatment of certain disorders provide the impetus for more sensitive analytical techniques. Compared to other analytical methods SIM offers the advantage of increased sensitivity coupled with specificity. In general it has been found to be 100 to 1000 times more sensitive than GC or GC-MS. Electron capture detectors can be as sensitive in some cases but their use is limited mostly to compounds containing halogen atoms. SIM compares favourably with the very sensitive immunoassay and haemagglutination methods. The latter two techniques have the advantage that a large number of analyses can be carried out in a short time (up to 400/day). However their specificity can be lowered as they are subject to cross-interference from substances having similar structures as the compound of interest, these methods giving little indication of the structure of the interfering substance. SIM can achieve the same sensitivity, is less adaptable for rapid analysis and requires expensive equipment. The technique of atmospheric pressure ionization (31) may considerably reduce the time required for SIM analysis. SIM is also particularly useful in determining the specificity and accuracy of simpler and less expensive analytical methods.

FACTORS INFLUENCING SENSITIVITY

The sensitivity of SIM measurements can be affected by a number of interrelated factors, the most important of which are briefly discussed below. Since most measurements are carried out with the GC-MS combination both instruments should be operating at optimal conditions if maximum sensitivity is required.

1. Gas chromatographic behaviour of the compounds: Although some drugs have good GC properties most have to be derivatized

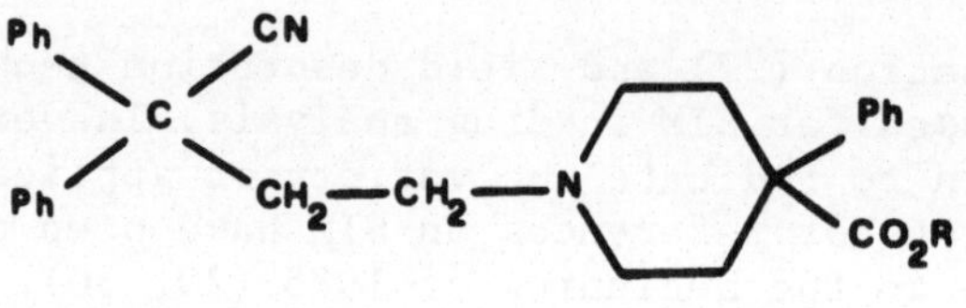

(18) R = C_2H_5

(19) R = H

(20) R = Me (MW 438)

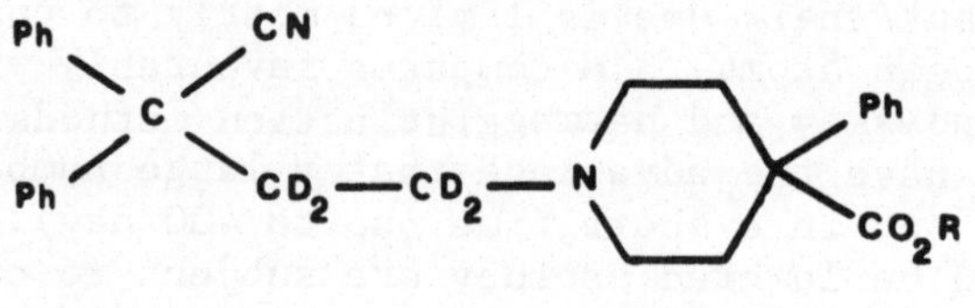

(21) R = H

(22) R = Me (MW 442)

FIG. 9. Structures of diphenoxylic acid (19), its derivatives and metabolites.

to enhance their volatility and stability and to minimize losses through absorption on the column. Other factors like phase, column temperature and gas flow rate must be selected so as to obtain optimum response for the compounds of interest. When small quantities of a compound are involved the addition of a stable isotope analogue can minimize losses due to absorption and degradation on the column.

This effect can be illustrated for clonidine, a potent hypotensive agent in man. Clonidine is normally administered in submilligram doses and investigation of its metabolism and kinetics requires a sensitive analytical method. The gas chromatographic behaviour of small quantities (< 10 ng/ml) of clonidine is variable. If quantities (40 ng) of $\underline{d}_4$-clonidine (Fig. 10) are added to the sample then levels down to 200 pg of clonidine could be reproducibly quantitated (Relative standard deviation 10%) by SIM (32). Similarly indoramin (Fig. 10) could be quantitated below its normal detectable levels of 100 ng if enough $\underline{d}_5$-indoramin was added before analysis to ensure that 1.5 ng of sample could be injected on the column (33). A linear calibration curve could be obtained in the range 5 - 100 ng/ml of plasma for endogenous indoramin.

The type of separator used can also be important. Generally one that will afford a higher yield of the substrate is better as long as the increased level of gas passing into the ion source does not adversely affect the ionization and mass analysis processes.

2. Ion source conditions. (a) Ionization voltage. The value which will give the highest relative abundance of the ion(s) of interest has to be determined. This value can lie between 16 and 70 eV. (b) Trap current. Sensitivity also increases with the levels of trap current used. However too high a trap current will shorten the lifetime of the filament. (c) Ion source temperature. The influence of this on the fragmentation pattern of a compound is well known. Decreasing the temperature can lead to an increase in the relative abundance of some ions. When a GC-MS combination is used however the ion source temperature should not be much below the column temperature since this can lead to contamination of ion source components. If this happens then the distortion of the electric fields adversely affects ion focusing.(d) Ion source pressure. Elevated ion source and analyzer pressure can decrease sensitivity and resolution due to "peak broadening". In one example (34) where the total effluent of a GC was passed into a differentially pumped MS, increase of the helium flow rate from 1.1 to 7.2 ml gave a two-fold decrease in sensitivity while a 12% decrease in resolution was observed.

3. Ion selection. (a) Intensity. The sensitivity available for SIM measurements is directly proportional to the percentage of the total ion current represented by the selected ion(s). It follows that the base peak should be preferred. Derivatization of

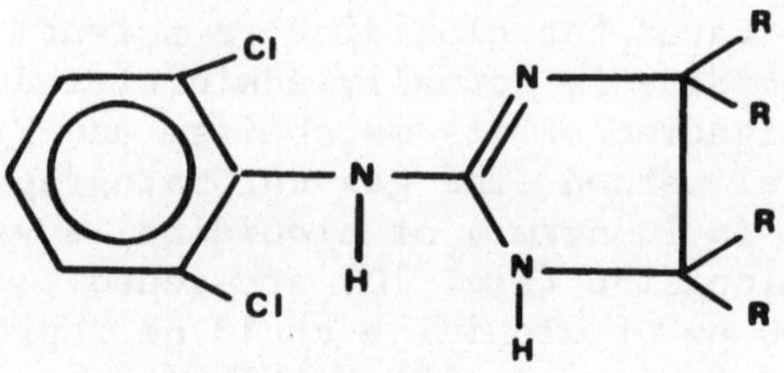

R = H CLONIDINE

R = D d_4-CLONIDINE

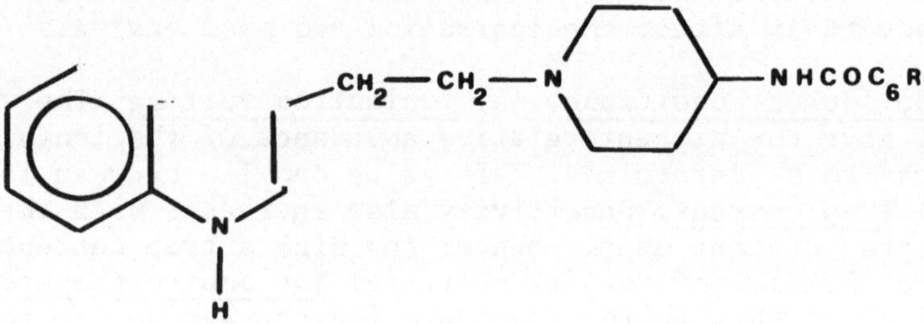

R = H INDORAMIN

R = D d_5-INDORAMIN

FIG. 10. Structures of clonidine, indoramin and their deuterated standards.

of a compound can result in directed fragmentation yielding intense ions characteristic of the compound. CI, FI and FD techniques give mass spectra in which only a few characteristic ions are present. As yet few experiments have been carried out to determine whether increased sensitivity is in fact achieved. Nevertheless these methods can be useful in reducing the number of signals from background contaminants which otherwise limit the sensitivity and specificity of SIM. (b) Number of ions. Maximum sensitivity results from monitoring a single ion. As more ions are selected each will be monitored for a shorter period with a subsequent decrease in the signal to noise ratio and increased contribution from random electrical noise. As a general approximation the overall GC-MS sensitivity for SIM decreases as a function of $1/\sqrt{n}$ where n is the number of ions monitored.

4. Background. The major sources of background in the mass spectrum are contributions from plastic containers, oil, column bleed, reagents and solvents used in the isolation and derivatization processes. As far as possible these should be checked for and eliminated. In this respect CI, FI and FD can be advantageous since the mass spectra of contaminants is reduced to a few recognizable peaks.

5. Internal standard. The advantages of using a labelled internal standard have been mentioned. In these cases however multiple ion monitoring must be used. The internal standard should give rise to ions well separated from the ions of the compounds to be analysed, and should not contribute to them. The purity and isotopic enrichment of the labelled standard can be checked in separate SIM experiments.

FACTORS INFLUENCING SPECIFICITY

This depends on the number of ions monitored. Clearly the more ions that are monitored the greater the specificity. On the other hand, monitoring a large number of ions leads to a poorer statistical representation of any single ion. In multiple ion monitoring at least two ions should be monitored for each compound. High resolution mass spectrometry has been used for increased specificity (35).

FACTORS INFLUENCING ACCURACY

(a) Calibration curves. Two methods are commonly used to derive calibration curves. These differ only in whether the varying amounts of compound and internal standard are extracted from water or biological fluids before analysis. Extracts from biological samples often show higher relative standard deviations (RSD) than standard samples of identical concentrations. Whereas the RSD in the measurement of 100 pg of N,N-dimethylamobarbital was 5.1%, injecting the same quantity obtained by extraction from plasma gave an RSD of 20% (36). The RSD value for a significant number of samples over the whole range of the calibration curve should be calculated. (b) Background. As random contributions increase so does the RSD of measurement. In this case the same considerations as discussed under sensitivity apply. (c) Internal standard. The work of Lee and Millard (37) on the quantitation of allobarbitone has shown that small quantities of this compound are most accurately measured by adding a labelled internal standard to act as carrier plus an internal standard giving an ion in common with the compound being measured. Using single ion monitoring they showed that 40 and 100 pg of methylated allobarbitone could be measured with an RSD of 11.5 and 2% respectively. When only the labelled internal standard was used the RSD values increased to 19.8 and 15.3% respectively.

REFERENCES

1) F.C. Falkner, B.J. Sweetman and J.T. Watson, Appl. Spectrosc. Rev., 1975, 10, 51.
2) B. Holmstedt and L. Palmer, Adv. Biochem. Psychopharmacol., 1973, 7, 1.
3) A.E. Gordon and A. Frigerio, J. Chromatogr., 1972, 73, 401.
4) D.J. Jenden and A.K. Cho, Annu. Rev. Pharmacol., 1973, 13, 371.
5) C. Fenselau, Appl. Spectrosc., 1974, 28, 305.
6) A.M. Lawson, Clin. Chem., 1975, 21, 803.
7) R. Carrington, A. Frigerio and R. Roncucci, in "Drug Fate and Metabolism: Methods and Techniques", E.R. Garret and J.L. Hirtz, Eds., to be published, 1976.
8) G. Vander Velde and J.F. Ryan, J. Chromatogr. Sci., 1975, 13, 322; and K.P. Evans, A. Mathias, N. Mellor, R. Silvester and A.E. Williams, Anal. Chem., 1975, 47, 821.
9) C.-G. Hammar, B. Holmstedt and R. Rhyage, Anal. Biochem., 1968, 25, 532.
10) T. Walle, J. Morrison, K. Walle and E. Conradi, J. Chromatogr., 1975, 114, 351.
11) F.B. Abramson and N.B. Fizette, in "22nd Ann. Conf. Mass Spectrometry Allied Topics", Philadelphia, 1974, p. 288.
12) R. Roncucci, M.-J. Simon, G. Jacques and L. Lambelin, Eur. J. Drug Metab. Pharmacokinet., 1976, 1, 9.
13) D.R. Knapp, T.E. Gaffney, R.E. McMahon and G. Kiplinger, J. Pharmacol. Exp. Ther., 1972, 180, 731.
14) D.R. Knapp and N.H. Holcombe, in "22nd Ann. Conf. Mass Spectrometry Allied Topics", Philadelphia, 1974, p. 289.
15) J.M. Strong, T. Abe, E.L. Gibbs and A. Atkinson, Neurology, 1974, 24, 250.
16) W.O.R. Ebbighausen, J.H. Mowat, H. Stearns and P. Vestergaard, Biomed. Mass Spectrom., 1974, 1, 305.

17) P.A. Clarke and R.L. Foltz, Clin. Chem., 1974, 20, 465.
18) M.G. Horning, W.G. Stillwell, J. Nowlin, K. Lertratanangkoon, D. Carrol, I. Dzidic, R.N. Stillwell and E.C. Horning, J. Chromatogr., 1974, 91, 413.
19) M.G. Horning, K. Lertratanangkoon, J. Nowlin, W.C. Stillwell, R.N. Stillwell, T.E. Zion, P. Kellaway and R.M. Hill, J. Chromatogr. Sci., 1974, 12, 630; and M.G. Horning, J. Nowlin, C.M. Butler, K. Lertratanangkoon, K Sommer and R.M. Hill, Clin. Chem., 1975, 21, 1282.
20) C. Pantarotto, A. Martini, G. Belvedere, A. Bossi, M.G. Donelli and A. Frigerio, J. Chromatogr., 1974, 99, 519.
21) G.C. Ford, N.J. Haskins, R.F. Palmer, M.J. Tidd and P.H. Buckley, Biomed. Mass Spectrom., 1976, 3, 45.
22) S.B. Matin and J.B. Knight, Biomed. Mass Spectrom., 1974, 1, 323.
23) W.A. Garland, W.F. Trager and S.D. Nelson, Biomed. Mass Spectrom., 1974, 1, 124.
24) S.B. Matin, J.H. Karam, P.H. Forsham and J.B. Knight, Biomed. Mass Spectrom., 1974, 1, 320.
25) I. Jardine, R. Brundrett, M. Colvin and C. Fenselau, Cancer Treat. Rep., 1976, 60, 403.
26) D.C.K. Lin, A.F. Fentiman, R.L. Foltz, R.D. Forney and I. Sunshine, Biomed. Mass Spectrom., 1975, 2, 206.
27) J.H. McReynolds, H. d'A. Heck and M. Anbar, in "22nd Ann. Conf. Mass Spectrometry Allied Topics", Philadelphia, 1974,p.195.
28) K.H. Maurer and U. Rapp, in "Advances in Mass Spectrometry in Biochemistry and Medicine",Vol. 1,A.Frigerio and N.Castagnoli, Eds.,Spectrum Publications, New York, 1976, p. 541.
29) E. Costa and B. Holmstedt, Adv. Biochem. Psychopharmacol., 1973, 7, 161.
30) L. Palmèr and B. Holmstedt, Science Tools, The LKB Instrument Journal, 1975, 22, 25 and 38.
31) E.C. Horning, M.G. Horning, D.I. Carrol, I. Dzidic and R.N. Stillwell, Anal. Chem., 1973, 45, 936.
32) G.H. Draffan, R.A. Clare, S. Murray, G.D. Bellward, D.S. Davies and C.T. Dollery, in "Advances in Mass Spectrometry in Biochemistry and Medicine", Vol. 2, A. Frigerio, Ed., Spectrum Publications, New York, 1976, in press.
33) G.H. Draffan, R.A. Clare, B.L. Goodwin, C.R.J. Ruthven and M. Sandler, in "Proceedings of the Sixth International Mass Spectrometry Conference", Edinburgh, Scotland, September 1973.
34) Conference Report by R.L. Foltz, Biomed. Mass Spectrom., 1975, 2, 227.
35) R.B. Parker, H.A. Simmonds, A.S. Jones and W. Snedden, Biochem. Pharmacol., 1973, 22, 2869 .
36) G.H. Draffan, R.A. Clare and F.M. Williams, J. Chromatogr., 1973, 75, 45.
37) M.G. Lee and B.J. Millard, Biomed. Mass Spectrom., 1975, 2, 78.

MASS SPECTROMETRIC IDENTIFICATION AND QUANTIFICATION OF COMPOUNDS INVOLVED IN THE METABOLISM OF PYRAZOFURIN

J. Roboz and T. Ohnuma

Mount Sinai School Medicine City University New York,

100th Street and Fifth Avenue New York, N.Y. 10029

SUMMARY

Gas chromatographic-mass spectrometric techniques were developed for the identification and quantification of pyrazofurin, orotic acid, and orotidine in human serum and urine. Free pyrazofurin was detected for the first time in both the serum and urine of patients receiving the drug. Plasma clearance studies established that the level of pyrazofurin decays exponentially with a half-life of the distribution phase of approx. 10 min. Urinary excretion yields only 10% of the total drug administered (in 10 days). Orotic acid and orotidine were positively identified in urine; quantification revealed that 20-100 mmole of orotic acid and 40-80 mmole of orotodine are excreted in a period of 8-10 days. Pyrazofurin 5'-phosphate was identified for the first time in cells (both T and B type) incubated with pyrazofurin. These findings together with the results of reversal experiments support the assumption that the mode of operation of the drug is the interruption of the conversion of orotidine 5'-monophosphate to uridine 5'-monophosphate in the course of the biosynthesis of pyrimidine nucleotides.

INTRODUCTION

Pyrazofurin (PF, 3-β-D-ribofuranosyl-4-hydroxypyrazol-5--carboxamide, Eli Lilly Compound # 47599, formerly called Pyrazomycin) is a carbon-linked nucleoside (Fig. 1) isolated from the broth filtrates of a strain of *Streptomyces candidus* (1). PF exhibits antiviral activity against several viruses, including herpes, rhino, measles, and influenza viruses *in vitro* (1), vaccinia

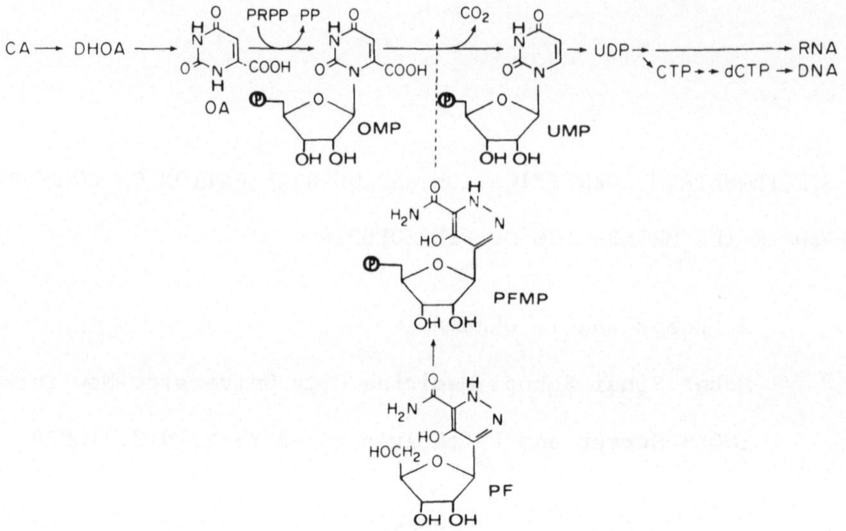

FIG. 1. Proposed mechanism of action of pyrazofurin (PF): converted into the 5'-monophosphate (PFMP) in cells, the drug inhibits the conversion of orotidine 5'-monophosphate (OMP) to uridine 5'-monophosphate (UMP). Other abbreviations are: CA-carbamylaspartase, DHOA-dihydroorotic acid, OA-orotic acid, PRPP-5'-phosphoribosyl pyrophosphate, PP-pyrophosphoric acid, UDP--uridine diphosphate, CTP-cytidine triphosphate, RNA-ribonucleate, DNA-deoxyribonucleate.

virus both *in vitro* and *in vivo* (2), and Friend virus leukemia (3). PF was also shown to exhibit activity of various degrees against several animal tumors, including Walker carcinosarcoma 256, mammary carcinoma 755, plasma cell myeloma (4) and DMBA (9,10-dimethyl--1,2-benzanthracene)-induced mammary carcinoma of the rat (5).

After toxicological studies in large animals, a Phase I study was started at the Lilly Laboratories for Clinical Research (6) with eight patients with a variety of cancers. Subsequently, Phase I studies were conducted in several research centers, including ours (7, 8). These reports provide data on drug doses and schedules used, observed systemic toxicity (mainly mucocutaneous toxicity and myelosuppression), and reasons for discontinuation of PF treatment. Our recommended dose is 250 mg/m^2 i.v. bolus, q 2-3 wk. Of the approximately 100 patients treated to-date, partial response was shown in patients with breast carcinoma (8) acute myelocytic leukemia (9) and Hodgkin's Diseases (5).

Relatively little is known about the metabolism and biological action of PF, mainly because of the lack of adequate analytical methodology. In reversal studies of the PF-induced inhibition of vaccinia virus *in vitro* (2), it was revealed that the activity of PF is reversed by uridine and uridylic acid, but not by cytidine or orotidine. This led to the suggestion (5) that the mode of action of PF might be the inhibition of the *de novo* biosynthetic pathway of pyrimidine nucleotides. The most likely mode of action of PF is shown in Fig. 1.

The first step of the biosynthesis of pyrimidine nucleotides is the irreversible carbamylation of L-aspartate by carbamyl-phosphate to form carbamylaspartate (catalyzed by the enzyme aspartate transcarbamylase). Next, carbamylaspartate is converted, by ring closure, to dihydro-orotic acid which, in turn, is reduced to orotic acid, catalyzed by the enzyme orotic acid dehydrogenase (OAD). Orotic acid (6-carboxyuracil) reacts with 5'-phosphoribosyl--1-pyrophosphate (PRPP) to form orotidine monophosphate (OMP). This reaction is catalyzed by the enzyme orotidylic pyrophosphorylase in the presence of Mg++. Here a β-glycoside linkage is formed between the N-atom at position 1 and the 1'C atom of the pentose of PRPP. (In PRPP the pyrophosphate radical replaces the OH group of ribose to form a β-glycosidic linkage). Orotidine 5'-phosphate is then decarboxylated to uridine 5'-phosphate (UMP) when catalyzed by the enzyme orotidylic acid decarboxylase (OAD). Although this step is irreversible, the reaction is subject to end product inhibition by uridine 5'-phosphate and also competitively inhibited by a number of other purine and pyrimidine nucleotides, including 6-azauridylic acid (10) and also pyrazofurin, or more likely phosphorylated PF. This suggestion is based upon the observation (2) that a 50% inhibition of OAD can be achieved with 5×10^{-8}M pyrazofurin 5'-monophosphate (PFP); pure PF shows no activity. It is noted for comparison, that a 20-fold excess of 6-azauridine 5'-phosphate is needed for the same effect. Cadman *et al.* (9) have recently shown that mono- and polyphosphate derivatives of PF can be detected in mouse L5178Y cells as well as in human leukemic cells by high pressure liquid chromatography, although no positive identification was made.

We have been studying the clinical, biochemical, and pharmacological properties of PF. A gas chromatographic-mass spectrometric technique was developed for the determination of free PF in blood and urine (11); clinical results, blood levels, and some biochemical and pharmacological results have been communicated (8). In the present paper we report additional observations concerning the biological actions of PF in man and *in vitro* experiments.

MATERIALS

Pyrazofurin was obtained courtesy of Eli Lilly Co., Indianapolis, Ind. (Lot # CT-2940-4B). Each ampoule contained 300 mg PF as lyophylized powder and was reconstituted with 10 ml of sterile 0.15 M NaCl prior to use; the vials were stored in a refrigerator. Pyrazofurin 5'-monophosphate (PFP) was obtained courtesy of Eli Lilly Co., Indianapolis, Inc. PFP was synthesized by the 5'--phosphorylation of a 2',3'-isopropylidene derivative of PF, followed by removal of the ketal protecting group (5). The field desorption mass spectrum of PFP revealed a strong signal at m/e 340, corresponding to the protonated molecular ion (12).

N,O-bis-trimethylsilyl-trifluoroacetamide+1% trimethylchlorosilane was purchased from Pierce Chemical Co.; constituents needed for the preparation of trimethylanilinium hydroxide were purchased from Aldrich Chemical Co. All solvents were of Nanograde quality.

PROTOCOLS AND METHODS

Doses and schedules. All patients in the study had histologic confirmation of diagnosed cancer, were without known curative treatment, and were beyond any palliative treatment thought to be of benefit. Informed consent was obtained from all patients in the study. Dose levels ranged from 100 to 300 mg/m^2. Bolus injections were given weekly, every 2-4 weeks. Twenty-four continuous infusions were scheduled for one to 6 days.

Sampling. Blood samples for PF levels were drawn from the contralateral arm at various intervals. The blood was allowed to clot at room temperature for about 25 min, the serum was separated by centrifuging (500 g, room temperature, 10 min) and stored at -80°C. Blood for OAD activity and leukocyte counting was obtained in heparin-containing syringes (final heparin conc: 7 µl/ml) both prior and after PF administration.

Urine samples were collected for 24 hr periods for as many as 20 days; toluene (5 ml) was used as preservative. Aliquots of urine samples were stored at -80°C.

Preparation of cell extracts. MOLT 4F is a human T-cell line established from a patient with acute lymphocytic leukemia (13) and has since been maintained in our laboratory. Characteristically, MOLT 4F cells form rosettes with sheep red blood cells (E-rosettes). They do not synthesize immunoglobulin and grow as a single cell suspension. RPMI 1788 is a human B-cell line established from the peripheral blood of a healthy volunteer. These cells produce immunoglobulin, have complement receptors and grow as clusters in suspension culture. Characteristics of these cell lines have recently been summarized by Moore (14). These cells were maintained in medium RPMI 1640 with 10% fetal calf serum and penicillin and streptomycin. The viability of the cells measured by a trypan blue exclusion study was about 95% at the time of their use.

The cells were incubated with 8 x 10^{-4} M of pyrazofurin for one hour at 37°C. This concentration is about 10 times that needed for the inhibition of 90% of the activity of orotidylate decarboxylase in L1210 cells (15). After incubation, cells were washed twice with 0.15 M NaCl to remove free pyrazofurin. Next 5 volumes of 1 N perchloric acid was added to the cell pellets and centrifuged for 15 min to remove the precipitate. The acid-soluble portion of the extract was neutralized with concentrated KOH and centrifuged to remove the precipitated potassium perchloride. The resulting extract was evaporated to total dryness under nitrogen stream (dry, room temperature), and the resulting residue (somewhat brownish in appearance, probably due to some charring) was derivatized as described later.

Determination of pyrazofurin in serum and urine. Details of this technique have been published (11, 16); here only a brief summary follows.

Serum sample was added to the internal standard (Ara-Virazole, 1-β-D-riboarabinosyl-1,2,4-triazole-3-carboxamide, ICN Corp., Irvine, Calif.). After diluting with equal volume of water, proteins were removed *via* molecular filtration (50,000 dalton exclusion limit, Amicon Co., Lexington, Mass), the "Centriflo" cones were centrifuged for 45 min at room temperature at 1100 g. Next, the filtrate was acidified (pH 6) with 0.5 N HCl and lipids were removed by extraction with dichloromethane.

A variety of interfering constituents were removed by filtering the lipid extracted serum through an ion exchange bed consisting of a Dowex-50 (acidic) and Biorex-5 (basic) resins placed into a small Pasteur pipette, one on top of the other (acidic on top). The filtrate was lyophylized and the solid residue silylated with BSTFA+1% TMCS at 60°C for 10 min. The silylated sample was analyzed in a combined gas chromatograph-mass spectrometer system (Finnigan, Model 3300) operated under the following conditions: Gas chromatographic column: 3% OV-17 or 3% SE-30, 1 m long, 2 mm ID; injection temperature is 270°C, temperature programming is from 175 to 240°C at 8°/min. The effluent entered the mass spectrometer *via* a heated 240°C interface (without molecular separator). Methane gas served as both the gas chromatographic carrier gas and the reagent gas in the chemical ionization source (at 1000 μ pressure). The mass spectrometer was operated in the selected ion recording mode, and the quasimolecular ions of PF and Ara-Virazole were monitored at m/e 692 and m/e 517, respectively; two fragment ions (loss of methyl group) were also monitored. A calibration curve was established using known amounts of PF added to pooled human serum. The observed areas of PF and the internal standard were used to calculate area ratios which, in turn, were used to determine the amount of PF in serum. Using this method minimum detectable quantity is 10 ng/ml serum; lowest level at which quantification is still reliable is 100 ng/ml serum.

Urinary PF was determined by adding urine to the internal standard, vortexing, adding the sample to ethyl alcohol (96%) dropwise, vortexing, centrifuging (770 g for 15 min), and evaporating the clear solution to dryness under nitrogen (dry, room temperature or 40°C water bath). Silylation and gas chromatographic-mass spectrometric conditions were the same as described above.

Determination of orotic acid and orotidine in urine by GC-MS. Urine samples were prepared as described above. In the gas chromatographic-mass spectrometric (GC-MS) analysis the total ion chromatogram was obtained and mass chromatographic search was made for the constituents. Other urinary constituents were identified by obtaining the mass spectrum at a given point in the total ion monitogram (reconstructed gas chromatogram) and identifying with comparisons to library data, followed by the analysis of authentic samples when warranted.

Identification of pyrazofurin 5'-phosphate in extracts. The dry residue of the cell extracts (see above) was derivatized with trimethylanilinium hydroxide at room temperature. A few drops of the solution (not clear because some components present will not dissolve in the reagent) were placed into the solid probe of the mass spectrometer and evaporated at slightly elevated temperature so that several droplets could be placed on the probe. Finally the probe was inserted into the ion source and slowly heated from room temperature to about 250°C. The mass spectrometer was operated in the chemical ionization mode using methane as the reagent gas (1000 μ pressure). Total ion current was monitored (reconstructed gas chromatogram) throughout the analysis, followed by interrogation for the mass chromatogram of the molecular ion of PFP (see Results).

Other methods employed. The activity of OAD in patient leukocytes was determined by a method described by Ohnuma and Holland (17). Quantification of orotic acid and orotidine was made by column chromatography (8). Urinary uric acid was measured by the technique of Pileggi *et al.* (18). Orotic acid, orotidine, and urinary uric acid were also separated by high pressure liquid chromatography (19) using a u-C_{18} column and 0.1M formic acid at a flow rate of 0.5 ml/min. Constituents were collected, evaporated to dryness with dry nitrogen, derivatized into the trimethylsilyl derivative as described above and positively identified by mass spectrometry.

RESULTS AND DISCUSSION

Pyrazofurin in serum and urine. Because of the lack of analytical techniques, Sweeney, *et al.* (4) measured "PF-like activity" in the plasma of rats by determining the inhibition of orotidylic acid decarboxylase. This activity reached a peak level 6-8 hr after dosing (p.o. or i.m.) with substantial levels remaining in the plasma for 24 hr. Activity was detected for at least 3 days; by 5 days all activity disappeared. This observed slow clearance was in accordance with the clinical observation that dosing at 3-4 days is adequate; indeed, daily dosing increases toxicity.

Our first result using the mass spectrometric technique is the proof that pyrazofurin does exist in the free form in blood for a considerable period of time after dosing. Attention is called to the fact that in the method described proteins are removed by molecular filtration. With this technique only the free drug is detected since protein-bound molecules stay intact and cannot pass the membrane.

Once the presence of free PF in blood was established, the technique was utilized for quantification of PF in blood as a function of dosage and also for the establishment of clearance curves. Only a summary of this work is presented here. Quantitative data were presented together with clinical and pharmacological observations on cancer patients (8), and also together with details of the analytical methodology developed for the drug (16).

Serum PF levels were determined in 8 patients. The quantity of PF found ranged from 0.5 μg/ml to 7.0 μg/ml and low levels were found up to 24 hr after administration. In general, the higher the dose the higher the concentration of PF for a given period of time after administration. Clearance curves indicate a biexponential decay (8). The initial part of the clearance curves (distribution phase) lasts on the average for 30 min and the average half-life (t 1/2 α) is about 10 min (range 4-18 min). The second part of the clearance curve (elimination phase) appears quite variable from patient to patient and the second half-life (t 1/2 β) could not be determined with reliability. Values range from 20 min to as high as 400 min. Assuming that 60% of the total body weight consists of water, extrapolation of the PF levels to zero-time yields about 1/2 of the expected value, or the central compartment is about twice that of the body water.

In the case of urinary PF, again the first objective was to determine if free PF was excreted. This fact was positively established by mass spectrometric identification. Next, similarly to blood levels, urinary levels were determined for various doses in 24 hr urine collections. PF was measured in urine for at least 5 days, with residual amounts detectable up to 10 days. Total PF recovered in urine is no more than 10% of the administered dose. The large apparent volume distribution implies either extensive tissue uptake or biotransformation.

Urinary pyrimidines and uric acid. Administration of PF was followed by the excretion of a large amount of orotic acid and orotidine. Fig. 2 shows the total ion chromatogram of the urine of a patient 2 days after PF infusion started (250 mg/m^2). The peaks corresponding to mass spectra # 230 and # 258 on the abscissa are hippuric and citric acid, respectively. In the control urine

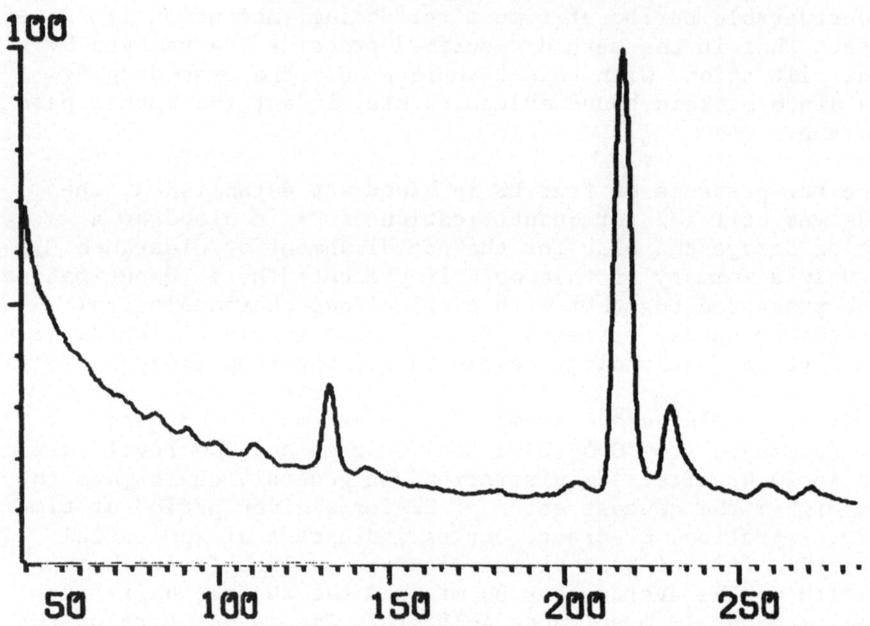

FIG. 2. Total ion chromatogram of patient urine extract, taken on the second day after administration of pyrazofurin. TMS-trimethylsilyl derivative. Temperature programming: 4°/min from 110 to 180°C. During the first min after injection, the excess derivatization reagent was diverted into waste. Abscissa: spectrum number, ordinate: intensity normalized to 100. Spectrum # 133 is the sugar moiety from orotidine, spectrum # 219 is orotidine, spectrum # 230 is hippuric acid, and spectrum # 258 is citric acid.

(taken before infusion) these peaks appear just as in figure shown.

Fig. 3 shows the mass spectrum of peak # 219. This peak is identified as orotic acid by comparing the mass spectrum to that of authentic orotic acid. Both mass spectra and chromatographic retention times were identical in two gas chromatographic columns (3% SE-30 and 3% OV-17). Orotic acid in urine was also identified in a similar way by collecting the corresponding peaks in a high

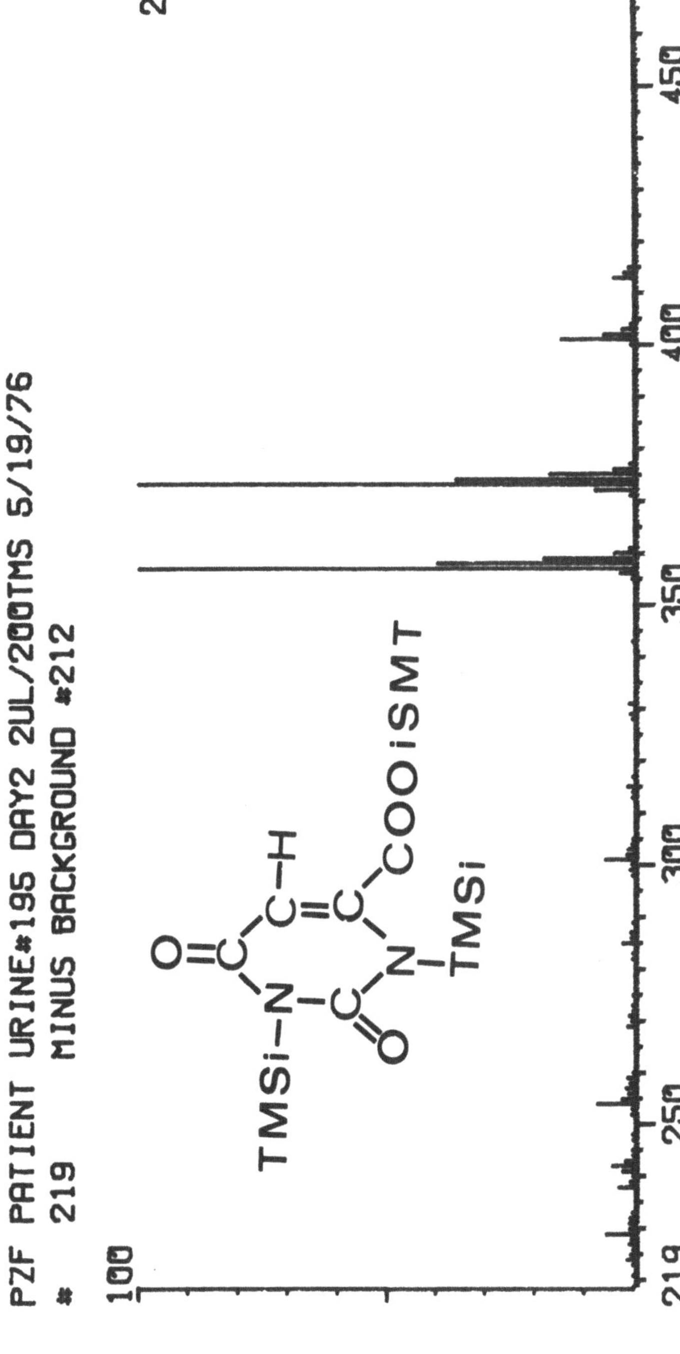

FIG. 3. Chemical ionization mass spectrum (methane reagent gas) of the trimethylsilyl derivative of orotic acid in the urine of patient receiving pyrazofurin.

pressure liquid chromatographic effluent.

Orotidine-TMS breaks down on the gas chromatographic columns, however, the fragments are quite characteristic and readily identifiable. The mass spectrum of peak # 133 of Fig. 2 show the characteristic fragmentation of the sugar moiety: a relatively weak peak for the fully trimethylsilylated sugar, and base peak corresponding to the loss of an OTMSi group (m/e 259).

In a typical case of a bolus injection of 250 mg/m^2 PF, orotic acid excretion lasts for 8 days and reaches a total of 20-100 mmole, and orotidine excretion lasts for 10 days and reaches a total of 40-80 mmole. Infusion (24 hr) consistently produces higher excretion of both orotic acid and orotidine with respect to bolus injection, but without increased clinical toxicity (8). The absolute amounts of orotic acid and oritidine excreted appear proportional to the amount of PF administered.

It is concluded from these experiments that the observance of orotic aciduria and orotidinuria is in accordance with the assumption of the proposed mode of action of PF. If it is true that PF interrupts the metabolic chain as shown in Fig. 1, it is expected that the levels of orotic acid and orotidine should increase.

The administration of PF also appears to produce an uricosuric effect, lasting for about 1 week, as evidence by increased uric acid excretion.

Orotidylate decarboxylase activity in leukocytes. Since mass spectrometry played no role in this phase of our work, results will only be summarized here for completeness. PF resulted in more than 90% inhibition of the enzyme activity in white blood cells in the first few days; within a week to 10 days activity returns to normal. Both bolus and infusion administration resulted in the same type and extent of inhibition.

In reversal experiments using the human B-cell line (RPMI 8422), pyrazofurin inhibited cell growth at a concentration of 10^{-7} M. This PF effect was completely reversed with the addition of 10^{-3} M to 10^{-4} M of uridine or cytidine but not by orotic acid or orotidine. This is additional confirmation of the assumption that the site of inhibition in the biosynthesis of pyrimidine nucleotides is at the site of orotidylic carboxylase.

Identification of pyrazofurin 5'-phosphate in cell extracts. Fig. 4 shows the chemical ionization mass spectrum of methylated (with trimethylanilinium hydroxide) pure synthetic pyrazofurin 5'-phosphate. It is seen that the base peak corresponds to the M+1 ion of the unbroken molecule which now contains 8 methyl groups: all hydrogens in the OH groups and those connected to nitrogen are replaced by methyl groups. The fully methylated compound does

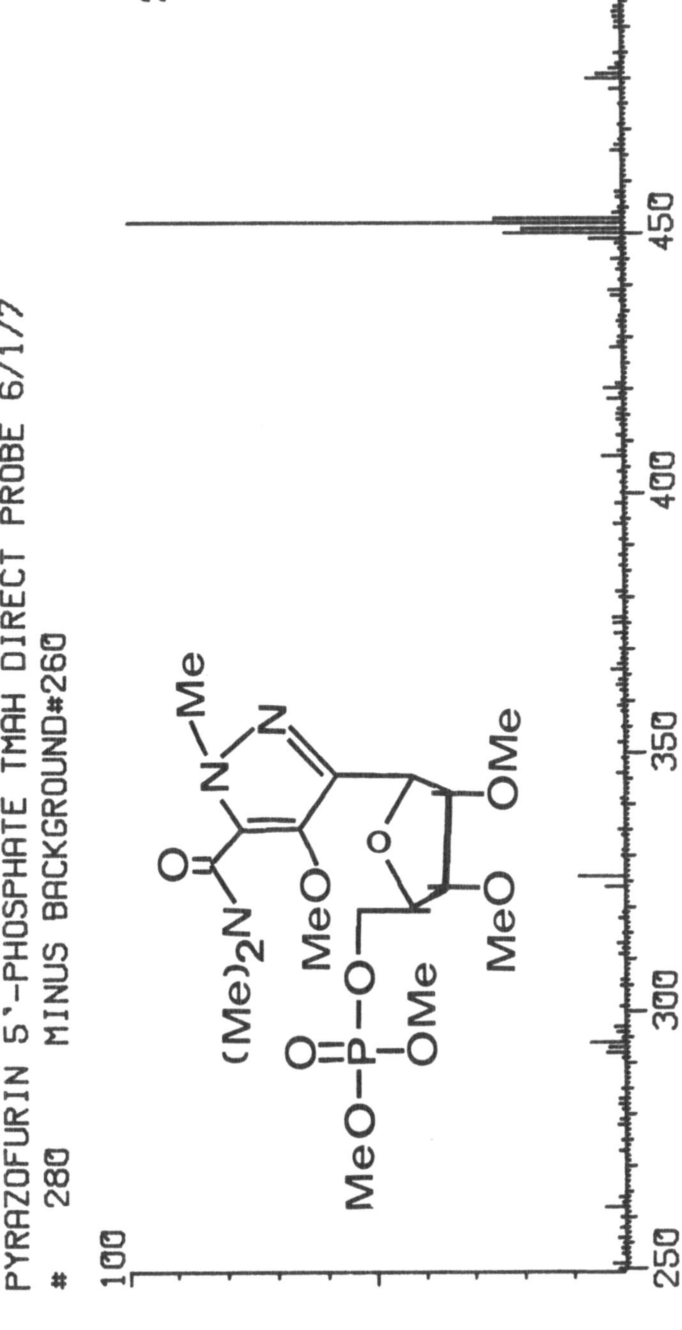

FIG. 4. Chemical ionization mass spectrum (methane reagent gas) of pyrazofurin 5'-monophosphate, derivatized with trimethylanilinium-hydroxide and introduced *via* the solid probe.

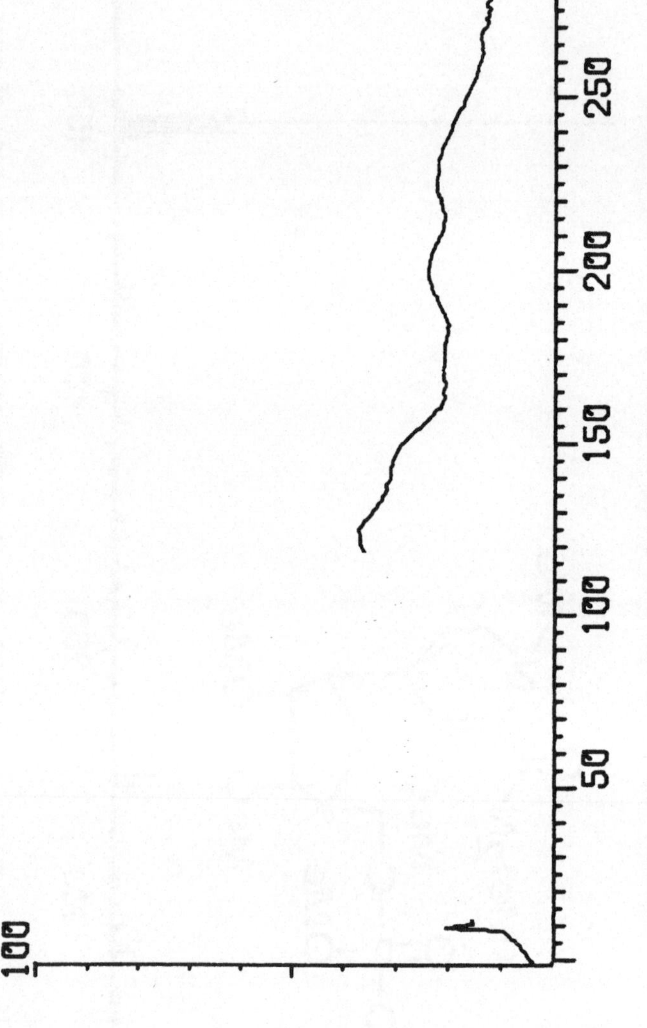

FIG. 5. Total ion chromatogram of extract from human cells incubated with pyrazofurin and methylated with trimethylanilinium hydroxide. Crude sample introduced *via* solid probe and heated from 50 to 300°C at a rate of 30°/min.

not survive passage through the gas chromatographic column and must be analyzed via the direct probe. The trimethylsilylated phosphate could not be detected either via the gas chromatograph or in the direct probe.

Fig. 5 shows the total ion chromatogram of all compounds evaporating from the crude cell extract heated in the solid probe. Based upon observation of the temperature at which the pure phosphate evaporates, one would expect to find the phosphate, if present, around the mass spectrum # 200. Fig. 6 is a mass chromatogram taken at m/e-452, corresponding to the quasimolecular ion of pyrazofurin

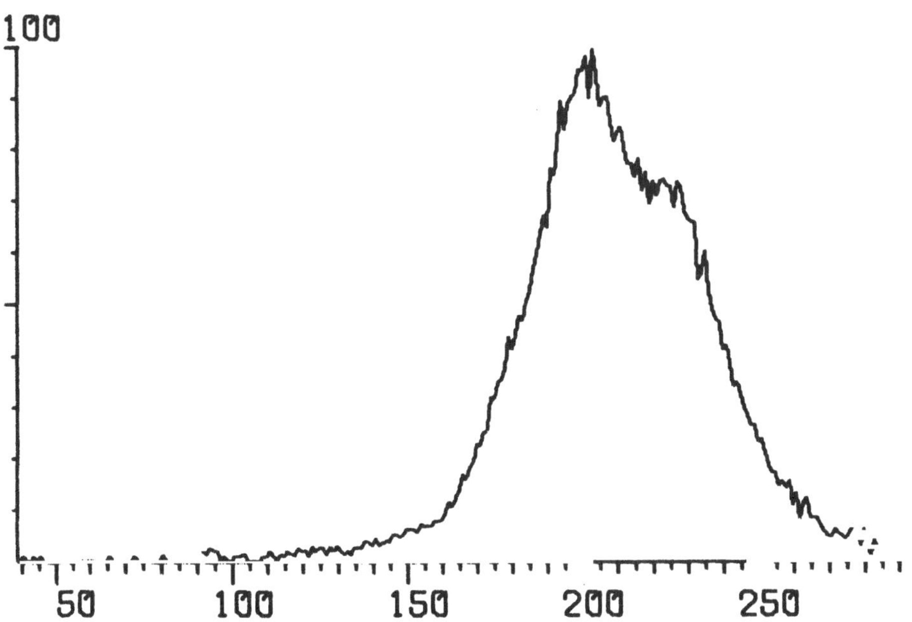

FIG. 6. Mass chromatogram of m/e 452 (the quasimolecular ion of methylated pyrazofurin 5'-phosphate) in the crude extract from human cells (see also Figs. 4 and 5). TMAH-trimethylanilinium hydroxide used to derivatize. Abscissa: spectrum number, ordinate: intensity normalized to 100. The approximate maximum of # 200 corresponds to the appearance of the phosphate in the reconstructed mass chromatogram (Fig. 5).

5'-phosphate. There is an abundant peak at around mass spectrum #200 indicating that the incubated cells do indeed contain phosphorylated pyrazofurin. The full mass spectrum taken at the indicated point agrees with that of the pure compound (Fig. 4).

It is concluded that the presence of phosphorylated pyrazofurin was established in cells incubated with pyrazofurin. Both the human T-cells and B-cells yielded essentially the same results. The amount of pyrazofurin used in the incubation was about 10 times higher than that needed for a 90% inhibition of OAD activity. The purpose of this was to assure adequate concentration in the initial experiments. The identification was made in a very crude cell extract, and it is clearly seen in Fig. 5 that the phosphate was only a very small fraction of all other components evaporating upon heating. Indeed, the phosphate could not have been detected without the aid of mass chromatography.

The positive identification of phosphate is considered an additional piece of evidence supporting the assumption that it is not PF but rather PFP that is the active ingredient in disrupting the biosynthetic pathway of pyrimidine nucleotides. Handschumacher suggested (20) that in mouse L5178Y cell system a major part of pyrazofurin is converted to polyphosphates. He also suggested that there appears to be a steady state between monophosphate and polyphosphate within the cell and since monophosphate is the only biologically active form the polyphosphate may play a role as an intracellular reservoir form to supply monophosphate. This mechanism is regarded as indicative of prolonged biological effects of pyrazofurin.

ACKNOWLEDGMENTS

The authors wish to thank G. Gutowski, Ph.D., M. Sweeney, Ph. D, and R. Dyke, M.D., all fromm Eli Lilly Co., for providing samples of pyrazofurin and pyrazofurin monophosphate and also for helpful discussions. Thanks are due to Mr. R. Suzuki for valuable technical assistance.

This work was supported by Grant # 5P01-CA15936-03 and also Contract # N01-CM-53837 from the National Cancer Institute, NIH, USA.

REFERENCES

1) R.H. Williams, K. Gerson, M. Hoehn, J. Gorman and D.C. DeLong, in "158th Amer.Chem.Soc.Natl.Mtg., New York, N.Y. (Abstr. MICR 38)", 1969.
2) F.J. Streightoff, J.D. Nelson, J.C. Cline, K. Gerson, M. Hoehn,

R.H. Williams, M. Gorman and D.C. DeLong, in "Ninth Conference on Antimicrobial Agents and Chemotherapy", Washington, D.C., 1969, p. 8.
3) D.C. DeLong, L.A. Baker, K. Gerson, G.E. Gutowski, R.H. Williams and R.L. Hamill, in "Proc. 7th Intern. Congr. of Chemother., Prague, Czechoslovakia", 1971.
4) M.J. Sweeney, F.A. Davis, G.E. Gutowski, R.L. Hamill, D.H. Hoffman and G.A. Poore, Cancer Res., 1973, 33, 2619.
5) G.E. Gutowski, M.J. Sweeney, D.C. DeLong, R.C. Hamill and K. Gerson, Ann. N.Y. Acad. Sci. , 1975, 255, 544.
6) Brochure, Compound 47599, Pyrazofurin, Eli Lilly Co., Indianapolis, Ind., 1973.
7) Minutes of the Phase I Working Group, Cancer Therapy Evaluation Program, National Cancer Institute, NIH, USA, Oct. 3, 1974 and Feb. 25, 1975.
8) T. Ohnuma, J. Roboz, M.L. Shapiro and J.F. Holland, Proc. Am. Assoc. Cancer Res. and Am. Soc. Clin. Oncol., 1976, 17, 184.
9) E.C. Cadman, D.E. Dix, S.C. Hill and R.E. Handshumacher, Proc. Am. Assoc. Cancer Res. and Am. Soc. Clin. Oncol., 1976, 17, 208.
10) R. Handschumacher, J. Biol. Chem., 1960, 235, 2917.
11) J. Roboz, R. Suzuki, M. Perloff and T. Ohnuma, in "Proc. 23rd Ann. Conf. Mass Spectrom., Dallas, Texas, 1975.
12) J.L. Occolowitz, M.R. Gleissner and G.E. Gutowski, personal communication, 1976.
13) J. Minowada, T. Ohnuma and G.E. Moore, J. Nat. Cancer Inst., 1972, 49, 891.
14) G.E. Moore, in "Cell lines from human with hematopoietic malignancies in human cells in vitro", Fogh J., Ed., Plenum Press, N.Y., 1975, p. 299.
15) T. Ohnuma, unpublished observation.
16) J. Roboz, R. Suzuki and T. Ohnuma, submitted for publication.
17) T. Ohnuma and J.F. Holland, Clin. Pharmacol. Ther., 1966, 7, 763.
18) V.J. Pileggi, J. DiGiorgio and D.R. Wybenga, Clin. Chim. Acta, 1972, 37, 141.
19) J. Roboz and L. Twanmoh, unpublished results.
20) R. Handschumacher, personal communication, 1976.

MASS SPECTROMETRY APPLIED TO METABOLISM STUDIES OF ORG GC 94, A NEW ANTI-MIGRAINE DRUG

J. Vink, J.J. de Ridder, C.J. Timmer and H. de Nijs

Organon International B.V., Scientific Development Group

Oss, Netherlands

INTRODUCTION

The growing interest in the annual "Mario Negri" Symposia demonstrates clearly the importance of mass spectrometry in the field of biochemistry and medicine. In drug metabolism research, mass spectrometry is an important tool in the identification of unknown metabolites, as well as in obtaining quantitative information on drug disposition and fate. The proliferation of mass spectrometry applications in this field is due to its sensitivity and specificity. Moreover, in combination with gas chromatography, extensive isolation and purification procedures, which are required for other analytical methods, can be avoided. Although mass spectrometric techniques are expensive, the general applicability makes it extremely useful for information gathering in the early stages of product development.

In the present paper, mass spectrometric applications are described in the development of Org GC 94 (Fig. 1), a new possible anti-migraine drug with a tetracyclic structure (1).

In particular the contribution of mass spectrometry to the determination of the drug pharmacokinetics and the evaluation of *in vitro* metabolism models are shown. The assay method developed for Org GC 94 is discussed in more detail.

ASSAY METHOD FOR ORG GC 94

The assay method for Org GC 94 includes a hexane extraction of the sample to which tetradeuterated Org GC 94 is added as an

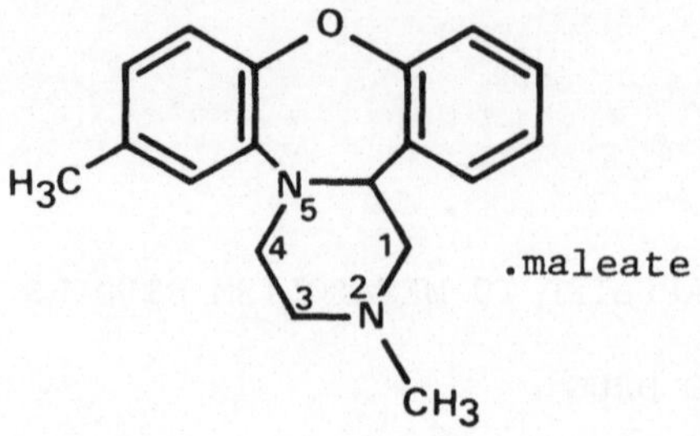

.maleate

FIG. 1. ORG GC 94.

internal standard. The crude extract is purified by high pressure liquid chromatography (HPLC) (2,3). The fraction of the eluate containing Org GC 94 and internal standard is collected and quantitated by gas chromatography-mass spectrometry (4). The flow chart of this procedure is shown in Fig. 2.

The HPLC clean-up requires some special comment. Because the amounts of drug and internal standard involved cannot be detected by the standard HPLC detectors, the retention times should be carefully determined to avoid loss of compound. This can be achieved by the use of detectable amounts of a marker compound added to the sample as illustrated in Fig. 3.

Once the relative retention time with respect to the marker compound (amitriptyline, e.g.) has been established, the eluate collection period can be checked for each sample. One should, however, ensure that both compounds, drug and internal standard, are trapped together. This is sometimes hampered by HPLC separation of the drug and its deuterated analogue as shown for the compound under study in Fig. 4 (5).

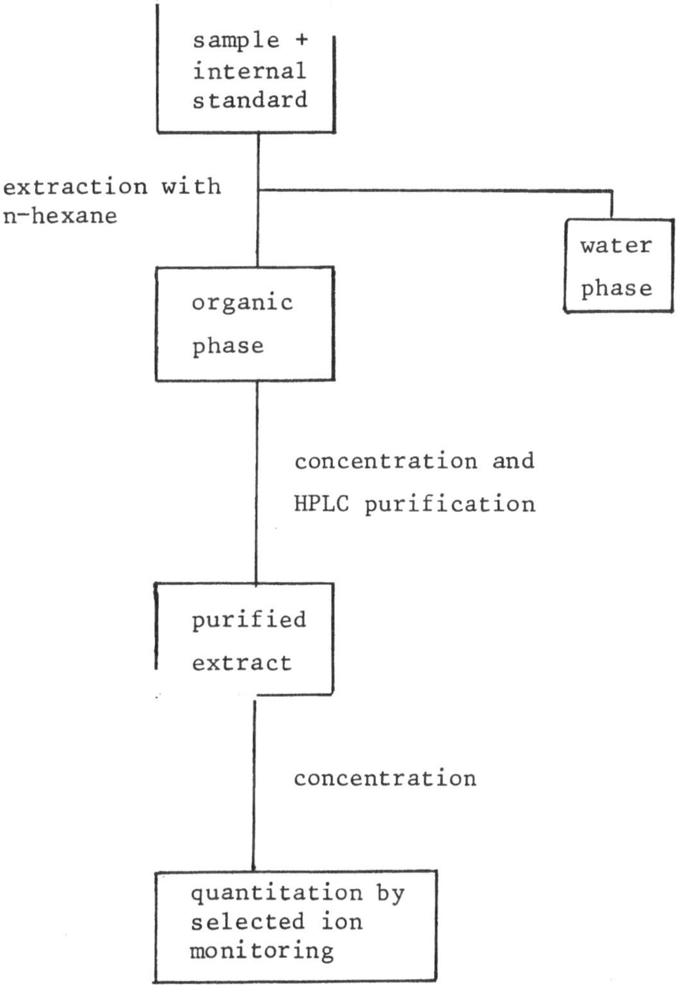

FIG. 2. Schematic procedure of sample preparation for the assay of ORG GC 94.

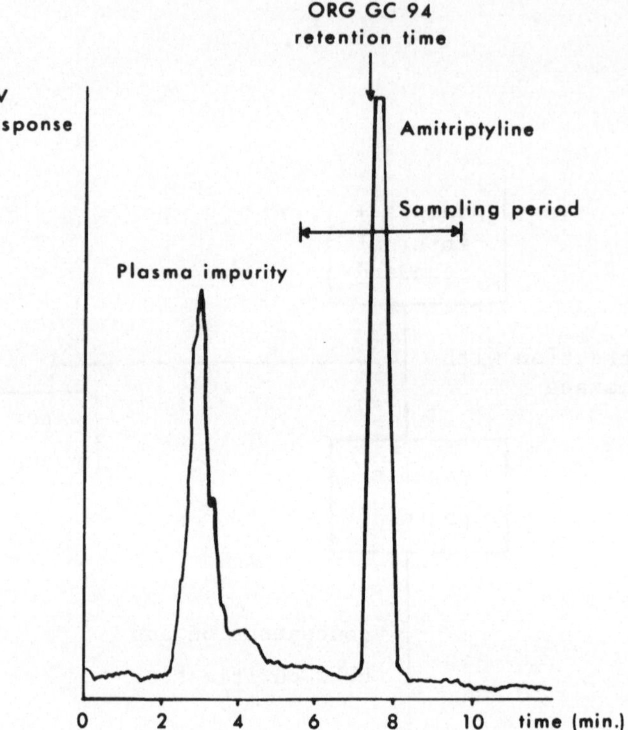

FIG. 3. High pressure liquid chromatogram of sample extract containing 5 µg of amitriptyline as marker.
UV detector 280 nm. Flow rate 2.5 ml/min.

Increase of sample collection period in this case warrants quantitative compound collection. The HPLC clean-up procedure turned out to be an extremely versatile and fast procedure, yielding sample devoid of interfering substances at low level determinations. The gas chromatographic-mass spectrometric quantitation with specific ion monitoring (SIM) of the molecular ions at m/e 280 and m/e 284 for Org GC 94 and tetradeuterated Org GC 94, respectively, yielded the following assay specifications: sensitivity <1 ng per ml; detection limit 50 pg with a signal to noise ratio of about 10; precision 6-7%; accuracy 1%. The precision and accuracy of the mean are determined at the 1 ng/ml level for ten samples. An example of the SIM trace of a 2 ng sample is shown in Fig. 5.

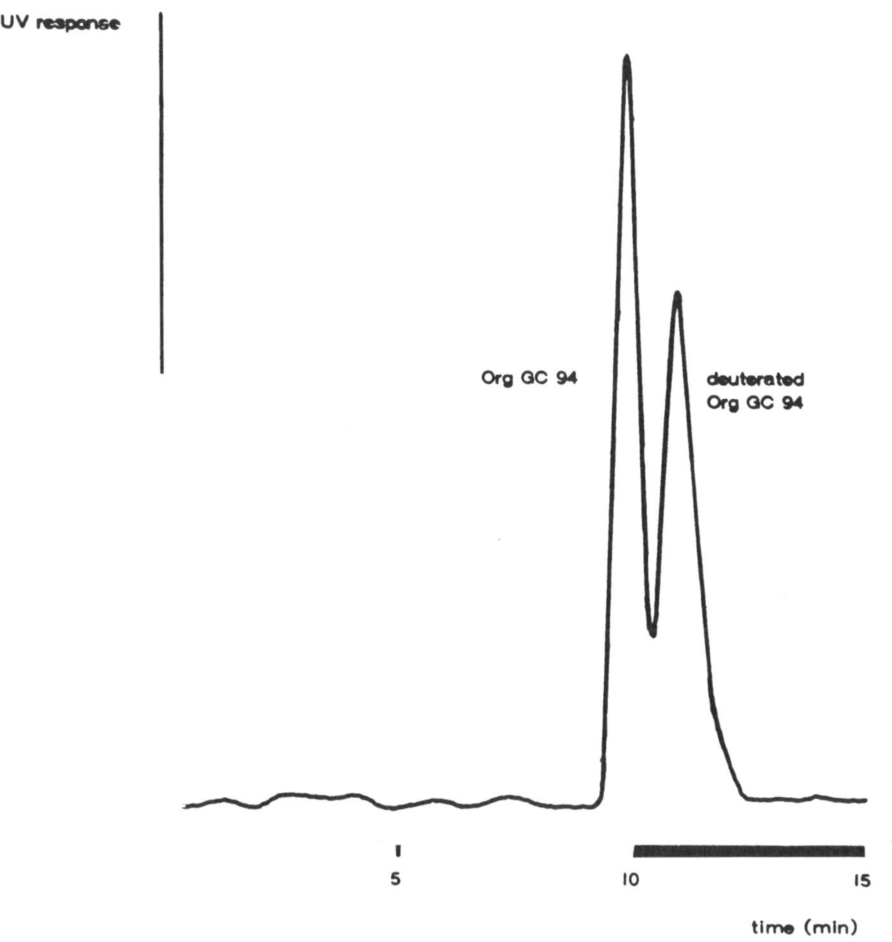

FIG. 4. High pressure liquid chromatogram of 3 µg of ORG GC 94 and tetradeuterated ORG GC 94. UV detector 280 nm. Flow rate 1 ml/min.

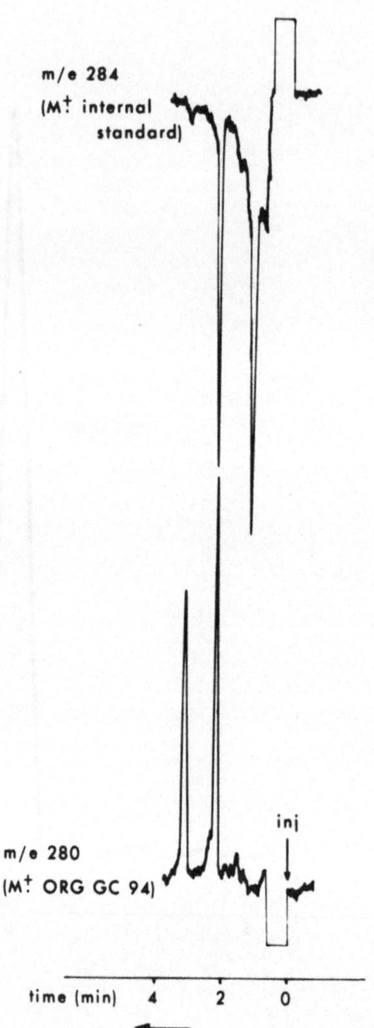

FIG. 5. Selected ion monitor traces of a 2 ng ORG GC 94 plasma sample.

APPLICATIONS OF THE ORG GC 94 ASSAY METHOD

I) Single dose and steady state studies in patients. With the assay method, the plasma level of the drug versus time is measured in patients after a single oral dose of 25 mg of Org GC 94, while after chronic therapeutic treatment with 3x5 mg/day the early morning levels were measured. An example of the plasma level curve after a single dose of 25 mg of Org GC 94 is shown in Fig. 6, demonstrating the sufficient sensitivity of the assay method to calculate the pharmacokinetic parameters of the drug.

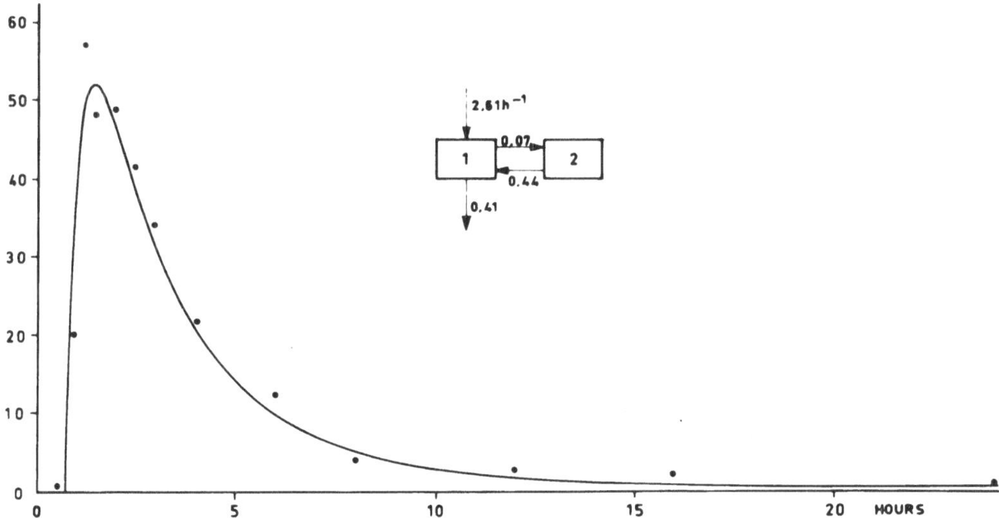

FIG. 6. ORG GC 94 plasma levels after a single oral dose of 25 mg.

The early morning Org GC 94 levels after chronic treatment amounted to 1-2 ng/ml, a level that requires the sensitivity of the assay method described.

II) In vitro metabolism model evaluation. Another application of the assay method was used for the evaluation of an *in vitro* metabolism model.

For many drugs, the liver is the main organ responsible for elimination of the drug from the circulation. For Org GC 94 the major routes of biotransformation were determined by mass spectral identification of the metabolites in rat urine and after incubation of Org GC 94 with 10,000 x g supernatant of rat liver homogenates. The similarity between the major routes of biotransformation in these in vivo and in vitro experiments could indicate that also for Org GC 94 metabolism predominantly took place in the liver. Therefore, the isolated perfused rat liver should offer a suitable simplified model for the study of Org GC 94 drug metabolism and kinetics. With some small alterations, the experimental set-up of the liver perfusion system was essentially that of Scholz (6) (Fig. 7).

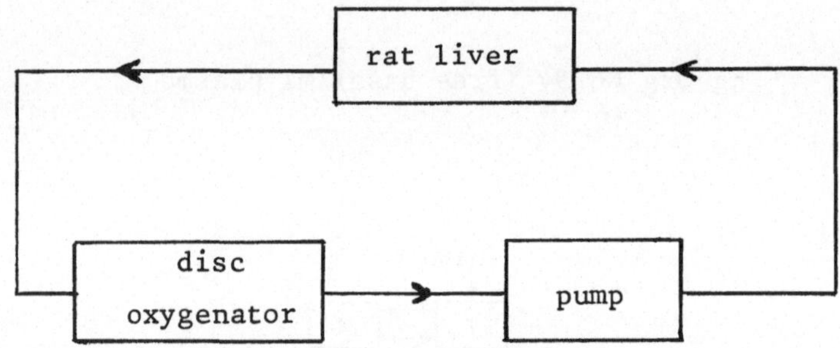

FIG. 7. Diagram of the recirculating rat liver perfusion system.

The recirculating perfusion medium was erythrocyte-free and consisted of a Krebs-Henseleit bicarbonate buffer pH 7.4, containing 2.7 g of bovine serum albumin per 100 ml perfusate and the drug Org GC 94. A disc oxygenator saturated the perfusion medium with sufficient oxygen. In this way, the isolated liver could be kept in good condition for at least two hours. With the assay method described, the Org GC 94 levels were measured in the perfusate of the rat liver perfusion system and in the plasma of the in vivo rat. In Figs. 8 and 9 the disappearance of Org GC 94 is shown from the perfusate during perfusion of the isolated rat liver, and from rat plasma after intravenous injection, respectively.

To compare the kinetic behaviour of a drug in the isolated, perfused rat liver system with that in the in vivo rat, the clearance of a drug is a meaningful index. The clearance is defined as the volume of blood or plasma which is completely cleared of the drug per unit of time and is calculated from the kinetic parameters of the disappearance curves obtained. De Nijs and Timmer (7) showed that the clearance values of Org GB 94 (a chemical analogue of Org GC 94) in the isolated perfused rat liver and in the anaesthetized rat were identical after correction

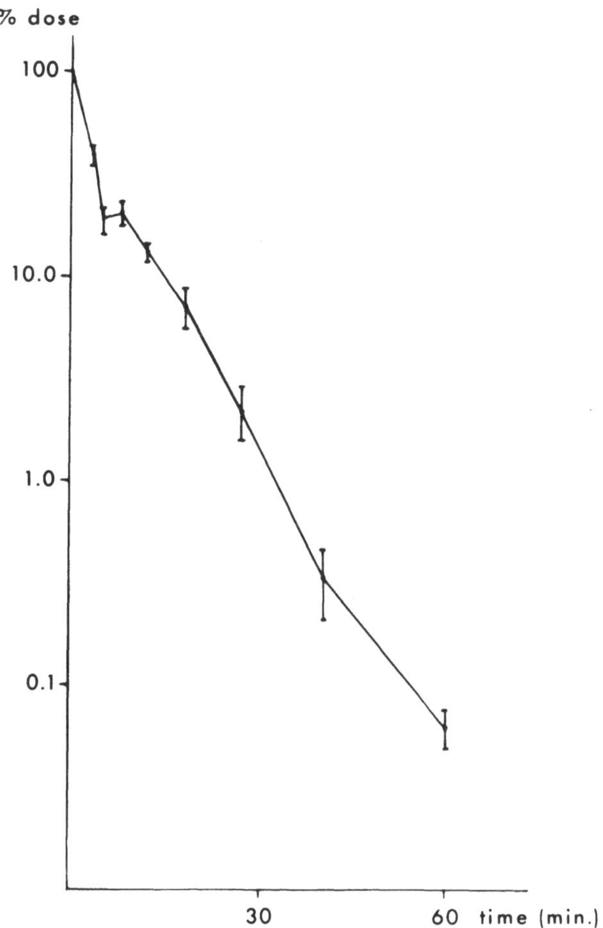

FIG. 8. ORG GC 94 levels during perfusion of the isolated rat liver. Initial ORG GC 94 concentration 44.5 μg/ml.

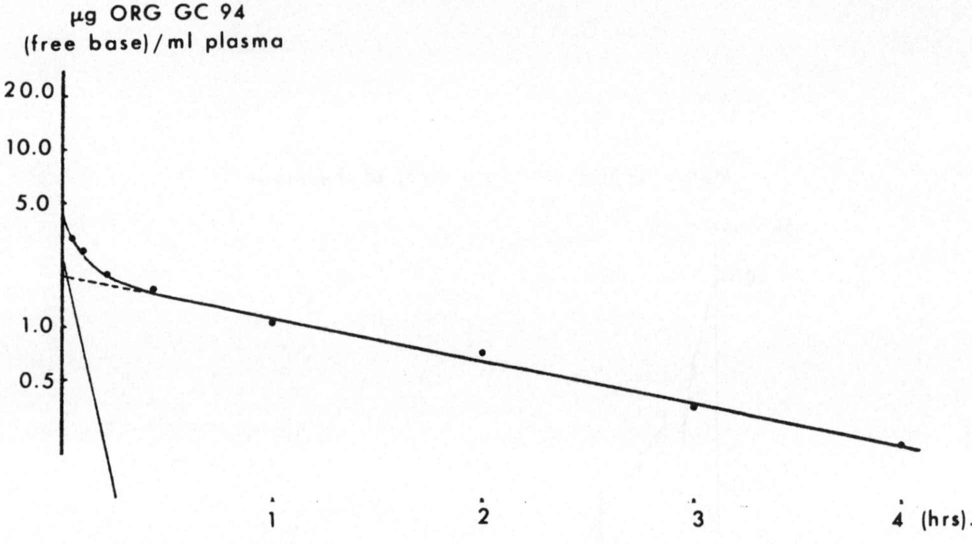

FIG. 9. ORG GC 94 rat plasma levels after intravenous injection of 8 mg/kg.

for body weight and blood flow rate. Similar values were found for nortriptyline and desmethylimipramine (Table 1).

The close similarity of these clearance values, presented strong evidence that in rats elimination of these drugs from the circulation is mainly the result of hepatic metabolism. From the kinetic parameters obtained in the perfusion experiment with

TABLE 1. *In vivo* and *in vitro* clearance values for some anti-depressant drugs, after correction for blood flow rate.

drug	clearance (ml/min/kg rat) perfused liver	*in vivo*
Org GB 94	45	43
Nortriptyline	50	56
Desmethylimipramine	50	54

Org GC 94 (Fig. 8), a high hepatic clearance of 108 ml/min/kg was derived, a value close to the flow rate of the perfusate through the liver (130 ml/min/kg). After correction of the isolated liver clearance for an *in vivo* blood flow of 60 ml/min/kg, a calculated hepatic clearance of 55 ml/min/kg was obtained. The plasma clearance of Org GC 94 in the anaesthetized rat appeared to be 36 ml/min/kg (Fig. 9). When the *in vivo* erythrocyte binding of 16% for Org GC 94 is taken into account, an *in vivo* blood clearance of 51 ml/min/kg was calculated for Org GC 94, a value that is almost the same as the calculated hepatic clearance:

$$Cl_{blood} = \frac{C_{plasma}}{C_{blood}} \; Cl_{plasma}$$

Consequently also Org GC 94 is mainly eliminated by the liver.

Correction for erythrocyte binding was not necessary for Org GB 94, because for Org GB 94 the plasma concentration equalled the blood concentration.

These results emphasize that not the *in vivo* plasma clearance is the correct parameter to compare with the hepatic clearance, but the *in vivo* blood clearance.

Acknowledgments. The work described was performed at the Drug Metabolism R&D Labs of Organon International B.V. The authors wish to thank all people involved in the Org GC 94 study, especially Mr. J. Wallaart and Mr. J.S. Favier for preparation of the deuterated Org GC 94 internal standard, Mr. G.D. de Jongh for performing the *in vivo* and *in vitro* metabolism studies, Mr. P.C.J.M. Koppens, Mr. H.J.M. van Hal and Mrs. W.J. de Jong-van Orsouw for technical assistance, Mr. H.P. Wijnand for providing pharmacokinetic data, and Dr. F. van der Veen for his everlasting encouragement.

Merck, Sharp & Dohme Nederland B.V. is thanked for the supply of amitriptyline.

REFERENCES

1) B. Anselmi, P.L. del Bianco, C.J. de Vos, P. Galli, J.C. Lamr, E. Schönbaum, F. Sicuteri and F. van der Veen, in "Clinical Pharmacology of Serotonin", F. Sicuteri and E. Schönbaum Eds., Karger, Basle, in press, 1976.
2) J.J. de Ridder, in "Advances in Mass Spectrometry in Biochemistry and Medicine", Vol. II, A. Frigerio Ed., Spectrum Publications, New York, 1976.

3) J.J. de Ridder, P.C.J.M. Koppens and H.J.M. van Hal, J. Chromatogr., submitted for publication, 1976 .
4) J.J. de Ridder, P.C.J.M. Koppens and H.J.M. van Hal, J. Chromatogr., submitted for publication, 1976 .
5) J.J. de Ridder and H.J.M. van Hal, J. Chromatogr., 1976, 121,96.
6) R. Scholz, in "Stoffwechsel der perfundierten Leber", W. Staib and R. Scholz Eds., Springer Verlag, Berlin, 1968, p. 25.
7) H. de Nijs and C.J. Timmer, 17th Dutch Federative Meeting, Amsterdam, abstracts symposia and communications, 1976, 312.

THE IDENTIFICATION AND MEASUREMENT IN URINE OF DIHYDRODIGOXIN, A CARDIOINACTIVE METABOLITE OF DIGOXIN

H. Greenwood and W. Snedden

Department of Chemical Pathology, St. Bartholomew's Hospital, London, EC1, U.K.

SUMMARY

The sensitivity of an established mass spectroscopic technique has been improved in order to measure dihydrodigoxin, the cardio-inactive metabolite of digoxin, in extracts of urine from paediatric patients maintained on the drug.

Aliquots were vapourized in a reproducible manner and the ion current at m/e 355.227 and 339.232 (characteristic of digoxin and dihydrodigoxin respectively) monitored at a resolving power of 5,000 throughout the evaporation process. The integrated ion current at each ion mass was proportional to the amount of each steroid admitted. The technique was validated by examining urine extracts containing known amounts of digoxin and dihydrodigoxin. Less than 0.01% of digoxin was converted to dihydrodigoxin during the extraction process.

The percentage of the daily maintenance dose excreted in urine in 24 h as this metabolite was low (mean 4.35%) compared to adults (mean 16.4%), with a wide variation between subjects. There was no significant difference between the results from subjects under one year of age and those over one year. Thus metabolism of digoxin to dihydrodigoxin does not appear to account for the increased dosage requirement of paediatric patients.

INTRODUCTION

The metabolism of digoxin in the body has been reported to be minimal (1, 2) and it has been estimated that a patient with normal renal function requires a daily maintenance dose sufficient only to replace digoxin lost from the body by excretion over this period (3).

Measurement of the urinary excretion of the drug by adults and young children receiving maintenance therapy, using radioimmunoassay, demonstrated that a low percentage of the daily dose was excreted by this route (4, 5). The antiserum employed in the assay was highly specific for digoxin, the cardioactive metabolites mono- and bis- digitoxoside and digoxigenin but did not cross-react with the cardioinactive metabolite dihydrodigoxin.

Dihydrodigoxin differs structurally from digoxin only in the saturation of the double bond (C20-22) in the lactone ring (Fig. 1) and cannot be separated from the parent compound by the normal solvent techniques. The metabolite has been identified and measured in morning urine samples by Clarke and Kalman (6) using gas chromatography. As the molecule is relatively involatile the method requires the glycoside to be converted to its heptafluorobutyryl derivative.

We have previously established a mass spectroscopic method for measuring the ratio of digoxin to dihydrodigoxin in extracted urine samples of adult patients. In this study a more sensitive technique was employed to directly identify and quantitate the lower concentrations of digoxin and dihydrodigoxin present in extracts from the 24 h urine collections of paediatric patients maintained on digoxin therapy.

METHODS

Patients. 24 h urine collections were made from 8 infants, under one year old and 8 young children, over one year old, using a urine collection bag where necessary. All had normal renal function and serum electrolytes, were receiving digoxin maintenance therapy and had been on the same digoxin dose for 5 days. The collection was abandoned if the bag leaked.

Urinary digoxin radioimmunoassay. This had been described previously (4). The urine digoxin concentration of all 24 h collections was determined directly by this method employing a specific antiserum for digoxin, raised in rabbits against a digoxin: bovine serum albumin conjugate, and ^{125}I-labelled digoxin (Burroughs Wellcome).

Dihydrodigoxin measurement. Sample preparation: It is not possible to separate dihydrodigoxin from digoxin using solvent separation or chromatographic techniques. Thus digoxin and dihydrodigoxin from a urine sample were extracted and purified together. 10 mℓ of each urine sample was extracted twice with an equal volume of ethyl acetate by mixing on a rotary mixer for 30 min.

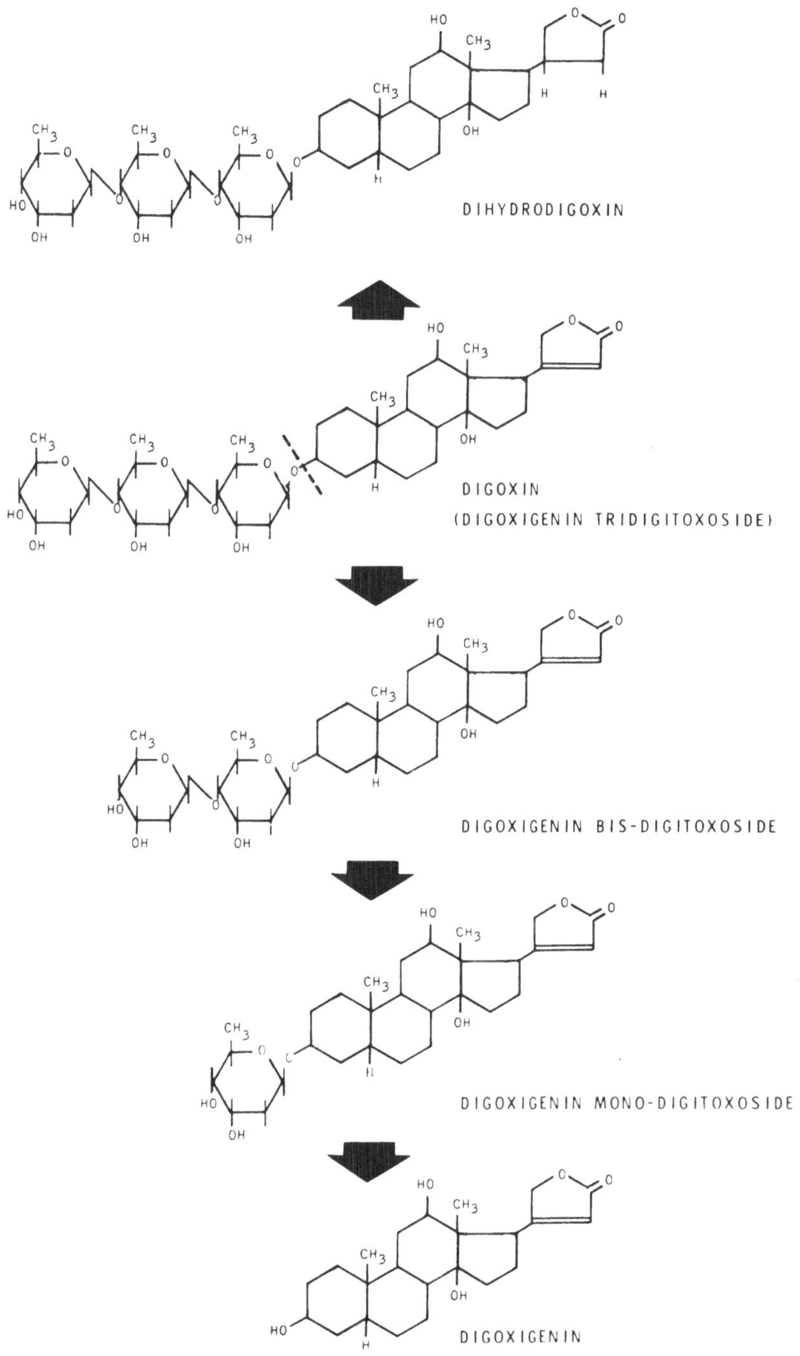

DIHYDRODIGOXIN

DIGOXIN
(DIGOXIGENIN TRIDIGITOXOSIDE)

DIGOXIGENIN BIS-DIGITOXOSIDE

DIGOXIGENIN MONO-DIGITOXOSIDE

DIGOXIGENIN

FIG. 1. The major metabolites of digoxin. The mass spectrum of digoxin shows facile cleavage of the

FIG. 1. (continued)
bond between ring A of the steroid and the glycosidic oxygen at position 3, to give a fragment of mass 373 which undergoes stepwise loss of 2 molecules of water to yield the intense fragments at masses 355 and 337. Dihydrodigoxin undergoes similar fragmentation to give the corresponding ions at masses 375, 357 and 339.

The solvent extract was dried under a stream of air in a water bath (60°) and the extract redissolved in ethanol (200 μℓ). This was applied to a thin layer plate (Merck, Silica Gel G). The extract together with marker spots of digoxin and dihydrodigoxin were run in the solvent system ethyl acetate:n-butanol (90:10 v/v) for 90 min and the marker spots detected by spraying with phosphomolybdic acid solution (5%). The appropriate area was scraped off and eluted in 10 mℓ ethanol. 5 mℓ of this was dried and the extract reconstituted in 100 μℓ ethanol for transfer to the direct insertion probe of the mass spectrometer.

Mass spectroscopy: The low resolution mass spectrum of digoxin and dihydrodigoxin are shown in Fig. 2. The glycosides fragment to give the characteristic ion peaks at 337, 355 and 373 for digoxin and those for dihydrodigoxin are shifted two atomic mass units to the higher mass of 339, 357 and 375, due to the saturation of the lactone ring.

In the low resolving power spectrum of an extracted adult urine sample the ions due to the glycosides are detectable above the background ions. The signal-background ratio is sufficient to allow the measurement of both steroids in the urine of adult patients. A more sensitive technique is required, however, for samples from paediatric patients where much lower concentrations of glycoside are involved.

The technique of high resolution mass fragmentography was employed for this study. The mass spectrometer (Varian MAT 311A) was adjusted to a resolving power of approximately 5000 and under these conditions the fragment ions from the glycosides were separated from the background ions. The intense fragment ions at masses 355.227 and 339.232, characteristic of digoxin and dihydro-digoxin respectively, were selected for quantitative measurement since cross contribution was minimal between them. Aliquots of the urine extract were admitted to the direct insertion probe and the response at each of the two selected masses monitored alternately, using the high resolution multi-ion detection unit. Meanwhile the sample was completely evaporated into the mass spectrometer by raising the inlet temperature according to a preselected standardised programme.

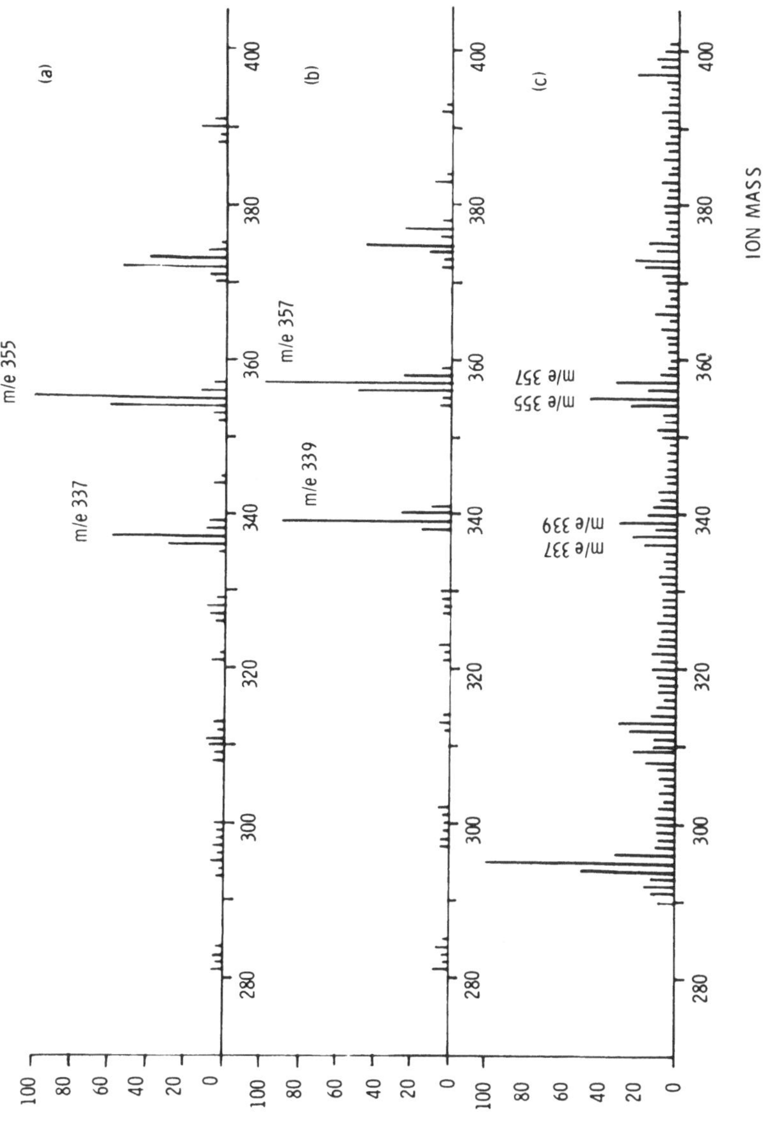

FIG. 2. Spectra of pure digoxin and dihydrodigoxin (a and b) and a typical urine extract containing both glycosides (c).

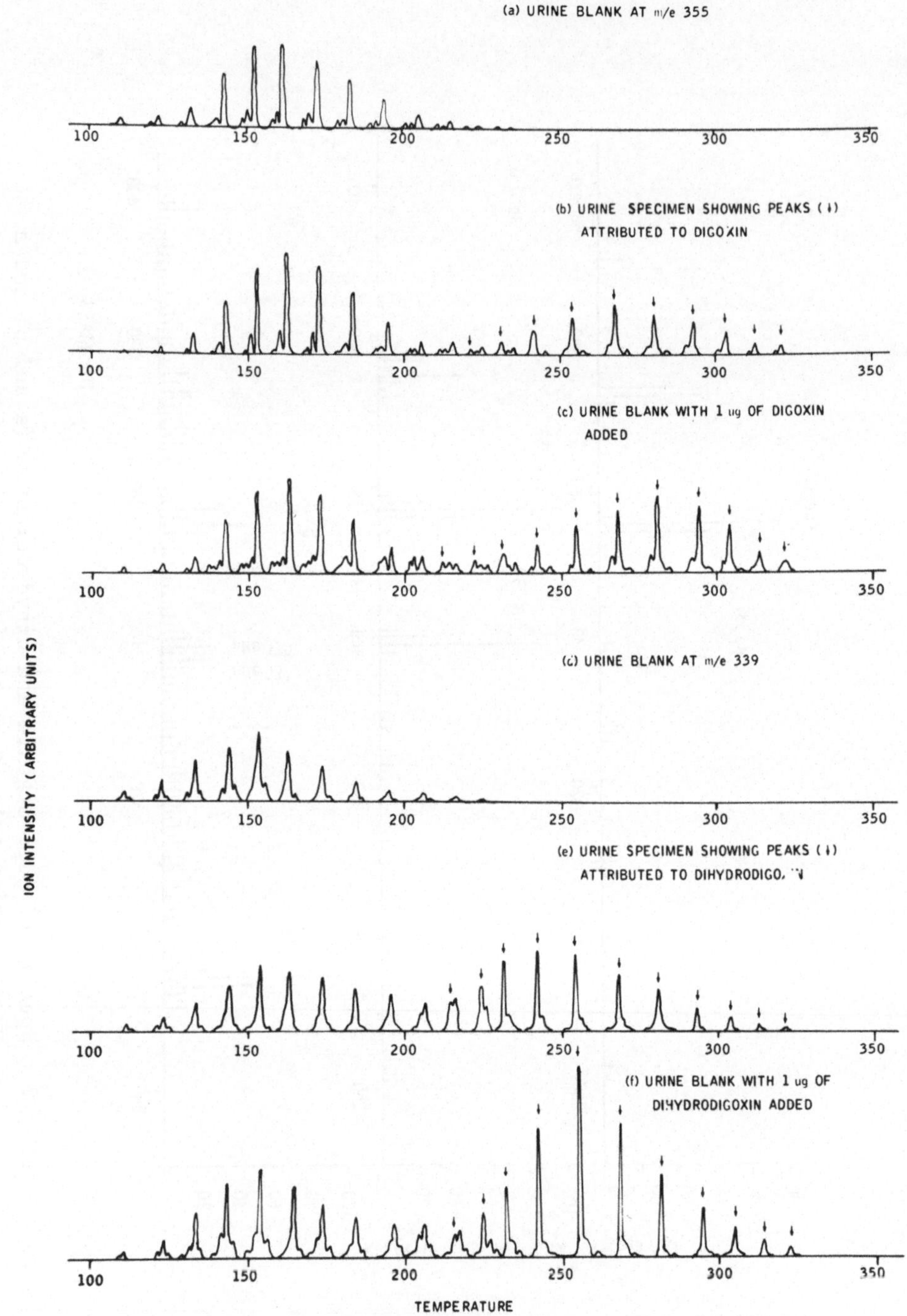

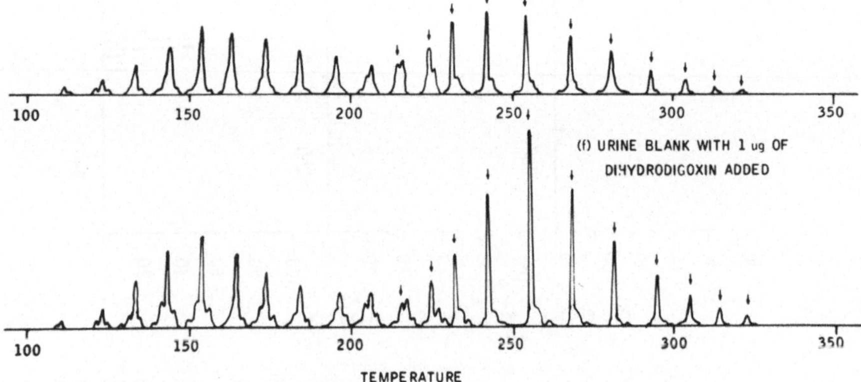

FIG. 3. High resolution mass fragmentogram at m/e 355 (a, b and c) and m/e 339 (d, e and f) of urine with and without added digoxin or dihydrodigoxin.

FIG. 3. (continued)
together with those of a urine sample from a patient maintained on digoxin therapy; showing the ion peaks arising from the fragmentation of digoxin or dihydrodigoxin.

Fig. 3 shows a plot of the ion intensity at masses 355 and 339 as a function of the inlet temperature. Each triplet corresponds to 3 ions of the same nominal molecular weight but of different atomic composition. The rise and fall of these ions results from their evaporation into and pump-out from the ion source, yielding an evaporation profile for each ion. Digoxin and dihydrodigoxin are completely separated from the background ions of the same nominal mass and also evaporate over a specific, reproducible temperature range of the inlet.

Quantitative analysis is by integration of the evaporation profile arising from the glycoside ions. This was achieved by summation of the component peak heights recorded throughout their lifetime in the inlet. The instrument was calibrated by admitting known amounts of the two glycosides separately and as mixtures to the mass spectrometer and recording their evaporation profiles under the same conditions. This takes into account the relative sensitivities of the two compounds and allows a correction to be made for the cross contribution of the two glycosides to each of their characteristic mass numbers. Sample recovery was estimated and corrected for in each batch by extracting control urine specimens to which had been added known amounts of the glycosides.

Although the method is capable of giving the absolute quantification of both glycosides separately, at this stage we were only concerned with the relative concentrations of dihydrodigoxin and digoxin, the concentration of the latter being measured absolutely by radioimmunoassay.

RESULTS

Standard amounts of digoxin and dihydrodigoxin admitted to the mass spectrometer either separately or as mixtures of known composition, gave integrated ion current ratios at masses 355.227 and 339.232 which were reproducible to within 10%. From the results of urine extracts to which known amounts of each glycoside had been added the recovery was 34.1% for digoxin and 39.8% for dihydrodigoxin. The system was capable of measuring as little as 10 ng of glycoside and the conversion of digoxin to dihydrodigoxin during the whole procedure was less than 0.01%.

The integrated ion intensity at each mass was converted to the absolute concentration of the corresponding glycoside using

TABLE 1. The ratio of Digoxin to Dihydrodigoxin in Urine Samples from Paediatric Patients Maintained on the Drug

Sample	Intensity m/e 355	Intensity m/e 339	Digoxin Concentration μg/mℓ	Dihydrodigoxin Concentration μg/mℓ	Ratio Dihydrodigoxin/ Digoxin
1	699	404	0.466	0.101	0.217
2°	381	189	0.252	0.047	0.187
3°	92	61	0.061	0.015	0.246
4	843	45	0.556	0.011	0.020
5°	249	12	0.164	0.003	0.018
6°	317	101	0.209	0.026	0.124
7	261	176	0.172	0.044	0.256
8°	216	171	0.143	0.043	0.301
9	179	109	0.118	0.027	0.229
10°	280	115	0.185	0.029	0.157
11	211	234	0.139	0.059	0.424
12	88	36	0.058	0.009	0.155
13	552	305	0.364	0.076	0.209
14°	49	46	0.032	0.012	0.375
15	71	19	0.046	0.005	0.109
16°	54	32	0.036	0.008	0.222

°Patient under one year of age.

the calibration coefficients obtained from the evaporation profiles of the pure compounds. Thus the ratio of dihydrodigoxin to digoxin was obtained (Table 1). The dihydrodigoxin concentration in the original sample was then obtained from this ratio and the digoxin concentration obtained by radioimmunoassay.

The percentage of the daily dose excreted in the urine as dihydrodigoxin by the paediatric patients is shown in Fig. 4. The mean of 4.35% (range 0.32-9.1%) was considerably lower than that shown by adult patients. There was no significant difference between the results for the eight subjects under one year (mean 3.8%) and the eight over one year old (mean 4.87%) with a large variation between subjects in both groups.

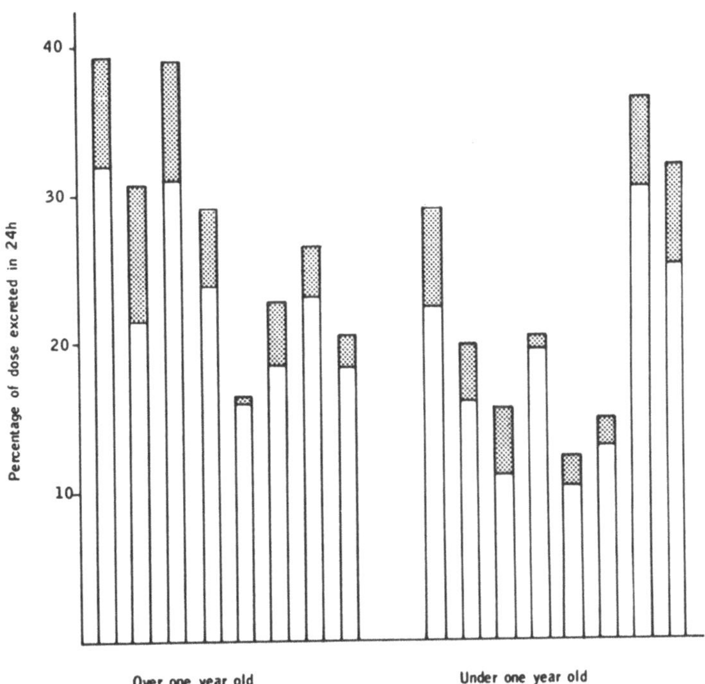

FIG. 4. The percentage of the daily dose excreted as digoxin ( ) and as dihydrodigoxin ( ) in 24 h by paediatric patients under and over one year of age.
Maintenance dose: under one year, 0.04-0.14 mg/24h
over one year, 0.02-0.2 mg/24 h.

DISCUSSION

The difficulty in separating digoxin from dihydrodigoxin probably accounts for the former not having been identified in earlier metabolic studies following administration of tritiated digoxin. Mass spectrometry permits the direct identification and measurement of this metabolite in extracts from urine samples. The high percentage of the daily dose found to be excreted in the urine by adults maintained on the drug demonstrates a significant metabolism of digoxin to this relatively cardioinactive form (4,6) and accounts for the low recovery of digoxin in the urine.

Infants and young children appear to receive and require in practice higher loading and maintenance doses of digoxin than adults (0.01-0.02 mg/kg body weight/24 h compared with 0.0054 mg/kg body weight/24 h for adults (7)) and those under one year old are able to tolerate serum concentrations of the drug which would be associated with toxicity in adults (8). Absorption, excretion and metabolism to the cardioactive metabolites following a single oral dose have been shown to be similar to that of adults (9) although we have found that paediatric patients on maintenance digoxin therapy excrete 20.4% of the daily dose in the urine in 24 h, slightly less than that previously found for adults.

This study was designed to determine whether an increased metabolism to the cardioinactive metabolite dihydrodigoxin could account for the higher dosage requirement. The highly sensitive mass spectrometric technique described here combined with a specific radioimmunoassay for digoxin which did not detect dihydrodigoxin, made possible the measurement of the very small concentrations of dihydrodigoxin excreted in the urine by paediatric patients. The percentage of the daily dose excreted as dihydrodigoxin by these patients (mean 4.35%) was considerably lower than that previously determined for adult patients (mean 16.4%) with considerable variation between individual subjects.

Studies in newborn infants have shown a decreased ability to metabolise drugs, resulting in a slow elimination rate and raised plasma concentrations. However, it has been reported that the urine of the newborn contains different conjugated metabolites than adults indicating that other metabolic pathways may exist (10). Thus the raised tolerance to digoxin in infants may be due to a difference in metabolism or to a decreased myocardial sensitivity but does not appear to be due to the production of unusually large amounts of dihydrodigoxin.

REFERENCES

1) J.E. Doherty, W.H. Perkins and G.K. Mitchell, Arch. Intern. Med., 1961, 108, 87.
2) F. Marcus, G.J. Kapadia and G.G. Kapadia, J.Pharmacol. Exp. Ther., 1964, 145, 203.
3) R.W. Jelliffe, Ann. Intern.Med., 1968, 69, 703.
4) H. Greenwood, W. Snedden, R.P. Hayward and J. Landon, Clin.Chim. Acta, 1975, 62, 213.
5) H. Greenwood, W. Snedden, J. Beardshaw and J. Landon, to be submitted for publication.
6) D.R. Clarke and S.M. Kalman, Drug Metab. Dispos., 1974, 2, 148.
7) D.J. Coltart, J.E. Cree and M.R. Howard, Br.J.Pharmacol., 1972, 44, 375.
8) H. Greenwood, M.R. Howard, J. Landon, B. Fraser and E. Shinebourne, Eur. J. of Cardiol. in press (1976).
9) A. Hernandez, R.M. Burton, R.D. Pagtakham and D. Goldring, Paediatrics, 1969, 44, 418.
10) M. Eriksson and S.J. Yaffe, Annu.Rev.Med., 1973, p. 29.

REFERENCES

1) [illegible] R.H. [illegible] and [illegible] Mitchell, [illegible] Chem., [illegible]
2) [illegible] and [illegible] Pharmacol. [illegible]
3) [illegible] Chem. [illegible], 1960, 69, [illegible]
4) [illegible] and [illegible], [illegible]
5) [illegible] and [illegible]
6) [illegible] Proc. [illegible], 197[illegible]
7) [illegible] and [illegible]
8) [illegible] and [illegible]
9) [illegible]
10) [illegible] 1977, [illegible]

DETERMINATION OF DIMETHYLNITROSAMINE IN EXTRACTS FROM CULTURES OF TRICHOMONAS VAGINALIS, USING GAS CHROMATOGRAPHY AND MASS SPECTROMETRY

M.Caramia, G.Poli°; M.Landi and L.F.Zerilli°°

°Ist.Patologia Gen. Università, Turin, Italy

°° Lepetit, Milan, Italy

INTRODUCTION

Recent findings suggest that chemical carcinogens may be formed in certain diseases of the human cervix (1). According to Harington _et al._ (2), dimethylnitrosamine (DMNA) might be present in the vaginal fluid from cervical and vaginal discharge. The report of these authors, based on the examination of vaginal fluid from a hundred African women, supports the idea that there is a link between the presence of DMNA in the vagina and the development of cancer of the cervix. Although it is generally agreed that DMNA's carcinogenic activity depends on prior conversion to the corresponding diazoalkane (3), a direct transformation of cells exposed to appropriate doses of DMNA has been found in one "_in vitro_" study (4). DMNA in the vagina could result from synthesis by microbes. It is possible in fact, that a large number of nitrate--reducing microorganisms, including _Trichomonas vaginalis_, can contribute at least partially to the production of DMNA from secondary amines and nitrates under neutral conditions (5, 6). The pH and temperature found in infected vaginas are thought to be conducive to nitrosamine formation from nitrates present in the urine and dimethylamine (DMA) formed from lecithin present in vaginal epithelium.

Since it has been epidemiologically demonstrated that there is increased risk of cancer of the cervix with _Trichomonas vaginalis_ infection (7, 8), we have examined under conditions as near as possible to those "_in vivo_", the possibility of a correlation between the production of DMNA and the presence of this Protozoa.

Trichomonas vaginalis was grown for 24 and 36 hours in a modified Diamond's medium (9), with or without sodium nitrate and DMA added. An aliquot of medium without Trichomonas was incubated for the same time, then tested for sterility. The samples were filtered through millipore filters at 2°C to separate medium from cells.

To detect the possible production of DMNA by Trichomonas vaginalis in modified Diamond's culture medium it was necessary to establish a method with a good recovery of very low quantities of DMNA. Known amounts of DMNA (4-8 ng/ml) were added to the medium for this purpose.

DMNA was distilled from the samples at an alkaline pH, then extracted from the distillate with dichloromethane. Interfering amines were removed from the dichloromethane extract by acid washing with a glycine-HCl acid buffer. A special apparatus for concentration of the organic extract was used in order to get a good percentage recovery. DMNA was looked for by gas liquid chromatography (GLC), using a flame ionization detector (10, 11), but the peak attributed to DMNA was poorly resolved. Therefore gas chromatography - mass spectrometry (GC-MS), operating in the single ion monitoring (SIM) mode, was applied (12, 13).

MATERIALS AND METHODS

Reagents. Solvents and reagents were of appropriate purity. ^{14}C-labelled products (DMNA and DMA) were supplied by SORIN, Saluggia Italy.

Sample preparation. To 250 ml of culture medium, filtered through a millipore filter (type XM 10-A, Hamicon Corp.) 30 g of sodium hydroxide were added and the mixture distilled in a glass apparatus, collecting about 150 ml of distillate. Fifteen g of potassium carbonate were dissolved in the distillate and the solution extracted twice with 150 ml portions of dichloromethane (extraction time 10 min). The combined extracts were shaken with 100 ml of glycine-HCl acid buffer, pH 2.1 $\pm$ 0.1 and the aqueous layer discarded (14). The dichloromethane layer was washed with 100 ml of 20% aqueous potassium carbonate solution and finally the organic layer was dried over anhydrous sodium sulphate. Half of the anhydrous extract was transferred into a 300 ml pear-shaped flask graduated from the bottom to indicate 15 ml and the chromatographic column, prepared as follows, was fitted. A glass column, 3 cm in diameter and 30-40 cm long, with a teflon stopcock, was used. A 10-15 cm layer of basic alumina (Aluminium oxide basic, activity grade 1 for chromatography) was poured into the column and a glass wool plug was inserted. The plug prevents the loss of aluminia during evaporation. The column was prepared just before use.

Vacuum was applied to the column and the flask maintained at 25°C

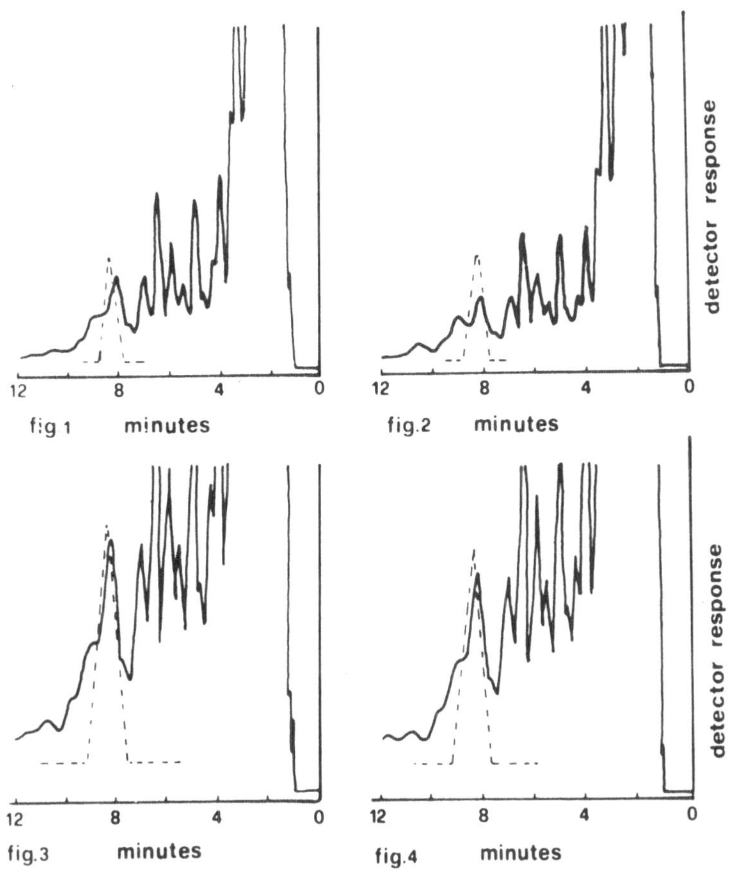

FIGS. 1-4. Gas liquid chromatographic research of DMNA: culture medium incubated with (1) and without (2) Trichomonas; culture medium with sodium nitrate and dimethylamine incubated with (3) and without (4) Trichomonas; dotted lines represent peaks obtained injecting DMNA alone. Column temperature 100°C.

in a water bath. In this way the solution was evaporated down to 15 ml. The vacuum was broken and the pear-shaped flask detached from the distillation unit and a graduated cylinder attached at the bottom of the column for the next operation. The basic alumina was eluted with sufficient dichloromethane to yield about 30 ml of eluate. The eluate was poured from the cylinder into the pear--shaped flask containing the 15 ml concentrate from the first evaporate and the remaining half of the anhydrous extract was added. At this point the flask was replaced on the distilling unit and the extract evaporated down to 15 ml under vacuum. Following the procedure outlined above the column was eluted again, collecting about 30 ml of eluate in a graduated cylinder and the eluate poured into the flask and evaporated down to 10 ml. The column was eluted once more and about 15 ml of eluate were collected and poured into the flask and evaporated down to 5 ml. With the flask in place, the column was eluted a final time with 3 ml of dichloromethane. The contents of the flask were transferred into a 10 ml graduated test tube and evaporated down to 0.5 ml under nitrogen at room temperature.

The percentage recovery of added DMNA was determined by using ^{14}C--labelled DMNA. The radioactivity of samples was determined with a liquid scintillation counter (Nuclear Chicago Co.). About 10^6 counts of ^{14}C-DMNA were added to 250 ml of culture medium corresponding to about 2 μg.

Apparatus. A fractovap model GV gas-chromatograph equipped with a flame ionization detector and 1 mV recorder was used. Conditions : stainless steel column 3 m long and 4 mm i.d. packed with 10% Carbowax 20 M on Chromosorb W (60-80 mesh) treated with HMDS; flow rate of nitrogen carrier gas 25 ml/min; injection port temperature 225°C; oven temperature 100°C; detector temperature 210°C; sample injections 0.5-1 μl. A Perkin Elmer 270 GC-MS system modified for SIM as described below was used for determining the sensitivity of the method and for quantitative measurements of any DMNA which might be present. The signal was taken from a driver stage of the oscillographic recorder galvanometer and fed into a multirange pen recorder.A direct current voltage stabilized source, providing zero suppression, was necessary in order to fully utilize the recorder span. Conditions : stainless steel column 3 m long and 2 mm i.d. packed with 15% Carbowax 20 M on Gas Chrom P (80-100 mesh) treated with HMDS; flow rate of helium carrier gas 25 ml/min; injection port temperature 250°C; oven temperature 115 or 130°C; separator temperature 200°C; ion source temperature 150°C; electron energy 70 eV; ionization current 100 μa.

RESULTS AND DISCUSSION

The percentage recovery at each step of the procedure was determined by using ^{14}C-labelled DMNA. These were : distillation 95%, extraction 92%, evaporation 60%. The overall percentage recovery was about 50%.

A typical gas chromatogram obtained from a extract of Diamond's medium incubated for 36 hours with T.vaginalis and processed as described above is shown in Fig. 1, while in Fig. 2, the gas chromatogram of a control extract is represented. In another series of experiments, T.vaginalis was grown in Diamond's medium (pH 6.5) containing sodium nitrate (0.295 g/l) and DMA (1 ml 40% aqueous solution) to see if the presence of these precursors would lead to synthesis of DMNA. Adding these precursors to the medium the maltose concentration must be reduced from 5 g/l to 0.5 g/l, in order to maintain the osmolarity. Under these conditions the growth of the Protozoa was very much accelerated. The results of GLC analysis of the media incubated with and without T.vaginalis are shown in Fig. 3 and 4 respectively.
The identification and quantification of DMNA appeared to be very difficult by GLC only, since the DMNA peak is not separated well from the other interfering peaks, as shown in Figs. 1-4.
Therefore gas chromatography-mass spectrometry provided with single ion monitoring technique (SIM) appeared to be the best way to solve the problem, because of its great sensitivity and specificity. In order to obtain the greatest sensitivity, the instrument was focused on the parent peak of DMNA at m/e 74, which corresponds to the most abundant ion in the spectrum, as shown in Fig. 5 (12). Typical total ion (TIC) and single ion (SIM) current profiles, obtained from culture medium extracts with and without added DMNA, are shown in Figs. 6A and 6B.

Repeated SIM analyses of many samples of Diamond's medium incubated with Trichomonas and with sodium nitrate and DMA gave no indication of the presence of DMNA, at least in amounts detectable by our procedure. The same result was obtained with a medium incubated with T.vaginalis, with sodium nitrate and $[^{14}C]$-DMA, added in the same amounts used in unlabelled experiments (Fig. 4). All the radioactivity was extracted into the aqueous phase, before sample concentration.

The limit of sensitivity of our procedure is 4 ng of DMNA per ml of culture medium. This was determined by extracting and analyzing culture media with known quantities of DMNA added.

In conclusion, the results of this study rule out the hypothesis that Trichomonas vaginalis participates in the induction of the carcinoma of the uterine cervix through production of DMNA. Of course the possibility that DMNA levels lower than 4 ng/ml may have biological activity can not be excluded.

Acknowledgments. Thanks are due to Mr. M. Zago for instrumental help and Dr. E. Beretta for helpful discussions.

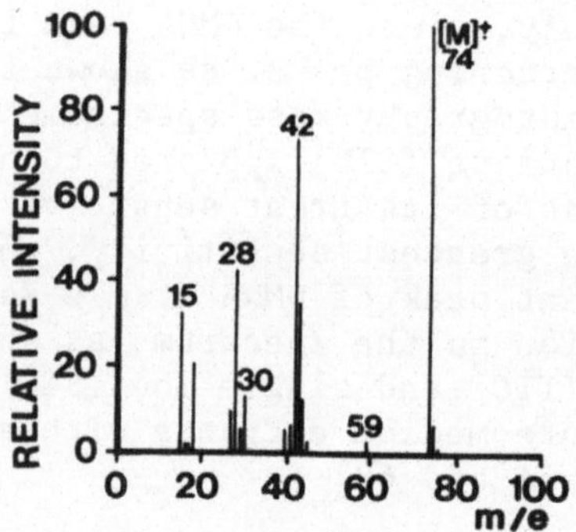

FIG. 5. 70 eV mass spectrum of DMNA obtained by GC-MS.

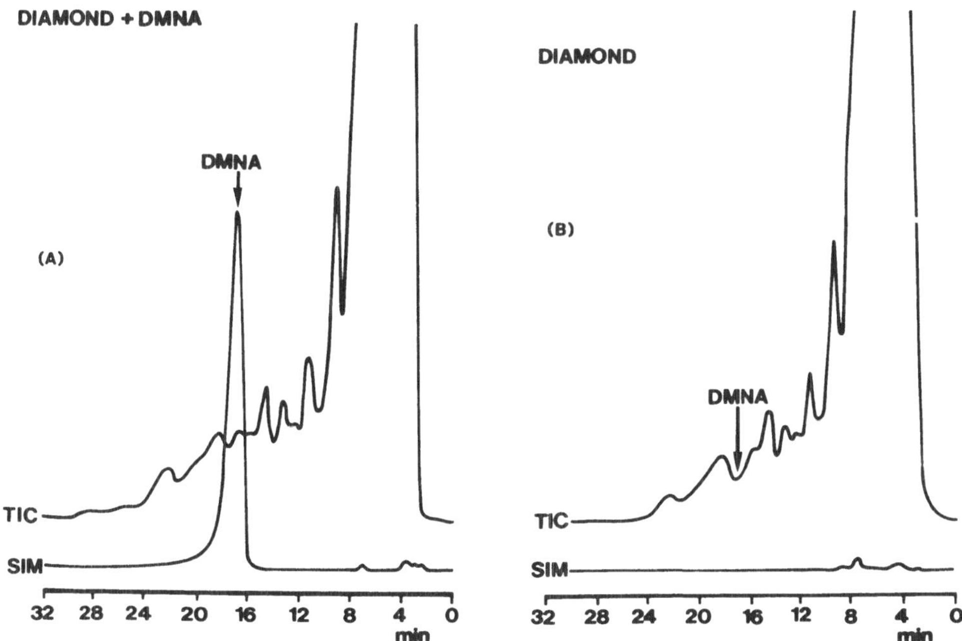

FIG. 6. Total ion current (TIC) and single ion monitoring (SIM) profiles obtained by GC-MS analysis of culture medium's samples incubated with <u>Trichomonas</u> with (a) and without (b) DMNA addition. Column temperature 115° C.

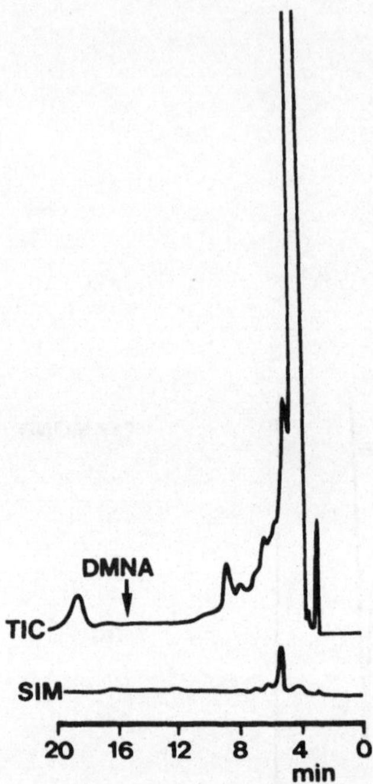

FIG. 7. TIC and SIM profiles of a sample of culture medium incubated with Trichomonas in presence of sodium nitrate and ^{14}C-dimethylamine. Column temperature 130°C.

REFERENCES

1) M.A. Robertson, J.S. Harington and E. Bradshaw, Brit. J. Cancer, 1971, 25, 377.
2) J.S. Harington, J.R. Nunn and L. Irwig, Nature, (London), 241, 49.
3) P.N. Magee, Biochem. J., 1956, 64, 676.
4) G. Di Mayorca, T. Greenblatt Trauthen, A. Soller and R. Giordano, Proc. Nat. Acad. Sci. U.S.A., 1973, 70, 46.
5) J. Sander, J. Assoc. Off. Anal. Chem., 1968, 52, 47.
6) A. Ayanaba and M. Alexander, Appl. Microbiol., 1973, 25, 826.
7) Barats, Am. Sov. Med., 1970, 33, 147.
8) B. Bertini and M. Hornstein, Acta Cytol., 1970, 14, 325.
9) P. Cappuccinelli, C. Lattes and I. Cagliani, Atti XVI Congresso Naz. Microbiol., 1972, 411, 411.
10) N.P. Sen, International Agency for research on cancer, 1973, p. 25.
11) P.N. Sen, D.C. Smith, L. Schwinghamer and J.J. Marleau, J. Assoc. Off. Anal. Chem., 1969, 52, 47.
12) B. Crathorne, M.W. Edwards, N.R. Jones, C.L. Walters and G. Woolford, J. Chromatogr., 1975, 115, 213.
13) G.M. Telling, T.A. Bryce and J. Althorpe, J. Agric. Food Chem., 1971, 19, 937.
14) N.P. Sen and C. Dalpe, Analyst, (London), 1972, 97, 216.

BLOOD CONCENTRATIONS OF MONOACYLCADAVERINES IN SCHIZOPHRENIA

H.Dolezalova,M.Stepita-Klauco,J.Kucera°,H.Uchimura[+] and M. Hirano[+];Univ.Connecticut,Connecticut,USA;Washington Univ., Missouri,USA°;Hizen Nat.Mental Hospital,Kanzaki,Japan[+]

INTRODUCTION

During the search for a chemical substance corresponding to the so called "pink spots" on chromatograms of urine from schizophrenic patients, both monoacetylcadaverine and monopropionylcadaverine were identified (1). Their connection, however, with mental illness was excluded in view of the then contemporary opinion on the exogenous origin of cadaverine.

It has been experimentally demonstrated that cadaverine in mammals originates from bacterial decomposition of food in intestines (2) or from tissue putrefaction (3). Monoacetylcadaverine and monopropionylcadaverine are believed to be the products of metabolic acylation of cadaverine (2). Moreover, it has been shown that the urine levels of monoacetylcadaverine and monopropionylcadaverine could be substantially reduced by administering large spectrum antibiotics, indicating that the suppression of intestinal bacterial flora resulted in less cadaverine formed exogenously and consequently in less cadaverine catabolites (1).

Some recent findings indicate that the actual role of cadaverine might be different from being simply an exogenous contaminant. Cadaverine is physiologically present in the mammalian brain and blood (4, 5). There is also an uptake system for cadaverine in mammalian brain which is inhibited by cyanide and some polyamines (6).Neither the blood nor the brain concentrations of cadaverine are lowered by the absence of bacterial flora in the intestine (7). Thus, even though there is no doubt that cadaverine is produced by bacteria, the brain concentration of cadaverine is maintained through mechanisms which are independent of bacterial decarboxylation.

TABLE 1.
Solvent systems for thin-layer chromatographic separation of dansylated cadaverine, monoacetylcadaverine and monopropionylcadaverine.

	First chromatography			Second chromatography		
Compound	Solvents	Runs	Eluant	Solvents	Runs	Eluant
bis-Dns-cadaverine	heptane-acetone (1:1)	one	ethyl acetate	chloroform-triethylamine (10:1)	one	ethyl acetate
Dns-acetylcadaverine Dns-propionylcadaverine	heptane-acetone (1:1)	one	methanol	benzene-methanol (14:1)	three	ethyl acetate-acetone (1:1)

METHODS

Subjects. The measurements were performed on blood samples obtained from two groups of subjects. The first group consisted of patients and controls from the United States, the second group originated from Japan. In both groups, the schizophrenic patients were hospitalized with acute or chronic schizophrenia. Two patients from the second group were newly admitted and had no medication. All other patients were taking phenothiazine drugs. The control subjects included laboratory personnel and neurological patients hospitalized for stroke. The patients and controls in the first group were not matched for age range or for sex ratio. The second group included only males, and the controls were selected for matched age range.

Samples. Cadaverine, monoacetylcadaverine and monopropionylcadaverine were measured in samples of the whole blood from the first group and in the blood plasma from the second group. The sample of the venous blood from the first group was transferred into a glass vial containing equal volume of 0.2 M perchloric acid, and the vial was immediately sealed. In the second group, the blood was collected into a vial containing EDTA (ethylenediaminetetraacetic acid disodium salt, 1 mg/ml of blood), centrifuged, the plasma was transferred into a glass vial containing equal volume of the perchloric acid, and the vial was sealed. With each sample in both groups, a second vial was simultaneously filled with the perchloric acid, sealed, and processed later as its blank.

Analytical procedures. A thin-layer chromatography-mass spectrometry method was used for determinations. Dansyl derivatives of amines were extracted into toluene and separated by thin-layer chromatography on silica gel coated plates (Merck HP). The fraction which co-chromatographed with standards of dansylated compound in question was scraped off, eluted, and separated with the second chromatography system. The thin-layer chromatography fractions were eluted again and their contents measured by high resolution mass spectrometry. The solvent systems for chromatography are shown in Table 1.

Quantitative mass spectrometry. A modified version of the 'integrated ion current technique' (8, 9) with the peak matching circuit (10) and the internal standard of a dansylated compound having the evaporation profile comparable to that of the compound in question (11, 12) was used. The quantitation of the recorded molecular ions was modified by correcting the signal for the spectrometer background, assuming a linear change of the background during the sample evaporation and by machine controlled determination of the beginning and the end of a part of the evaporation curve used for quantitation. The main advantages of this procedure are

TABLE II.
Mass spectrometric internal standards for quantitative determination of dansylated cadaverine, monoacetylcadaverine and monopropionylcadaverine.

Compound	Composition	m/e	Standard	Composition	m/e	Ratio
bis-Dns-cadaverine	$C_{29}H_{36}N_4O_4S_2$	568.2178	*bis*-Dns-hexamethylenediamine	$C_{30}H_{38}N_4O_4S_2$	582.2344	1.024666
Dns-acetyl-cadaverine	$C_{19}H_{27}N_3O_3S$	377.1773	Dns-acetylhexamethylenediamine	$C_{20}H_{29}N_3O_3S$	391.1930	1.037159
Dns-propionyl-cadaverine	$C_{20}H_{29}N_3O_3S$	391.1930	Dns-propionylhexamethylenediamine	$C_{21}H_{31}N_3O_3S$	405.2086	1.035828

that a higher reproducibility can be obtained and a larger number of biological samples can be measured in one session without having to wait until the background is reduced. A known quantity of an internal standard was added to each eluted chromatographic fraction, and the dried mixture was introduced via the probe into the mass spectrometer (AEI MS 902). The list of internal standards and their molecular ions are shown in Table II. The sample was evaporated using the full power of the probe heating system and ionized with the electron beam energy 70 eV. The molecular ions corresponding to the dansylated compound in question and to its internal standard were recorded with use of the peak matching circuit of the spectrometer at a resolving power between 2000 and 8000. Their molecular ratio was preset with an accuracy of 2ppm. The peaks corresponding to the two molecular ions were consecutively displayed on the screen of a storage oscilloscope and, simultaneously, each individual peak was digitized in a 14-bit analog-to-digital converter and recorded digitally on a magnetic tape. From a ratio between the intensity of the ion of interest and that generated by an internal standard substance, the quantity of the substance of interest in a sample was calculated.

The matching of the two peaks (intensities of ions or fragments) has been evaluated only by the operator in all previous systems for this type of application. In our method, both the intensity and the mass identity (with the accuracy of 10 ppm) are objectively recorded. An interference in the spectrum due to impurities or contamination of the sample is therefore demonstrated and such samples can be eliminated. Moreover, the method provides proof of the identity of the studied compound during the whole evaporation of the substance in mass spectrometer.

It should be noted that the term peak matching is used in the above description for a procedure which is slightly different from the classical peak matching and could be called, more accurately, reversed peak matching. While the former implies a stepwise adjustment of a ratio between the two masses in an attempt to find an accurate numeric value for the actual mass ratio; in the latter procedure, a fixed, calculated six digit ratio is preset, and the difference between the preset (theoretical) ratio and that observed during evaporation of the sample in the mass spectrometer is objectively evaluated by a computer. Thus, the disagreement between the expected and observed mass values is instantaneously evaluated for each pair of two matched mass regions, dynamically, during the evaporation of the whole sample. In contrast the subjective evaluation by an operator can be done only for a few sweeps during evaporation of one sample and rests on the operator's impression rather than on a recorded measurement.

Calculations of the amount of a substance in each sample were performed after substracting the background, by comparing the sums of the molecular ion maxima collected during the evaporation, generated by the known amount of the internal standard and by the unknown amount of the compound in question. The amplitude of each peak was established and an increase or decrease of the signal _per_ step for each two consecutive displays was calculated. The first and the last peak of the evaporation curve used for the quantitative evaluation were defined as those in which the increase of the signal _per_ step was for the first time larger than 5% of the maximal change of the signal _per_ step found during the whole course of the evaporation of that sample, when searching forward for the first peak and backward for the last one. The amplitude of the signal prior to the first peak and after the last were considered as background. In those cases where these two background values differed, a linear change of the background during the evaporation was assumed, and the corresponding background value was subtracted from each peak. From the corrected ratio between the sums of the molecular ion maxima the actual amount of the compound in question was calculated.

It is assumed in the course of the calculation procedure that for the same molar quantities of the two compounds the intensities of both molecular ions are identical. This is obviously not true for most compounds. To establish the quantitative relationship between the intensity of the molecular ion of an internal standard and that of the compound in question, a calibration procedure based on measuring standard samples is performed. These standard samples contain both the internal standard and the compound in question in known quantities (two standard samples for each of about 5 different amounts within the expected operating range). Assuming the above quantitative relationship, the content of the compound in question is calculated for each standard sample giving a calculated and an observed (actual) value. A regression line is calculated for the whole calibration set of pairs of expected and observed quantities. From the equation for the regression line a corrected amount of the internal standard is calculated. The corrected internal standard is subsequently used in all experimental samples to calculate the amount of the compound in question.

If the mass ratio between the two matched peak maxima differed at any time during the evaporation of the sample by more than 40 ppm (due to drift of the instrument, sample contamination, or electrical interference), the sample was disregarded. The calibration functions calculated by linear regression for 12 calibration samples containing known picomole quantities of each compound had the correlation coefficients 0.9920, 0.9914 and 0.9865, for cadaverine, monoacetylcadaverine and monopropionylcadaverine, respectively. The reported concentrations were measured as quantities more than three times higher than their blanks. They were not corrected for losses during extraction and thin-layer chromatography.

RESULTS

As shown in Table III, the concentrations of the measured compounds were lower in the samples of blood plasma from the second group of subjects than in the whole blood of the first group. With the exception of cadaverine in schizophrenic patients, all other concentrations in the second group were much lower than in the first group.

There was no significant difference between the mean concentrations of cadaverine in the blood of controls and schizophrenic patients in the first group. In the plasma samples from the second group, there was a significant increase of cadaverine in schizophrenic patients against controls.

Monoacetylcadaverine and monopropionylcadaverine concentrations in both the blood and plasma of schizophrenics showed a larger variance than those in controls (Fig. 1). While several values were within the region of control concentrations, the others were almost one order of magnitude higher. The mean values were significantly higher in schizophrenic than in controls for both monoacetylcadaverine and monopropionylcadaverine in both the groups (Table III).

DISCUSSION

The presence of reported substances in the human blood was established by our measurements. This report is based on a relatively small number of subjects, and the conclusion of an increase in the blood concentrations of monoacetylcadaverine and monopropionylcadaverine in some schizophrenic patients should not be generalized. Further analysis of elevated monoacylcadaverine in the blood of schizophrenic patients will require an extensive and detailed study.

The lower values of all measured compounds in the plasma of the second group are puzzling. They are not caused by relatively higher concentrations in the blood cells, for in some samples in which both blood cells and plasma were analyzed the concentrations in the blood cells were not significantly higher than in plasma. The presence of EDTA and one additional step in preparing the samples from the second group could contribute to the difference. However, the fact that there are different concentration ratios between cadaverine and monoacetylcadaverine or monopropionylcadaverine in the two groups, and that there is significantly higher mean cadaverine concentration in the plasma of schizophrenic than in controls from the second group while there is no difference in those of the first group, seem to indicate that the observed differences between the two groups might have some biological importance.

TABLE III.
Concentrations of cadaverine, monoacetylcadaverine and monopropionylcadaverine in blood samples from two groups of control subjects and schizophrenic patients ($x10^{-12}$mole/g of wet weight, mean ± standard error, the numbers of subjects are given in parentheses).

Compound	Group	Control subjects	Schizophrenic patients	t	P
Cadaverine	I	14.13 ± 2.78 (10)	15.53 ± 1.74 (18)	0.45101	>0.6
	II	7.59 ± 1.32 (14)	16.70 ± 2.73 (21)	2.58470	<0.05
Monoacetylcadaverine	I	3.28 ± 0.37 (13)	33.30 ± 7.93 (23)	2.82916	<0.01
	II	0.48 ± 0.10 (13)	4.14 ± 0.91 (21)	3.14762	<0.01
Monopropionylcadaverine	I	1.70 ± 0.67 (14)	15.75 ± 5.66 (17)	2.23594	<0.05
	II	1.09 ± 0.16 (15)	2.64 ± 0.31 (20)	4.07265	<0.001

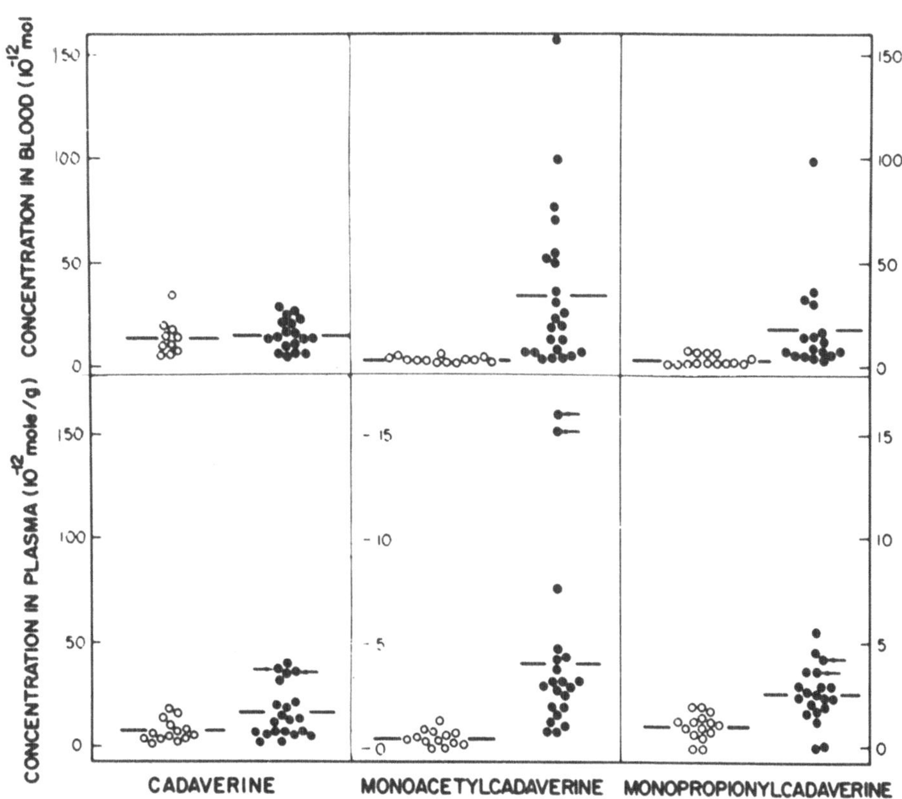

FIG. 1. Concentrations of cadaverine, monoacetylcadaverine and monopropionylcadaverine in the blood samples from the two groups of control subjects and schizophrenic patients. Group I - in the upper, group II - in the lower part of the diagram. Note the expanded concentration scale for monoacetylcadaverine and monopropionylcadaverine in group II. (Empty circles - controls; filled circles - schizophrenics).

TABLE IV.

The effect of monoamine oxidase inhibitors on concentrations of monoacetylcadaverine and monopropionylcadaverine in the mouse brain (in 10^{-12} mole/g of wet weight, mean $\pm$ standard error, the numbers of subjects are given in parentheses).#

	Controls	Nialamide	Pargyline
Monoacetylcadaverine	2.7 ± 0.7 (8)	19.8 ± 7.8 (7) ^{x}	12.5 ± 2.8 (8) ^{xx}
Monopropionylcadaverine	3.4 ± 1.0 (8)	17.7 ± 4.4 (6) ^{xx}	30.4 ± 4.9 (8) ^{xxx}

The mice were injected for 7 days with 25 mg/kg/day of Nialamide or Pargyline intraperitoneally. Controls were injected with saline.

Levels of significance: ^{x} $P<0.05$; ^{xx} $P<0.01$; ^{xxx} $P<0.001$.

The elevated blood levels of monoacylcadaverines in schizophrenic patients do not seem to be pharmacologically induced. In contrast, it is possible that phenothiazines are decreasing the elevated concentrations of monoacylcadaverines in the blood of schizophrenics because the concentrations found in samples from the two schizophrenic patients without medication (Fig. 1, arrows) were among the highest in that group.

Both the role and origin of cadaverine, monoacetylcadaverine and monopropionylcadaverine in human blood are unknown. Cadaverine concentrations in the blood and brain have been reported to fluctuate during sleep in mice and during hibernation in molluscs (4, 13).

It has been demonstrated that 1,4-diaminobutane is preferentially acetylated by the rat brain (14). It is therefore reasonable to consider the possibility that similar mechanisms might also metabolize 1,5-diaminopentane (cadaverine) in humans. The increase of monoacetylcadaverine and monopropionylcadaverine in blood could be caused by a higher rate of acetylation, or by a lowered catabolism of monoacylcadaverines. It is probable that monoacetylcadaverine and monopropionylcadaverine are substrates for monoamine oxidases because their four-carbon analog (monoacetylputrescine) is metabolized by monoamine oxidase in the rat (15). To test this assumption, we have measured brain concentrations of monoacylcadaverines in mice treated with monoamine oxidase inhibitors. The results in Table IV indicate that monoacetylcadaverine and monopropionylcadaverine are indeed the substrates for monoamine oxidases in the mouse brain.

The observed higher values of monoacetylcadaverine and monopropionylcadaverine in the blood of some schizophrenic patients could therefore be caused by an inefficient monoamine oxidizing system. Several studies have indicated that an altered activity of blood platelet monoamine oxidase accompanies mental illness (16, 24).

The investigators were also searching for alterations of monoamine oxidase activity in the brain corresponding to those found in blood platelets (18). Disappointingly, no changes were found in brain monoamine oxidase activity of mental patients as evaluated _post mortem_ (25, 26). The attempts to search for an endogenous substrate for monoamine oxidase displaying concomitant changes with lowering of blood platelet monoamine oxidase activity were also unsuccessful (27). The blood levels of monoacylcadaverines might be the first promising step in that direction. Irrespective whether the blood concentrations of monoacylcadaverines are actually dependent or reflecting brain concentrations of the same compounds, they might be sensitive indicators of the functional activity of monoamine oxidizing systems. Ultimately, it would be most interesting to see whether by using monoacylcadaverines as substrates, changes in brain monoamine oxidase activity could be detected in mental patients.

The use of blood levels of acylcadaverines in screening for a predisposition to mental illness could be twofold: 1) Monoacetylcadaverine and monopropionylcadaverine could be simply blood metabolites having no connection with the etiopathogenesis of mental illness or even with the physiology of the central nervous system. Their concentration may or may not depend on the overall monoamine oxidase activity in tissues of the body. Yet, if a significantly high correlation is found between their blood levels and some forms of mental illness, they could be successfully used for a more "objective" form of clinical diagnosis. 2) Monoacetylcadaverine and monopropionylcadaverine could be metabolites of cadaverine, preferentially formed in the brain, their blood levels reflecting a steady-state equilibrium with the corresponding concentrations in the brain tissue. Their increase in the blood of schizophrenic patients would be caused by a lowered activity of a monoamine oxidase isoenzyme specific for monoacylcadaverines in the brain. This type of enzymatic anomaly could be directly connected with the etiology of some forms of mental illness.

The two types of conditions described above are delimiting the extremes of a range of different possibilities in connections between the elevated blood levels of monoacylcadaverines and schizophrenia. The actual relationship will be probably somewhere between these two extremes.

Acknowledgments. We thank Dr. Robert Fairweather and Marvin Thompson for their help in operating the mass spectrometer. This work was supported by PHS grants NS 11716, NS 12482, and MH 27158.

REFERENCES

1) T.H. Perry, S. Hansen and L. MacDougall, Nature (London), 1967, 214, 484.
2) H. Tabor and C.W. Tabor, Advan. Enzymol. Relat. Areas Mol. Biol., 1972, 36, 203.
3) F. Dreyfuss, R. Chayen, G. Dreyfuss, R. Dvir and J. Ratan, Isr. J. Med. Sci., 1975, 11, 785.
4) H. Dolezalova, M. Stepita-Klauco and R. Fairweather, Brain Res., 1974, 77, 166.
5) H. Dolezalova and M. Stepita-Klauco, in "Advances in Mass Spectrometry in Biochemistry and Medicine", Vol. 1, A. Frigerio and N. Castagnoli, Eds., Spectrum Publications, Inc., New York, 1976, p. 207.
6) F. Picolli and A. Lajtha, Biochim. Biophys. Acta, 1971, 225, 356.
7) M. Stepita-Klauco and H. Dolezalova, Nature (London), 1974, 252, 158.

8) J.R. Majer and A.A. Boulton, Nature (London), 1970, 225, 658.
9) A.A. Boulton and J.R. Majer, J. Chromatogr., 1970, 48, 322.
10) N.M. Frew and T.L. Isenhour, Anal. Chem., 1972, 44, 659.
11) A.A. Boulton, S.R. Philips and D.A. Durden, J. Chromatogr., 1973, 82, 137.
12) N. Seiler and B. Knodgen, Org. Mass Spectrom., 1973, 7, 97.
13) H. Dolezalova, M. Stepita-Klauco and N. Seiler, Brain Res., 1974, 67, 349.
14) N. Seiler and M.J. Al-Therib, Biochim. Biophys. Acta, 1974, 354, 206.
15) N. Seiler and M.J. Al-Therib, Biochem. J., 1974, 144, 29.
16) A. Nies, D.S. Robinson, J.M. Davis and C.L. Ravaris, Psychosom. Med., 1971, 33, 470.
17) D.L. Murphy and R. Weiss, Am. J. Psychiatry, 1972, 128, 1351.
18) D.L. Murphy and R.J. Wyatt, Nature (London), 1972, 238, 225.
19) R.J. Wyatt, D.L. Murphy, R. Belmaker, S. Cohen, C.H. Donnelly and V. Pollin, Science, 1973, 179, 916.
20) H.Y. Meltzer and S.M. Stahl, Res. Commun. Chem. Pathol. Pharmacol., 1974, 7, 419.
21) A. Nies, D.S. Robinson, L.S. Harris and K.R. Lamborn, Adv. Biochem. Psychopharmacol., 1974, 12, 59.
22) I.B. Cookson, F. Owen and A.P. Ridges, Psychol. Med., 1975, 5, 314.
23) E.A. Zeller, B. Boshes, J.M. Davis and M. Thorner, Lancet, 1975, 1, 1385.
24) J.J. Schildkraut, J.M. Herzog, P.J. Orsulak, S.E. Edelman, H.M. Shein and S.H. Frazier, Am. J. Psychiatry, 1976, 133, 438.
25) M.S. Schwartz, A.M. Aikens and R.J. Wyatt, Psychopharmacologia, 1974, 38, 319.
26) M.A. Schwartz, R.J. Wyatt, N.Y.T. Yang and H.H. Neff, Arch. Gen. Psychiatry, 1974, 31, 557.
27) K.L. Murphy, R. Belmaker and R.J. Wyatt, in "Catecholamines and Their Enzymes in the Neuropathology of Schizophrenia", S. Matthysse and S.S. Kety, Eds., Oxford, Pergamon, 1975, p. 221.

MHPG, AMITRIPTYLINE AND DEPRESSION: A COLLABORATIVE STUDY

E.Sacchetti,E.Smeraldi,E.Allaria°;M.Cagnasso,P.A.Biondi°°

°Department of Psychiatry and °°School of Veterinary

Medicine, University of Milan, Milan, Italy

INTRODUCTION

A large body of clinical pharmacological data seems to be consistent with the existence of a relationship, perhaps causal, between abnormalities of catecholamine metabolism and affective disorders. This point of view has been formally organized into a "catecholamine hypotheses of affective disorders" (1 - 6), which differ in detail, but all present the view that depression is associated with a functional deficit in the amount of catecholamines available at the site of the receptor of critical central synapses.

Kinetic studies of the catecholamine levels in the spinal fluid would be the most direct means for exploring the function of the catecholaminergic system in the brain. However, it would be difficult in such studies to obtain repeated samples at small enough intervals of time to permit proper evaluation of the catecholaminergic system and its homeostasis.

Therefore, studies of catecholamines in peripheral fluids must be used, even though valid interpretation is difficult and caution must be applied. These studies may or may not reflect truly what is occurring in the CNS, because the two systems may or may not be in equilibrium with one another and because there is a selective barrier which influences the movement of catecholamines in and out of the brain.

It may be possible to get around some of these difficulties by studying 3-methoxy-4-hydroxy-phenylethyleneglycol (MHPG), a deaminated-O-methylated metabolite of catecholamines which is naturally occurring in man. MHPG is, in fact, the major metabolite of brain norepinephrine, even though there is still some doubt as to the exact proportion of the urinary MHPG which originates in the catecholamine pools of the CNS (6-16). Nevertheless, changes in the urinary excretion of this compound may serve as an indicator of changes in the metabolism of norepinephrine in the brain.

On this supposition, many studies are currently under-way on the urinary excretion of MHPG in patients with affective disorders (for reviews see 5, 6, 15, 17, 18, 19).

Our laboratories have been developing experimental approaches for the study of MHPG in depressive patients, with the aim of clarifying the role which catecholamines play in causing affective disorders. In this paper we bring together all our results in this area, adding new data to a summary of our previous findings.

ANALYTICAL PROCEDURE FOR MHPG

Gas chromatography with flame ionization detection (GC-FID). Total MHPG, free and conjugated, was extracted from urine with ethyl acetate after enzymatic hydrolysis. Quantitative analysis was carried out with the GC-FID technique. Some difficulties were encountered in the purification of the urinary extracts and in choosing a derivative which would give satisfactory resolution on GC. The following efforts were made toward improving the feasibility and the specificity of the method.

In our first experiments (20) the urinary extract was chromatographed on a Sephadex LH-20 column, according to the method of Anggard *et al.* (21) to separate the MHPG from interfering urinary components. The volatile derivative used was the tri-trimethylsilyl one (MHPG-triTMS). Although this procedure gave good results, it was very time consuming.

We then turned our attention to using new reagents, the boronic acids, $RB(OH)_2$, taking advantage of the distinctive glycol moiety of MHPG (22). The MHPG side chain derivatives with methane- and butane-boronic acids, $MeB(OH)_2$ and $BuB(OH)_2$, were prepared, with the phenolic hydroxyl protected as the TMS ether (MHPG-MeB-TMS and MHPG-BuB-TMS). Even though the reaction with the boronic acids was more specific than silylation, the preliminary purification before derivative formation could not be always avoided.

In order to overcome this difficulty, we followed the procedure of Sharman (23). The extracted MHPG was reacted with acetic anhydride in slightly alkaline solution, giving acetylation of the phenolic group. This acetylation in aqueous phase was both a derivatization and a purification step. Since it involved only the phenolic compounds, the acidic interfering components could be removed by a subsequent extraction of acetyl-MHPG (Ac-MHPG) into dichloromethane. Ac-MHPG was then reacted with the boronic acids to yield the two possible derivatives, Ac-MHPG-MeB and Ac-MHPG-BuB. With this procedure it was no longer necessary to perform the time--consuming column chromatographic purification and the whole procedure became suitable for use in routine analysis.

Among the stationary phases tested (polyester, SE and OV types), OV-101 gave the best resolution of the MHPG peak in urinary extracts.

Mass spectrometry (MS). The identities of the MHPG derivative peaks in the urinary GC patterns were determined by comparison of their mass spectra with those of standard MHPG derivatives. Molecular and base peaks of the derivatives used are summarized in Table 1.

The acetyl-MHPG-boronates have some MS features which make them suitable for use in the analysis by mass fragmentography of MHPG in biological samples. The mass fragmentographic (MF) technique is the most recent advance in the analysis of monoamine metabolites. Examples of studies by this technique are those in ventricular fluid (24), in urine, plasma, CSF and tissues (25) and in amniotic fluid (26). In those studies the molecular ion or others which are typical of the parent molecule are monitored. If these ions do not predominate in the mass spectrum, the advantage of MF cannot be fully utilized. In our experiments, for instance, we tried to apply MF to the analysis of MHPG, using the MHPG-BuB--TMS derivative, but the ion at m/e 292 (M^+-30) had a relative abundance of 4% and the method was therefore not sufficiently sensitive.

However, in the mass spectra of the Ac-MHPG-boronates, the peaks of the fragments which are obtained after the loss of acetyl moiety as ketene are both distinctive and very predominant. The values of $\%\Sigma_{40}$ for the methyl and butyl derivatives are 18 and 22 respectively (27).

At present we can state that the Ac-MHPG-boronates can be the derivatives of choice for the routine analysis of MHPG in urine, either by the GC-FID technique, generally used in clinical laboratories, or by MF, which is often necessary for obtaining the highest specificity.

STUDIES IN THE HUMAN

Criteria for selection of subjects. All the twelve depressive patients studied in the different phases of our investigation were diagnosed, in accordance with the criteria of Robins and Guze (28),

TABLE 1

MASS SPECTROMETRIC DATA OF MHPG DERIVATIVES

	MHPG-triTMS	MHPG-MeB-TMS	MHPG-BuB-TMS	Ac-MHPG-MeB	Ac-MHPG-BuB
M^+(%)	400 (1)	280 (15)	322 (15)	250 (10)	292 (5)
Base peak	297	73	73	208	250
	(M^+-CH_2OTMS)	(TMS^+)	(TMS^+)	(M^+-CH_2CO)	(M^+-CH_2CO)
	not specific	not specific	not specific	specific	specific

as having primary affective disorders of either the unipolar and bipolar type. The healthy volunteers had no first or second degree relatives with history of primary affective disorders. All the groups were balanced for age and sex.

None of the subjects had taken any medication for at least three weeks before beginning the study. The subjects, including the controls, were maintained on a normocaloric constant diet throughout the study (29).

After a basal period, the depressive patients were given 1.5 mg/kg body weight of amitriptyline, i.v., every day.

A quantitative evaluation of the depressive state was made by means of the self-rating scale of Aitken (30), with the criterion of improvement being a decrease in the rating score of at least 60%. The MHPG values given for each subject under each experimental circumstance are means of values measured on four consecutive days.

Urinary MHPG levels in controls and in patients with primary affective depression. (a) Base-line values. The mean values ± standard errors (S.E.) for the two groups are shown in Fig. 1.

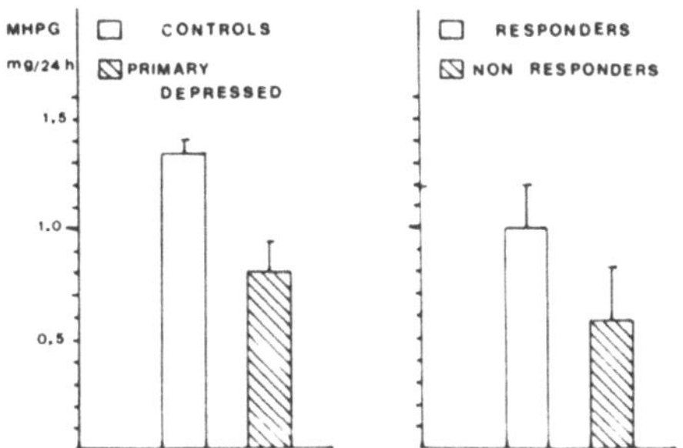

FIG. 1. MHPG in baseline conditions.

The patients' values are significantly lower, with a P value <0.01 (Student *t* test).

There was no significant difference between the values found in the unipolar patients and those found in the bipolar patients.

These findings agree with those published earlier (6, 15, 17, 31, 32), and are in accord with the "catecholamine hypothesis" of affective disorders. However, closer inspection of the data shows that there is a considerable inter-patient variability, and that the low values of MHPG represent a group-tendency, on the average, and are not necessarily found in individuals who are still undoubtedly suffering from the disorder. It would appear that the values for the group of patients with primary affective disorders are shifted toward the lower end of the distribution pattern for the normal population.

These data indicate that there is a heterogeneity among the patients classified as having primary affective disorders (6, 29, 33, 34). One group corresponds to the "catecholamine hypothesis" and has low MHPG levels. Another group has higher MHPG (equal to or greater than those in the controls). In these the catecholaminergic system is either operating normally or there is a slight hyposensitivity of the post-synaptic receptors, with a consequent compensatory increase in catecholamine synthesis.

(b) Correlation of clinical response to amitriptyline with basal MHPG levels. In an attempt to define the characteristics of the groups with "high" and "low" MHPG, we have decided to compare the responses of the two groups to treatment with an anti-depressive agent, amitriptyline, and to see how the MHPG levels correlate with "responders" and "non-responders".

The "responders" as a group had MHPG levels higher than those of the "non-responders", which agrees with the findings of other investigators (18, 19, 34, 35), although our results were different from those of Beckmann and Goodwin (34) in that the difference was not statistically significant. This discrepancy probably is due to the fact that in our patients there was a considerable overlap between the MHPG values of the two groups. On the basis of our data, it appears that subdivision into "low" and "high" MHPG groups and into "responders" and "non-responders" are only partially coincident, and that MHPG levels are not completely predictive for responsiveness to amitriptyline. This lack of correlation of MHPG levels to responsiveness to amitriptyline is not surprising in view of the overlap in the clinical responsiveness of depressive patients to amitriptyline and imipramine. The latter is supposed (32, 34, 36) theoretically to be of help only in those with low catecholamine levels.

MHPG levels after amitriptyline treatment. To elucidate the relationship between the pharmacological activity of amitriptyline and the catecholamines, we have compared the levels of MHPG before and after 3 weeks of treatment of primary depressive patients with amitriptyline.

When the patients are considered as a pool, without being separated into "low" and "high" basal level groups, it appears that amitriptyline had no effect on the urinary MHPG (before treatment 0.742+0.120 mg/24h, after treatment 0.808+0.100 mg/24h, not significantly different by the Student t test).

When, on the other hand, the groups are divided according to "high" or "low" MHPG values (discrimination value 0.9 mg/24 h), or according to whether they were "responders" or "non-responders", it could be seen that amitriptyline causes variable but definite changes in excretion in some of these (Fig. 2).

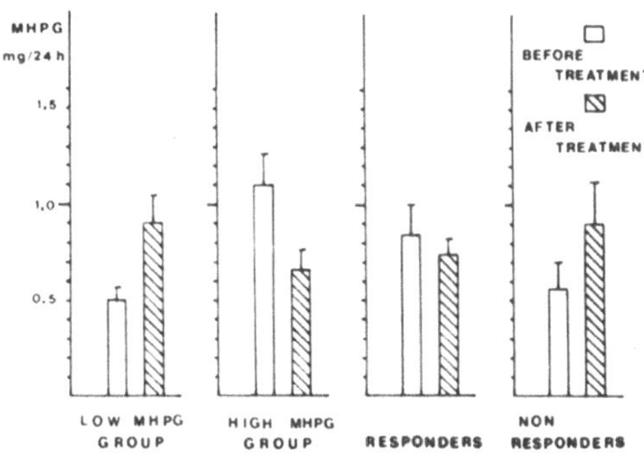

FIG. 2. MHPG excretion before and after treatment with amitriptyline.

In the group with "low" MHPG values, amitriptyline caused a significant increase in the MHPG value ($P < 0.05$). In the group with "high" values, amitriptyline caused a decrease, which was at the limit of significance ($t = 2.35$; for $P < 0.05$, t must be 2.4). Probably an increase in the number of subjects would establish this as significant.

When the patients are divided into "responders" and "non-responders", the levels in "responders" decreased on treatment, and those in the "non-responders" increased on treatment. These are not significant, and would not be expected to be in the light of our previous demonstration that the biochemical and clinical criteria did not completely correlate with each other.

The apparent lack of effect of amitriptyline on the pooled group of patients, therefore, arose from the balancing out of these various factors.

Temporal relationship between MHPG modification and improvement in the depression. We have previously reported (29) and have continued to observe in the clinic that there is a dissociation between changes in MHPG levels under treatment and improvement in the mood of the patients. In fact, 1) there is usually a lapse of about one week between the appearence of changes in the MHPG and changes in the mood; 2) significant changes in MHPG are sometimes seen in patients who do not improve clinically; 3) a few patients respond clinically in a manner unrelated to the changes in MHPG.

In addition, although our studies have shown that the basal levels of MHPG are relatively constant, during the first week of amitriptyline treatment, they become extremely variable, with some dampening of this variability when treatment is prolonged. At the moment we think that these fluctuations indicate that amitriptyline has its effect on some homeostatic mechanisms of the catecholamine system which may be unstable in these patients.

CONCLUSIONS

The data presented here lend further support to the idea that PAD patients are a heterogeneous group, and that the "catecholamine hypothesis" is a useful but over-simplified model for a very complex pathometabolic picture, in which all the factors are closely interconnected. Depression is the final common phenomenological expression of all these factors.

REFERENCES

1) W.E. Bunney,Jr. and J.M. Davis, Arch. Gen. Psychiatry, 1965, 13, 483.
2) J.I. Schildkraut, Am. J. Psychiatry, 1965, 112, 509.
3) J.I. Schildkraut, in "Biochemistry, schizophrenias and affective illness", Himwich, Ed., Williams and Wilkins, Baltimore, 1970, p. 198.
4) F.K. Goodwin and W.E. Bunney,Jr., Psychiatr. Ann., 1973, 3, 19.
5) F.K. Goodwin and R.L. Sack, in "Frontiers in catecholamine research", Usdin and Snyder, Eds., Pergamon Press, New York, 1973, p. 1157.
6) J.W. Maas, Arch. Gen. Psychiatry, 1975, 32, 1357.
7) E. Mannarino, N. Kirshner and B.S. Nashold, J. Neurochem., 1963, 10, 373.
8) J. Glowinsky, I.J. Kopin and J. Axelrod, J. Neurochem., 1965, 12, 25.
9) J.W. Maas and D.H. Landis, J. Pharmacol. Exp. Ther., 1968, 163, 147.
10) S.M. Shamberg, J.I. Schildkraut, G.R. Breese, E.K. Gordon and

I.J. Kopin, Biochem. Pharmacol., 1968, 17, 2006.
11) S.M. Schamberg, J.I. Schildkraut, G.R. Breese and I.J. Kopin, Biochem. Pharmacol., 1968, 17 , 247.
12) G.R. Breese, A.J. Prange, J.L. Howard, M.A. Lipton, W.Y. McKinney, R.E. Bowman and P. Bushnall, Nature, London, 1972, 240, 286.
13) J.W. Maas, H. Dekirmenjian, D. Garver, D.E. Redmond Jr. and D.H. Landis, Brain. Res., 1972, 41, 507.
14) S. Wilk and E. Watson, in "Frontiers in catecholamine research", Usdin and Snyder, Eds., Pergamon press, New York, 1973.
15) J.W. Maas, H. Dekirmenjian and J.A. Fawcett, Pharmacopsychiatry, 1974, 9, 14.
16) F. Karoum, R.J. Wyatt and E. Costa, Neuropharmacology, 1975, 13, 302.
17) J.W. Maas, H. Dekirmenjian and F. Jones, in "Frontiers in catecholamine research", Usdin and Snyder, Eds., Pergamon Press, New York, 1973, p. 1091.
18) J.I. Schildkraut, Am. J. Psychiatry, 1973, 130, 695.
19) J.I. Schildkraut, in "Frontiers in catecholamine research", Usdin and Snyder, Eds., Pergamon Press, New York, 1973, p. 1165.
20) M. Cagnasso and P.A. Biondi, Ital. J. Biochem., 1974, 23 , 345.
21) E. Anggard, B. Sjoquist and R. Siostrom, J. Chromatogr., 1970, 50, 251.
22) M. Cagnasso and P.A. Biondi, Anal. Biochem., 1976, 71, 597.
23) D.F. Sharman, Br. J. Pharmacol., 1969, 36, 523.
24) F. Karoum, J.C. Gillin, D. McCullough and R.J. Wyatt, Clin. Chim. Acta, 1975, 62, 451.
25) B. Sjoquist, B. Lindstrom and E. Anggard, J. Chromatogr., 1975, 105, 309.
26) F. Zambotti, K. Blau, G.S. King, S. Campbell and M. Sandler, Clin. Chim. Acta, 1975, 61, 247.
27) P.A. Biondi and M. Cagnasso, Anal. Lett., in press, 1976.
28) E. Robins and S.B. Guze, in "Recent advances in the psychobiology of the depressive illness", Williams, Katz and Shield, Eds., US Govt. Printing Office, Washington, 1972, p. 283.
29) E. Sacchetti, E. Smeraldi, M. Cagnasso, P.A. Biondi and L. Bellodi, Int. Pharmacopsychiatry, in press, 1976.
30) R.C.B. Aitken, Proc. R. Soc. Med., 1969, 62, 989.
31) J.W. Maas, H. Dekirmenjian and J.A. Fawcett, Int. Pharmacopsychiatry, 1974, 9, 14.
32) J. Fawcett, J.W. Maas and H. Dekirmenjian, Arch. Gen. Psychiatry, 1972, 26, 246.
33) H. Beckmann, J. St-Laurent and F.K. Goodwin, Psychopharmacologia, 1975, 42, 277.
34) H. Beckmann and F.K. Goodwin, Arch. Gen. Psychiatry, 1975, 32, 17.
35) J.I. Schildkraut, P.R. Draskoczy, E.S. Gerhon, P. Reich and E.L. Grab, J. Psychiatr. Res., 1972, 9, 173.

36) J.W. Maas, J.A. Fawcett and H. Dekirmenjian, Arch. Gen. Psychiatry, 1972, 26, 252.

DRUGS AND DRUG METABOLITES AS ENVIRONMENTAL CONTAMINANTS: DETECTION OF CHLOROPHENOXYISOBUTYRATE AND SALICYLIC ACID BY GC-MS

C. Hignite° and D.L. Azarnoff°°

° Kansas City, Missouri

°° Kansas City, Kansas

INTRODUCTION

The annual consumption of many drugs in the United States is measured in thousands of kilograms and many of these drugs pass through the body unchanged or are eliminated as biologically active metabolites. Drugs, therefore, represent a potentially serious source of biologically active environmental contamination since the compounds which are excreted may eventually enter the public water supplies. To evaluate this problem we have been screening the effluent of the Big Blue River sewage treatment plant which serves a population of approximately 600,000 people in the greater Kansas City, Missouri, area.

Sewage water was extracted under acidic, neutral, and basic conditions with methylene chloride and these extracts, both free and methylated, were screened by gas chromatography-mass spectrometry. The presence of 2-(4-chlorophenoxy)-2-methylpropanoic acid (CPIB), the active metabolite of the widely used hypolipidemic drug, clofibrate, and 2-hydroxy benzoic acid (salicyclic acid), a metabolite of aspirin, was established in these screens. This is to our knowledge the first report of detection of drugs in water samples. A review describing the detection of other organic compounds has recently appeared (1).

MATERIALS AND METHODS

Samples. Sewage water samples were collected in glass bottles with teflon-lined caps previously cleaned by a solution of potassium dichromate in sulfuric acid. All samples were collected

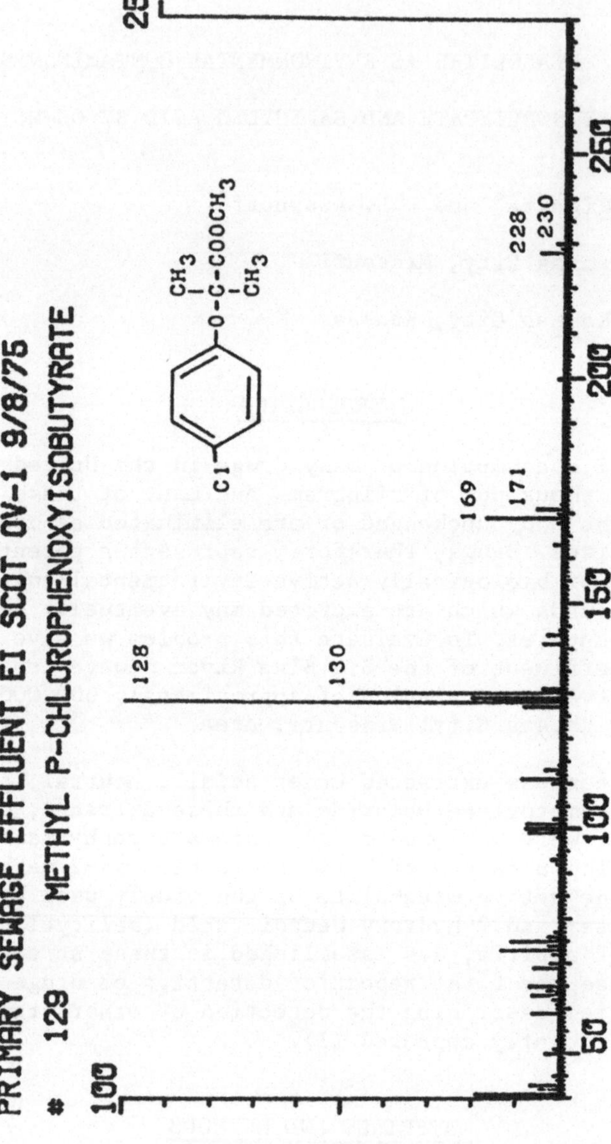

FIG. 1. Mass spectrum of methyl p-chlorophenoxyisobutyrate.

by Big Blue River sewage treatment plant personnel and were composite 24 hr samples (1/24th of the total volume was collected each hr for a 24 hr period). Samples were obtained after primary treatment (removal of solids) and immediately prior to discharge from the treatment plant. Water distilled from glass served as procedural blanks for these analyses.

Quantitation procedure. Ion exchange chromatography was used in preference to extraction due to the higher recovery of carboxylates (2). Anion exchange chromatography was carried out on 80-100 mesh Dowex 1-X1 (Bio-Rad Laboratories, Richmond, California) columns 25 x 120 mm. The columns were generated prior to each analysis with 100 ml 1N HCl saturated with NaCl and were then washed with distilled water until neutral. Sewage samples to which internal standards were added and blanks (500 ml) were then passed through the column and followed by 100 ml distilled water. The column was then eluted with 50 ml 1N HCl. This eluant was extracted with 50 ml nanograde methylene chloride and evaporated (rotary evaporator) to a volume of 3-5 ml. This condensed solution was treated with excess diazomethane (until yellow color persists) and then further condensed to a volume of approximately 100 µl. A 2-5 µl aliquot of this solution was injected directly into a gas chromatography-mass spectrometry system for analysis.

Internal standards. 4-Chlorophenoxyacetic acid was synthesized by the Williamson reaction (3) and involved reaction of sodium 4-chlorophenoxylate with ethylbromoacetate in ethanol, and the ester obtained was converted to the carboxylic acid by hydrolysis in 1N KOH.

Gas chromatography-mass spectrometry. A Finnigan 3300 gas chromatograph-mass spectrometer system operated on line with a Finnigan 6000 data system was used for all analyses. Thirty m x 0.3 mm support-coated open tubular (SCOT) gas chromatographic glass columns with OV-1 as the liquid phase were used for these studies. The columns were programmed from 50° C to 300° C at 10° C/min and held at the upper temperature until no further peaks were eluted. The injection temperature was 270° C. The effluent from the gas chromatograph passed through a single-stage glass jet separator before entering the ion source of the mass spectrometer. Mass spectra were recorded continuously throughout the analyses at either 2 or 4 sec intervals and were stored on magnetic discs. After the data were recorded the total ionization plot, mass spectra, and mass chromatograms were examined on an oscilloscope or plotted in hard copy form.

RESULTS

CPIB, salicylic acid and their internal standards were detected in these analyses as their methyl esters. The structure and mass spectrum of the methyl ester of CPIB is shown in Fig. 1, and the structure and mass spectrum of the methyl ester of the

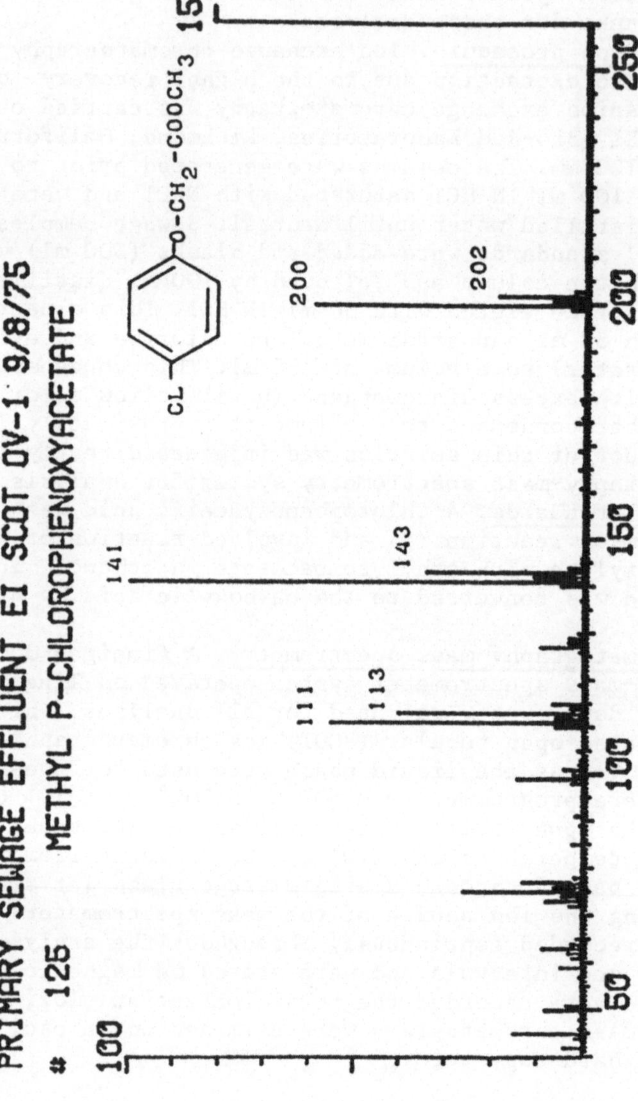

FIG. 2. Mass spectrum of methyl p-chlorophenoxyacetate.

internal standard, methyl 4-chlorophenoxyacetate, is shown in Fig. 2 . The total ionization plot from a typical run is shown in Fig. 3. The mass chromatogram of m/e 200, the molecular ion of methyl 4-chlorophenoxyacetate, is shown in Fig. 4 and the mass chromatogram of m/e 228, the molecular ion of the methyl ester of CPIB, in Fig. 5. The areas of the appropriate peaks in the mass chromatograms were calculated, and the ratio of these areas was used to determine the concentration of CPIB in the original sewage water sample. The same procedure employing molecular ions at m/e 152 of methyl salicylate and at m/e 170 of methyl 4-chlorobenzoate was used to quantitate salicylate levels in the sewage water. The choice of the proper peak in the mass chromatogram of m/e 152 must be done with care since 4-hydroxybenzoic acid is also present in many samples at significant concentrations.

The levels of CPIB and salicylic acid in sewage water are given in Table 1. The average amount of CPIB discharged for a 24 hr period from August 1975 to May 1976 was 2.1 kg/day (range 0.76--2.92). The average amount of salicylic acid discharged in the sewage effluent during the same period was 8.64 kg/24 hr (range 0.55-28.69).

In addition to analyzing the sewage water being discharged we also analyzed the raw sewage entering the sewage treatment plant. The concentration of CPIB was less than 20% higher in raw sewage (2.8 kg/day) indicating that most is not removed by the treatment plant. The concentration of salicylic acid, however, was substantially higher and approximately 90% of this compound is removed by the treatment plant.

We also analyzed the raw sewage and treated sewage water before and after acid hydrolysis. The results of these experiments indicate that there are negligible amounts of conjugated CPIB and only a minor amount (less than 25%) of salicylic acid is conjugated.

DISCUSSION

The Big Blue River Sewage Treatment Plant was placed in operation in 1966 to comply with a federal and state mandate to remove suspended solids from sewage prior to discharge into the Missouri River. This plant now provides primary treatment for 75-80% of the sewage generated in Kansas City, Missouri, and this treatment removes essentially all solids and about 35% of the total organic matter from the raw sewage. The volume of liquid passing through the plant averages 300 million liters/day, and the U.S. Army Corps of Engineers report the average flow of the Missouri River at Kansas City at approximately 100 billion liters/day. The effluent of the Big Blue Sewage Treatment Plant is therefore diluted more than 300 fold after entering the Missouri River.

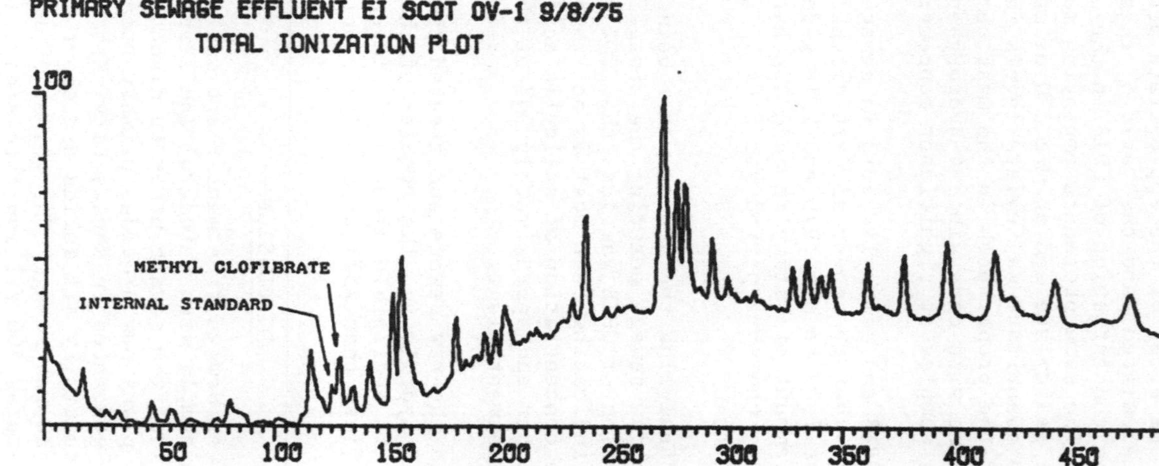

FIG. 3. Total ionization current profile of methylated extracts from sewage effluent.

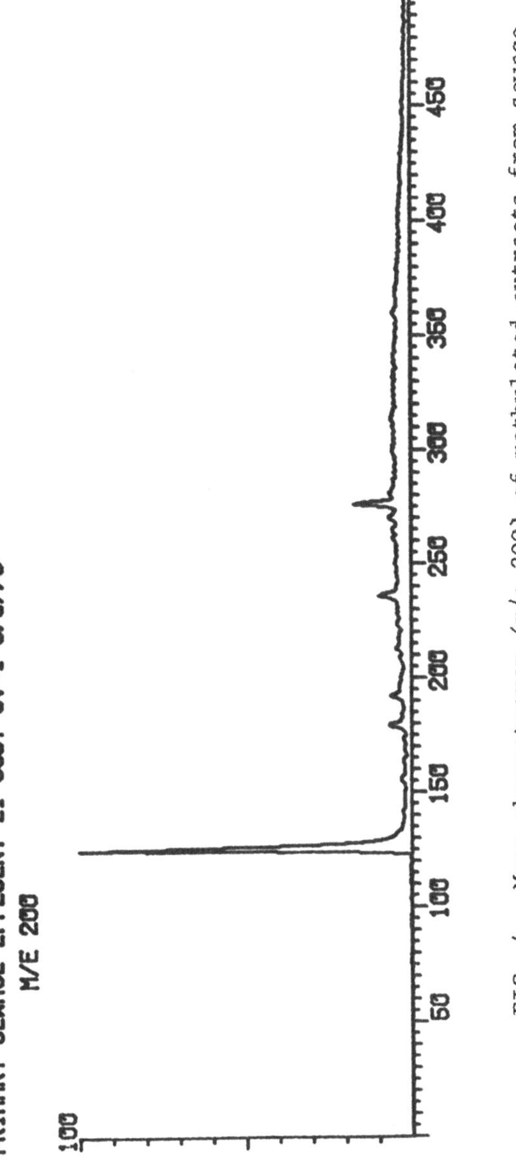

FIG. 4. Mass chromatogram (m/e 200) of methylated extracts from sewage effluent.

TABLE 1.
Concentration and 24 hr discharge of CPIB acid and salicylic acid in effluent of Big Blue River Sewage Treatment plant.

Date	CPIB		Salicylic acid	
	ppb	kg/day	ppb	kg/day
8/75	9.71	2.91	4.90	1.47
9/75	6.48	1.94	-	-
10/75	5.21	1.56	-	-
11/75	2.54	0.76	1.83	0.55
12/75	6.99	2.10	95.62	28.69
3/76	9.74	2.92	39.19	11.76
5/76	8.95	2.69	2.41	0.72
average	7.09	2.13	28.79	8.64

CPIB = 2-(4-chlorophenoxy)-2-methyl propanoic acid.

Because this dilution factor could be accurately estimated, we concentrated our efforts on analysis of the effluent of the sewage treatment plant rather than on analyses of Missouri River water. In this manner, our analyses are determined at higher concentrations, and the problem of procedural artifacts is subsequently minimized.

Clofibrate (ethyl 2-(4-chlorophenoxy)-2-methylpropanoate) is a drug widely used to lower plasma triglycerides and cholesterol concentrations in humans. It is administered in oral doses of 1.5-2.5 gm/day. It undergoes rapid metabolism at the ester linkage to produce a free carboxyl group, the biologically active form. Very little, if any, additional destruction of the molecule takes place. It has a plasma half-life of approximately 12 hrs and is excreted in the urine as the glucuronide derivative. The biologically active moiety of the molecule, therefore, remains intact. The amount of this drug which is discharged into the Missouri River is large, an average of 2.13 kg/day (Table 1). The best sales figures we could obtain were for the greater Kansas City area (1.3 million inhabitants) and suggest a daily consumption of approximately 10 kg/day in this area. A crude estimate suggests that half of the population of this area has its sewage treated by the Big Blue River plant. If we assume an equal distribution of the drug throughout the area, approximately 5 kg/day is consumed by the inhabitants of the area served by the treatment plant and 2.13 kg or 40% of the drug consumed is discharged in the sewage effluent.

Aspirin, a drug widely used, is hydrolyzed in the body to salicylic acid and then is excreted in the urine as either an ester glucuronide (20%), as a glycine conjugate (69%), as free salicylate (10%), and as a hydroxylated metabolite, gentisic acid (1%) (4). The amount of salicylic acid we found to be discharged in the sewage effluent, 8.64 kg/day, represents free salicylate and is even more than the amount of CPIB. We have not been able to obtain any reasonable information about consumption of aspirin in this area since it is supplied by numerous distributors. It can be seen from Table 1 that the daily amounts of salicylic acid discharged vary dramatically. This is not too surprising since the major use of aspirin is in an erratic manner for acute conditions rather than on a chronic basis by a relatively constant number of individuals as is clofibrate. The largest amounts discharged occur during the winter months when nonspecific use for upper respiratory infections would be greatest. Salicylate may also enter the sewage from various industrial uses or be formed by microorganisms.

The studies reported here show that kilogram amounts of drugs and drug metabolites may be discharged into our rivers in sewage effluent. In the instance of Kansas City, the dilution of these

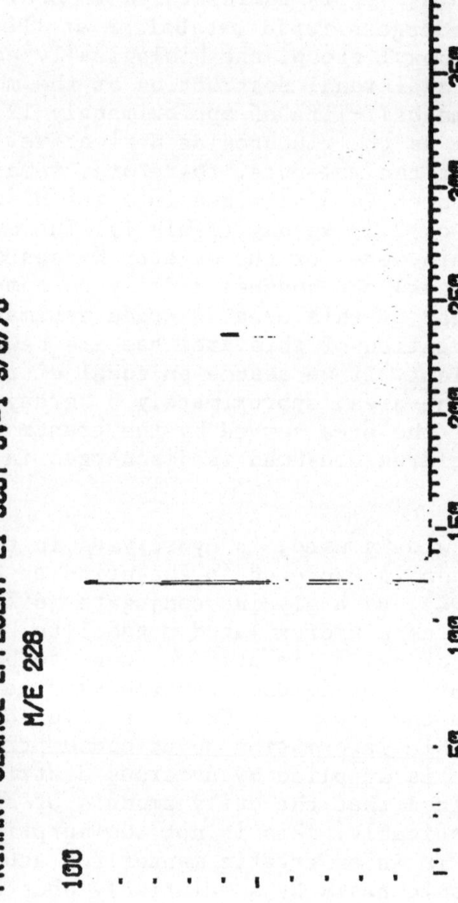

FIG. 5. Mass chromatogram (m/e 228) of methylated extracts from sewage effluent.

compounds upon entering the river is very large. We were unable to detect either compound in the drinking water in Kansas City or downstream in a sample of drinking water obtained in St. Charles, Missouri. Thus, at the present time, there is no detectable contamination of public water supplies with these drugs. However, in allergic individuals, minute amounts of a drug may be all that is required to initiate a reaction.

Acknowledgments. The authors wish to thank Mr. Raymond Newman for his excellent technical assistance. The authors also thank the personnel of Big Blue Sewage Treatment Plant for collecting the samples and also for providing data and discussion concerning the operation of the plant. We especially thank Jim Davis, John Reece and Huguette Woodrum.

This study was supported by grant GM 15956 from the United States Public Health Service.

REFERENCES

1) A.L. Burlingame, B.J. Kimble and P.J. Derrick, Anal. Chem. 1976, 48, 368R.
2) J.A. Thompson and S.P. Markey, Anal. Chem., 1975, 47, 1313.
3) A.W. Williamson, J. Chem. Soc., 1852, 4, 229.
4) R.T. Williams, in "Detoxification Mechanisms", 2nd ed., John Wiley, New York, 1959, p. 360.

APPLICATION OF CHEMICAL IONIZATION MS AND THE TWIN-ION TECHNIQUE IN THE ANALYSIS OF REACTIVE INTERMEDIATES IN DRUG METABOLISM

S.D. Nelson,[a] J.R. Mitchell, and L.R. Pohl

Laboratory Chemical Pharmacology, National Heart and Lung Inst., National Inst. Health, Bethesda, Maryland, 20014 USA

INTRODUCTION

Since the pioneering work of the Millers (1) and Magee and Barnes (2), it has become increasingly evident that many stable chemicals are metabolized to electrophilic intermediates that alkylate and arylate tissue macromolecules. This laboratory (3 - 5) has shown that drugs, such as acetaminophen (paracetamol), phenacetin, furosemide (frusemide), isoniazid, and iproniazid, also are oxidatively activated by microsomal enzymes to electrophilic intermediates that covalently bind to tissue macromolecules and cause massive tissue necrosis.

Because these drug metabolites are highly reactive and usually represent only a small proportion of the administered dose, their analysis by classical means is extremely difficult. In order to facilitate the characterization of these metabolites, we have trapped the intermediates *in vitro* with the naturally occurring sulfhydryl compounds, cysteine and glutathione. The addition of these compounds to incubation mixtures containing microsomes (prepared from the organ manifesting the tissue damage), NADPH and drug substrate decreases the binding of reactive metabolites to microsomal protein by combining with the electrophilic drug metabolites. The cysteine and glutathione conjugates thus formed are purified by a combination of Sephadex, ion exchange and thin-layer chromatography and identified by chemical ionization (CI) mass spectrometry and electron ionization (EI) mass spectrometry.

This paper describes the formation of reactive intermediates from acetylhydrazine and isopropylhydrazine, hepatotoxic metabolites of the tuberculostatic drug, isoniazid, and the antidepressant drug, iproniazid. In addition to the trapping experiments, we have utilized stable isotope labeling combined with a CI mass spectrometry twin--ion technique as aids in defining the nature of the reactive intermediates.

METHODS

Syntheses. Acetylhydrazine and trideuteroacetylhydrazine were prepared as previously described (6). An EI mass spectrum of trideuteroacetylhydrazine showed the following isotope abundances: 84% d_3, 15% d_2, and 1% d_1 - acetylhydrazine.

Isopropyl-[$2-^{14}C$]-hydrazine was prepared by tris-triphenylphosphorhodium catalyzed hydrogenation of N^2-isopropylidene--tertiary-butyl-carbazate prepared by condensing acetone [$2-^{14}C$]-and tertiary-butyl-carbazate. The product, N^2-isopropyl [$2-^{14}C$]-tertiary butylcarbazate was hydrolyzed in dilute methanolic hydrochloric acid to yield isopropyl-[$2-^{14}C$]-hydrazine hydrochloride which was purified by recrystallization from methanol: ether.

Isopropyl-[$2-^{2}H$]-hydrazine was prepared by hydrogenation of N^2-isopropylidene-tertiary-butylcarbazate with deuterium gas. In addition, this compound was prepared by reductive alkylation of the tertiary-butylcarbazate with sodium cyanoborodeuteride in acetone followed by mild acid hydrolysis. Deuteration of the isopropyl-[$2-^{2}H$]-hydrazine in the above syntheses occurred specifically on the C-2 carbon atom as determined by NMR (lack of C-2 hydrogen septet at 3.25 ppm downfield from TMS) and EI mass spectral analyses. The parent ion at m/e 75 indicated greater than 98% deuterium incorporation. The base peak at m/e 60 (M-15) with no contribution at m/e 59 indicated loss of a methyl group with no loss of deuterium. This interpretation of the fragmentation pattern was supported by the presence of a substantial metastable transition at m/e 48.

Trapping experiments. Liver microsomes were obtained from male Fischer rats (180-200 g) and incubations were carried out as described by Nelson *et al*. (6). Experiments routinely consisted of four sets of incubations with microsomes at a concentration of 2 mg protein/ml. One set contained substrate and NADPH-generating system and was used to determine total covalent binding of reactive metabolite. A second set contained substrate, NADPH-generating system, and either cysteine or glutathione (1 mM). A control set contained cysteine and substrate but lacked the NADPH-generating system in order to demonstrate that no direct reaction occurred between the sulfhydryl reagents and the substrate. The final set of incubations contained substrate but no NADPH-generating system to determine the amount of nonspecific binding of radiolabel to microsomes.

Reactions were run for 15 minutes and stopped by the addition of 4 volumes of ice-cold methanol. The precipitated protein was collected and covalent binding of radiolabel to microsomal protein was determined as described by Potter et al.(7). The supernatant from each set of incubations was lyophilized, dissolved in 2 ml of water and applied to a Sephadex G-10 column (1.5 x 90 cm), which was equilibrated and eluted with deionized water. Those incubations containing microsomes, substrate, NADPH and sulfhydryl component showed peaks of activity not found in controls. These fractions were combined, lyophilized and purified further by a combination of ion-exchange and thin-layer chromatography as described previously (6).

Mass spectral studies. Mass spectra were obtained on a single focusing VG Micro Mass 16F instrument. Samples were introduced with a temperature-controlled direct insertion probe and spectra recorded at a resolution (10% valley) of about 1000. EI mass spectrometry was performed at an accelerating voltage of 4 kV, an electron energy of 70 eV, and an ionizing current of 100 μA. CI mass spectrometry was performed at an accelerating voltage of 4 kV, an electron energy of 100 eV, and an ionizing current of 200 μA. Ion source pressure was 0.3-0.5 Torr (isobutane reactant gas). Probe and source temperatures were variable as described for each spectrum.

The N-trifluoroacetyl (TFA) methyl esters of the glutathione conjugates were prepared by trifluoroacetylation of the isolated glutathione conjugate (∿40 μg) in 100 μl of trifluoroacetic anhydride for 15 minutes. The mixture was evaporated under a gentle stream of nitrogen and the residue taken up in 0.5 ml anhydrous methanol to which was added 0.2 ml of freshly distilled diazomethane in ether. After a 5 minute reaction period the solvents and excess diazomethane were evaporated under a stream of nitrogen and the residue was dissolved in 10 μl of methanol. Samples of 1, 2 and 5 μl were evaporated in capillary tubes for mass spectral analysis.

N-Acetylacetonyl methyl ester derivatives were prepared as described by Schier et al.(8) by dissolving the peptide in a methanolic solution of trimethylanilinium hydroxide and adding acetylacetone. After a 3 hour reaction period, the mixture was evaporated under nitrogen and the sample inserted into the mass spectrometer using a direct insertion probe. The probe was heated until ions appeared at m/e 122 for protonated dimethylaniline indicating methylation of the peptide and then spectra were recorded.

RESULTS

Glutathione conjugates. Glutathione decreased the covalent binding of a radiolabeled acetylhydrazine metabolite to microsomal protein by 65%, with the concomitant formation of a radiolabeled

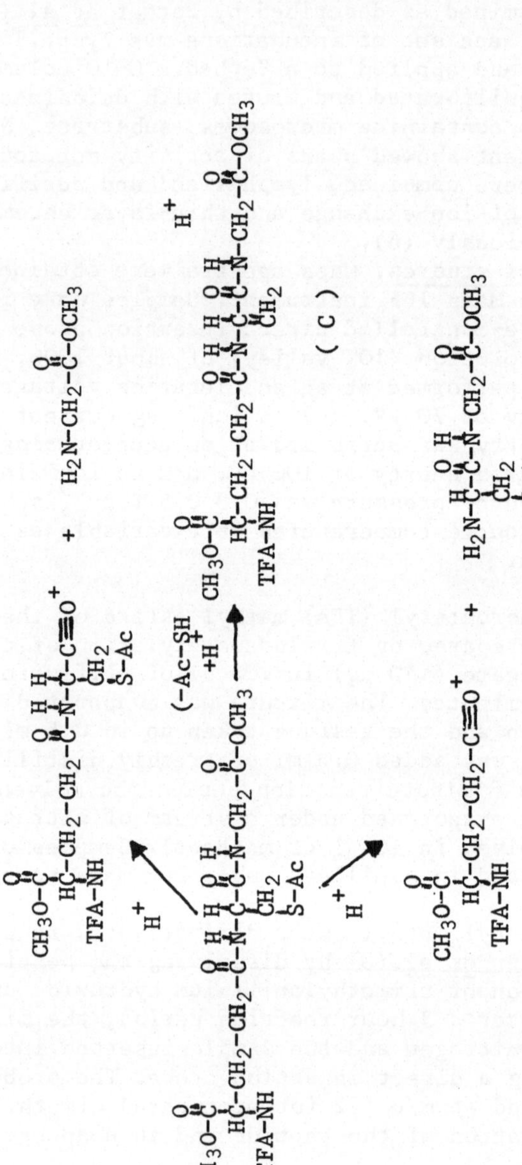

(A)

$CH_3O\text{-}C(=O)\text{-}HC(NH\text{-}TFA)\text{-}CH_2\text{-}CH_2\text{-}C(=O)\text{-}NH\text{-}CH(CH_2\text{-}S\text{-}Ac)\text{-}C{\equiv}O^+$ + $H_2N\text{-}CH_2\text{-}C(=O)\text{-}OCH_3$

H^+ ↗

$CH_3O\text{-}C(=O)\text{-}HC(NH\text{-}TFA)\text{-}CH_2\text{-}CH_2\text{-}C(=O)\text{-}NH\text{-}CH(CH_2\text{-}S\text{-}Ac)\text{-}C(=O)\text{-}NH\text{-}CH_2\text{-}C(=O)\text{-}OCH_3$ $\xrightarrow{(-Ac\text{-}SH^+ \; +H^+)}$ $CH_3O\text{-}C(=O)\text{-}HC(NH\text{-}TFA)\text{-}CH_2\text{-}CH_2\text{-}C(=O)\text{-}NH\text{-}C(=CH_2)\text{-}C(=O)\text{-}NH\text{-}CH_2\text{-}C(=O)\text{-}OCH_3 \; H^+$

(C)

H^+ ↘

$CH_3O\text{-}C(=O)\text{-}HC(NH\text{-}TFA)\text{-}CH_2\text{-}CH_2\text{-}C{\equiv}O^+$ + $H_2N\text{-}CH(CH_2\text{-}S\text{-}Ac)\text{-}C(=O)\text{-}NH\text{-}CH_2\text{-}C(=O)\text{-}OCH_3$

(B)

glutathione conjugate. A CI mass spectrum of the acetylacetonyl-methyl ester derivative, obtained by Schiff base formation and pyrolytic methylation, showed a quasimolecular ion (MH^+) at m/e 460 and a major ion at m/e 428 corresponding to the loss of methanol. However, other major ions in the spectrum (m/e 400, 374, 333, 225, 164, and 144) provided little information concerning the peptide sequence or position of acetyl group substitution.

However Kiryushkin *et al.* (9) have shown that CI mass spectra of acylated methyl esters of simple peptides provide ions which are important in determining the sequence of amino acids in the peptides and these sequence-determining ions are of enhanced intensity when compared to the EI mass spectra. We, therefore, prepared the TFA methyl ester of the glutathione conjugate. A CI mass spectrum of this derivative (Fig. 1) provided substantial information concerning the sequence of amino acids in the peptide as well as the position of reactive metabolite conjugation.

Analysis of the spectrum reveals a substantial quasimolecular ion (MH^+) at m/e 474 for the protonated glutathione derivative from which methanol is eliminated to give an ion at m/e 442. Additional loss of 207 mass units to produce an ion at m/e 235 indicates removal of the entire α-methoxy-N-TFA-γ-glutamyl residue. The base peak at m/e 90 corresponds to the quasimolecular ion of methyl glycinate which indicates the additional loss of an acetylated cysteine residue. The sequence of the glutathione (γ-glutamyl-cysteinyl-glycine) peptide is confirmed since major ions are present at m/e 385 for the elimination of methyl glycinate from the quasimolecular ion to produce the acyl carbonium ion (A), as well as at m/e 240 due to cleavage of the glutamyl-cysteine C-N bond with formation of the acyl carbonium ion (B).

Additional information about the glutathione conjugate is found in the spectrum from the ions at m/e 398, 309 and 159. The elimination of thiolacetic acid from S-acetylglutathione produces a dehydroalanine (Δ-ala) containing peptide (C) which when protonated forms a quasimolecular ion at m/e 398. Heterolysis of the C-N bond at the dehydroalanyl C-terminus produces a γ-glutamyl-dehydroalanyl acylonium ion at m/e 309. Alternatively, cleavage with hydrogen transfer at the N-terminus, as described by Kiryushkin *et al.* (9), would form a dehydroalanylglycinate fragment with m/e 159. These ions describe a fragmentation pathway which showed that the acetyl group was bound to the cysteinyl sulfydryl group. Thus, S-acetylglutathione was the conjugate formed from glutathione and a reactive intermediate in acetylhydrazine metabolism.

Glutathione also decreased the binding of a metabolite of isopropylhydrazine to microsomal protein by 50%. A glutathione conjugate was isolated and purified as described for S-acetyl-

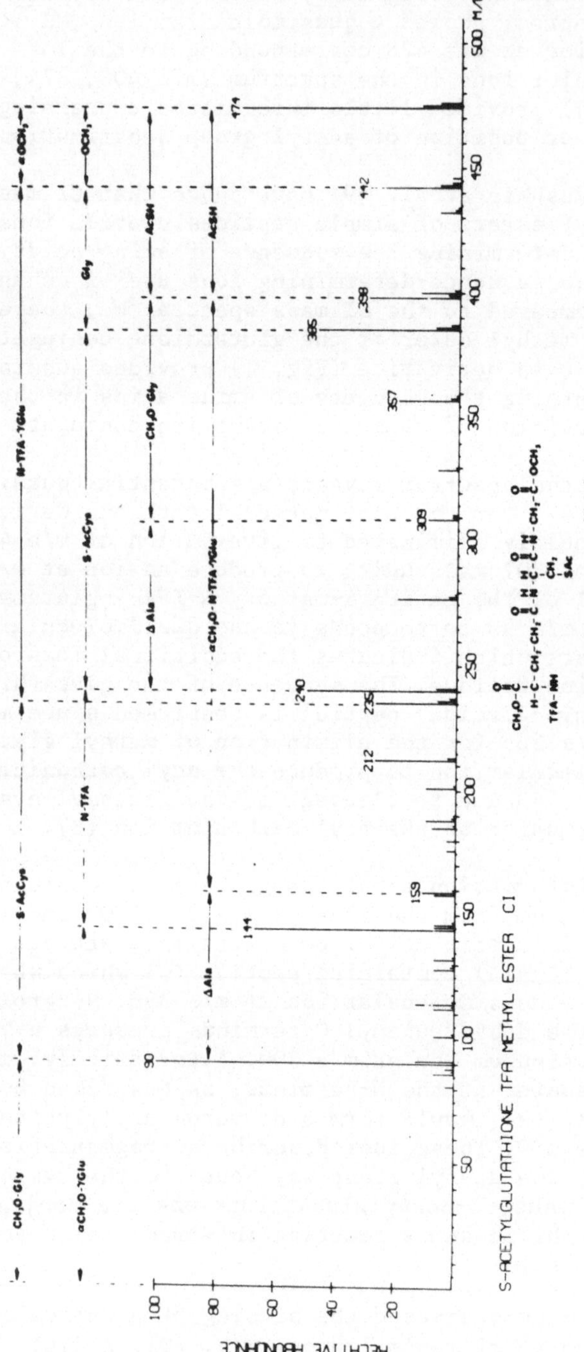

FIG. 1. Chemical ionization (CI) mass spectrum of the methylated TFA derivative of S-acetylglutathione. The reagent gas was isobutane at a source pressure of ∿ 0.4 Torr; source temp., 300°C; probe temp., 150°C. The term Δ-ala refers to a dehydroalanyl structure.

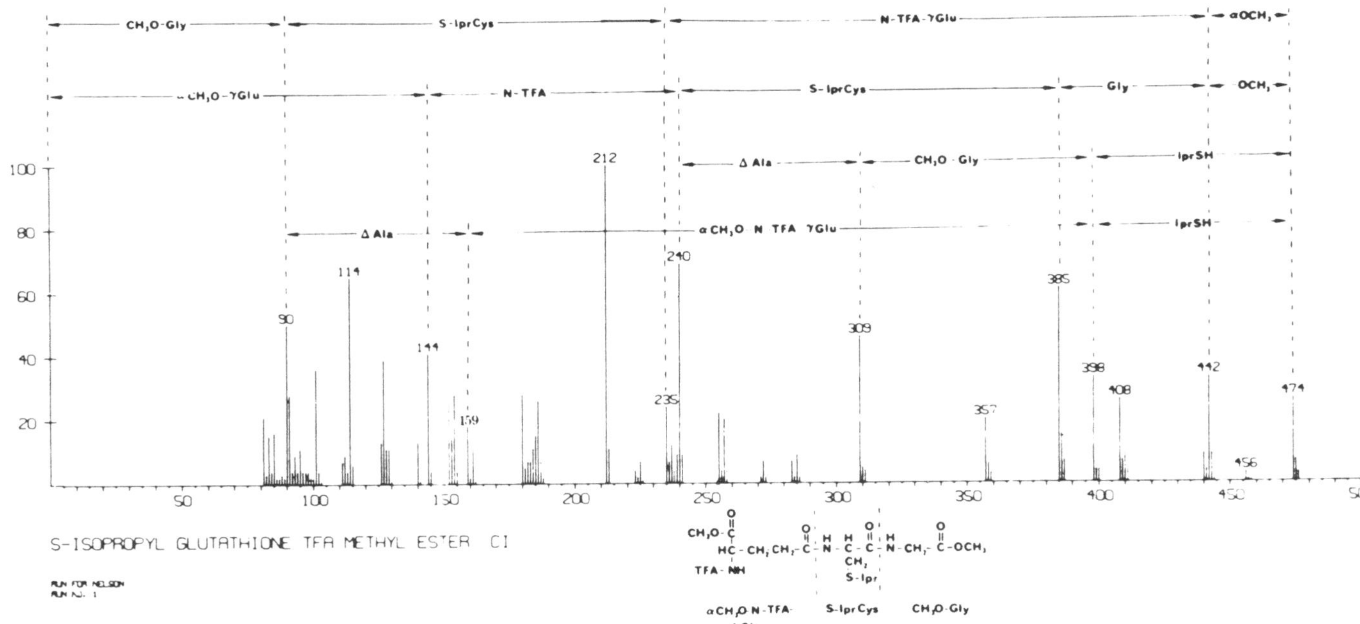

FIG. 2. Chemical ionization (CI) mass spectrum of the methylated TFA derivative of S-isopropylglutathione. The reagent gas was isobutane at a source pressure of ~ 0.4 Torr; source temp., 280°C; probe temp., 150°C. The term Δ-ala refers to a dehydroalanyl structure.

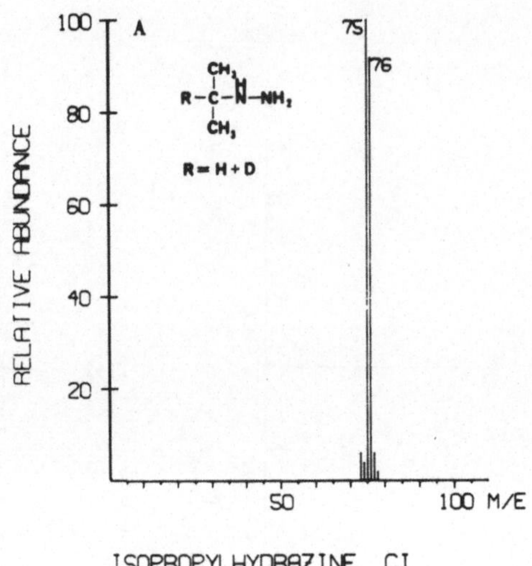

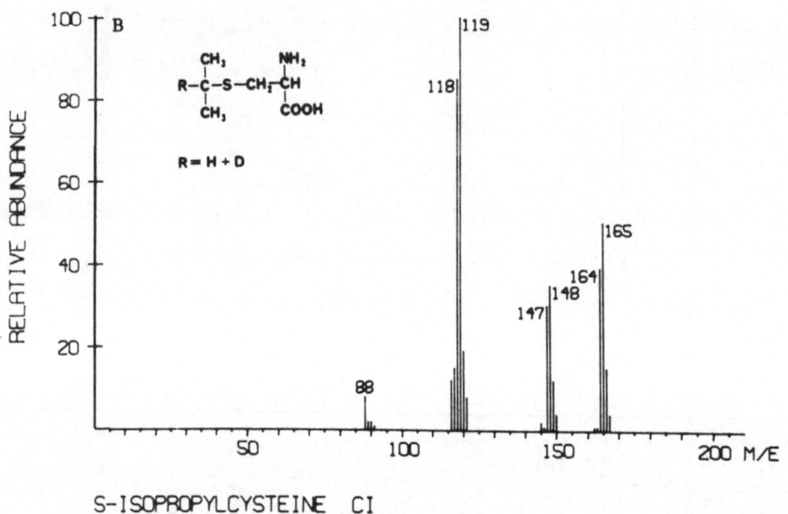

FIG. 3. Chemical ionization mass spectra of a sample of the substrate mixture of isopropylhydrazine and isopropyl/$2-{}^2H$/-hydrazine as their hydrochloride salts (A) and of the cysteine adduct isolated from a microsomal incubation containing the substrate mixture, NADPH cofactor and cysteine (B). The reagent gas was isobutane at a source pressure of ∿ 0.3 Torr; source temp., 200°C (A) and 250°C (B); probe temp., ambient (A) and 110°C (B).

glutathione. A CI mass spectrum of the N-acetylacetonyl methyl ester derivative showed a quasimolecular ion at m/e 460 and a major ion at m/e 428 due to the elimination of methanol. As determined for the N-acetylacetonyl methyl ester of S-acetylglutathione, other ions in the spectrum provided little information concerning the structure of the glutathione derivative. A CI mass spectrum of the N-TFA methyl ester derivative (Fig. 2) revealed many of the ions found in the spectrum of the N-TFA methyl ester of S-acetylglutathione from which the same fragmentation pattern can be described. Since the unit molecular weight (43) of the acetyl group is the same as the isopropyl group, this suggests that the conjugate formed was S-isopropylglutathione. However, other major ions occur in the spectrum of m/e 408, 357, 212, 101, etc. which cannot be readily interpreted. They may indicate degradation of the peptide during derivatization or possibly incomplete purification of the peptide. Conclusive evidence for the structure will require spectral comparison with synthetic compound, an experiment in progress.

Cysteine conjugates. Additional trapping experiments were carried out with stable isotope derivatives of acetylhydrazine and isopropylhydrazine and the sulfhydryl containing amino acid, cysteine. When an equimolar mixture of acetylhydrazine and trideuteroacetylhydrazine was incubated with rat liver microsomes. NADPH and cysteine, N-acetyl-cysteine was isolated which contained the same ratio of hydrogen and deuterium as the initial hydrazine substrate mixture (6). This study eliminated ketene as the possible reactive acylating intermediate formed by the microsomal oxidation of acetylhydrazine.

In a similar manner, an equimolar mixture of isopropylhydrazine and isopropyl-[$2-^2H$]-hydrazine (Fig. 3A) was incubated with rat liver microsomes, NADPH and cysteine. A CI mass spectrum of the cysteine derivative isolated from these reactions (Fig. 3B) showed MH^+ ion doublets at m/e 164 and 165, as well as ion doublets at m/e 147 and 148 and m/e 118 and 119. These ions correspond to the loss of ammonia and formic acid, respectively, and are fragments typically found in the CI mass spectra of amino acids (10). A low intensity ion at m/e 88 corresponds to the elimination of isopropylmercaptan and isopropyl-[$2-^2H$]-mercaptan from the quasimolecular ion. This evidence suggests that S-isopropylcysteine is the product formed from a reactive metabolite of isopropylhydrazine and cysteine and that the C-2 hydrogen is retained, thus eliminating acetone as the intermediate.

DISCUSSION

Postulated mechanism for hydrazine toxicity. Previous work (11) has provided evidence that acetylhydrazine is the metabolite of ioniazid and that isopropylhydrazine is the metabolite of

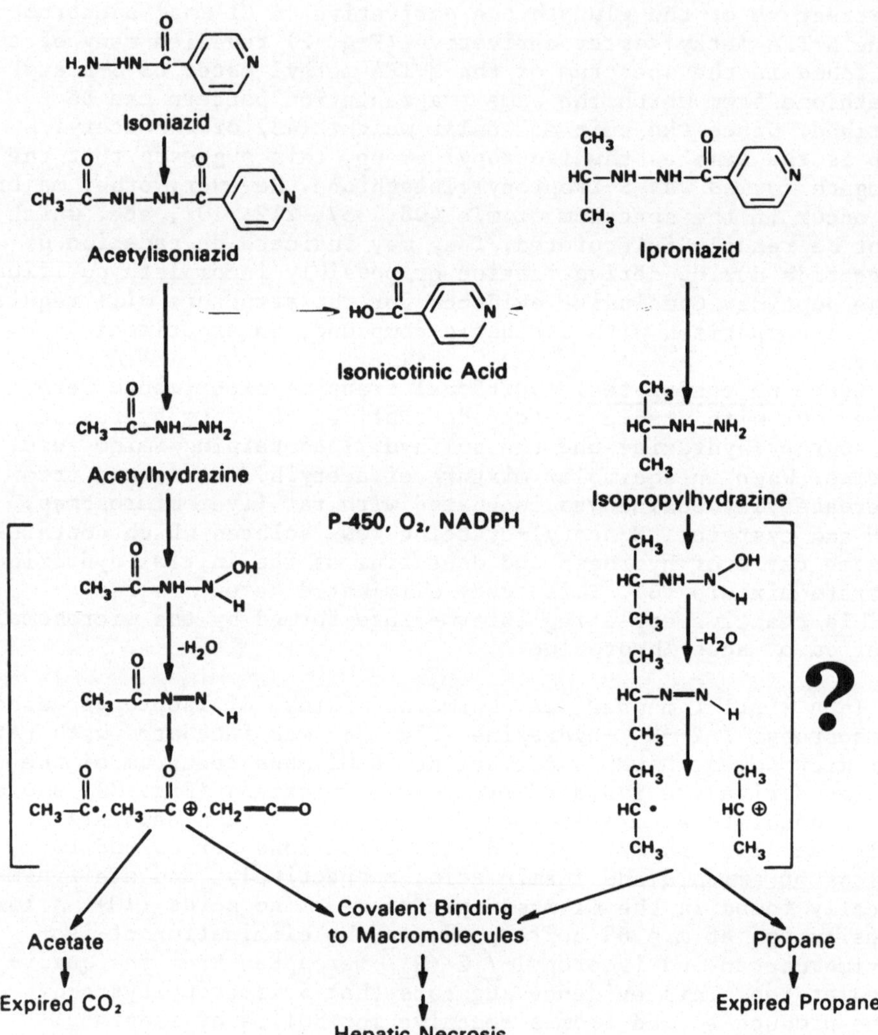

FIG. 4. Postulated reaction scheme for the metabolic activation of isoniazid and iproniazid.

iproniazid that are responsible for the hepatic necrosis caused by these drugs in man (Fig. 4). These metabolites are oxidized by a microsomal P-450 enzyme to reactive intermediates that covalently bind to tissue macromolecules (12). Furthermore the covalent binding of acetylhydrazine is paralleled by carbon dioxide production, reflecting the formation of a reactive acylating agent, and the binding of isopropylhydrazine is paralleled by the evolution of propane, reflecting the formation of a reactive alkylating agent.

Glutathione conjugates. This paper has described the use of chemical ionization mass spectrometry in the analysis of the glutathione and cysteine conjugates of reactive intermediates formed by the microsomal oxidation of these hydrazine metabolite. Although Polan et al. (13) have described EI mass spectra of permethylated glutathione, the sequence-determining ions were of relatively low intensity compared to the ions we observed in the CI mass spectra of the N-TFA-methyl esters. We also determined the CI and EI spectra of the silyl derivatives of the glutathione conjugates as reported by Hoppen et al. (14) for 2-hydroxyestrogens. Although these spectra revealed some information about the molecular weight of the conjugate and losses of silyl groups, they were uninformative in terms of the backbone peptide structure. We, therefore, prefer using the TFA-methyl esters even though some degradation occurs during the derivatization procedure.

These initial studies on the analysis of glutathione conjugates of reactive intermediates in drug metabolism are now being expanded in several directions. First, we are attempting the purification and analysis of reactive intermediates formed during the metabolism of other drugs such as acetaminophen (paracetamol) and furosemide (frusemide). Secondly, to verify the proposed fragmentation pathways for the glutathione conjugates, we are preparing glutathione specifically labeled with stable isotopes in the constituent amino acids. High resolution CI mass spectrometry coupled with metastable defocusing will complement this study. Finally, we hope to apply this technique to ensure specificity in the analysis of conjugates isolated from in vitro reactions as well as those isolated from plasma and bile samples, since the analysis of such conjugates will be necessary for determining the role of enzymes such as glutathione transferases in detoxification mechanisms.

Cysteine conjugates and the twin-ion technique. In addition to studies of glutathione conjugates, we have utilized the twin-ion technique to study reaction mechanisms in reactive metabolite formation from acetylhydrazine and isopropylhydrazine. This technique has proven to be a useful tool when applied to biotransformation studies. Hammar and Holmstedt (15) used the naturally occurring isotopic doublet of chlorine to study the metabolism of chlorpromazine. Morfin et al. (16) were the first to use stable isotope standards in studies of the pathways of testosterone metabolism and several other stable isotope studies have followed. Recently, Pohl *et al.* (17) coupled the twin-ion technique with chemical ionization mass spectrometry to determine a new pathway of warfarin metabolism. In the present studies this technique again proved useful because no derivatization was required to obtain good spectra by chemical ionization mass spectrometry of the cysteine derivatives formed in the trapping experiments. In addition, the isotopic doublets formed by the deuterium containing derivatives enabled us: 1) to establish the structure of the cysteine conjugates formed in the microsomal incubations containing NADPH, oxygen, cysteine and acetylhydrazine or isopropylhydrazine; 2) to determine that the conjugates arise from the substrates without apparent loss of deuterium, which allowed us to eliminate certain reaction pathways in the formation of the reactive intermediates.

Thus, mass spectrometry is a powerful technique for studying not only drug metabolite detoxification pathways but also those pathways which lead to metabolic activation of the parent drug to chemically reactive, hepatotoxic intermediates.

REFERENCES

1) E.C. Miller and J.A. Miller, Pharmacol. Rev., 1966, 18, 805.
2) P.N. Magee and J.M. Barnes, Adv. Canc. Res., 1967, 10, 163.
3) J.R. Mitchell, D.J. Jollow, J.R. Gillette and B.B. Brodie, Drug Metab. Dispos., 1973, 1, 418.
4) J.R. Mitchell and D.J. Jollow, Israel J. Med. Sci., 1974, 10, 339.
5) J.R. Gillette, J.R. Mitchell and B.B. Brodie, Annu. Rev. Pharmacol., 1974, 14, 271.
6) S.D. Nelson, J.A. Hinson and J.R. Mitchell, Biochem. Biophys. Res. Commun., 1976, 69, 900.
7) W.Z. Potter, D.C. Davis, J.R. Mitchell, D.J. Jollow, J.R. Gillette and B.B. Brodie, J. Pharmacol. Exp. Ther., 1973, 187, 203.
8) G.M. Schier, D. Halpern and G.W.A. Milne, Biomedical Mass Spectrom., 1974, 1, 212.
9) A.A. Kiryushkin, H.M. Fales, T. Axenrod , E.J. Gilbert and G.W.A. Milne, Org. Mass Spectrom., 1971, 5, 19.
10) G.W.A. Milne, T. Axenrod and H.M. Fales, J. Amer. Chem. Soc., 1970, 92, 5170.
11) J.R. Mitchell, U.P. Thorgeirsson, M.A. Black, J.A. Timbrell, W.R. Snodgrass, W.Z. Potter, D.J. Jollow and H.A. Keiser, Clin. Pharmacol. Ther., 1975, 18, 70.
12) S.D. Nelson, J.R. Mitchell, J.A. Timbrell, W.R. Snodgrass and G.B. Corcoran, Science, in press, 1976.
13) M.L. Polan, W.J. McMurray, S.R. Lipsky and S. Lande, Biochem. Biophys. Res. Commun., 1970, 38, 1127.

14) H.O. Hoppen, L. Siekmann and H. Breuer, Z. Anal. Chem., 1970, 252, 299.
15) C.G. Hammar and B. Holmstedt, Anal. Biochem., 1968, 25, 532.
16) R.F. Morfin, I. Leav , P. Ofner and J.C. Orr, Fed. Proc., 1970, 29, 247.
17) L.R. Pohl, S.D. Nelson, W.A. Garland and W.F. Trager, Biomed. Mass Spectrom., 1975, 2, 23.

a) Staff Fellow in the Pharmacology Research Associate Program of the National Institute of General Medical Sciences, Bethesda, Maryland 20014.

RELATIONSHIP BETWEEN STRUCTURE AND METABOLISM OF TRICYCLIC DRUGS: STABLE EPOXIDE-DIOL PATHWAY

A. Frigerio and C. Pantarotto

Istituto Ricerche Farmacologiche "Mario Negri"

Milan, Italy

INTRODUCTION

Aromatic and olefinic compounds can be metabolized by mammalian microsomal monooxygenases to arene oxides and epoxides respectively. These oxygenated species represent electrophilic intermediates in the formation of phenols and dihydrodiols, which are considered the real detoxification products of these substances (1, 4).

In order to explain the toxic, carcinogenic and mutagenic effects of these "bioactivated intermediates" many factors have to be taken into consideration: (i) the rate of formation, (ii) the high reactivity towards nucleophilic agents such as DNA, RNA and proteins and (iii) sufficient stability to reach target sites in the organism (5, 8). It becomes therefore important to identify the presence of these intermediates and to study their effects in the different sites where their toxic action may occur.

As far as the direct identification of these epoxide metabolites is concerned, it has to be pointed out that for polycyclic hydrocarbons there is some direct evidence for the formation of relatively stable epoxides in the "K" region of these molecules (9); however these data were obtained with microsomal preparations (10) or from tissue homogenates (11, 12). For other substances such as alkylbenzenes (13) and drugs like diphenylhydantoin, glutethimide and antipyrine (14, 16), the high chemical reactivity of the intermediate epoxide precluded its isolation. However the chemical nature of the metabolites identified can only be reasonably explained assuming the intermediacy of an epoxide.

The concern that clinically useful drugs containing aromatic and/or olefinic groups may be transformed to bioactivated intermediates led us to examine the metabolism of several tricyclic compounds used in psychopharmacology. The drugs that have been chosen for this study belong to two different classes of compounds: the derivatives of 5H-dibenzo(a,d)cycloheptene and the derivatives of 5H-dibenzo(b,f)azepine as reported in Scheme 1.

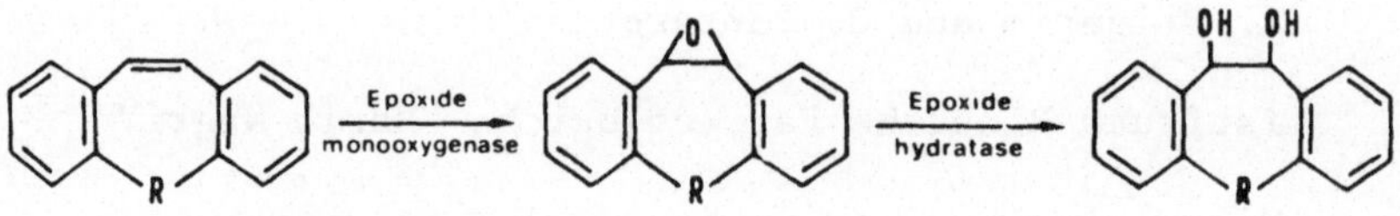

<u>5H-dibenzo(a,d)cycloheptene and its derivatives:</u>

R= $>CH_2$	5H-dibenzo(a,d)cycloheptene
R= $>CHCONH_2$	Cytenamide
R= $>CHCH_2CH_2CH_2NHCH_3$	Protriptyline
R= $>C{=}CHCH_2CH_2N(CH_3)_2$	Cyclobenzaprine
R= $>C{=}CHC{\equiv}CCH_2N(CH_3)_2$	Intriptyline
R= $>C{=}C\quad NCH_3$	Cyproheptadine

<u>5H-dibenzo(b,f)azepine and its derivatives:</u>

R= $>NH$	5H-dibenzo(b,f)azepine
R= $>NCONH_2$	Carbamazepine
R= $>NCH_2CH_2CH_2N\quad NCH_2CH_2OH$	Opipramol

Scheme 1. Major sequence of biotransformation of the tricyclic drugs.

Biotransformation studies of these compounds were carried out both "<u>in vitro</u>", by incubation with rat liver microsomes, and "<u>in vivo</u>", in rat and, for clinically used drugs, also in man. The original hypothesis of the involvement of an "epoxide-diol" metabolic pathway common to all the listed drugs has been confirmed be means of chromatographic and mass spectrometric techniques (17, 22).

As an example of the procedure that we used for the identification of this oxidative metabolic step, we report here the study of the metabolism of 5H-dibenzo(a,d)cycloheptene and

5H-dibenzo(b,f)azepine, the parent compounds of the two classes of drugs (23, 24).

ANALYTICAL ASPECTS

The first approach involved the use of gas chromatography in order to obtain better separation of the compounds of interest from endogenous material.

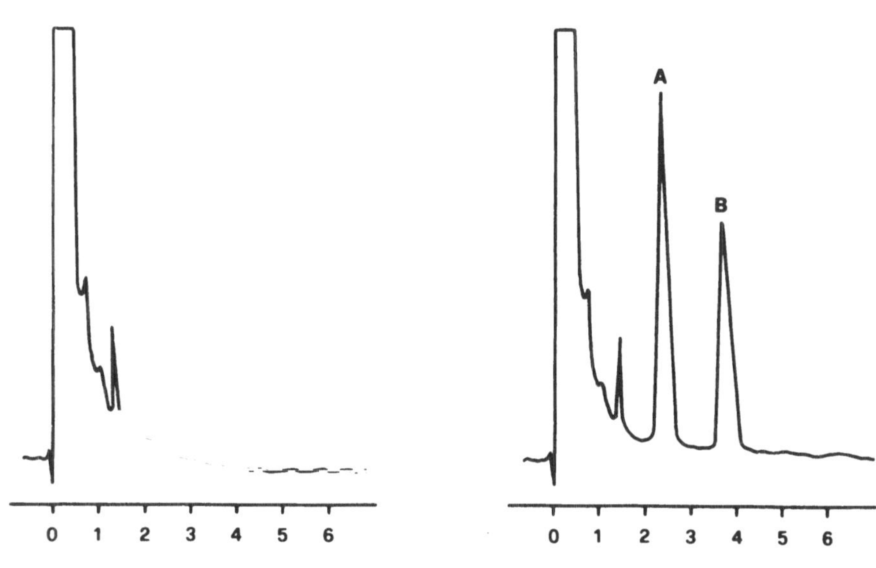

FIG. 1. Gas chromatograms of urine extracts. On the left, the chromatogram corresponding to the extract of control urine, and on the right, the chromatogram corresponding to the urine extract of rats treated with 5H-dibenzo (a,d)cycloheptene.
Peak A: 5H-dibenzo(a,d)cycloheptene.
Peak B: 5H-dibenzo(a,d)cycloheptene-10,11-epoxide.

In Fig. 1 are shown the gas chromatograms of the urine extracts of rats that had been treated intraperitoneally with a single dose of 5H-dibenzo(a,d)cycloheptene. Two gas chromatographic

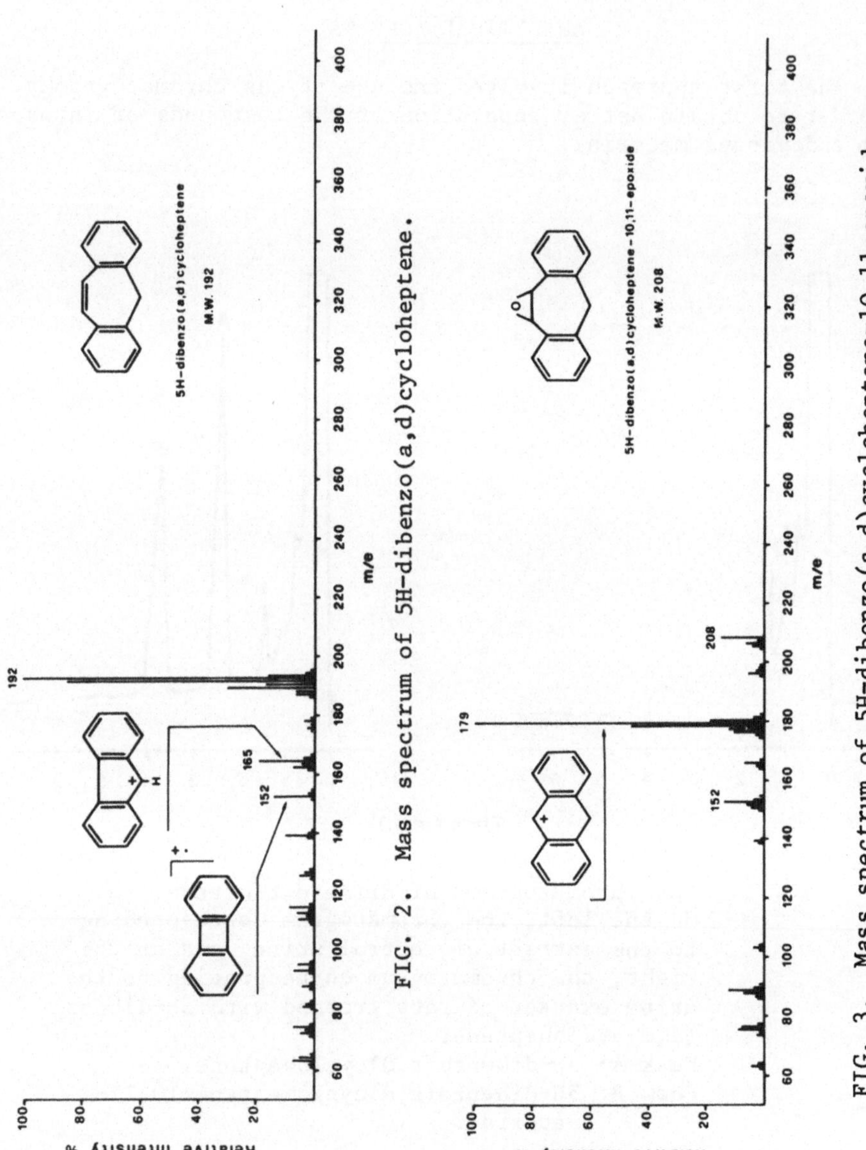

FIG. 2. Mass spectrum of 5H-dibenzo(a,d)cycloheptene.

FIG. 3. Mass spectrum of 5H-dibenzo(a,d)cycloheptene-10,11-epoxide.

peak (A,B) not present in the urine of the control animals were observed. Mass spectrometric analysis of peak (A) gave the spectrum reported in Fig. 2. The molecular ion at m/e 192 and the occurrence of peaks at m/e 165 and 152, were consistent with the structure of the unchanged 5H-dibenzo(a,d)cycloheptene.

The mass spectrum of peak (B) is reported in Fig. 3. The molecular ion at m/e 208 (an increase of 16 mass units in comparison to the spectrum of the parent compound) and the subsequent loss of 29 mass units from the molecular ion to give the base peak at m/e 179, suggested the presence of an epoxide at the 10,11-position. The loss of 29 amu has been rationalized as shown in Scheme 2. This type of rearrangement followed by the loss of 29 mass units was common to all the epoxide metabolites studied and therefore it has been considered as one of the more characterictic parameters for the structural elucidation of these intermediates.

The diol metabolites were also identified by gas chromatographic and mass spectrometric techniques, following derivatization with n-butyl boronic acid, a specific reagent for compounds containing vicinal hydroxy groups (25). This derivatization reaction was performed by co-injecting the reagent and the urine extracts at a relatively higher column temperature. In Fig. 4 are shown the gas chromatograms of the urine extracts of both control and treated rats obtained after the co-injection with n-butyl boronic acid. A peak (C) was observed, not present in the chromatogram corresponding to the urine extracts of the control animals. The nature of the compound giving this peak was examined by mass spectrometry: the spectrum is shown in Fig. 5. The molecular ion and base peak at m/e 292 and the occurrence of the ions at m/e 235, 207, 191 and 179 arising as shown on the spectrum, confirmed the compound as the butyl boronate of the 10,11-dihydro-10,11--dihydroxy-5H-dibenzo(a,d)cycloheptene.

The metabolism of the other "model compound", the 5H-dibenzo (b,f)azepine was found to occur by a similar "epoxide-diol" pathway. The fragmentation pattern of this compound and its derivatives under electron impact was interesting. The mass spectrum of 5H-dibenzo(b,f)azepine-10,11-epoxide (Fig. 6), shows the characteristic loss of 29 mass units from the molecular ion to form the ion at m/e 180 and loss of dihydrogen cyanide radical species from the molecule, a loss that in general has been described as occurring <u>via</u> two distinct processes (26). The ion at m/e 180 in the spectrum of 5H-dibenzo(b,f)azepine-10,11-epoxide loses a H_2CN radical as has been demonstrated by high resolution measurements and metastable defocusing data. Scheme 3 shows a possible mechanism rationalizing a rearrangement of the protonated acridinium ion to the benzoquinoline species. 5H-Dibenzo(b,f) azepine dihydrodiol has been identified in the urine of the treated rats as the butyl boronate derivative. The mass spectrum (Fig. 7)

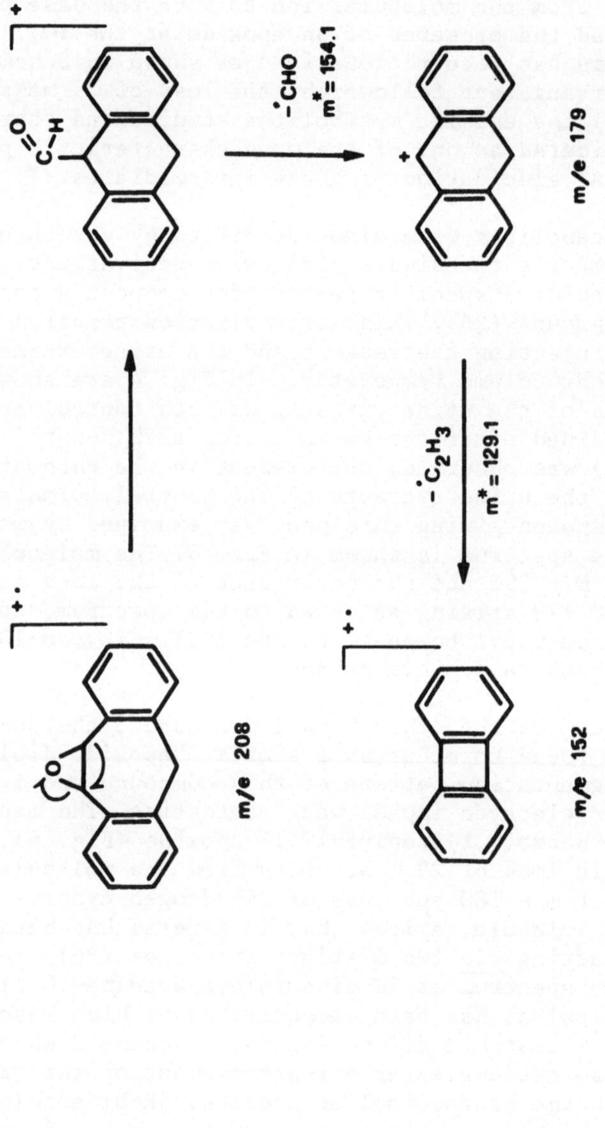

SCHEME 2. Fragmentation pathway of 10,11-epoxides.

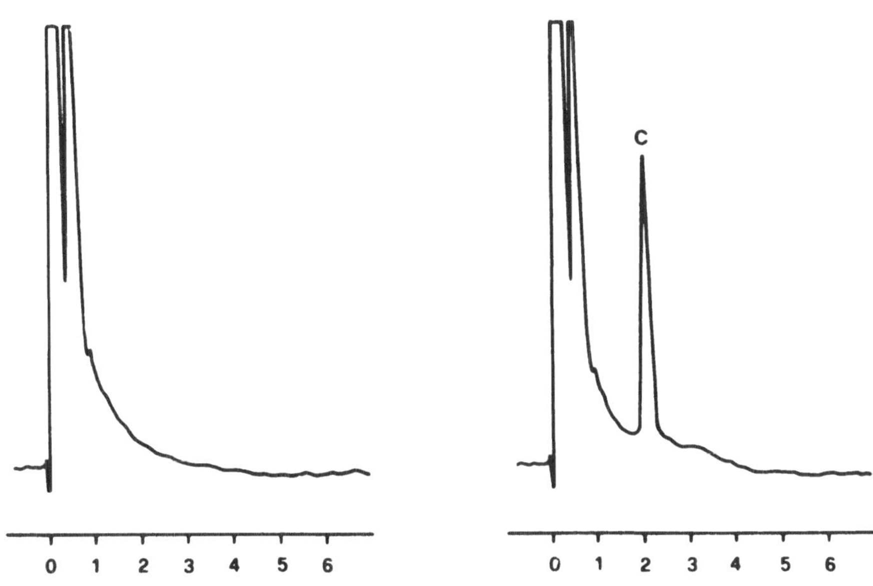

FIG. 4. Gas chromatograms of urine extracts of both control (left) and 5H-dibenzo(a,d)cycloheptene treated rats (right) obtained after co-injection with n-butyl boronic acid.
Peak C: 10,11-dihydro-10,11-dihydroxy-5H--dibenzo(a,d)cycloheptene-butyl-boronate.

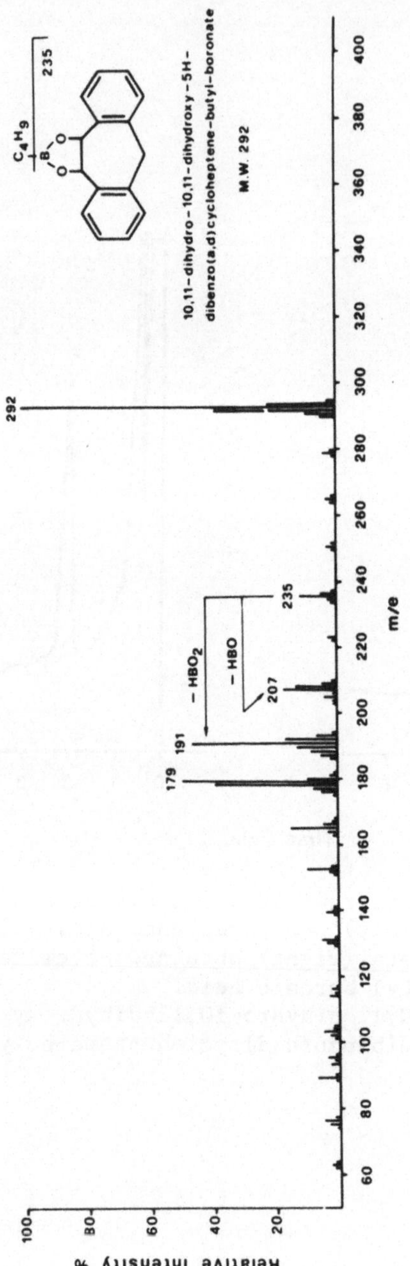

FIG. 5. Mass spectrum of 10,11-dihydro-10,11-dihydroxy-5H-dibenzo (a,d)cycloheptene-butyl-boronate.

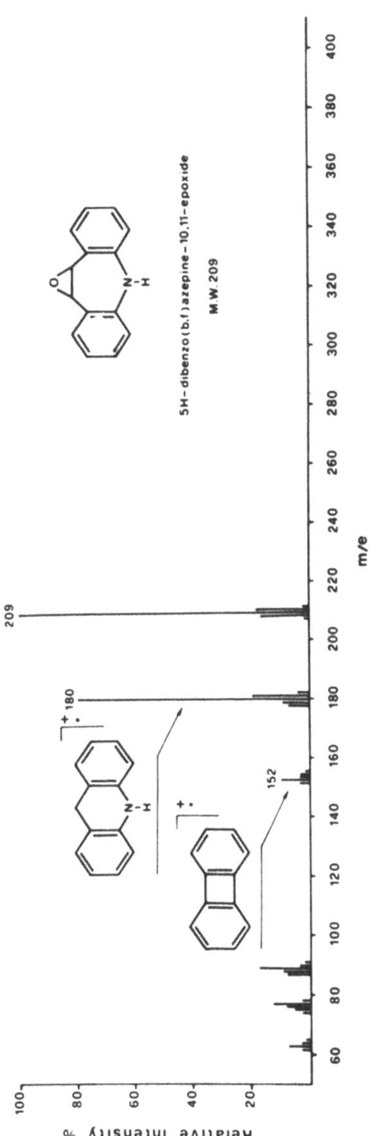

FIG. 6. Mass spectrum of 5H-dibenzo(b,f)azepine-10,11-epoxide.

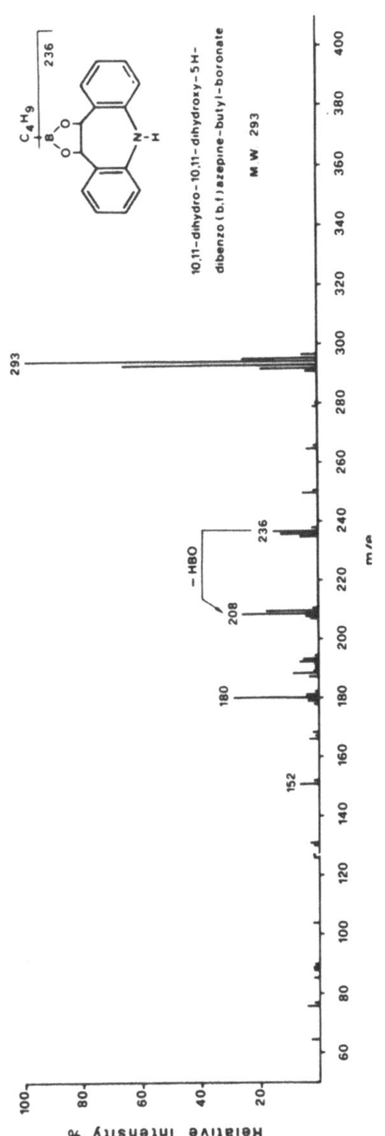

FIG. 7. Mass spectrum of 10,11-dihydro-10,11-dihydroxy-5H-dibenzo-(b,f)azepine-butyl-boronate.

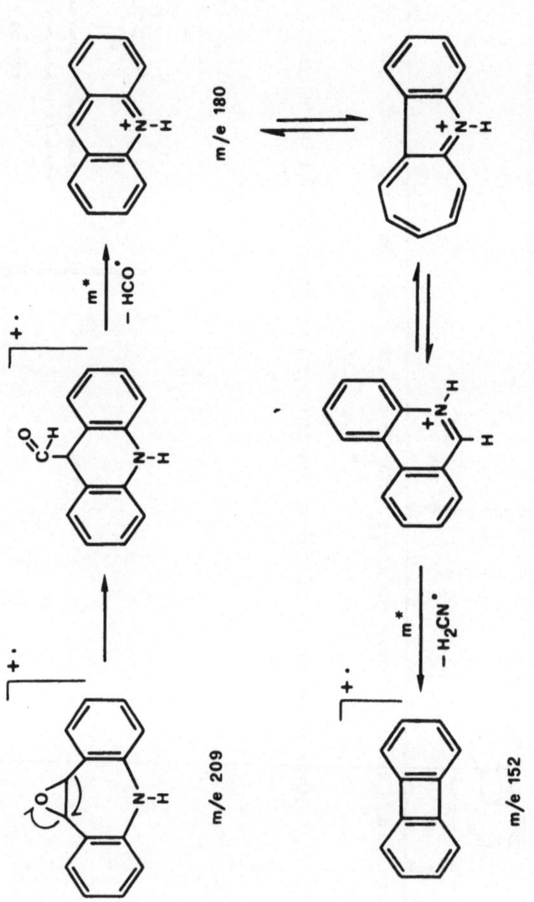

SCHEME 3. Loss of H_2CN from 5H-dibenzo(b,f)azepine derivatives.

shows the molecular ion at m/e 293 and other ions at m/e 236, 208, 180 and 152 that are consistent with the proposed structure. In the spectrum, the loss of 28 mass units from the ion at m/e 180 to form the ion at m/e 152 can be attributed to the loss of a dihydrogen cyanide radical species.

That these epoxides were not artifacts formed from the parent drugs, and the stability of the chemically synthetized epoxides was investigated by incubating the drugs and the epoxides for 48 hours at 37°C with rat urine. No degradations of the drugs or chemical modifications of the epoxides were found under these conditions. Furthermore, to ascertain that the formation of these metabolites was due to an enzymatic process, "in vitro" experiments were performed by incubating the drugs with rat liver microsomes under various experimental conditions as summarized in Table 1.

TABLE 1. Incubation of the drugs, under various experimental conditions with rat liver microsomes, showing the enzymatic formation of epoxide and diol metabolites.

Experimental conditions	Epoxide	Diol
Microsomes + cofactors	+	+
Denatured microsomes + cofactors	-	-
Microsomes + cofactors - NADP	-	-
Cofactors only	-	-

With inactivated microsomes, after boiling, or in the absence of NADP in the incubation mixture, epoxidation and successive formation of the dihydrodiols was not observed, a further confirmation for the absence of chemical artifacts.

In general, for the "in vitro" metabolism, the metabolites were identified by multiple ion detection mass fragmentography. Typical mass fragmentograms of extracts from microsomal preparations are shown in Fig. 8, 9. This technique was used since the dihydroxy derivatives were, in general, present only in trace amounts; the 10,11-epoxides seem to have a relatively poor affinity as epoxide hydratase substrates (27).

This, in general, has been our methodological approach in the study of the metabolism of the above mentioned classes of compounds. As we reported, the procedures used were similar in all our studies that were performed by chromatographic and spectrometric techniques. Particular attention was paid to the gas chromatographic behaviour of some of these drugs and their metabolites, in relation to column decomposition and, hence, misidentification, Carbamazepine, for instance, degrades

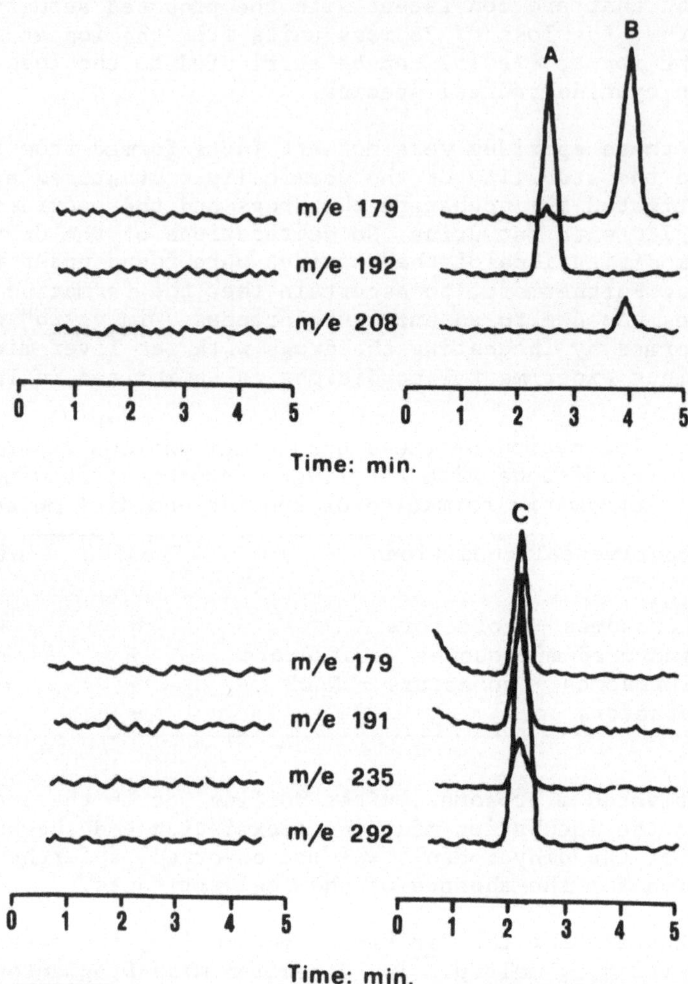

FIG. 8. Mass fragmentograms of extracts from microsomal preparations. On the left, extracts from control microsomes and on the right, extracts from microsomes incubated with 5H-dibenzo(a,d) cycloheptene.
Peak A: 5H-dibenzo(a,d)cycloheptene.
Peak B: 5H-dibenzo(a,d)cycloheptene-10,11-epoxide.
Peak C: 10,11-dihydro-10,11-dihydroxy-5H-dibenzo (a,d)cycloheptene-butyl-boronate.

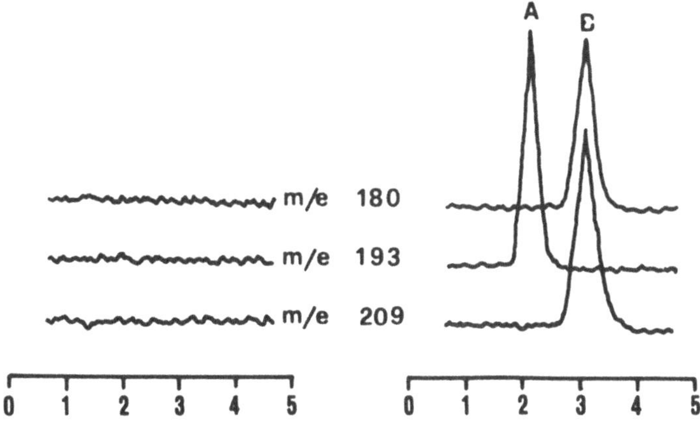

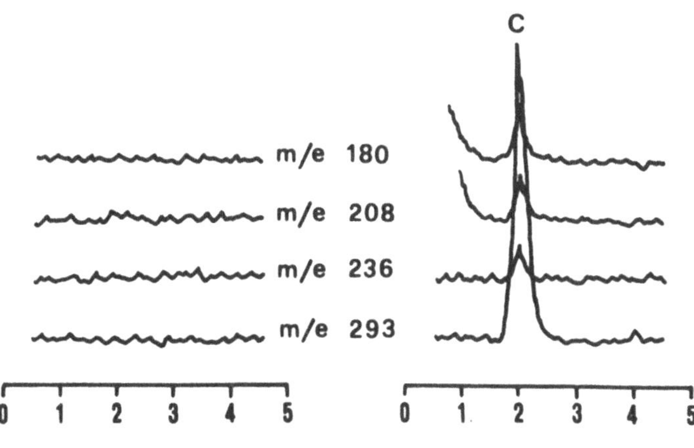

FIG. 9. Mass fragmentograms of extracts from microsomal preparations. On the left, extracts from control microsomes and on the right, extracts from microsomes incubated with 5H-dibenzo(b,f) azepine.
Peak A: 5H-dibenzo(b,f)azepine.
Peak B: 5H-dibenzo(b,f)azepine-10,11-epoxide.
Peak C: 10,11-dihydro-10,11-dihydroxy-5H-dibenzo(b,f)azepine-butyl-boronate.

(28) when injected as a methanol solution onto an OV 17 glass column (Fig. 10). Besides a large peak due to carbamazepine, there

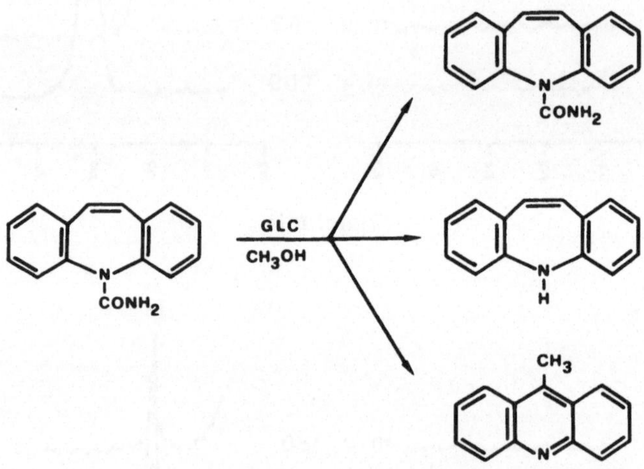

FIG. 10. Gas chromatographic degradation of carbamazepine to 5H-dibenzo(b,f)azepine and 9-methylacridine.

are two minor components of lower polarity, which have been identified as 5H-dibenzo(b,f)azepine and 9-methylacridine. Their structure was elucidated by mass spectrometry and by comparison of their mass spectra with those of authentic standards.

This degradation takes place only when carbamazepine is injected onto the column as a methanol solution; injection as an ethanol or acetone solution does not give any degradation reaction. The methanol present in the system, increases the acidity of the column, which produces the hydrolysis reaction to form iminostilbene and the extensive rearrangement to give 9-methylacridine. Carbamazepine-10,11-epoxide also decomposes when injected onto an

OV 17 glass column showing a clean gas chromatogram with only one component which has been identified as 9-acridine carboxyaldehyde (28). Fig. 11 shows a mechanism rationalizing this

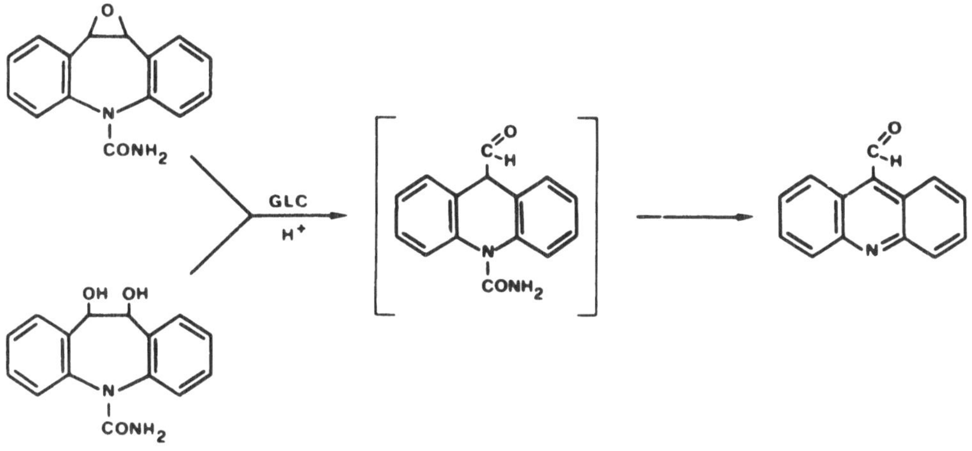

FIG. 11. Mechanism for the acid-catalyzed rearrangements of carbamazepine-10,11-epoxide and 10,11-dihydro--10,11-dihydroxy-carbamazepine, during the GC analysis, to give 9-acridinecarboxyaldehyde.

rearrangement during the gas chromatographic analysis; the process is probably catalyzed by the slightly acid column at high temperature, the intermediate amide then decomposes to give the more stable, fully conjugated system, 9-acridine carboxyaldehyde, 10,11-dihydro-10,11-dihydroxy carbamazepine also undergoes a degradation to the same 9-acridine carboxyaldehyde; this rearrangement to give the aldehyde is an example of the pinacol type. In Fig. 12 is shown another example of column degradation, referred to the intramolecular cyclization of 10,11-dihydro-10,11--dihydroxy-cytenamide to form cytenamide-*syn*-hydroxylactone (22).

These results indicate the care that must be taken in identifying these metabolites; often the quantities available for identification may be very small and gas chromatography-mass spectrometry must be used, however alternative methodologies should be employed in conjunction, wherever possible, to check that no

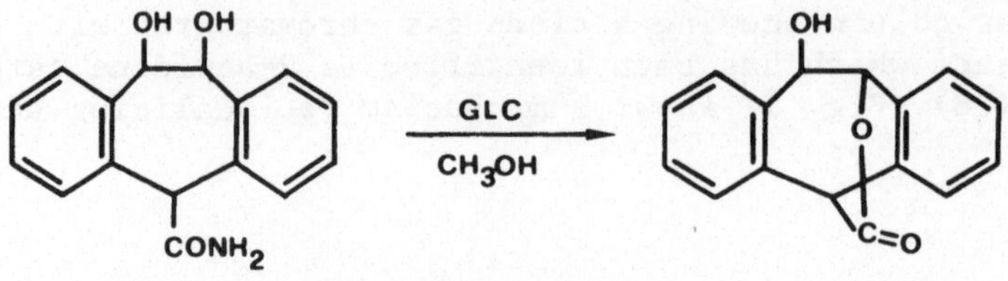

FIG. 12. Intramolecular cyclization of 10,11-dihydro--10,11-dihydroxy-cytenamide to cytenamide--*syn*-hydroxylactone.

rearrangements or degradations take place during one of the physical measurements.

For all the drugs studied the major biotransformation process has been found to be epoxidation of the 10,11-double bond, although other important metabolic pathways are involved and these include N-demethylation, N-oxidation and hydroxylation of the aromatic rings (18 -20, 29 -33). It can now be stated that epoxidation is a general biotransformation process for these compounds as is the conversion of the epoxides to the corresponding diols and their "*in vivo*" conjugation mainly with glucuronic acid.

BIOLOGICAL SIGNIFICANCE OF THE 10,11-EPOXIDES

The metabolism of the compounds studied shows the common feature of yielding epoxides which can be measured in blood and tissues and are excreted in the urine. Whether these metabolites maintain the pharmacological activity of the parent compound is therefore of great interest in the interpretation of the drug action.

Carbamazepine-10,11-epoxide, when injected into rats, crosses the blood-brain barrier similarly to carbamazepine; it is remarkable that the epoxide has the same anticonvulsant activity as carbamazepine (34). Table 2 shows a comparison of activity between carbamazepine and its epoxide on three different tests in mice; the convulsions induced by strychnine, pentylenetetrazol and electroshock. The epoxide was nearly as active as its parent compound, indeed with respect to convulsions induced by electroshock it was found even more active.

In Table 3 are reported the results of a study on the antihistamine and antiserotonin activity of cyproheptadine and cyproheptadine-10,11-epoxide (35). A comparison of the inhibitory activity of the two drugs was carried out on the enhanced vascular permeability induced by serotonin (5HT) or histamine (H) injected intradermally in rats. The antagonistic activity was expressed as percent inhibition of the amount of Evans blue, previously injected intravenously, that was found in the skin of treated animals as compared to control animals. The antiserotonin and antihistamine activity of cyproheptadine-10,11-epoxide did not differ from that of cyproheptadine. Also, there was no difference between the inhibition measured 30 minutes or 2 hours after the administration of the compounds. These epoxides seem therefore to have pharmacological importance since they may contribute to the total pharmacological activity of their parent compounds.

In the light of these results, the important question is now raised concerning the possibility that these epoxides may share the properties of other epoxides which are considered responsible for several toxic effects including carcinogenesis, teratogenesis and mutagenesis. The fact that these epoxides can be detected in the urine is indirect evidence that these chemical species are not highly reactive and therefore not available for the binding to macromolecules (proteins and nucleic acids) which is considered the basis of toxic effects. Moreover the LD_{50} of these epoxides was found similar to that of their parent compounds and a comparison between the drugs and the epoxides for their cytotoxic effects "in vitro" on three different strains of human cells, KB (carcinoma of the rhinopharinx), EUE (normal embryonic epithelium) and HeLa (uterine carcinoma) showed the epoxides to be not as toxic as the parent compounds (36). Table 4, as an example, reports the cytotoxicity "in vitro" of carbamazepine compared to that of the 10,11-epoxide. In this test, performed at concentrations ranging from 1 to 1000 μg/ml, carbamazepine-10,11--epoxide at higher concentrations was found even less toxic than the parent compound carbamazepine (34).

Both the drugs and the epoxide metabolites did not show cytotoxic activity "in vivo" on the survival of $B_6D_2F_1$ mice bearing leukemia L1210 at various doses and various schedules of treatment (37). A consequence of the high stability of these epoxides could

TABLE 2. Anticonvulsant effect of carbamazepine and its epoxide.

Compound dose (mg/kg/os)	Protection from convulsions by: Strychnine	Pentylenetetrazol	Electroshock
Carbamazepine			
5	-	-	0
10	-	-	0
20	0	-	100
25	20	40	-
50	80	80	-
100	80	90	-
200	100	100	-
300	100	100	-
400	-	-	-
Carbamazepine-10,11-epoxide			
5	-	-	20
10	-	-	33
20	-	-	100
25	0	60	-
50	0	80	-
100	40	100	-
200	60	100	-
300	100	-	-
400	-	-	-

TABLE 3. Antihistamine and antiserotonine activity of cyproheptadine and cyproheptadine-10,11-epoxide at 30 minutes and 2 hours after i.p. injection.

No. of animals	Treatment	Dye extracted (µg) ± S.E. after			
		Histamine	%	5-HT	%
8	Controls	12.67 ± 1.4		18.21 ± 0.9	
4	Cyproheptadine 30' 800 µg/kg	2.73 ± 0.2	21.5	4.30 ± 0.29	23.6
4	Cyproheptadine 120' 800 µg/kg	2.33 ± 0.3	18.4	4.98 ± 0.5	27.3
4	Cyproheptadine epoxide 30' 800 µg/kg	2.09 ± 0.26	16.5	3.08 ± 0.18	16.9
3	Cyproheptadine epoxide 120' 800 µg/kg	2.34 ± 0.3	18.5	6.04 ± 0.5	33.2

TABLE 4. "*In vitro*" cytotoxicity of carbamazepine and its epoxide in mice.

Compound and concentration (μg/ml)	Effect on the cell strains					
	KB		EUE		HeLa	
	A	B	A	B	A	B
Carbamazepine						
1	147	0	220	0	146	0
10	117	0	210	0	115	0
100	0	23	109	0	48	0
1,000	0	42	0	11	0	29
Carbamazepine-10,11-epoxide						
1	95	0	117	0	37	0
10	114	0	137	0	92	0
100	98	0	100	0	91	0
1,000	76	0	0	19	1	0

A = % of cell growth (controls = 100).
B = % of cell destruction (controls = 0).

TABLE 5. Effects of parent compounds and epoxide metabolites on the number of His^+ revertant colonies.

Compound	Amount (ug per plate)	His^+ revertant colonies per plate		
		TA 1535	TA 1537	TA 1538
None		21,24,24,26	13,18,23,24	16,18,18,23
Benzo(a)pyrene	2	20,24	21,24	18,21
	10	17,23	19,19	17,21
	50	20,26	23,28	19,19
Benzo(a)pyrene-4,5-oxide	0.025	26,29	36,40	15,17
	0.1	26,27	75,90	32,50
	0.25	24,26	245,260	67,80
	1	25,31	615,630	317,358
	2.5	16,25	850,870	740,890
Carbamazepine	30, 100	23(22-26)	19(17-21)	20(16-23)
Carbamazepine-10,11-epoxide	1-500	26(18-33)	18(21-25)	20(17-24)
Cyproheptadine	30, 100	23(12-32)	23(19-27)	24(19-26)
Cyproheptadine-10,11-epoxide	1-100	27(21-35)	17(5-23)	22(16-26)
Cyclobenzaprine	30, 100	26(24-27)	18(12-24)	20(17-23)
Cyclobenzaprine-10,11-epoxide	1-500	26(17-40)	18(11-24)	19(14-25)

be reflected in a lower level of mutational potential and there is evidence that these metabolites do not possess mutagenic activity unlike the 4,5 ("K" region) epoxide of benzo(a)pyrene towards some *Salmonella* strains (38).

In Table 5 are shown the results of mutagenicity tests using histidine-dependent *Salmonella typhinurium* strains. The tests were performed by scoring for revertant colonies able to grow on a histidine-poor medium. Strains which were sensitive to frameshift (TA 1537 and TA 1538) and substitution mutations (TS 1535) were used. Benzo(a)pyrene-4,5-oxide was extremely mutagenic for the frameshift strains, while the epoxide metabolites of the drugs: carbamazepine, cyproheptadine and cyclobenzaprine proved completely non-mutagenic even at doses 1000 fold higher than benzo(a)pyrene--4,5-oxide.

CONCLUSION

According to these data we think that these 10,11-epoxide metabolites could be real detoxification products of drugs biotransformation (39). Furthermore, the considerable stability of these epoxides and their relative non-reactivity as epoxide hydratase substrates, one of the most important enzymes involved in the metabolism of epoxides, makes these drugs a suitable tool for studying the monooxygenases responsible for epoxide formation (27).

Recently these epoxide metabolites have been found to possess epoxide hydratase inhibitory effect (Table 6) (40, 41).

TABLE 6. Inhibition of epoxide hydratase by metabolically produced epoxides. Styrene oxide was used as substrate.

Inhibitor	Inhibition (%)
Carbamazepine-10,11-epoxide	11
Cyproheptadine-10,11-epoxide	25
Cyclobenzaprine-10,11-epoxide	67

On the other hand, if not *per se*, these epoxides may lead to potentiation of toxic effects of epoxide metabolites derived from ubiquitously present pollutants, such as benzo(a)pyrene, by interaction with enzymes responsible for their further biotransformation. Since drugs will normally have to be present at much

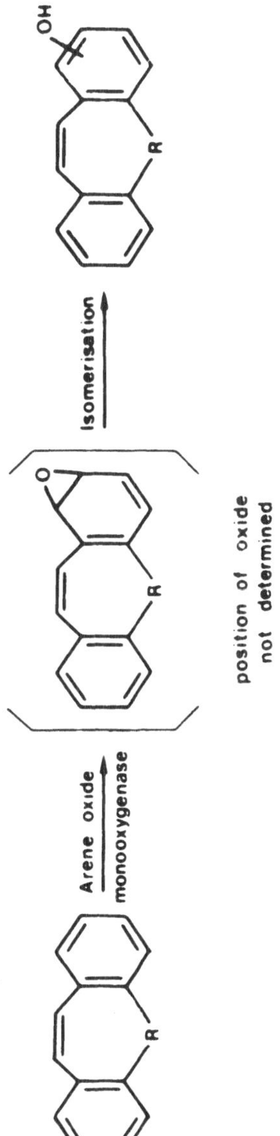

FIG. 13. Isomerization of arene oxides to phenols.

higher concentrations than are pollutants, they may lead to very effective competitive inhibition of processes which normally inactivate potentially harmful epoxide metabolites of polycyclic hydrocarbons.

All these compounds however possess other sites where epoxidation or other reactions leading to reactive metabolites could also take place.

Recently an irreversible protein binding of /^{14}C/-imipramine with rat and human liver microsomes has been demonstrated (42). The results support the concept that irreversible protein binding of imipramine is catalyzed by a cytochrome P-450 dependent hydroxylation of the aromatic ring *via* an epoxidation step. The similarity of this antidepressant with the drugs considered here suggests that also for our compounds hydroxylation of the aromatic rings could be related to the intermediacy of an oxide as shown in Fig. 13 (43). This could provide an explanation for the toxic effects that have been observed for some of these drugs, including pancreatic toxicity, hepatic necrosis and agranulocytosis (44, 45).

To conclude, binding to subcellular structures and structure-activity relationship for the drugs as substrates of epoxide synthetase and of the epoxides as substrates or inhibitors of epoxide hydratase (46), using hepatic microsomal preparations, are also under study to try to give us a clearer understanding of the mechanisms of action, formation and inactivation of these intermediates.

REFERENCES

1) J.W. Daly, D.M. Jerina and B. Witkop, Experientia, 1972, 28, 1129.
2) P.L. Grover, A. Hewer and P. Sims, Biochem. Pharmacol., 1972, 21, 2713.
3) G.R. Keysell, J. Booth, P. Sims, P.L. Grover and A. Hewer, Biochem. J., 1972, 129, 41P.
4) D.J. Jollow, J.R. Mitchell, N. Zampaglione and J.R. Gillette, Pharmacology, 1974, 11, 151.
5) D.M. Jerina and J.W. Daly, Science, 1974, 185, 573.
6) J.R. Gillette, Biochem. Pharmacol., 1974, 23, 2785.
7) M.J. Cookson, P. Sims and P.L. Grover, Nature New Biol., 1971, 234, 186.
8) T. Kuroki, E. Huberman, H. Marquardt, J.K. Selkirk, C. Heidelberger, P.L. Grover and P. Sims, Chem. Biol. Interactions, 1972, 4, 389.
9) P.L. Grover, A. Hewer and P. Sims, FEBS Lett., 1971, 18, 76.
10) P. Sims, Biochem. Pharmacol., 1970, 19, 795.

11) P.L. Grover, A. Hewer and P. Sims, Biochem. Pharmacol., 1974, 23, 323.
12) P.L. Grover, Biochem. Pharmacol., 1974, 23, 333.
13) G.J. Kasperek, T.C. Bruice, H. Yagi, N. Kaubisch and D.M. Jerina, J. Am. Chem. Soc., 1972, 94, 7876.
14) T. Chang, A. Savory and A.J. Glazko, Biochem. Biophys. Res. Commun., 1970, 38, 444.
15) W.G. Stillwell, M. Stafford and M.G. Horning, Res. Commun. Chem. Pathol. Pharmacol., 1973, 6, 579.
16) M. Stafford, G. Kellermann, R.N. Stillwell and M.G. Horning, Res. Commun. Chem. Pathol. Pharmacol., 1974, 8, 593.
17) A. Frigerio, R. Fanelli, P. Biandrate, G. Passerini, P.L. Morselli and S. Garattini, J. Pharm. Sci., 1972, 61, 1144.
18) A. Frigerio, N. Sassi, G. Belvedere, C. Pantarotto and S. Garattini, J. Pharm. Sci., 1974, 63, 1536.
19) G. Belvedere, V. Rovei, C. Pantarotto and A. Frigerio, Xenobiotica, 1975, 5, 765.
20) G. Belvedere, C. Pantarotto, V. Rovei and A. Frigerio, J. Pharm. Sci., 1976, 65, 815.
21) V. Rovei, G. Belvedere, C. Pantarotto and A. Frigerio, J. Pharm. Sci., 1976, 65, 810.
22) A. Frigerio, J. Lanzoni, C. Pantarotto, E. Rossi, V. Rovei and M. Zanol, Xenobiotica, 1976, in press.
23) C. Pantarotto, L. Cappellini, A. De Pascale and A. Frigerio, Xenobiotica, 1976, in press.
24) C. Pantarotto, P. Negrini, L. Cappellini and A. Frigerio, J. Chromatogr., 1976, in press.
25) C.J.W. Brooks and I. Maclean, J. Chromatogr. Sci., 1971, 9, 18.
26) K.M. Baker and A. Frigerio, J. Chem. Soc. (Perkin Trans. II), 1973, 648.
27) J. Pachecka, M. Salmona, L. Cantoni, E. Mussini, C. Pantarotto, A. Frigerio and G. Belvedere, Xenobiotica, 1976, in press.
28) A. Frigerio, K.M. Baker and G. Belvedere, Anal. Chem., 1973, 45, 1846.
29) J.W. Faigle, S. Brechbühler, K.F. Feldmann and W.J. Richter, in "Epileptic Seizures Behaviour Pain", by W. Birkmayer, Ed., Hans Huber Publishers, Wien, 1976, p. 127.
30) H.B. Hucker, A.J. Balletto, S.C. Stauffer, A.G. Zacchei and B.H. Arison, Drug Metab. Dispos., 1974, 2, 406.
31) C.C. Porter, B.H. Arison, V.F. Gruber, D.C. Titus and W.J.A. Vandenheuvel, Drug Metab. Dispos., 1975, 3, 189.
32) H.B. Hucker, A.J. Balletto, J. Demetriades, B.H. Arison and A.G. Zacchei, Drug Metab. Dispos., 1975, 3, 80.
33) G. Belvedere, V. Rovei, C. Pantarotto and A. Frigerio, Biomed. Mass Spectrom., 1974, 1, 329.
34) A. Frigerio and P.L. Morselli, in "Advances in Neurology", Vol. 11, J.K. Penry and D.D. Daly, Eds., Raven Press, New York, 1975, p. 295.
35) A. Bonaccorsi, R. Franco and P.L. Morselli, unpublished results.

36) L. Morasca, unpublished results.
37) F. Spreafico, unpublished results.
38) H.R. Glatt, F. Oesch, A. Frigerio and S. Garattini, Int. J. Cancer, 1975, 16, 787.
39) G. Belvedere, C. Pantarotto and A. Frigerio, Biomed. Mass Spectrom., 1975, 2, 115.
40) J. Pachecka, M. Salmona, G. Belvedere, L. Cantoni, E. Mussini and S. Garattini, Experientia, 1976, in press.
41) F. Oesch, P. Bentley and H.R. Glatt, in "Active Intermediates: Formation, Toxicity and Inactivation", J.R. Gillette, D. Jollow, J.J. Kocsis, H. Remmer, R. Snyder, H. Vainio and A. Hänninen, Eds., Plenum Publishing Company, New York, 1976, in press.
42) H. Kappus and H. Remmer, Biochem. Pharmacol., 1975, 24, 1079.
43) A. Frigerio and C. Pantarotto, J. Pharm. Pharmacol., 1976, 28, 665.
44) W.E. Crill, Ann. Intern. Med., 1973, 79, 844.
45) J.S. Wold and L.J. Fischer, J. Pharmacol. Exp. Ther., 1972, 183, 188.
46) F. Oesch, H. Thoenen and H. Fahrlaender, Biochem. Pharmacol., 1974, 23, 1307.

A STUDY OF PROTEIN METABOLISM IN CHRONIC-ENTEROCOLITIS PATIENTS AND THE EFFECT OF PHARMACOLOGICAL AGENTS USING MASS SPECTROMETRY

A.S. Belousov, N.A. Bredikhina and L.T. Zhuravlev

Centre of the Lenin Order

Moscow, USSR

INTRODUCTION

A study of the functions of the human small intestine, both normal and diseased, is rather difficult but is of great interest from the point of view of physiology and pathology. A number of investigations of the fundamental functions of the intestinal tract (digestive, absorbing, motor and excretory) have broadened and deepened our understanding of the importance of the intestinal tract in the organism. However many questions have remained unanswered, particularly that concerning the interrelation of the processes of hydrolysis,absorption and interstitial protein metabolism, and the effect of pharmacological agents on these processes under many pathological conditions. Early detection of disturbances of protein digestion, amino acid absorption and other metabolic disorders of protein metabolism is of diagnostic importance since not only do they complicate the disease but they are an important link in the pathological development of the process. The study of the absorption and metabolism of endogenous substances, amino acids, sugars and fats, which are rapidly metabolised is very difficult and requires the application of modern analytical techniques such as mass spectrometry. We have used this method in combination with the twin-ion technique to study some aspects of amino acid metabolism in patients with chronic-enterocolitis. Taking into account the fundamental clinical problem of our work, we have chosen the amino acid lysine, labelled with ^{15}N (Fig. 1) as a carrier of labelled nitrogen. In nature the ^{14}N and ^{15}N isotopes are in a strictly stable ratio and the MS determination of the ratio of these isotopes in biological materials (blood, urine and faeces) reflects the fate of the amino acid introduced, i.e. the process of absorption. The parallel determination of the

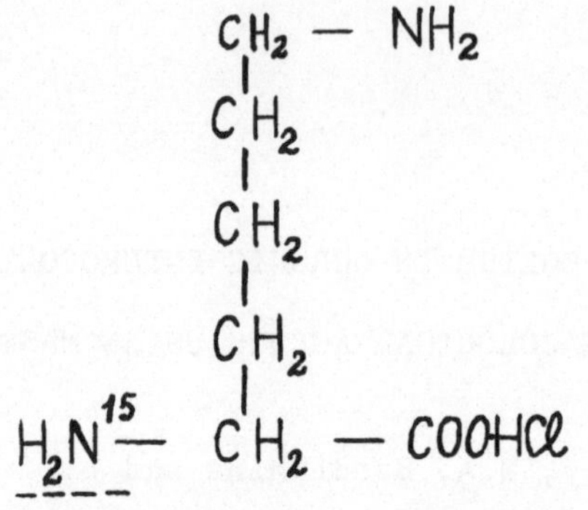

$$
\begin{array}{l}
\quad\quad\; CH_2 - NH_2 \\
\quad\quad\; | \\
\quad\quad\; CH_2 \\
\quad\quad\; | \\
\quad\quad\; CH_2 \\
\quad\quad\; | \\
\quad\quad\; CH_2 \\
\quad\quad\; | \\
H_2N^{15} - CH_2 - COOHCl
\end{array}
$$

FIG. 1. L - Lysine Hydrochloride (α,E -Diaminokapron acid).

label in blood, urine and faeces characterizes the functional condition of the small intestine.

METHODS

Subjects were given 20 mg of $[^{15}N]$-lysine perorally and then after 15 minutes and subsequently every 30 minutes blood samples were taken from the fingers of subjects during 2-3 hour periods. Samples of urine and faeces were taken during the first day after administration. The amount of biological material needed was such as to yield 0.7-1.0 mg of total nitrogen. Before MS analysis all biological samples were treated by the Rittenberg method (1948). The ^{15}N concentration in biological material was estimated using a mass spectrometer (MX 1303; USSR) and the results are expressed as atomic percentages (at %). Twenty healthy people and sixty chronic-enterocolitis patients were examined by the method developed. Over 2,000 MS analyses were carried out.

RESULTS AND DISCUSSION

The results of the investigation show that in healthy people the isotope concentration in blood increased during the first 90 minutes after oral administration of ^{15}N-lysine (Fig. 2). This was followed by a sharp fall until at 150 minutes the minimum isotope concentration was noted in all cases. The general pattern for the changes in ^{15}N level was the same in all healthy subjects. The concentrations during the first 90 minutes essentially reflect the absorption of the labelled amino acid, while the subsequent values represent its utilization in the biochemical processes

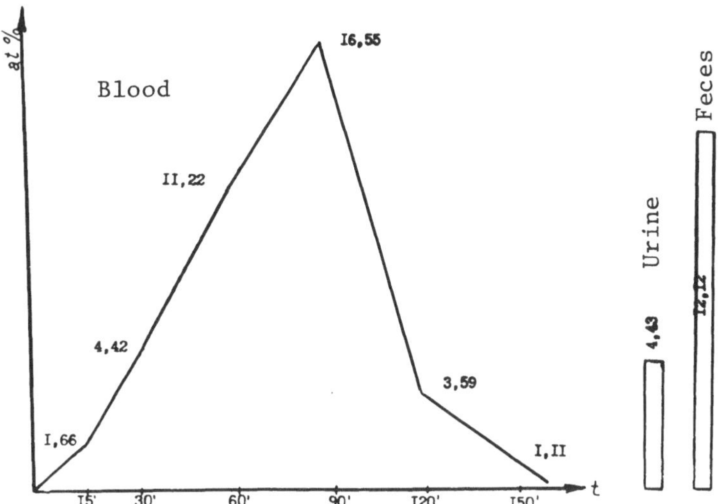

FIG. 2. Dynamics of isotope ^{15}N concentration in blood and the daily excretion of the isotope with urine and faeces in healthy people.

of the organism. However in patients with chronic-enterocolitis a different pattern for isotope concentration was observed after oral administration of labelled lysine. The maximum level reached was lower than in healthy people, the time to reach peak concentration was different and the amount excreted in the urine and faeces during the first day was increased. The extent of these differences correlates with the severity of the disease (Fig. 3). It is worth noting that the maximum isotope concentration is recorded at 90 minutes in both healthy people and patients in the first stage of enterocolitis whereas it occurs at 60 and 30 minutes in patients at the second and third stages respectively. It is possible that this shift is due to the rapid uptake and metabolism of the amino acid by protein deficient tissues. In addition extensive isotope excretion is noted in the urine of diseased patients. In the first day healthy people excrete in urine 27.3% of the maximum isotope concentration recorded in blood whereas patients at the first stage of chronic-enterocolitis excrete 44.7%, those at the second stage 71.2% and those in the third stage 99.8%. This fact probably reflects either the lower utilization of the amino acid introduced for protein biosynthesis and its greater-decomposition or the predominance of protein catabolism especially in patients with pronounced symptoms of chronic lesion of the small intestine.

The use of both mass spectrometry and a labelled amino acid

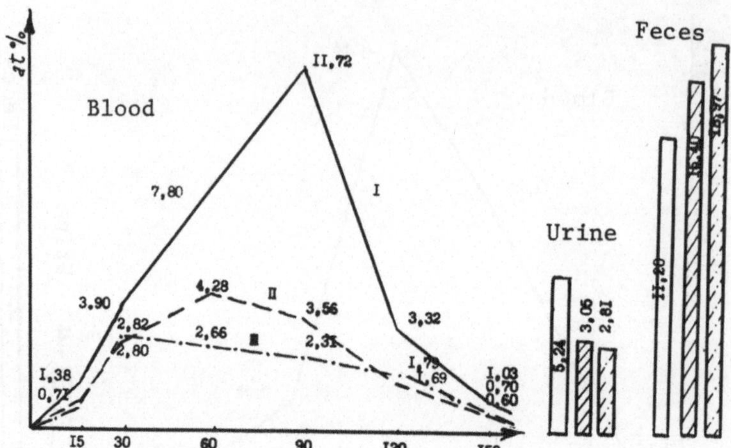

FIG. 3. Dynamics of isotope ^{15}N concentration in blood and the daily excretion of the isotope with urine and faeces in chronic-enterocolitis patients. Notations: I, II, III Extent of seriousness.

allowed us to carry out different investigations in healthy people and patients with chronic- enterocolitis. Thus the effect of insulin on both groups perorally treated with a complete set of amino acids was studied. The results show that a balanced set of amino acids considerably improves absorption and metabolism of amino acids favourably affecting the course of the pathological process (Fig. 4). Mass spectrometry was used to study the effect

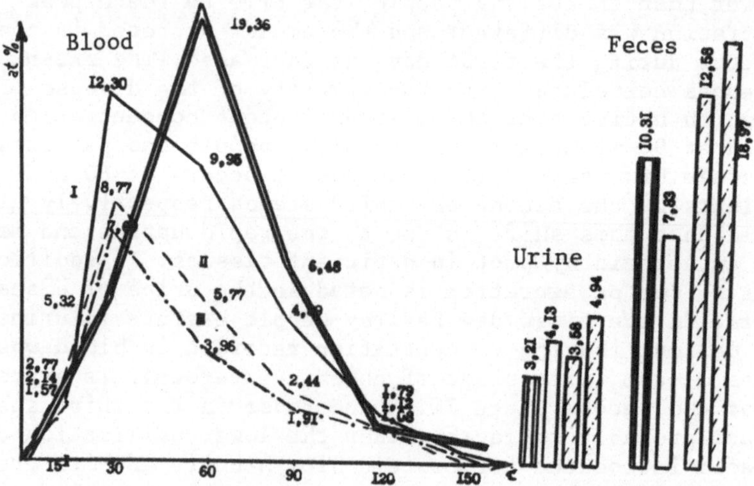

FIG. 4. Insulin effect on the absorption process of lysine ^{15}N in grastrointestinal tract in chronic-enterocolitis patients. Notations: ━ Healthy people; Chronic-enterocolitis patients I, II, III.

of enzyme preparations on protein hydrolysis and amino acid absorption. For this purpose we have undertaken an investigation where labelled lysine was injected into a piece of boiled chicken egg white. In the gastroenteric tract the chicken egg white is subject to the action of digestive juices, the result being the liberation of the labelled amino acid which, in combination with other amino acids formed during hydrolysis, is absorbed in the small intestine.

The investigations have shown that in all chronic-enterocolitis patients at the second and third stages of the disease one observed a decrease of the level of isotope concentration in blood within 15-30 minutes and a reduction of the maximum ^{15}N level, especially in patients of the third stage of the disease (Fig. 5).

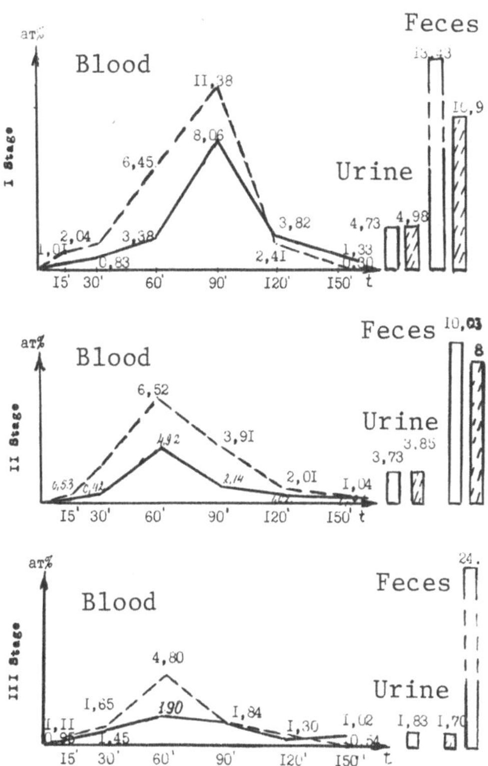

FIG. 5. Effect of enzyme preparation (Mexasae) on the processes of protein hydrolysis and absorption of amino acids in the gastrointestinal tract of chronic-enterocolitis patients.
—— Egg-white + lysine ^{15}N.

These changes are indicative of the impairment of absorption of labelled lysine in the gastroenteric tract of chronic-enterocolitis patients when given the whole egg white. This fact seems to be explained by the disturbances of the processes of the enzymes involved in protein hydrolysis, because of that the yield of labelled lysine into the intestinal lumen is insufficient. The disturbance of protein hydrolysis is likely to be indirect evidence of the enzyme deficiency which aggravates the disturbances of absorption still more.

But while in all patients at the second and third stages of the disease these features are observed to different extents, in 50% of the patients in the first stage of the disease the differences in the changes of the label concentration in blood are absent when only lysine or whole protein are given. This indicates that no disturbances in the processes of enzyme hydrolysis are present.

The same method has been used to study the effect of enzyme preparations on protein hydrolysis and amino acid absorption. For this purpose the investigation was repeated over several days and the subjects were administered a piece of chicken egg-white and an enzyme preparation (mexasae fermento duodenale, panzynorm forte). Both investigations were performed on the same patient and the effect of the enzyme preparation was estimated.

We have found that the application of enzyme preparations influenced the process of hydrolysis of the egg-white favourably, and hence contributed to more rapid absorption and excretion of labelled amino acid (Fig. 5). Consequently it is not sufficient to give chronic-enterocolitis patients a diet of full protein value but it is also necessary to prescribe enzyme preparations to improve cavitary hydrolysis of protein.

Mass spectrometric investigations have also shown that in chronic-enterocolitis patients nicotinic acid does not affect considerably lysine absorption but contributes to more intensive decrease of the label concentration in blood and its lesser excretion in urine which indicates that the nitrogen is retained in the organism (Fig. 6).

Thus mass spectrometry, in combination with stable isotope techniques, is of great value to clinical gastroenterology. It allows stable (safe) isotopes to be used for the dynamic study of the absorption processes and amino acid metabolism in healthy and diseased people. The method is of diagnostic importance helping to discover some new aspects of the pathogenesis of disturbances in patients with abnormal pathology of the intestinal tract. Furthermore it provides the possibility of estimating the effect of pharmacological preparations on the disease and the results of the treatment.

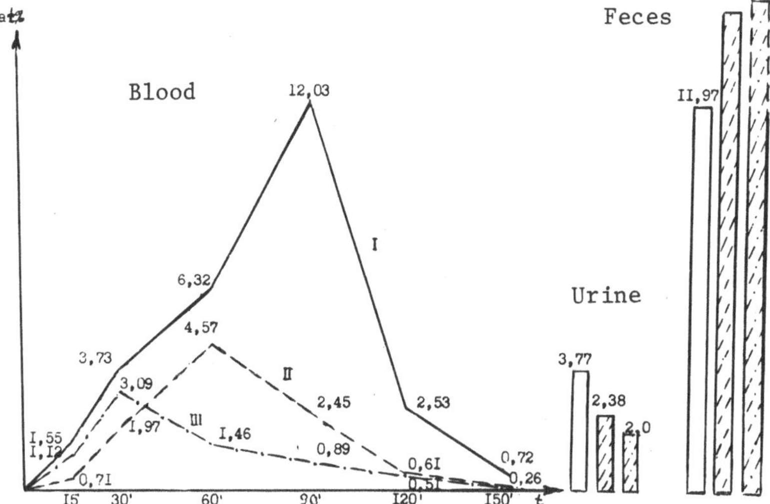

FIG. 6. The effect of Nicotinic acid on lysine ^{15}N absorption in gastrointestinal tract of chronic-enterocolitis patients.

BIOCHEMICAL APPLICATIONS OF HIGH PRECISION ISOTOPE ANALYSIS BY COMBINED GAS CHROMATOGRAPHY-MASS SPECTROMETRY

E.J. Heron°, E.J. Bonelli° and C.T. Gregg°°

° Sunnyvale, California, USA

°° Los Alamos, New Mexico, USA

INTRODUCTION

When labelled sugars are produced by *in vivo* photosynthesis in the presence of $^{13}CO_2$ the resulting material is often a combination of at least two types of isotopic species, the endogenous material containing ^{13}C in natural abundance and the enriched material. Before such materials can be used in clinical applications where quantitative studies are being carried out, it is necessary to determine the enrichment of the various isotopic components. The analyses described in this work were designed to measure: 1. The relative amounts of unenriched and uniformly labelled material. 2. The uniformity of the ^{13}C label in the enriched material. 3. The enrichment in the labelled material. These measurements were done with a combined gas chromatograph-mass spectrometer (GC-MS) operating in the selected ion monitoring mode. This technique is especially useful for isotopic analyses because it can be successfully used to obtain quantitative data even on microgram quantities of complex mixtures.

METHODS AND MATERIALS

All mass spectrometry was done with a Finnigan 3300F combined gas chromatograph-mass spectrometer (GC-MS), with a Finnigan 6100 Data System. The GC-MS interface in the electron impact (EI) was a glass jet separator maintained at 220°C. The source temperature was 120°C. For the analyses in the EI mode, the electron energy was 70 eV. In the chemical ionization (CI) mode, the electron energy was 160 eV and the reagent gas was methane at a source pressure of 0.32 torr.A 3 foot 3% OV-1 column was used for the gas chromatography. Helium was used as the carrier gas in the EI mode while methane was used in the CI mode. The column oven was maintained at 180°C throughout the analysis and the injector was held at 200°C. Glucose was analysed as the poly-O-acetylaldonic nitrile (1). Approximately one microgram of this derivative was used for each injection.

RESULTS AND DISCUSSION

Table 1 shows the results obtained from measurements of the

TABLE 1.
Analysis of natural abundance glucose by electron impact

m/e	Relative Abundance (measured)	Standard Deviation	Relative Abundance (calculated)
314.2	100.00	-	100.00
315.2	14.98	0.04	15.50
316.2	2.77	0.02	2.75

relative intensities of m/e 314.2, 315.2 and 316.2 arising from the unenriched glucose derivative in the electron impact mode. The means and standard deviations are the result of five independent measurements. The ratios obtained from these measurements may be compared with those obtained by calculation of theoretically ratios based on average natural abundances. For such calculations, the empirical formula $C_{13}H_{16}O_{8}N$ was used for the ion at m/e 314.

The data obtained from the analysis of a mixture of enriched and unenriched glucose is given in Table 2. Again, the standard deviations are the result of five independent measurements. The glucose sample was also analyzed in the chemical ionization mode and the results are presented in Table 3. Comparison of the measured and calculated ion ratios given in Table 1 gives an indication of the accuracy of this method of isotopic analysis. Since significant variations in natural abundances are known to exist, it is difficult to evaluate accuracy obtainable by this method, but this data shows that ratios with less than 1% deviation from the theoretical value can be obtained. This is consistent with results reported elsewhere (2). In order to determine the amounts of enriched and unenriched material in the glucose sample, the relative abundances of the ions in the m/e 314 to m/e 322 region of the EI spectrum of the glucose derivative were studied. The

TABLE 2.
Analysis of glucose sample containing a mixture of ^{13}C enriched and unenriched material

m/e	Relative Abundance	Standard Deviation
314.2	6.70	0.21
315.2	2.43	0.86
316.2	6.00	0.22
317.2	25.95	0.15
318.2	74.82	0.23
319.2	100.00	-
320.2	10.56	0.05
321.2	2.24	0.02
322.2	0.11	0.02

TABLE 3.
Analysis of a glucose sample containing a mixture of ^{13}C enriched and unenriched material by chemical ionization

m/e	Relative Abundance	Standard Deviation
326.2	0.50	.022
327.2	2.78	.063
328.2	100.00	-
329.2	20.62	.056
330.2	28.78	.342
331.2	132.95	1.512
332.2	509.83	3.76
333.2	1215.83	7.25
334.2	1359.74	6.62
335.2	149.90	.243
336.2	29.96	.171
337.2	2.67	.086
338.2	.46	.065

ions in this region would be expected to contain only 5 of the 6 original carbon atoms in the glucose molecule. Initially it was assumed that all of the intensity at m/e 314.2 was a result of the unenriched glucose. The intensities at m/e 315.2 and 316.2 were corrected for isotopic contributions on the basis of this assumption. Each of the intensities in this cluster were also corrected for the ^{15}N,^{13}C,^{17}O,^{18}O and ^{2}H contributions from all atoms except the carbon atoms arising from the glucose. This leaves intensities for monoisotopic species at each nominal mass. These intensities were normalized so that they have a sum of unity because the probability that the ion in question will appear at one of these masses is 1. Expression for the probability (i.e. relative intensity) that a particular ion is at each of these masses were derived with the assumption that the material was uniformly labelled. These normalized intensity values and expressions are given in Table 4. The normalized intensity at m/e 319 was used

TABLE 4.
Comparison of labelled glucose E.I. results with theoretical uniform distribution

m/e	Relative Intensity[a]	Expression	Calculated	Difference
314.2	(0.000)[b]	$(1-x)^5$ [c]	.000055	.000055
315.2	.0071	$5x(1-x)^4$	.0017	.0054
316.2	.0288	$10x^2(1-x)^3$	.0206	.0082
317.2	.1287	$10x^3(1-x)^2$	.1257	.0030
318.2	.3669	$5x^4(1-x)$	.3837	.0168
319.2	.4684[d]	x^5	(14684)[d]	-

[a] Corrected for intensities from isotopes other than the ^{13}C from the enriched glucose.

[b] Assumed to be 0.00

[c] x is defined as the probability that a particular carbon atom in the glucose molecule will be ^{13}C

[d] This value was used to calculate x by the equation $x = \sqrt[5]{.4684}$ i.e. x = .8592

to calculate X which is the probability that a particular carbon atom is ^{13}C. This value for X, 0.8592 was substituted into the other probability expressions and relative intensities were calculated for the theoretical uniform distribution. Comparison of the measured and calculated abundances shows that the material is very nearly uniform and that the assumption that the enriched material makes only a slight contribution to m/e 314.2 was valid. These results are given in Table 4. To illustrate the precision of the method, the numerical values of the expression were also calculated for x = 0.9 and x = 0.8. these are given in Table 5 and

TABLE 5.
Comparison of measured and calculated isotope ratios for different enrichments

m/e	Measured Abundance	Calculated Relative Abundance x = .8592	x = .90	x = .80
315.2	.0071	.0017	.00045	.0064
316.2	.0288	.0206	.0081	.0512
317.2	.1287	.1257	.0729	.2048
318.2	.3669	.3837	.3280	.4096
319.2	.4684	.4684	.5905	.3277

show that x = .8592 gives a good fit to the data, indicating that the material is at least very nearly uniformly labelled with an enrichment of 85.92% in each carbon atom. In order to find the percentage labelled and unlabelled material, the ion intensities from each of these species were summed and compared. This result indicated that the material contained 96.54 atom % of the uniformly labelled material.

The relative intensities of the ions in the m/e 326.2 to 338.2 region of the CI spectrum of the glucose derivative were analyzed in the same manner. These ions are of particular interest because they expected to contain all the carbon atoms in the original glucose molecule. The intensities for these ions were corrected for the contributions from the unlabelled material and the natural abundance contribution from ^{13}C, ^{2}H, ^{17}O, ^{18}O, and ^{15}N. These corrected intensities were normalized to a total of one and the results are given in Table 6. Again there is excellent agreement between the measured distribution and the calculated for the theoretical uniform distribution. The value for the enrichment in each carbon atom was calculated to be 85.95. This compares well with the data obtained for the ion containing only five of the six carbon atoms. This data was also used to calculate the percentage of labelled material. It was found that the material was 96.80 atom % of the uniformily labelled material.

TABLE 6.
Comparison of labelled glucose CI results with theoretical uniform distribution.

m/e	Relative Intensity [a]	Expression	Calculated
328.2	(.00000) [b]	$(1-x)^6$ [c]	.000008
329.2	.00275	$6(1-x)^5 (x)$	.000238
330.2	.00849	$15(1-x)^4 (x)^2$	.00432
331.2	.04260	$20(1-x)^3 (x)^3$	.03523
332.2	.1624	$15(1-x)^2 (x)^4$	.1534
333.2	.3806	$6(1-x) (x)^5$	.3754
334.2	.4031 [d]	x^6	(.4031)

[a] Corrected for intensities from isotopes other than ^{13}C from enriched glucose.

[b] Assumed to be 0.00

[c] x is defined as the probability that a particular carbon atom in the glucose molecule will be ^{13}C.

[d] This value was used to calculate x by the equation $x = \sqrt[6]{.4031}$ i.e. $x = .8595$

The agreement between the results obtained for the two groups of ions measured shows that GC-MS can be used to give a detailed characterization of a sample containing two isotopic species. In one analysis it is possible to accurately determine how much labelled material is present and to determine its enrichment. In addition the same analysis gives information on the uniformity of the label.

REFERENCES

1) J. Szafranek, C.D. Pfaffenberger and E.C. Horning, Anal. Lett., 1973, 6, 479.
2) R.M. Caprioli, W.F. Fies and M.S. Story, Anal. Chem., 1974, 45, 453A.

CHEMICAL IONIZATION MASS SPECTROMETRY IN THE IDENTIFICATION OF DRUGS AND DRUG METABOLITES

E.L. Ghisalberti

Department of Organic Chemistry, University of Western Australia, Nedlands, 6009, Western Australia

INTRODUCTION

In the ten years since its development, chemical ionization mass spectrometry (CIMS) has undergone remarkable growth to become an important analytical tool for the identification, structural elucidation and quantitation of organic compounds. Its utility in pharmacological and biochemical research has been firmly established over the last few years.

The interest in CIMS arose from the initial observations that, as a "softer" method of ionization, it often provided enhanced ions related to the molecular weight of the sample under analysis. The comparability of chemical ionization techniques (CI) with the already established electron impact systems was a major factor in the rapid development of CIMS. In fact the combined gas chromatography-CI mass spectrometry systems offered some advantages. In this combination the ion source could take most or all of the column effluent and the carrier gas could often be used as the reagent gas. The advent of computer interfacing facilitated rapid data collection and processing, an important aspect in the accurate analysis of complex mixtures in pharmacological problems. In this paper the recent advances in the application of CIMS to the identification of drugs and drug metabolites will be reviewed. The coverage is meant to be illustrative rather than complete. Various reviews have dealt with the theory (1-9) and general applications (7-11) of CIMS and the reader is referred to these for a comprehensive reference list.

TABLE 1.
Chemical ionization reactions

1) Proton transfer

M (sample molecule) + XH^+ (reagent gas) $\longrightarrow MH^+ + X$

2) Addition reaction

$M + XH^+ \longrightarrow MXH^+$

3) Proton abstraction

$M + XH^+ \longrightarrow M\text{-}H^+ + H_2 + X$

4) Elimination of a neutral molecule

$M + XH^+ \longrightarrow M\text{-}Y^+ + HY + X$

5) Charge exchange reactions

(a) Direct charge exchange

$M + X^{+\cdot} \longrightarrow M^{+\cdot} + X$

$E_{int} = R.E.(X^{+\cdot}) - IP(M)$

E_{int} = energy interval
R.E. = recombination energy
IP = ionization potential

(b) Combined charge exchange/chemical ionization (for a proton containing second gas)

$X_1 + e^- \longrightarrow X_1^{+\cdot} + 2e^-$
$X_1^{+\cdot} + X_2H \longrightarrow X_1 + X_2H^{+\cdot}$
$X_2H^{+\cdot} + X_2H \longrightarrow X_2H_2^+ + X_2^{\cdot}$
$\quad\quad n\,X_2H \longrightarrow X_2H_2^+(X_2H)_n$
$X_2H_2^+ + M \longrightarrow MH^+ + X_2H$

6) Negative ion formation

$M + X^- \longrightarrow [MX]^- \longrightarrow M\text{-}H^- + XH$

PRINCIPLES OF CHEMICAL IONIZATION

CIMS can be divided into three main classes. The first includes the use of reagent gases which can transfer ionic species to the sample molecule giving rise to even electron ions. The second, more appropriately referred to as charge exchange (CE), involves direct electron exchange with the sample molecule giving odd electron ions. A combination of these two approaches is one of the significant advances in CIMS and will be discussed later. A third class is the recently-developed atmospheric pressure ionization (API). Since the background and applications of this technique have been discussed by Professor Horning at this Symposium it will not be considered here. The theory of CIMS has been discussed and reviewed in a number of articles (1-9). Here it is useful to summarize the more common reactions which can occur between the reagent gas (ionization gas) and the molecule of the sample being analysed (Table 1).

Proton transfer occurs when the proton affinity of the sample, or of a functional group in the sample molecule, is greater than that of the reagent gas, i.e. PA(M) > PA(XH). The resultant ionic species $(MH)^+$ is an even electron species which, if unstable, can further fragment to lose a neutral molecule. In general, the degree of fragmentation increases as the difference between the two parameters increases. For a given molecule more fragmentation will be observed as the proton affinity of the gas decreases. Thus H_2 will cause more fragmentation than will CH_4 or NH_3 (see Table 2).

Proton abstraction occurs mostly with compounds of low proton affinity e.g. alkanes. Addition reactions between functional groups and the reagent gas can also be observed. The extent of this reaction depends on the temperature, pressure and concentration conditions in the ion chamber.

Elimination of a neutral fragment from the molecule is believed to occur via an unstable addition product. Charge exchange (CE) reactions arise when the ionized reagent gas cannot donate a proton and on the condition that the ionization potential (IP) of the molecule is less than the recombination energy (RE) of the ionized gas (12). The degree of fragmentation will depend on the energy difference (E_{int}) between the two parameters. Combined charge exchange -chemical ionization can occur (13) when a second gas (X_2H) is used to "quench" the CE gas. The second gas once ionized can then ionize the sample molecule. The spectra of molecules ionized in this way will present a situation between that obtained for true EI and CI spectra. Negative ion formation should also be considered. In conventional ion sources negative ions which are produced are normally trapped. A reversal of the potentials

TABLE 2.
Common reagent gases for CIMS.

Gas	Proton affinity (a) (Kcal/mole)	Major ions formed Composition	(m/e)	% Total ion current	Pressure (torr)
H_2	100	H_3^+ (b)	(3)	-	0.7 - 1.0
CH_4	127	CH_5^+ $C_2H_5^+$ $C_3H_7^+$	(17) (29) (43)	47 41 6	0.5 - 2.0
H_2O	165	H_3O^+ $(H_2O)_nH_3O^+$	(19)	 (c)	0.25
$\underline{i}$-C_4H_{10}	195	$C_4H_9^+$ $C_3H_7^+$ $C_3H_3^+$	(57) (43) (39)	93 4 3	0.6 - 0.8
NH_3	207	NH_4^+ $NH_3NH_4^+$ $(NH_3)_nNH_4^+$	(18) (35)	97 3 (c)	0.5 - 1.0
CH_3NH_2	216	$CH_3NH_3^+$ $(CH_3NH_2)_2H^+$	(32) (63)	83 6	0.5

(a) Approximate values only.

(b) Ions derived from residual air (H_3O^+,N_2H^+,O_2^+,O_2H^+) are also present.

(c) Amount of cluster ion formation is pressure dependent.

of the ion repeller and accelerator voltage plates can result in the acceleration and focusing of these negative ions.

In all these cases it can be seen that the typical ions which are formed are predictably related in mass to the molecular ion. Furthermore the occurrence of both proton transfer and addition reactions on the same compound can often provide further evidence for the molecular weight assigned. The reagent gases most often used are shown in Table 2.

APPLICATIONS

The analytical problems in drug research can be summarized as being concerned with:
(a) The identification of drugs and their known metabolites,
(b) the detection and structural elucidation of new metabolites and
(c) the quantitation of these in biological tissues and fluids.
In clinical and forensic laboratories time is usually an important factor. In the past more emphasis has been placed on (a) and (c) but the importance of (b) cannot be overstressed. In most cases the percentage of drug administered accounted for by known metabolites is very low. Secondly, a proper understanding of drug action requires as complete a picture as possible of its biotransformation products. Not uncommonly a metabolite may be more active or toxic than the original drug. In the following pages are presented some examples of the use of CIMS applied to the solution of these problems either individually or in combination.

DIRECT CIMS ANALYSIS

In dealing with cases of drug abuse the problem of rapidly identifying the compounds involved is of paramount importance. Since CI often provides abundant quasimolecular ions and relatively simpler mass spectra it is ideally suited for the analysis of multi-drug mixtures. In one of the first applications of CIMS to the analysis of multicomponent mixtures, the CI isobutane mass spectra of 48 commonly-used drugs were obtained (14). Of these 36 compounds gave spectra which had no ions with a relative abundance of more than 5% with respect to the base peak, other than the MH^+ ion. Thirty drugs had unique molecular weights and therefore unique $CI_{isobutane}$ mass spectra. This background information was used in the investigation of several cases of drug overdoses. In one example (14) the extract of the gastric contents were applied directly to the probe of the mass spectrometer. The $CI_{isobutane}$ spectrum obtained (Fig. 1) clearly indicated the presence of a number of common drugs. It should also be mentioned

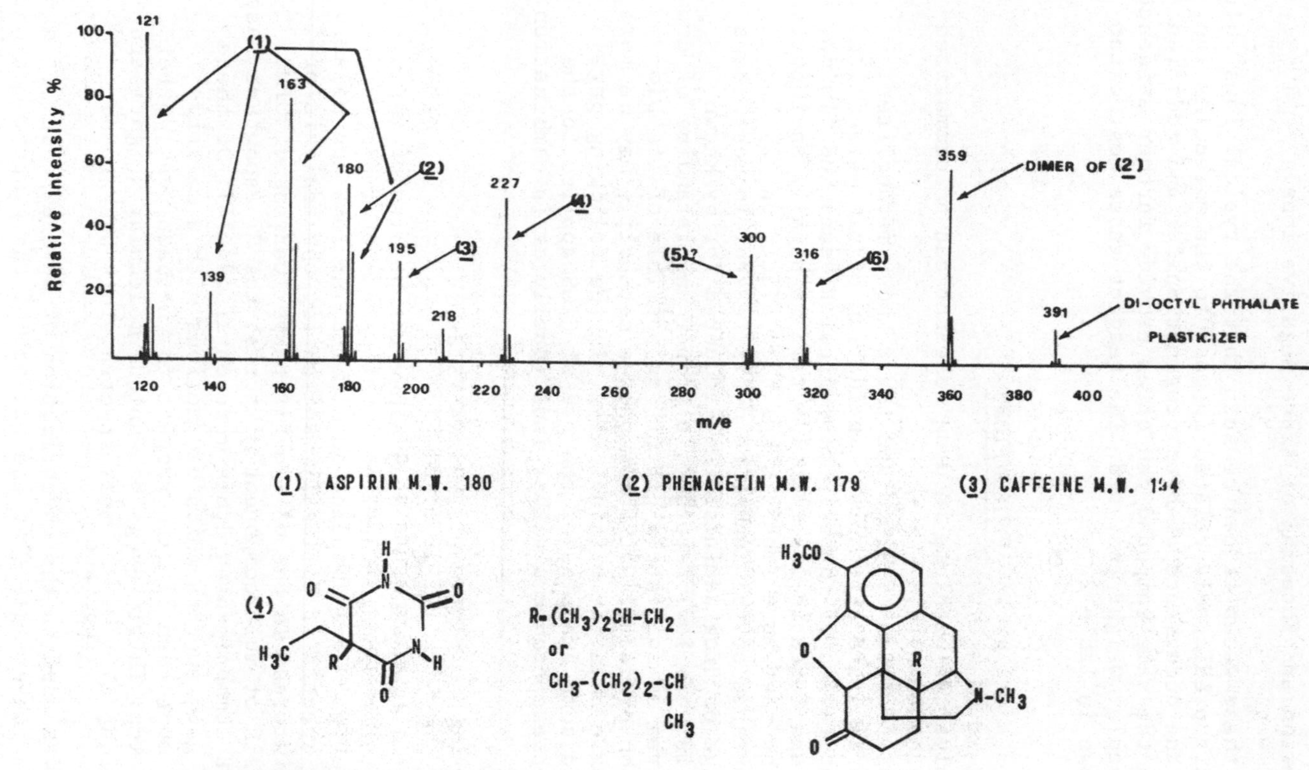

FIG. 1. $CI_{isobutane}$ mass spectrum of extract of gastric contents in an overdose case.

that the identification in this case was aided by prior indications of the preparation ingested (Percodan). Nevertheless homoatropine, known to be present in small (0.05%) amounts was not detected. Furthermore the peak at m/e 300 was initially wrongly assigned to codeine which should also have a peak due to the facile loss of H_2O at m/e 282. One possibility is that the peak represents dihydrocodeinone, a probable contaminant.

Another compilation (15) lists the $CI_{methane}$ mass spectra of over 300 drugs, drug metabolites and other compounds encountered in body fluid extracts. Over 60% of these compounds gave mass spectra in which the MH^+ ion appeared as the base peak. These reference data systems should greatly aid the specific identification of drugs and drug metabolites in direct analyses of mixtures. This technique has also been used (16) to identify heroin (7) and its diluents, acetyl codeine, quinine, caffeine, procaine, methapyriline, mannitol and sugars, in illicit heroin preparations.

A similar method of direct analysis has been used to identify and quantitate the antiarrhytmic agents, quinidine (8) and lidocaine (11) in human plasma (17). In one experiment a known amount of d_2-dihydroquinidine (9) was added to plasma samples from patients who had received quinidine gluconate, and the extract was applied directly to the probe of the mass spectrometer. The resulting $CI_{isobutane}$ mass spectrum obtained (Fig. 2) clearly shows the ion pattern due to quinidine and d_2-dihydroquinidine. The peak at m/e 327 is due to dihydroquinidine originally present in the quinidine sample (7.8%) and that (17.8%) present in the deuterated sample. The relative intensities of the ions at m/e 325 and 329 allows quantitation. Although quinidine can be determined in plasma by a GLC method a distinction between quinidine and dihydroquinidine is not possible (18). Lidocaine (11) and monoethylglycinexylidide (12), a known metabolite of 11 were identified and quantified in human plasma in a similar way. For quantitation the corresponding deuterated standards (11-d_4 and 12-d_3) were added to the crude extract from human plasma. The CI mass spectrum obtained is shown in Fig. 3. Again the ion doublet observed for each compound allows ready identification and quantitation. The range of detection was from 5ng to 4μg and 0.1 to 1μg respectively. No evidence for the presence of glycinexylidide (13) as a metabolite was obtained, although the peak for the MH^+ of the deuterated standard (13-d_2) at m/e 181 was seen to levels of 0.25μg.

This method of direct analysis is rapid and in cases of drug overdose does not require identification of all the components. The disadvantages however are: (a) the inability to differentiate between some pairs of drugs e.g. amo- and pentobarbital (see Fig. 1), (b) relatively large quantities of material are needed, and

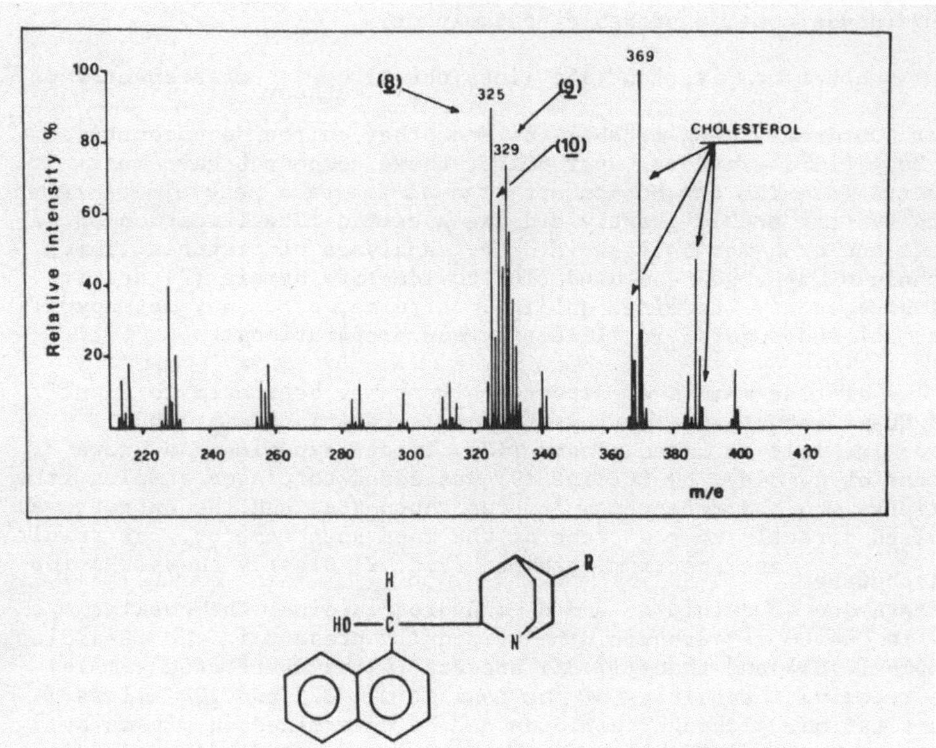

R (8) CH=CH$_2$ (9) CHD-CH$_2$D (10) CH$_2$-CH$_3$

FIG. 2. CI$_{isobutane}$ mass spectrum of plasma extract of a subject treated with quinidine gluconate.

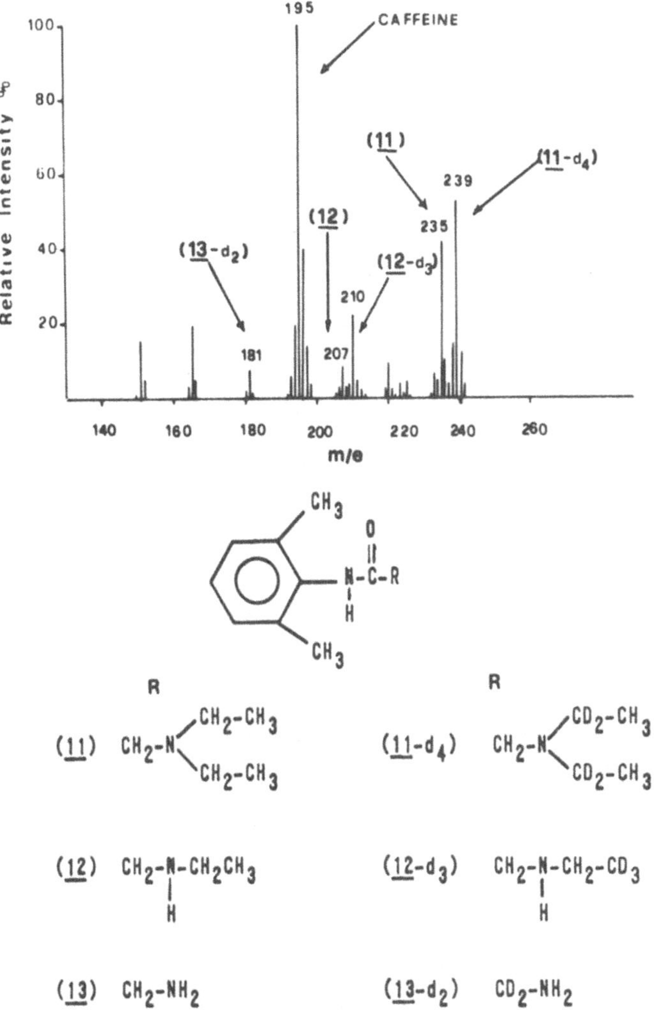

FIG. 3. $CI_{isobutane}$ mass spectrum of plasma extract of a subject treated with lidocaine hydrochloride.

(c) the presence of non-drug components could be falsely interpreted.

CIMS was found (19) useful in the structural elucidation of the glucuronides of the antidepressant drugs, oxazepam (14) and lorazepam (15) (Fig. 4). Oxazepam is also a metabolite of a number

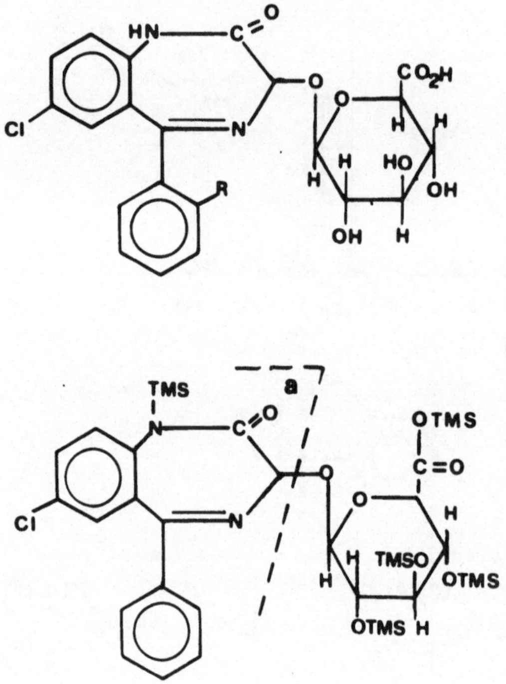

	R
(14)	H
(15)	Cl

M.W. 822

PENTA SILYLATED OXAZEPAM GLUCURONIDE

FIG. 4. Structures of oxazepam and lorazepam glucuronide (14 and 15).

of benzodiazepine drugs. The urinary metabolites, after separation on XAD-2 resin were purified by charcoal and DEAE-cellulose chromatography. Direct CIMS with isobutane as the reagent gas gave a spectrum which clearly included an MH^+ peak at m/e 823 consistent with a pentasilylated oxazepam glucuronide (Fig. 4). Diagnostic peaks were noted at m/e 341 (the ion derived from the oxazepam portion), 269, 359 and 431. The last two were absent from the EI spectrum. As a result the location of the glucuronide moiety on the carbon α-to the carbonyl was established. The structure shown for lorazepam glucuronide was determined in a similar way (19).

CIMS has been used for the identification of macrolide antibiotics and related model substances (20). In developing methods for the analysis and rapid microidentification of β-lactam antibiotics, Mitscher et al. (21) have recorded the CI mass spectra of the free acids and methyl esters of eight clinically-significant penicillins and their breakdown products. Isobutane and ammonia were used separately as the reagent gases and high resolution mass measurements were obtained to assign peaks.

The CI spectra of all the penicillin esters gave an MH^+ ion and showed a base peak at m/e 174 arising from a [2+2] cycloreversion reaction (Fig. 5). These two peaks provide ready

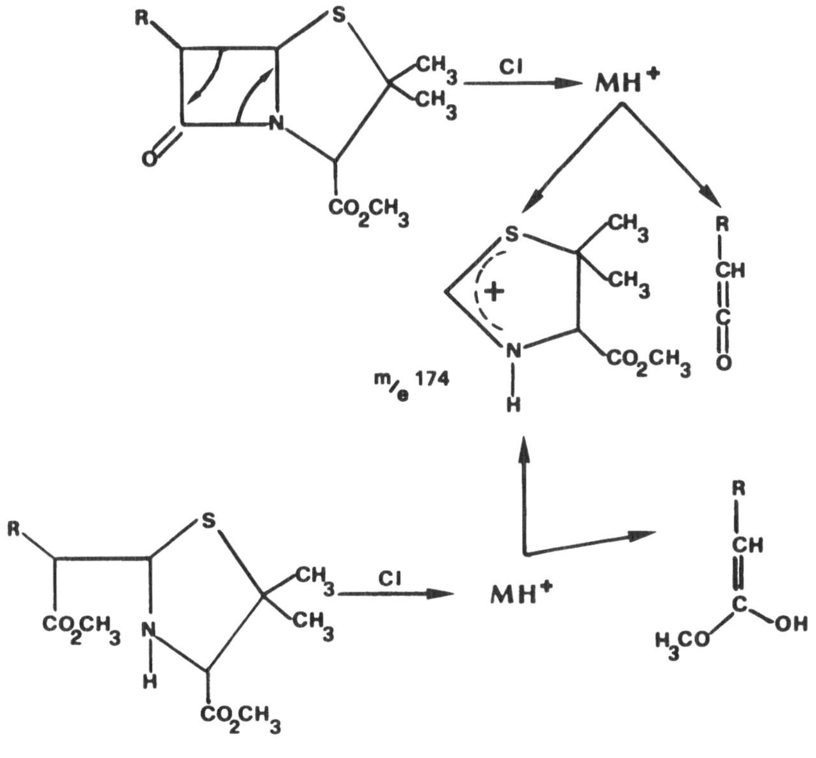

FIG. 5. CI mass spectral fragmentations of penicillin and penicilloic acid methyl esters.

recognition of the type of side chain present. In contrast, the penicilloic acid esters gave MH^+ as the base peak with m/e 174 still relatively abundant. The $CI_{isobutane}$ mass spectra of the free acids lacked MH^+ ions and showed more extensive fragmentation. Because of their instability and lack of volatility some of these fragmentations may be thermally induced prior to ionization.

In these cases some of the fragmentations could be eliminated by exposing the sample to the ion plasma in a CI source. In this way, Baldwin and McLafferty (22) showed that mass spectra of relatively non-volatile compounds can be obtained with temperatures approximately 150° below those normally required to vapourize the sample. This method could be useful particularly when derivatization of the sample is incomplete, gives a number of by-products, or when the derivative decomposes on the column. For example, in an attempt (23) to quantitate the oral hypoglycemic agent, tolbutamide (<u>18</u>) and its metabolites (<u>19</u> and <u>20</u>) in human plasma by GC-MS, it was found that even when derivatized these sulphonyl ureas partly decompose to sulphonamides (fig. 6). This decomposition is thermally

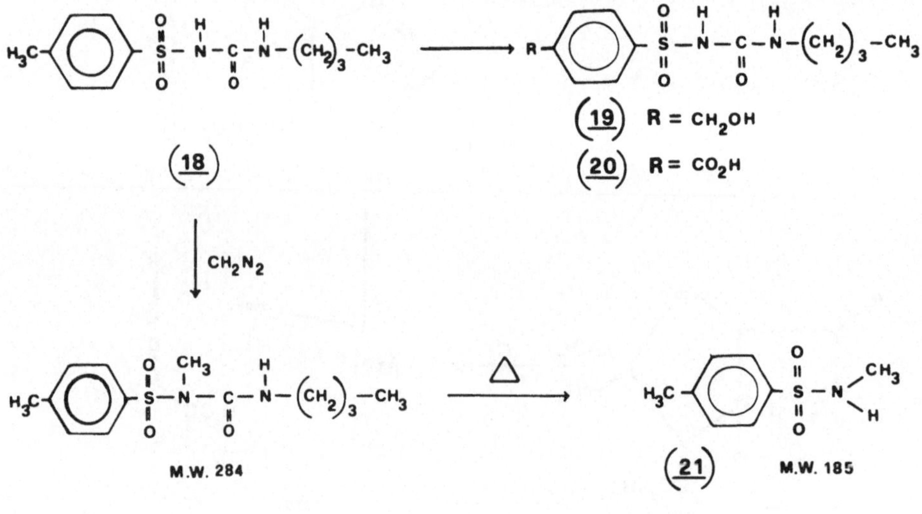

FIG. 6. Structures of tolbutamide (<u>18</u>), its metabolites and derivatives.

induced and occurs both in the injector and ion source. When the derivatized compounds were evaporated directly from the probe in the presence of methane as the CI reagent gas decomposition was not observed. For quantitation the appropriately deuterated standards were employed and the levels of each compound was determined by selected ion monitoring (SIM) using the MH^+ ion which carries over 60% of the total ion current (22). Sulphonyl urea levels of 200ng/ml could be estimated. In this study it was found that the rates of removal of tolbutamide and its metabolites were similar and that the ratio of the acid to hydroxy (19) metabolite was 2 : 1.

CI was used in the structural elucidation of a new metabolite of the anticoagulant warfarin (22). The metabolism of warfarin was known to give the alcohol, by reduction of the methyl ketone group, and phenols, by hydroxylation of either phenyl ring (Fig. 7). In this work Pohl *et al*. (24) also identified benzylically

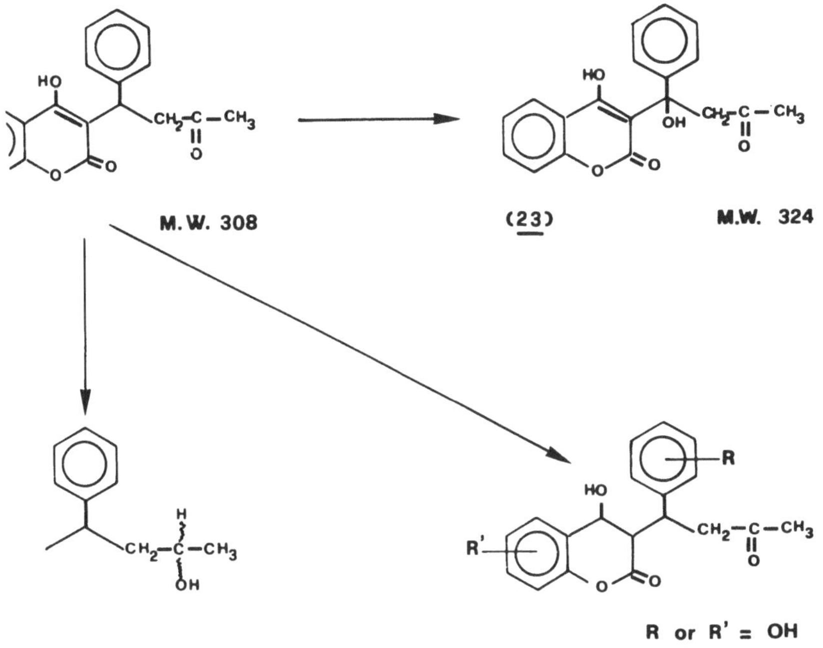

FIG. 7. Partial structure of warfarin (22) and some of its metabolites.

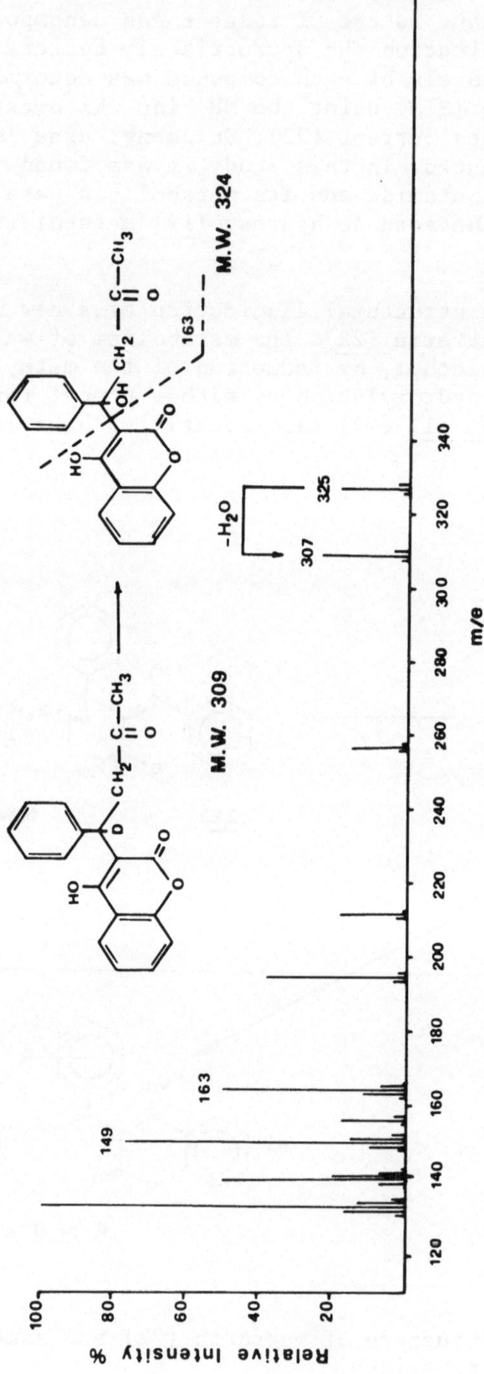

FIG. 8. $CI_{methane}$ mass spectrum of a metabolite of warfarin. A sample of warfarin, deuterated at the benzylic position (85%), was incubated with rat liver microsomes.

hydroxylated warfarin as a metabolite in rat liver microsomes. A detailed examination of its mass spectrum indicated that the extra hydroxyl group was located either at the benzylic position or α- to the carbonyl group. To distinguish between these two possibilities the benzylically deuterated warfarin was incubated. The CI mass spectrum (Fig. 8) of the metabolite obtained indicated that the deuterium had been replaced by a hydroxyl group. The ion at m/e 163 was assigned to the non-coumaran part of the molecule. This information was used to detect this compound as a metabolite of warfarin in humans. The urinary extract of patients treated with an equimolar mixture of warfarin and benzylically deuterated warfarin gave the mass spectrum shown in Fig. 9. This clearly shows the presence of monohydroxylated metabolites which include 6-and 7-hydroxymetabolites (m/e 325, 326), warfarin (m/e 309, 310), reduced warfarin metabolites (m/e 311, 312) and ions arising from the loss of H_2O at m/e 293 and 294. The ion at m/e 163 implied the presence of the benzylically hydroxylated metabolite. Separation of the components gave a compound identical to that obtained from the microsomal incubation.

GAS CHROMATOGRAPHY - CI MASS SPECTROMETRY (GC-CIMS)

The increased sensitivity and specificity provided by the combination of the GC and MS systems have given impetus to the study of drug metabolism and drug quantitation (25-27). GC-MS used in the selected ion monitoring mode becomes one of the most sensitive detection systems known (28). In all these applications CI is being increasingly used. The relatively simpler spectra obtained by CI are especially useful in the analysis of mixtures often encountered in pharmacological work. For SIM also CI often simplifies the spectra of contaminating compounds thus minimizing their contributions to the ions being monitored.

The CI mass spectrometer is more easily combined than its EI counterpart with the gas chromatograph (7, 30). The ability of the CI source to handle high gas loads decreases the pressure differential between the two instruments. Furthermore the GC carrier gas can often be used as the reagent gas. The reagent gas can also be added to the gas stream within the GC-CIMS interface (31) or introduced into the ion source via a separate inlet port (32) (e.g. via the liquid inlet system). When reagent gases (H_2O, NO, O_2) that may adversely affect the GC stationary phase are used these should be introduced as far as possible from the GC interface. In this respect a coaxial dual gas interface has been developed which allows the gas to be introduced very close to the ion source. This is particularly useful for thin-film capillary columns. The small diameter of the interface capillary allows columns to be changed without turning off the vacuum controls of the spectrometer.

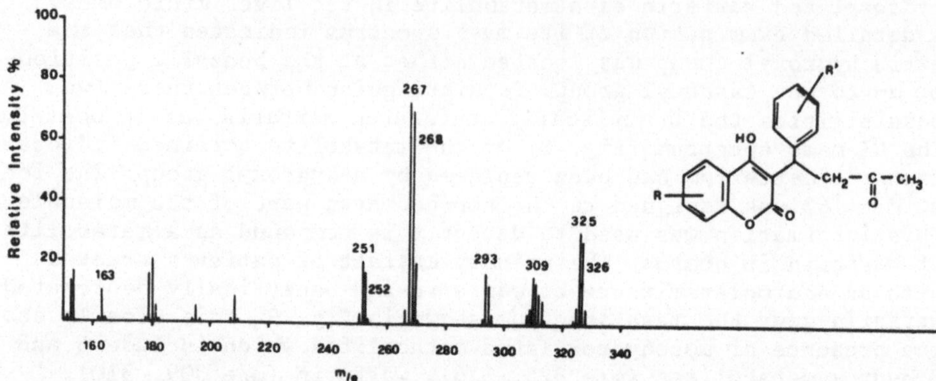

FIG. 9. $CI_{isobutane}$ mass spectrum of the urine extract of a subject administered a mixture of warfarin and benzylically deuterated warfarin.

In those cases where the reagent gas is introduced independently of the column effluent conversion between CI and EI operational modes is rapid. Several reviews have discussed the applications of GC-CIMS from different aspects. Only a selection of recent applications is given here.

Phencyclidine (24) (Fig. 10), an analgesic anaesthetic with potent side effects has become one of the drugs of abuse. The quantitation of this drug in body fluids and the structural elucidation of two of its metabolites were achieved by GC-CIMS (34). Blood and urine samples from individuals intoxicated with phencyclidine on extraction gave only unchanged drug. The metabolites were present as conjugates and could be extracted after enzymatic hydrolysis with β-glucuronidase.

Quantitation of phencyclidine was achieved by SIM using 1-(1-phenyl-$[^2H_5]$-cyclohexyl)piperidine as the internal standard. Levels of 1 ng/ml in body fluid could be measured. Interestingly, it was established that 1-phenylcyclohexene, previously reported as a metabolite, in fact arises from the thermal degradation in the injection port of the GC even at temperatures as low as 150°. GC separation of the metabolites could be achieved only after silylation. The $CI_{methane}$ mass spectra of each metabolite (25 and

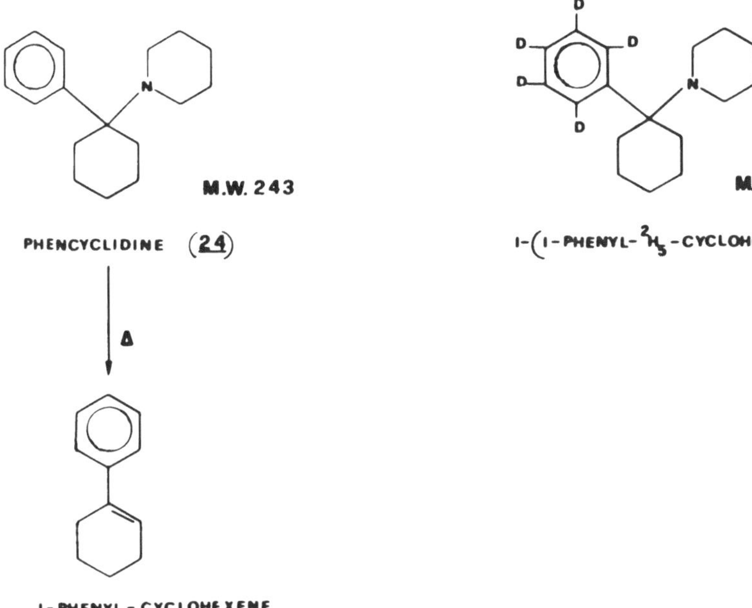

FIG. 10. Structures of phencyclidine (24) and $\underline{d}_5$-phencyclidine.

26) indicated the molecular weight and interpretation of their mass spectra, by comparison with their EI spectra, allowed tentative structures to be postulated (Fig. 11). The structure of the metabolites was confirmed by comparison with synthetic material.

Phenformin (27) used clinically as an oral hypoglycemic agent, in common with the biguanides, degrades on volatilization. Treatment of phenformin with trifluoroacetic anhydride yields the diamino-S-triazine (28) which is more amenable to GC analysis. Quantitation of phenformin in human plasma was achieved (35) by SIM using the dideuterated phenformin as internal standard and by monitoring the MH^+ ions, m/e 284 and 286 respectively, which are the base peaks in $CI_{methane}$ mass spectra of their corresponding derivatives (Fig. 12). Levels of 20 ng/ml could be measured.

A similar method was used (36) to estimate cyclophosphamide (29) in the serum, urine and tissues of tumour-bearing mice using isophosphamide (30) as the internal standard. Both compounds

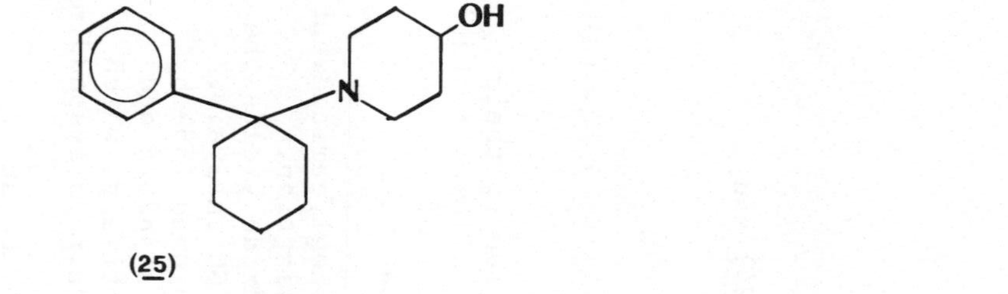

(25)

(26)

EI		CI		EI		CI	
m/e	ASSIGNMENT	m/e	ASSIGNMENT	m/e	ASSIGNMENT	m/e	ASSIGNMENT
331	$M^{+\cdot}$	332	MH^+	331	$M^{+\cdot}$	332	MH^+
288	[structure] OTMS, N^+, CH=CH_2	174	[structure] OTMS, $H-\overset{+}{N}-H$	200	[structure] N^+, CH=CH_2	157	[structure] +
		154 (bp)	[structure] +			86 (bp)	[structure] $H-\overset{+}{N}-H$

FIG. 11. Important peaks in the mass spectra of the silyl ether derivatives of the metabolites (25 and 26) of phencyclidine.

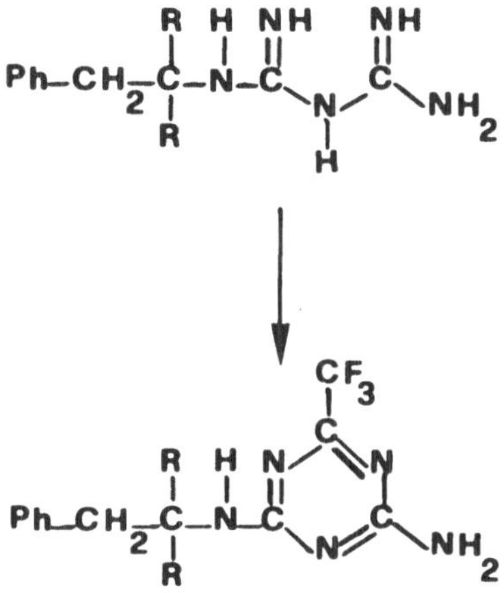

(27)

(28)

	R = H		R = D
$CI_{methane}$	284 (b.p.)	$[MH]^+$	m/e 286 (b.p.)
	264	$[M-19]^+$	266
	312	$[M+29]^+$	314
	326	$[M+43]^+$	328

FIG. 12. Major peaks in the $CI_{methane}$ mass spectra of derivatized phenformin (28) and its deuterated standard.

normally undergo intramolecular alkylation under GC conditions (Fig. 13). This can be avoided by forming their trifluoroacetyl

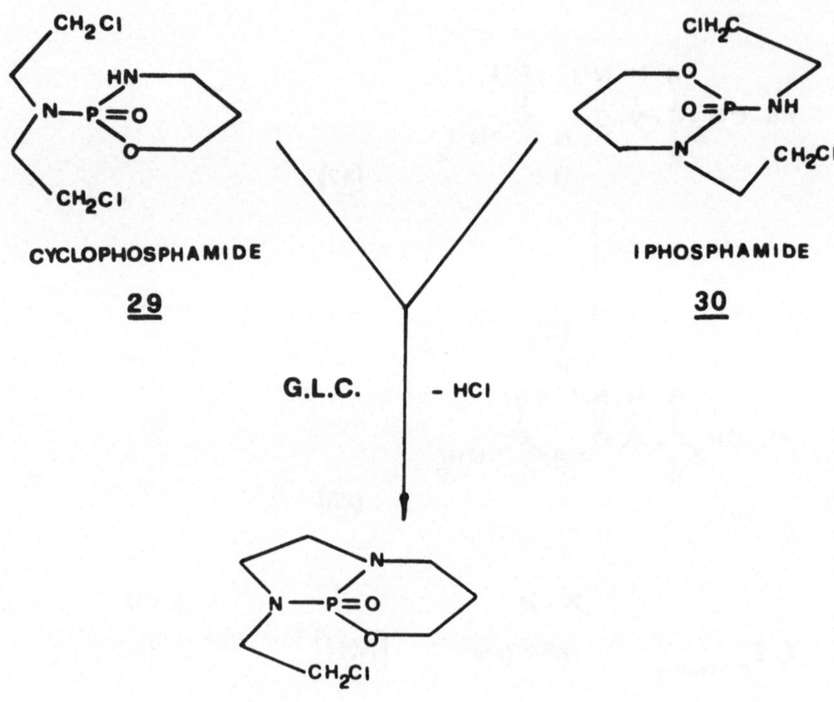

FIG. 13. Intramolecular cyclization of cyclophosphamide (29) and isophosphamide (30).

derivatives (31 and 32) before analysis. The $CI_{methane}$ mass spectra of the trifluoroacetate derivatives are very simple with the MH^+ ion being the base peak. The minimum detectable amount was 500 pg/injection.

As previously mentioned, use of CI can often simplify the detection and measurement of drugs and their metabolites in biological fluids. This is illustrated in the following example. In a study of the metabolism of cyclophosphamide in humans it was important to determine whether N,N-bis(2-chloroethyl)phosphoro-

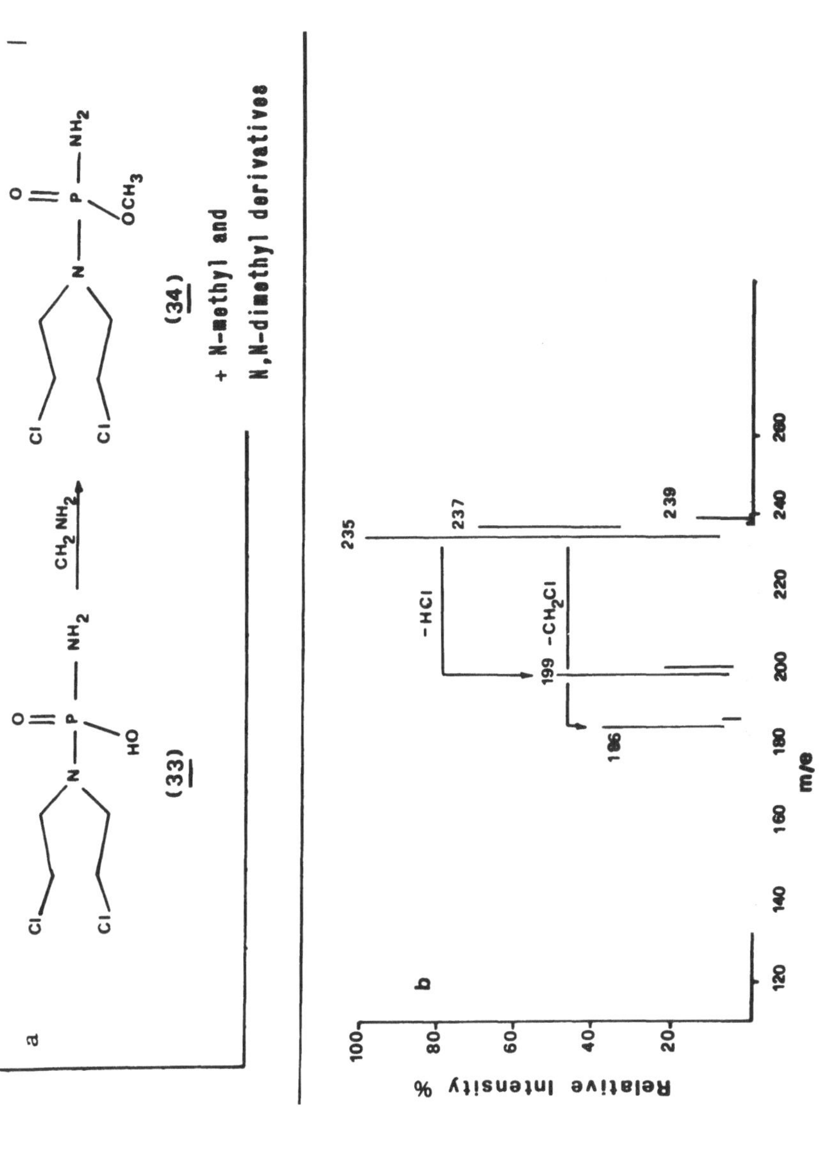

FIG. 14. (a) Methylation of 33 with diazomethane. The O-methyl derivative (34) is formed in over 80% yield, (b) $CI_{isobutane}$ mass spectrum of 34.

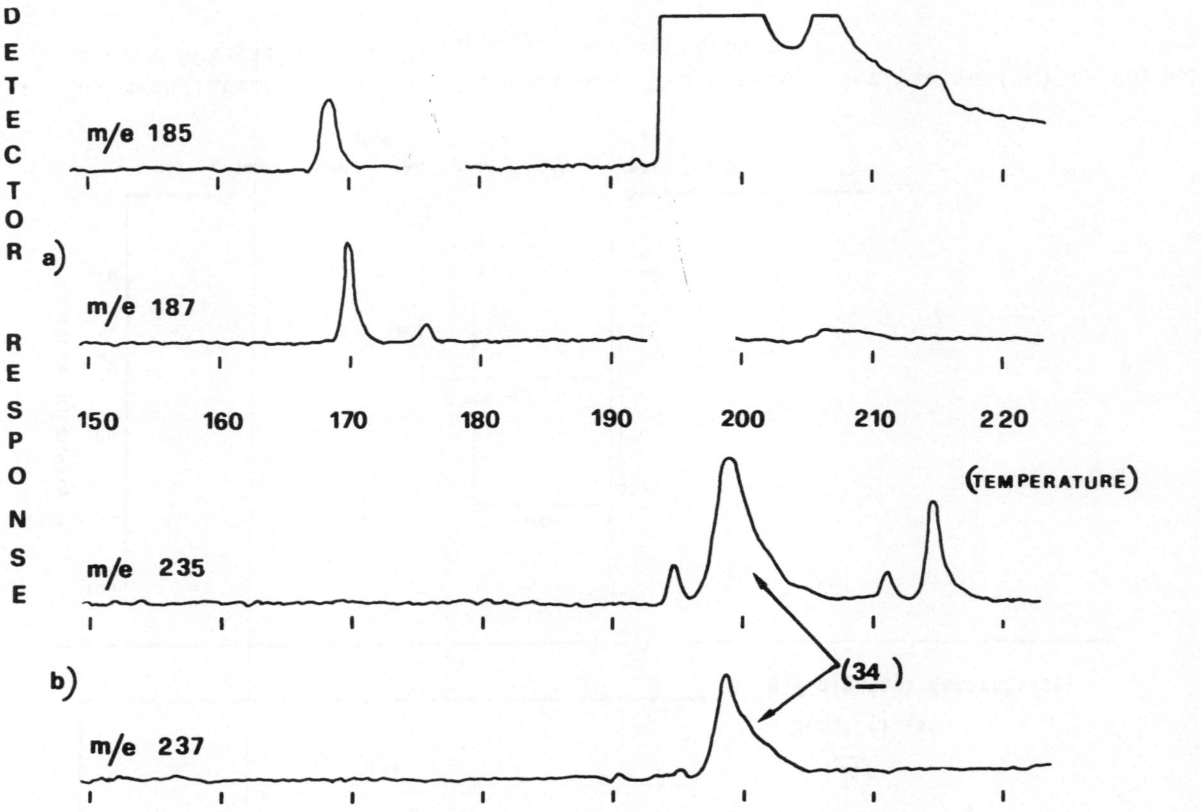

FIG. 15. Selected ion profiles of methylated plasma extract of subject treated with cyclophosphamide. (a) The ions monitored (m/e 185 and 187) are two isotopic peaks of the base peak ($M^{+}-CH_2Cl$) in the EI mass spectrum of 34. (b) The ions monitored are two isotopic peaks of the base peak (MH^{+}) in the $CI_{isobutane}$ mass spectrum of 34.

diamidic acid (33) was formed (37). Previous work had shown that this metabolite could be detected in GC-MS after treatment with diazomethane. This reaction yields a mixture of mono-, di- and trimethylated derivatives in which the monomethylated compound (34) predominates (80%). The $CI_{isobutane}$ mass spectrum of 34 is shown in Fig. 14. The MH^+ is clearly the base peak in contrast to the EI spectrum which does not show a molecular ion peak and in which the base peak is given by the $M-CH_2Cl$ ion. Although phosphoroamidic acid could be detected in methylated extracts from urine by SIM using EI methods, analysis of the plasma extracts, similarly derivatized, gave the selected ion profiles shown in Fig. 15a. The contributions of a number of compounds to these ions is evident. When SIM was used in the CI mode the selected ion profiles shown in Fig. 15b were obtained. From this and comparison of the retention time with that of a standard the presence of this compound as a circulating metabolite in plasma was confirmed. Quantitation of this compound in plasma using CIMS has recently been reported (38).

The metabolism of etidocaine (35), a long-acting anesthetic, has been studied using GC-CIMS (39). GC analysis of the urinary extracts from subjects treated with etidocaine showed the presence of the drug and six basic metabolites (Fig. 16). Two other metabolites were separately identified as 2,6-dimethyl-and 4-hydroxy--2,6-dimethylaniline. GC-CIMS analysis of these metabolites allowed the identification of unchanged drug (35, peak 6 in Fig. 16) and four other compounds (Fig. 17) corresponding to peaks 3, 4, 5 and 7. Interpretation of the $CI_{methane}$ mass spectrum was facilitated by prior knowledge of the fragmentation pattern for etidocaine. The major ions formed arise by cleavage of the $\overset{O}{\overset{\|}{C}}$-CH bond giving a peak at m/e 128 (side chain) and m/e 148 (phenyl ring). The metabolite (36, peak 7 in Fig. 16) instead showed only a peak at m/e 147, indicating that the amide nitrogen did not carry a hydrogen atom. Support for this structure came from the CE mass spectrum (He) although confirmation must await comparison with an authentic sample. CIMS methods are particularly useful in problems of toxicology, pharmacokinetics and perinatal pharmacology as demonstrated by the work of M.G. Horning *et al.* (40, 41). Particularly interesting is the study of the exposure of the foetus/neonate to pharmacologically active agents taken by gravid/lactating females. In most cases the drug is transferred across the placenta and *via* breast milk to the foetus and neonate respectively. Using SIM techniques, quantitative studies can also be carried out and the high sensitivity of this method has made possible studies of the metabolism of drugs in the human neonate.

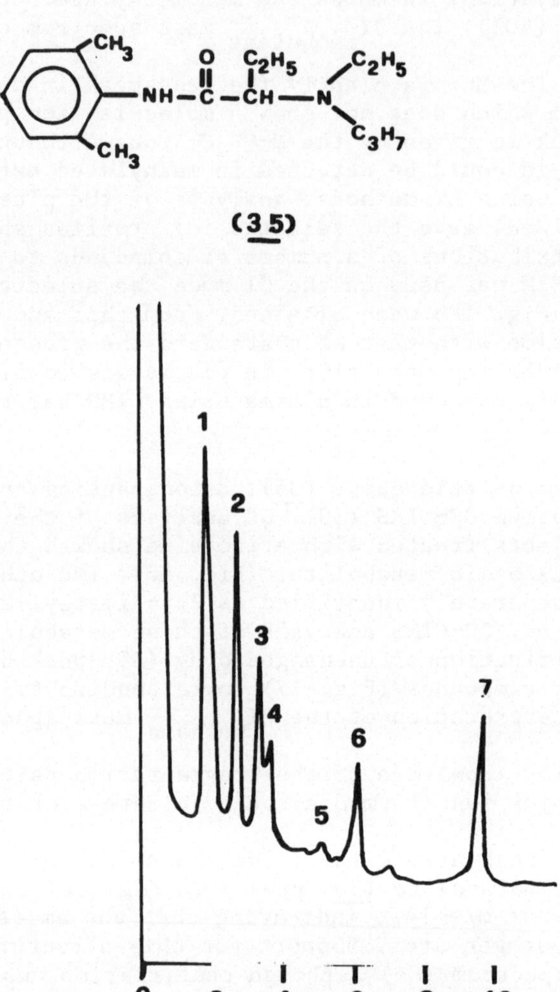

FIG. 16. Gas chromatogram of the basic metabolites of etidocaine (35). Structures corresponding to the peaks are given in Fig. 17.

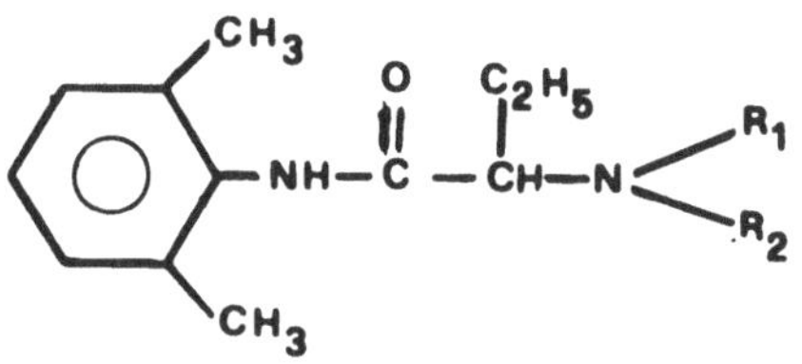

	R_1	R_2
	H	H
	C_2H_5	H
	H	C_3H_7
(35)	C_2H_5	C_3H_7

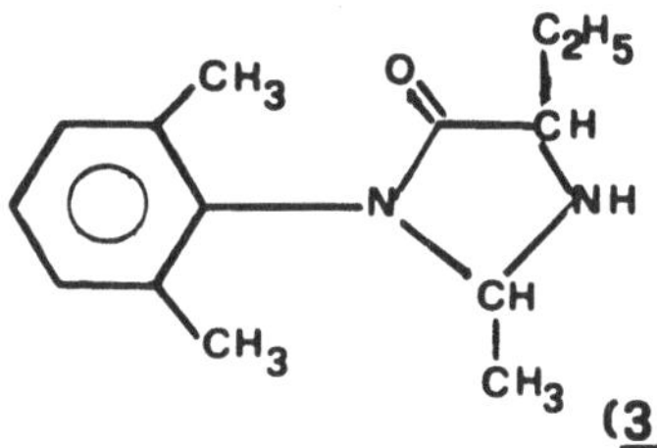

(36)

FIG. 17. Structures of etidocaine (35) and its metabolites.

TABLE 3.
Chemical ionization gases.

Reagent gas	Characteristics
H_2	Enhances MH^+ ions, More fragmentation, Can be used for G.C.
CH_4	Enhances MH^+ ions, Can be used for G.C.
H_2O	Enhances MH^+ ion, Gives abundant M+19 ions, Allows analysis of aqueous solutions.
D_2O	Determination of active hydrogens.
_i_-C_4H_{10}	Enhances MH^+ ions, less fragmentation.
NH_3	Selectively protonates amines, amides, α,β-unsatured ketones; Adds to ketones, aldehydes, ester and acids.
C_4H_{10} modifed with CH_2-NH_2 / CH_2-NH_2 or CH_2-OH / CH_2-NH_2	Can be used in place of NH_3

RECENT DEVELOPMENTS

This section deals with some of the recent developments in CI methodology. It should be pointed out that in many cases the new methods and techniques have been tested only with model substances. Their application to real-life situations awaits more investigation to determine the potential of these methods.

Reagent gases. One of the great potentials of CIMS is the possibility of varying the reagent gas used in the ionization step. Traditionally methane and isobutane have been used since the difference in their proton affinities and therefore the Bronsted activity of the reactant species covers a wide range. As shown earlier a number of other gases have been used for specific purposes. Some of these and the advantages possible are listed in Table 3. Water has been shown to be useful for the direct determination of organic compounds in aqueous solution down to 1 ppm (42). Typical ions produced are MH^+ and $M+H_3O^+$. Deuterated water appears to offer some advantages in the determination of the number of active hydrogens in a compound (43, 44). Importantly, amide hydrogens can be exchanged, whereas hydrogens α-to carbonyl groups are exchanged only in amounts less than 15% (44).

Ammonia and alkylamines selectively protonate amine, amide and α,β-unsaturated ketones and form addition products $(M+NHR)^+$ from ketones, aldehydes, acids and esters (44). To circumvent the use of ammonia, isobutane modified with ethanolamine and ethylenediamine has been used (45). Both additives have a pk_b comparable to that of ammonia. CI mass spectra obtained by using these mixtures as the reagent gas showed less fragmentation than if isobutane was used alone. Trial experiments with some biological compounds indicate that this technique may be useful in the analysis of mixtures from biological extracts. The comparative behaviour of a series of reagent gases with nucleosides, purine and pyrimidine bases has been studied to determine the mechanisms of ion formation and the relative proton affinities (46). As an example, the variation in the relative abundance of the MH^+ for adenosine is shown in Table 4. Other work has involved an estimation of the effectiveness of different functional groups on steroid molecules for directing attack by CH_5^+ and $C_2H_5^+$. The following order was estimated: carboxyl > methoxy ≃ carbonyl > bromo ≃ chloro > hydroxy (47).

A proper understanding of the protonating ability of a reagent gas is potentially of great use for detecting specific basic groups in a molecule of unknown structure. On the other hand if the functional groups present in a molecule are known then a reagent gas can be chosen which will selectively protonate one of these. This can lead to fragmentations which are initiated by, and are therefore specific for, that functional group.

Charge exchange gases. If the reagent gas does not contain

TABLE 4.
Variation in the relative abundance of MH^+ of adenosine with different reagent gases.

Gas	P.A.	% of MH^+
CH_4	127	34
$\underline{i}$-C_4H_{10}	195	36
NH_3	207	66
CH_3NH_2	216	60
$(CH_3)_2NH$	222	92
$(CH_3)_3N$	226	95

hydrogen then charge exchange or addition reactions may occur to produce ions of the compound (12). Charge exchange gases (He, Ar, N_2) yield spectra which do not differ significantly from those obtained by EI, since the recombination energies of these gases are much greater than the IP of most organic molecules. Nitric oxide (RE 8.3-9.3 eV) is one exception (44,48) and mass spectra obtained with this gas show relatively abundant $M+NO^+$ ions when groups containing π-electrons are present in the molecule (44,49). Alkanes, cycloalkanes and alkenes can be distinguished since only the latter give $M+NO^+$ ions (44, 50). Nitric oxide CI may be used to differentiate between the 1°, 2° and 3° alcohols. Primary alcohols give spectra in which M-2, M-3 and M-2+NO ions are present. The major ions present in the spectra of secondary alcohols correspond to M-1, M-17 and M-2+NO whereas M-17 ions only are produced from tertiary alcohols. Oxygen appears to be particularly useful for negative CI studies on polyaromatic and polyhalogenated compounds (44). Use of oxidizing gases however reduces filament life time. When mixtures of NO and common GC carrier gases are used, no deleterious effects are noted (48).

Mixtures of NO/N_2 (2-15%) were shown to enhance the M^+ ion in the mass spectra of steroidal alcohols and the trimethylsilyl-ethers of biologically-important polyols and glucuronides (48). The mass spectra of twelve morphine and tropane alkaloids obtained by using N_2/NO as the reagent gas all showed the M^+ as the base peak (51). In general, acetoxy, hydroxy, carbonyl and aromatic groups will not be involved in CE since their IP is greater than the RE of the NO^+ ion. Consequently fragmentations characteristic of these groups will not occur to the same extent as in the EI spectra.

In an experiment designed to compare the relative sensitivities of EI, CI and CE ionization methods the relative intensity of the molecular or MH^+ ion of heroin was measured by single ion monitoring (51). In this study EI was shown to be four times more sensitive than CI and twenty times more sensitive than CE. The use of mixtures of CI and CE gases also appears to offer some advantages. A list of useful CE gases and other gases commonly used in mixtures is given in Table 5. Most of the common GC carrier gases are good CE gases and addition of a small percentage of the CI gas can be simply effected. Mass spectra obtained with these mixtures combine the features of those obtained by either CI or EI methods. Intense MH^+ ions are observed whereas the daughter ions produced appear to arise from fragmentations directed by specific functional groups. Mixtures of Ar/H_2O (52) and Ar/CH_4 (53) have been shown to distinguish between the two isomeric compounds amobarbital and pentobarbital on the basis of their mass spectra. He/H_2O mixtures proved useful in the elucidation of the structure of the naturally-occurring isoprenoid fatty acid 6,10,14--trimethylpentadecanoic acid (13). Use of the combined CI/CE gases

TABLE 5.
Charge exchange/chemical ionization gases.

Charge exchange gas	R.E. eV	Second gas	Composition (pressure)	Major ions	% $I_t/\Sigma I_t$
Ar	15.8			Ar^+ Ar_2^+	87 6
		NO	2-15% (1 Torr)	Ar^+ NO^+	10 90
		H_2O	5% (1 Torr)	Ar^+ H_3O^+	30 70
		CH_4	5-10%	$C_2H_5^+$ CH_3^+ Ar^+	60 22 18
He	24.6			He^+	99
		H_2O NO	13%	H_3O^+ NO^+	100 100
N_2	15.3			N_2^+	67
		NO	10% (0.5-1 Torr)	N_2^+ NO^+	8-20 80-92
NO	8 - 9		(1 Torr)	NO^+	97
O_2	10 -17			O_2^+	96
				O_2^- O^-	80 20

appear be promising particularly in structural elucidation although much work needs to be done to determine the full potential of this approach.

Developments in instrumentation. Continuing efforts to develop the analytical potential of CIMS are being made. One of the major problems in pharmacological research is the degradation of some compounds when analysis is attempted by GC methods. Although this drawback can sometimes be overcome by derivative formation, it continues to limit the range of compounds which can be analysed and militates against the detection of relatively unstable metabolites. The combination of high pressure liquid chromatography (HPLC) and CIMS appears promising as a less drastic method of sample introduction into the MS (54, 55). A system in which about 1% of the liquid chromatographic effluent solution is introduced directly into the ion source of the CIMS has been developed. The solvent is evaporated and the gas acts as the ionizing agent. The solute is left as a microcrystalline structure in contact with the ion plasma, a situation which aids the ionization into the gas phase of molecules of otherwise insufficient vapour pressure. All the common liquid chromatographic solvents (H_2O, CH_3OH, CH_3CN, $CHCl_3$, n-heptane) can be used. Despite the 1% eluent split single ion monitoring allows the detection of subnanogram quantities of solute introduced onto the column (55). Also important is that in this way the MS acts as a "non-destructive" detector.

Ionization of the CI gas by Townsend discharge can be used to generate both positive and negative CI spectra with comparable sensitivity (56). In this source, oxidizing and reducing agents can be used without affecting hot metal filaments. Ionization can be carried out over a pressure range of 1 Torr to atmospheric pressure. Recent developments in Ion Cyclotron Resonance instrumentation seem promising for the study of gaseous ion--molecule reactions of low vapour pressure compounds (57). The problem of low resolution in these instruments have been overcome. Since pressures as low as 10^{-6} Torr can be used no clustering of polar reagent gases is observed. In contrast to the ion collection efficiency of high pressure CI instruments of 0.1% the trapped ICR analyser cell collects and detects all ions produced.

An instrument incorporating a new CI source has been used for the determination of molecular weights of compounds such as vitamin B_{12}, neurotoxins and other non-volatile high molecular weight compounds (58, 59). The technique relies on the generation of high localized temperatures which vapourize impurities e.g. H_2 and sodium, in the nickel foil supporting the sample. These impurities react with elemental ions formed from fission fragments of californium 252 to generate secondary ions which behave as CI agents. A time-of -flight instrument is used to analyse the ions produced.

There is little doubt that CI techniques will become more

widely used as commercial instruments become more accessible. The value of CI in the identification and quantitation of drugs and their metabolites has been established. Recent developments suggest that CI may prove particularly useful as an aid for the structural elucidation of metabolites available only in submicrogram quantities. However before the potential of this approach can be fully utilized a better understanding of the factors which determine the fragmentations observed in CI mass spectra is required. It also seems clear that CIMS is best utilized in conjunction with other ionization techniques (60), EI, FI, and FD, and an instrument incorporating all four ionization methods appears to be feasible (61).

REFERENCES

1) F.H. Field, Acc. Chem. Res., 1968, 1, 42.
2) F.H. Field, in "Mass Spectrometry", A. Maccol, Ed., M.T.P. International Review of Science, Series 1, Vol. 5, Butterworths, London, 1972.
3) B. Munson, Anal. Chem., 1971, 43, 28A.
4) G.W.A. Milne and M.J. Lacey, Crit. Rev. Anal. Chem., 1974, 45.
5) J.M. Wilson, in "Mass Spectrometry", Vol. 1, D.H. Williams, Ed., (Specialist Periodical Reports), The Chemical Society, London, 1971, Chapter 1.
6) J.M. Wilson, in "Mass Spectrometry", R.A.W. Johnstone, Ed., (Specialist Periodical Reports), The Chemical Society, Vol. 3, London, 1975, Chapter 3.
7) G.P. Arsenault, in "Biochemical Applications of Mass Spectrometry", G.R. Waller, Ed., Wiley, New York, 1971, Chapter 31, p. 817.
8) A.L. Burlingame, R.E. Cox and P.J. Derrick, Anal. Chem., 1974, 46, 248R.
9) E.C. Horning, M.G. Horning, D.I. Carrol, I. Dzidic and R.N. Stillwell, in "Advances in Biochemical Psychopharmacology", Vol. 7, E. Costa and B. Holmstedt, Eds., Raven Press, New York, 1973, p. 15.
10) J.M. Wilson, Biochem. Soc. Trans., 1976, 3, 453.
11) G. Vander Velde and J.F. Ryan, J. Chromatogr. Sci., 1975, 13, 322.
12) N. Einolf and B. Munson, Int. J. Mass Spectrom. Ion Phys., 1972, 9, 141.
13) G.P. Arsenault, J. Am. Chem. Soc., 1972, 94, 8341.
14) G.W.A. Milne, H.M. Fales and T. Axenrod, Anal. Chem., 1971, 43, 1815.
15) B.S. Finkle, R.L. Foltz and D.M. Taylor, J. Chromatogr. Sci., 1974, 12, 304.
16) J.M. Chao, R. Saferstein and J. Manura, Anal. Chem., 1974, 46, 296.
17) W.A. Garland, W.F. Trager and S.D. Nelson, Biomed. Mass

Spectrom., 1974, 1, 124.
18) J.L. Valentine, P. Driscoll, E.L. Halburg and E.D. Thompson, J. Pharm. Sci., 1976, 65, 96.
19) T.T.L. Chang, Ch.F. Kuhlman, R.T. Schillings, S.F. Sisenwine, C.O. Tio and H.W. Ruelius, Experientia, 1973, 29, 653.
20) R.S. Egan, S.L. Mueller, L.A. Mitscher, I. Kawamoto, R. Okachi, H. Kato, S. Yamamoto, S. Takasawa and T. Nara, J. Antibiot., 1974, 27, 544; and references therein; J.D. Wander and R.L. Foltz, Anal. Biochem., 1974, 59, 452.
21) L.A. Mitscher, H.D.H. Showalter, K. Shirahata and R.L. Foltz, J. Antibiot., 1975, 28, 668.
22) M.A. Baldwin and F.W. McLafferty, Org. Mass Spectrom., 1973, 7, 1353.
23) S.B. Matin and J.B. Knight, Biomed. Spectrom., 1974, 1, 323.
24) L.R. Pohl, S.D. Nelson, W.A. Garland and W.F. Trager, Biomed. Mass Spectrom., 1975, 2, 23.
25) D.J. Jenden and A.K. Cho, Annu.Rev. Pharmacol., 1973, 13, 371.
26) A.M. Lawson and G.H. Draffan, in "Progress in Medicinal Chemistry", Vol. 12, G.P. Ellis and G.B. West, Eds., North Holland Publishing Co., Amsterdam and Oxford, 1975, p. 1; and A.M. Lawson, Clin. Chem., 1975, 21, 803.
27) C.J.W. Brooks and B.S. Middletich, in "Mass Spectrometry", Vol. 3, R.A.W. Johnstone, Ed., (Specialist Periodical Reports), The Chemical Society, London, 1975, p. 296; and B.J. Millard, ibid., p. 339.
28) F.C. Falkner, B.J. Sweetman and J.T. Watson, Appl. Spectros. Rev., 1975, 10, 51.
29) C. Fenselau, Appl. Spectrosc., 1974, 28, 305.
30) D.M. Schoengold and B. Munson, Anal. Chem., 1970, 42, 1811.
31) W. Blum and W.J. Richter, Tetrahedron Lett., 1973, 835.
32) J. Ynon and H. Boettger, Chem. Instrum., 1972, 4, 103.
33) W. Blum and W.J. Richter, Finnigan Spectra, 1975, 5.
34) D.C.K. Lin, A.F. Fentiman, R.L. Foltz, R.D. Forney and I. Sunshine, Biomed. Mass Spectrom., 1975, 2, 206.
35) S.B. Matin, J.H. Karam, P.H. Forsham and J.B. Knight, Biomed. Mass Spectrom., 1974, 1, 320.
36) C. Pantarotto, A. Martini, G. Belvedere, M.G. Donelli and A. Frigerio, Cancer Treat. Rep., 1976, 60, 493.
37) C. Fenselau, M.-N.N. Kan, S. Billets and M. Colvin, Cancer Res., 1975, 35, 1453.
38) I. Jardine, R. Brundett, M. Colvin and C. Fenselau, Cancer Treat. Rep., 1976, 60, 403.
39) J. Thomas, D. Morgan and J. Vine, Xenobiotica, 1976, 6, 39.
40) M.G. Horning, J. Nowlin, C.M. Butler, K. Lertratanangkoon, K. Sommer and R.M. Hill, Clin. Chem., 1975, 21, 1282; and references therein.
41) M.G. Horning, C.M. Butler, J. Nowling and R.M. Hill, Life Sci., 1975, 16, 651.
42) P. Price, D.P. Martinsen, R.A. Upham, H.S. Swofford and S.E. Buttril, Anal. Chem., 1975, 47, 190.

43) D.F. Hunt, C.N. McEwen and R.A. Upham, Anal. Chem., 1972, 44, 1292.
44) D.F. Hunt, Finnigan Spectra, 1976, 6.
45) D.V. Bowen and F.H. Field, Org. Mass Spectrom., 1974, 9, 195.
46) M.S. Wilson and J.A. McCloskey, J. Am. Chem. Soc., 1975, 97, 3436.
47) J. Michnowicz and B. Munson, Org. Mass Spectrom., 1974, 8, 49.
48) B.L. Jelus, B. Munson and C. Fenselau, Biomed. Mass Spectrom., 1974, 1, 96.
49) D.F. Hunt and M.T. Harvey, Anal. Chem., 1975, 47, 1730 and 2136.
50) B.L. Jelus, B. Munson, K.A. Babiak and R.K. Murray, J. Org. Chem., 1974, 39, 3250.
51) I. Jardine and C. Fenselau, Anal. Chem., 1975, 47, 730.
52) D.F. Hunt and J.F. Ryan, Anal. Chem., 1972, 44, 1306.
53) D.P. Beggs, Hewlett Packard, Application Note, AN176-19.
54) F.W. McLafferty and B.G. Dawkins, Biochem. Soc. Trans., 1975, 3, 856; and P.J. Arpino, B.G. Dawkins and F.W. McLafferty, J. Chromatogr. Sci., 1974, 12, 574.
55) F.W. McLafferty, R. Knutti, R. Venkataraghavan, P.J. Arpino and B.G. Dawkins, Anal. Chem., 1975, 47, 1503.
56) D.F. Hunt, C.N. McEwen and T.M. Harvey, Anal. Chem., 1975, 47, 1730.
57) R.T. McIver, E.B. Ledford and J.S. Miller, Anal. Chem., 1975, 47, 692.
58) D.F. Torgeson, R.P. Skowronski and R.D. Macfarlane, Biochem. Biophys. Res. Commun., 1974, 60, 616.
59) T.H. Maugh, Science, 1975, 187, 529.
60) H.M. Fales, G.W.A. Milne, H.U. Winkler, H.D. Beckey, J.N. Damico and R. Barron, Anal. Chem., 1975, 47, 207.
61) H.H. Gierlich, A. Heindrichs and H.D. Beckey, Rev. Sci. Instrum., 1974, 45, 1208.

FIELD DESORPTION MASS SPECTROMETRY OF PHARMACOLOGICALLY ACTIVE SALTS

W.D. Lehmann and H. -R. Schulten

Institute of Physical Chemistry, University of Bonn

5300 Bonn, W. Germany

INTRODUCTION

In the mass spectrometric analysis of drugs and their metabolites new ionisation techniques such as chemical ionisation or field ionisation/field desorption are of great interest. Compared to the conventional electron impact ionisation these techniques involve a relatively low energy transfer during the process of ion formation. Furthermore, the field desorption method offers the unique advantage that no separate evaporation process for the sample is required. Thus field desorption mass spectrometry (FDMS) seems to be especially suited for the analysis of thermally labile and/or highly polar compounds. The utility of FDMS for the analysis of a wide variety of organic and inorganic compounds has been demonstrated in the last years (1, 3). FDMS of organic and inorganic salts (4, 22, 32, 33) is an expanding field since many of these intact salts cannot be analysed successfully by EI or CI mass spectrometry.

The FD spectra of inorganic and many organic salts are characterised by high ion intensities of the salt cations. In addition cluster ions with m/e values above the molecular weight (M = C + A; M=molecular weight, C=cation, A=anion) of the salt are observed. These cluster ions are produced by field induced reactions on the emitter surface and normally enable the determination of the anion of the salt. Thus FDMS can be used for molecular weight determination of inorganic and organic salts. Since drugs are frequently synthesized and applied as salts we have investigated a number of representative hydrochlorides, sulphates and sulphonates which have pharmacological activity in order to explore the use of the technique for studies in drug research.

Table 1. The three most intense ions (except the cations) in the FD mass spectra of a number of inorganic salts.

Salt	
LiF	$(LiF)Li^+$, $(LiF)_2Li^+$, $(LiF)_3Li^+$
LiCl	$(LiCl)Li^+$, $(LiCl)_2Li^+$, $LiCl^+$
LiJ	$(LiJ)Li^+$, J^+, $(LiJ)_2Li^+$
Li_2CO_3	$(Li_2CO_3)Li^+$, $(Li_2CO_3)_2Li^+$
NaCl	$(NaCl)Na^+$, $(NaCl)_2Na^+$, $NaCl^+$
NaN_3	$(NaN_3)Na^+$
$NaNO_3$	$(NaNO_3)Na^+$, $(NaNO_3)_2Na^+$, $(NaNO_3)_2Na^+$
Na_2SO_4	$(Na_2SO_4)Na^+$
$NaClO_3$	$(NaClO_3)Na^+$, $(NaClO_3)_2Na^+$, $(NaCl)Na^+$
$Na(CH_3CO_2)$	$Na(CH_3CO_2)\ Na^+$, $Na(CH_3CO_2)\ _2Na^+$
KF	$(KF)K^+$, $(KF)_2K^+$
KCl	$(KCl)K^+$, $(KCl)_2K^+$, KCl^+
KBr	$(KBr)K^+$, Br^+
KJ	$(KJ)K^+$, $(KJ)_2K^+$, J^+
AgCl	$AgCl^+$, $(AgCl)Ag^+$
$AgNO_3$	$(AgNO_3)Ag^+$, $(AgNO_3)_2Ag^+$
$CaCl_2$	$CaCl^+$, Ca^+, $CaCl_2^+$

EXPERIMENTAL

All FD spectra were run on a double-focussing CEC 21-1108 mass spectrometer at a resolution of 15000 (at half peak width). The ions were detected on vacuum-evaporated AgBr photoplates (Ionomet, Waban, Mass.,). Field anodes employed were 10 μm tungsten wires activated at high temperature with microneedles of 30 μm length on average (5). The ionisation efficiency of the emitters was checked in the field ionisation mode using the signal at m/e 58 of acetone. The FD spectra were produced by emission-controlled desorption (18, 23) with a threshold of 2×10^{-8} A. For the organic hydrochlorides and the sulphate salts the thermograms (25) show that a constant total emission is achieved for values between 12 and 18 mA emitter heating current.

RESULTS AND DISCUSSION

<u>Inorganic salts</u>. When organic compounds containing small amounts of an inorganic salt, e.g. a sodium or potassium salt, are analysed by FDMS the salt cations are displayed with very high intensity in the FD spectrum. In the FD spectra of inorganic salts the salt cations give by far the most intense ion signals. Additionally, a series of cluster ions is observed that follow the general formula $[nM + C]^+$,where M is the intact salt molecule and C is the salt cation. Table 1 shows the characteristic ions in the FD spectra of a number of inorganic salts.

Sometimes, with lower relative intensity, even the salt molecule (e.g. sodium and potassium chloride) or the salt anion (e.g. bromine, iodine) can be observed as positively charged anion. The utility of FD mass spectrometry for the detection of inorganic salts in genuine samples has recently been demonstrated. Inorganic salts have been found in natural aerosol samples (18), and commercial products such as drugs (3, 31), pesticides (22) and dyestuffs (26).

<u>Organic hydrochlorides</u>. A number of basic organic compounds are available as the reaction products with hydrochloric acid. The organic hydrochlorides I - IX, structures are shown in Fig. 1, were investigated by FDMS.

The characteristic FD ions of these organic hydrochlorides are exhibited in Table 2.

All compounds display high cation and $[C-H]^+$ ion intensities. Some of the hydrochlorides show $[C+(C-H)]^+$ cluster ions and the compounds VI - VIII give doubly charged ions of low relative intensity as frequently observed in the FD mass spectra of organic compounds containing an aromatic system. In addition to the ions displayed in Table 2 only novocaine (IX) shows a FD cluster ion

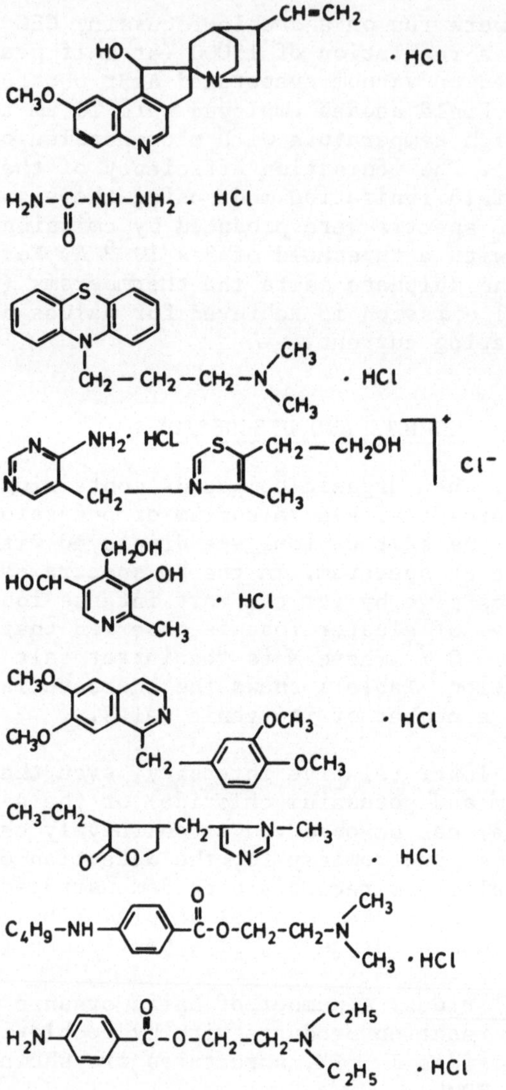

(I)

(II)

(III)

(IV)

(V)

(VI)

(VII)

(VIII)

(IX)

FIG. 1. Structures of the investigated hydrochlorides: (I) chinine hydrochloride, (II) semicarbazide hydrochloride, (III) chlorpromazine hydrochloride, (IV) thiamine chloride hydrochloride, (V) pyridoxine hydrochloride, (VI) papaverine hydrochloride, (VII) pilocarpine hydrochloride, (VIII) pantocaine hydrochloride, (IX) novocaine hydrochloride.

Table 2. Characteristic ions in the FD mass spectra of the organic hydrochlorides I-IX. The relative intensities of the signals are given in four degrees of blackening of the photoplate (high rel.int.= + + + +. medium rel.int.= + + +, low rel.int.= + +, weak rel.int.= +).

		$C-H^{++}$	C^{++}	C - H	C	C + (C-H)
(I)	chinine hydrochloride			++++	++	
(II)	semicarbazide hydrochloride			+++	++++	+++
(III)	chlorpromazine hydrochloride			++++	+	
(IV)	thiamine chloride hydrochloride			++	++++	
(V)	pyridoxine hydrochloride			+++	++++	++++
(VI)	papaverine hydrochloride	++		++++	+++	
(VII)	pilocarpine hydrochloride	++	+	++++	+++	+++
(VIII)	pantocaine hydrochloride	+		++++	++++	
(IX)	novocaine hydrochloride			+++	++++	

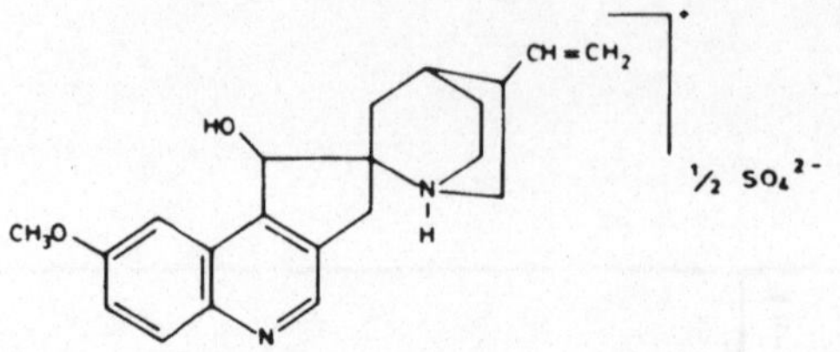

FIG. 2. Structures of the investigated organic sulphate salts: (X) chinine sulphate, (XI) strychnine sulphate, (XII) atropine sulphate, (XIII) bamethane sulphate, (XIV) isoprenaline sulphate, (XV) orciprenaline sulphate, (XVI) amphetamine sulphate. (Ref. 23)).

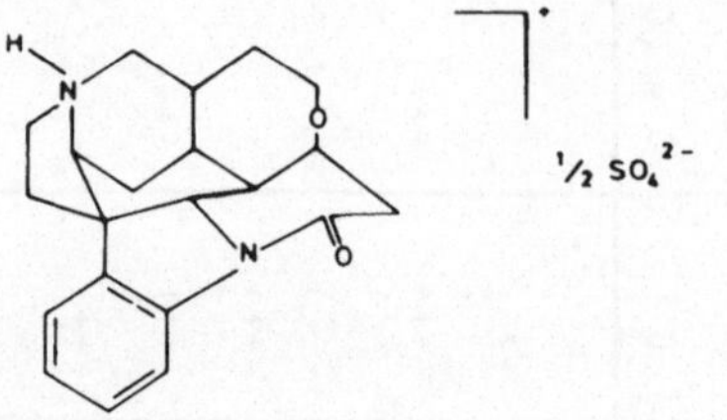

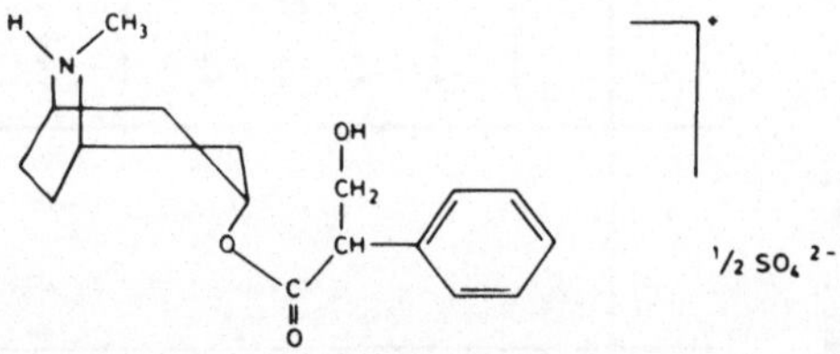

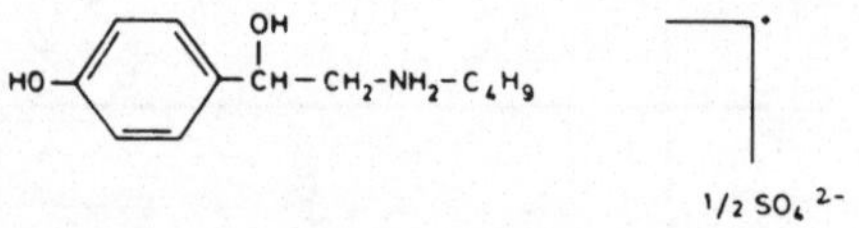

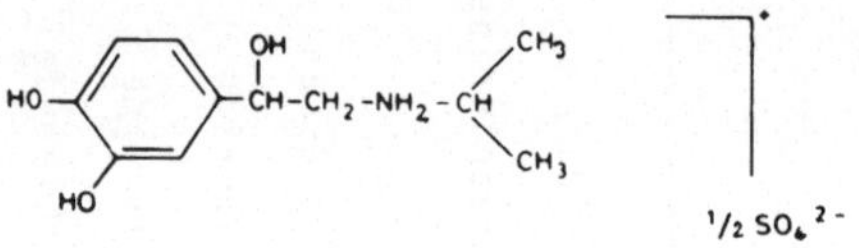

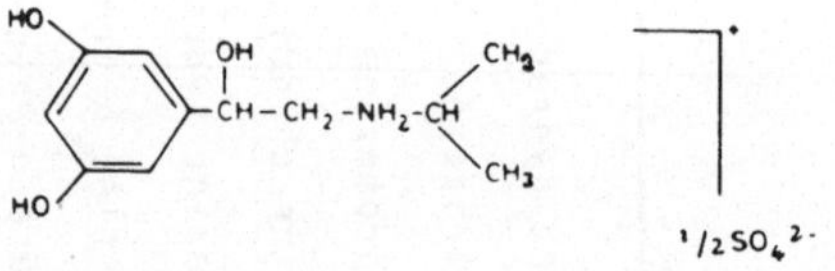

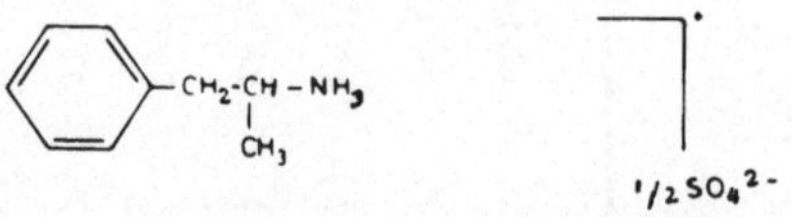

$[M + C]^+$ at m/e 509 that contains the intact hydrochloride. As observed with other ionization techniques during field desorption the organic hydrochlorides readily undergo HCl elimination. This is confirmed by signals at m/e 36 and m/e 38 for hydrochloric acid.

Organic sulphate salts. Many basic drugs are applied as their sulphate salts. In Fig. 2 the structure of seven organic sulphate salts (X - XVI) with large organic cations are shown.

Corresponding to the FD spectra of the hydrochloride salts the most intense ions in the spectra of the sulphate salts are the $[C - H]^+$ and $[C]^+$ ions. However, cluster ions and multiply charged ions are observed with strongly enhanced intensities. In Table 3 the characteristic singly charged ions in the FD spectra of the sulphate salts X - XVI are displayed.

The salt cation in all cases is by far the most intense ion species. In the region of higher m/e values than the molecular unit of the salt a set of cluster ions containing at least one molecule of H_2SO_4 is observed. From these cluster ions, e.g. the $[C + H_2SO_4]^+$ ion was present in all spectra, the anionic moiety of the sulphate salt can be derived. The extent of formation of organic cluster ions, purely inorganic ions, and multiply charged ions during the field desorption of organic sulphate salts is strongly dependent on the emitter heating program employed. Fig. 3 shows the FD spectrum of amphetamine sulphate (XVI) produced by emission-controlled desorption at a threshold of 2x 10^{-8} A. The ions produced with an emitter heating current between 18 and 30 mA were detected.

Using the high resolution data, ions of organic and purely inorganic composition could easily be discerned. In order to simplify the understanding of the spectrum in Fig. 3 the ions with inorganic composition are plotted downwards whereas all other ions are displayed on the upper half of the Figure. Since it was confirmed by C, H, N, O elemental analysis and $[SO_4]^{2-}$ analysis that the sample of amphetamine sulphate contained no inorganic impurities the set of inorganic cluster ions is a product of the field desorption process. Among the organic ions in the FD spectrum of amphetamine sulphate the salt cation is the most intense ion species. In the region of higher m/e values a set of cluster ions containing at least one molecule of sulphuric acid is observed. The complex spectrum in Fig. 3 (emitter heating current 18-30 mA) can be considerably simplified when only the fraction of ions produced with an emitter heating current between 18 and 21 mA is recorded. Fig. 4 shows that under these conditions only a few significant ions are detected that enable the identification of cationic and the anionic moiety of the sulphate salt.

In the emission-controlled FD spectra of sulphate salts that

Table 3. Characteristic ions in the FD mass spectra of the organic sulphate salts X-XVI. The relative intensities are given in four degrees as in Table 2.

	C-H - H_2O	C-H_2O	C-2H	C-H	C	C-H-H_2O + H_2SO_4	C-H_2O +H_2SO_4	C+H_2SO_4	C+2H_2SO_4	2C-H +H_2SO_4
(X) chinine sulphate				++	++++			+++	++	
(XI) strychnine sulphate	+			++++	+++	+++		++		+
(XII) atropine sulphate	+++	+		++++	+++			+		
(XIII) bamethane sulphate	+++	++		+++	++++			++		
(XIV) isoprenaline sulphate		+++	+++	+++	++++		+++	++		
(XV) orciprenaline sulphate		+++	+++	+++	++++		+++	++		
(XVI) amphetamine sulphate			+++	+++	++++			+++		+++

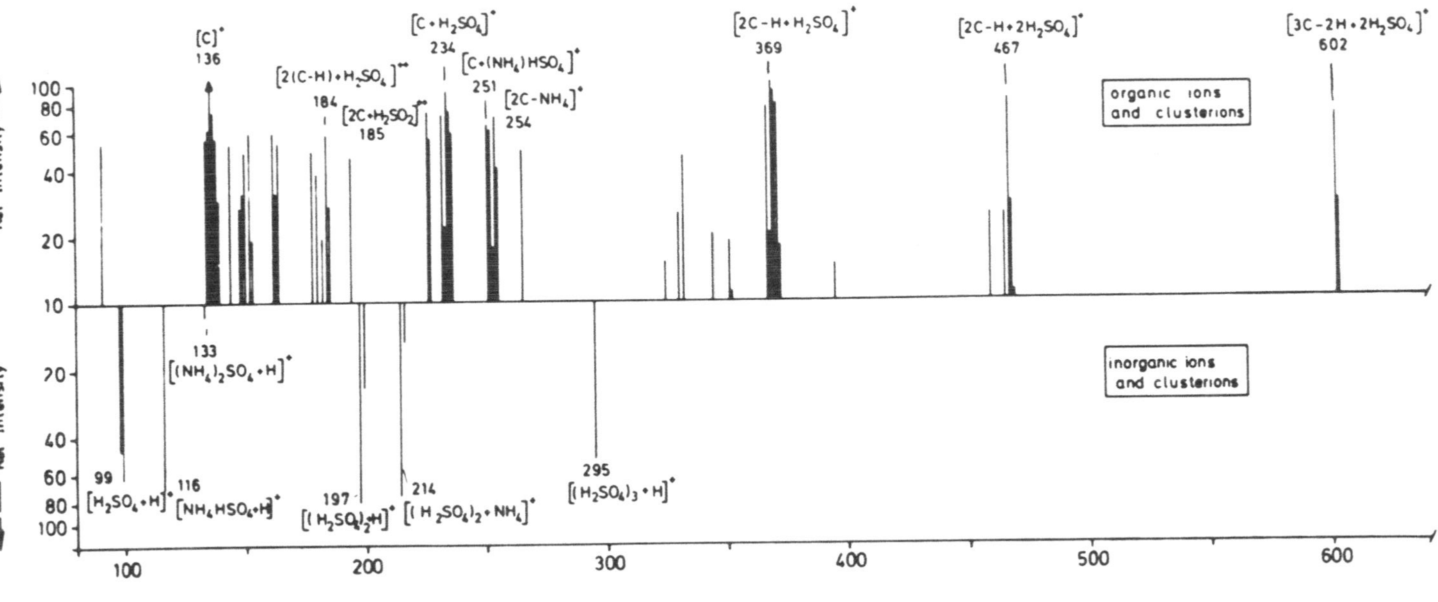

FIG. 3. FD spectrum of amphetamine sulphate (XVI). Exposure time of the photoplate: 11 min; emitter heating current: 18-30 mA; threshold: 6 x 10 8 A; recorded mass range: m/e 30 - m/e 960. (Ref. 23)).

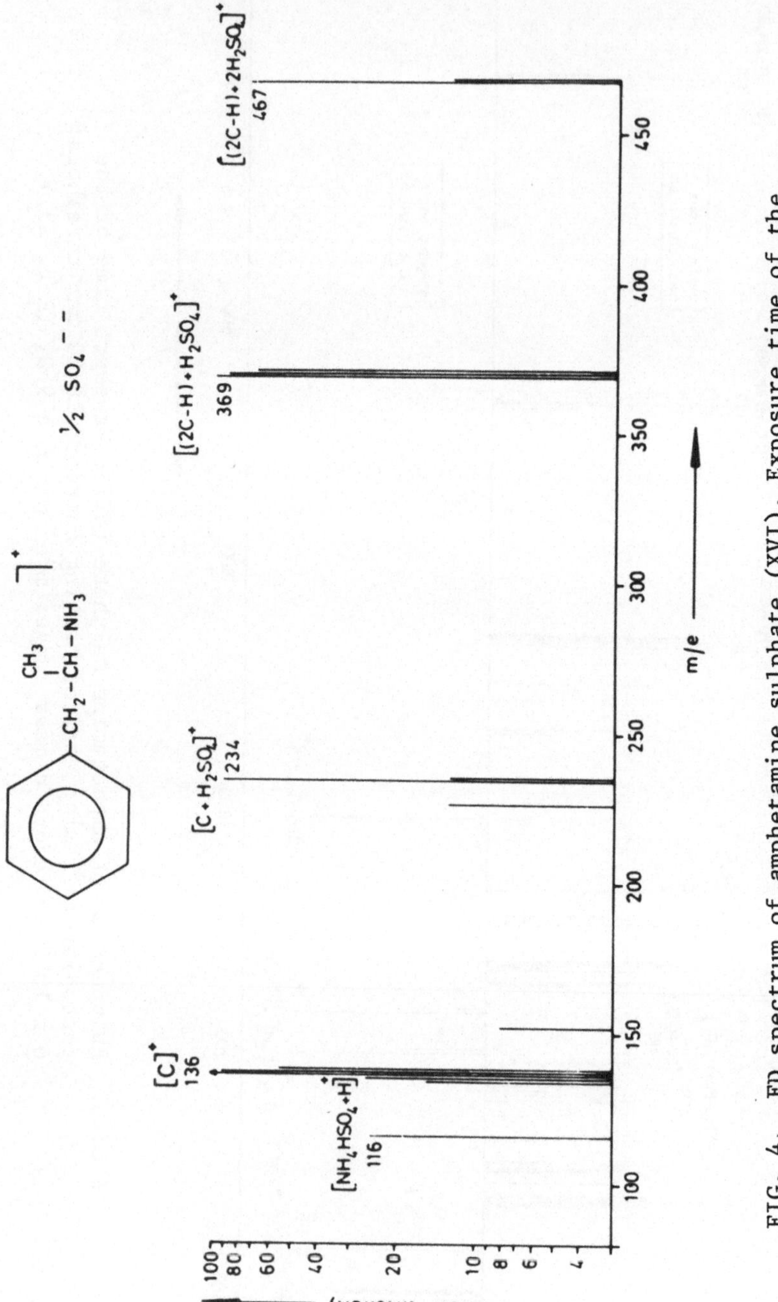

FIG. 4. FD spectrum of amphetamine sulphate (XVI). Exposure time of the photoplate: 6 min; emitter heating current 18-21 mA; threshold: 2×10^{-8} A; recorded mass range: m/e 17 - m/e 560. (Ref. 23)).

include the ions produced during the total desorption process the occurrence of relatively intense doubly and triply charged ions is a common feature. The building blocks of these multiply charged cluster ions in general are also observed as intense singly charged ions. This is demonstrated for isoprenaline sulphate (XIV) in Table 4.

It is suggested that the multiply charged ions of large organic sulphate salts are generated under FD conditions <u>via</u> aggregation of singly charged ions bound in the adsorbed layer on the emitter surface.

<u>Organic alkali salts</u>. Organic salts containing large organic anions and alkali cations such as Li^+, Na^+, K^+ can also be investigated by FDMS. Fig. 5 shows the FD spectrum of the potassium salt of phenethicilline.

Besides a number of structurally significant fragment ions no molecular ion but an intense signal at m/e 441 for $[M + {}^{39}K]^+$ is recorded. This type of ion parallels the $[M + C]^+$ ion always observed in the FDMS of inorganic salts.

The FD spectra of a number of sulphonic acid alkali salts have been reported to show intense $[M + C]^+$ ions besides $[M]^+$ ions (26). Protonation can also be an ion generating process in the field desorption of an organic alkali salt as is demonstrated in Fig. 6 for 3-deoxyfluoro-D-glucose-6-disodium phosphate.

The ion $[M + Na]^+$ though not displayed in Fig. 6 is also present with high relative intensity. A partial decomposition of the phosphate ester salt is indicated by a set of inorganic phosphate cluster ions. To our present experience ions of the type $[M + C]^+$ are always observed in the FD spectra of organic alkali salts (sometimes higher cluster ions $[nM + C]^+$ also occur). Additionally the intact alkali salt molecule can be ionised or protonated and thus a $[M]^+$ or $[M + H]^+$ ion is detected.

Ions of the type $[M + C]^+$ are also observed when an organic compound is mixed with a salt, preferably alkali halides (27). For a mixture of adenosine with LiCl it has been found that with increasing concentrations of LiCl the $[M]^+$ and $[M + H]^+$ ions of adenosine are suppressed in favour of an intense signal for $[M + Li]^+$. Simultaneously the intensity of the organic ions decreases and the fluctuations of the ion current are enhanced. Moreover the fragmentation pattern is governed by the attachment of alkali cations as has been shown for the FD spectra of oligonucleotides containing sodium chloride as an impurity (28).

Efforts have been made to eliminate the influence of the inorganic salts by addition of macrocyclic ligands to the sample (29) prior to the FD analysis. It is also possible to remove the

Table 4. Doubly and triply charged ions in the FD spectrum of isoprenaline sulphate (XIV). The relative intensities are given in four degrees as in Table 2. (Ref. 23).

doubly charged ions	rel.int.	triply charged ions	rel.int.
$[2C-H_2O]^{++}$	++	$[2(C-H_2O) + C]^{+++}$	+++
$[2C-H]^{++}$	+++	$[3(C-H_2O) + (C-H) + H_2SO_4]^{+++}$	++
$[2(C-H_2O) + H_2SO_4]^{++}$	+++		
$[(2C-H_2O) + H_2SO_4]^{++}$	++		
$[2(C-H) + H_2SO_4]^{++}$	++		
$[2C + H_2SO_4]^{++}$	++		

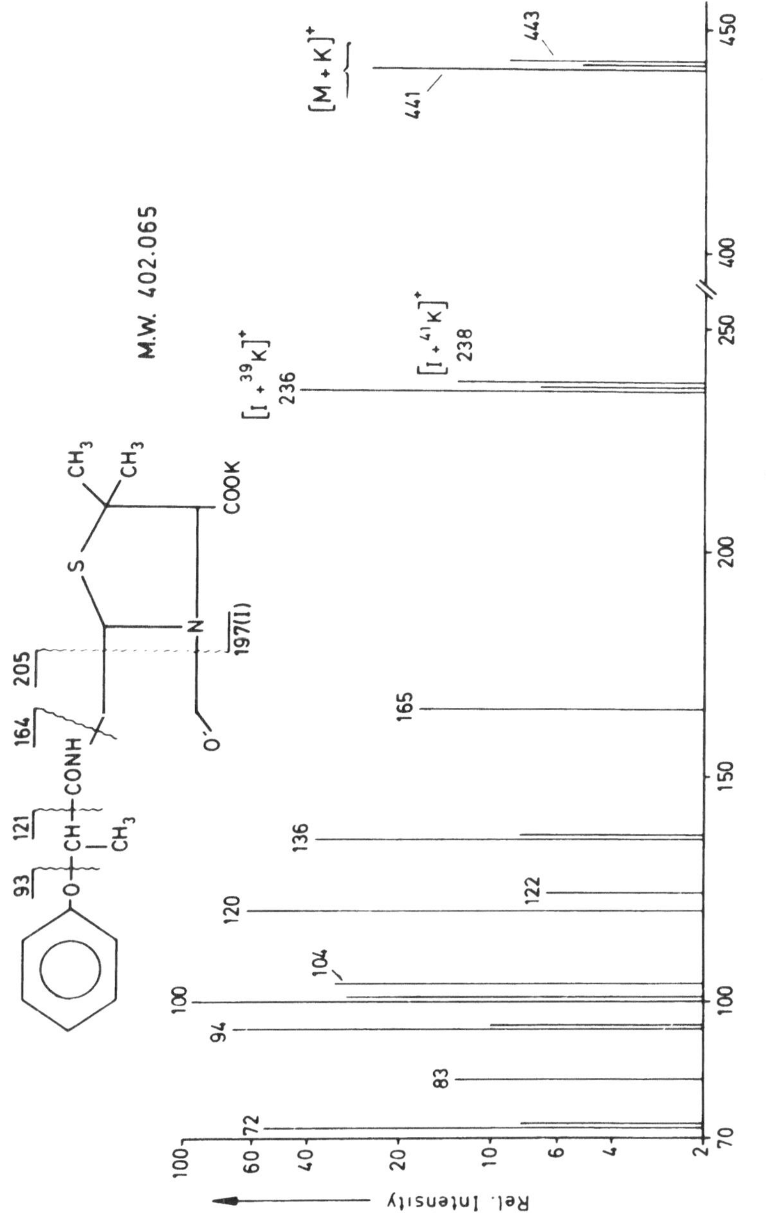

FIG. 5. FD mass spectrum of phenethicilline potassium salt.

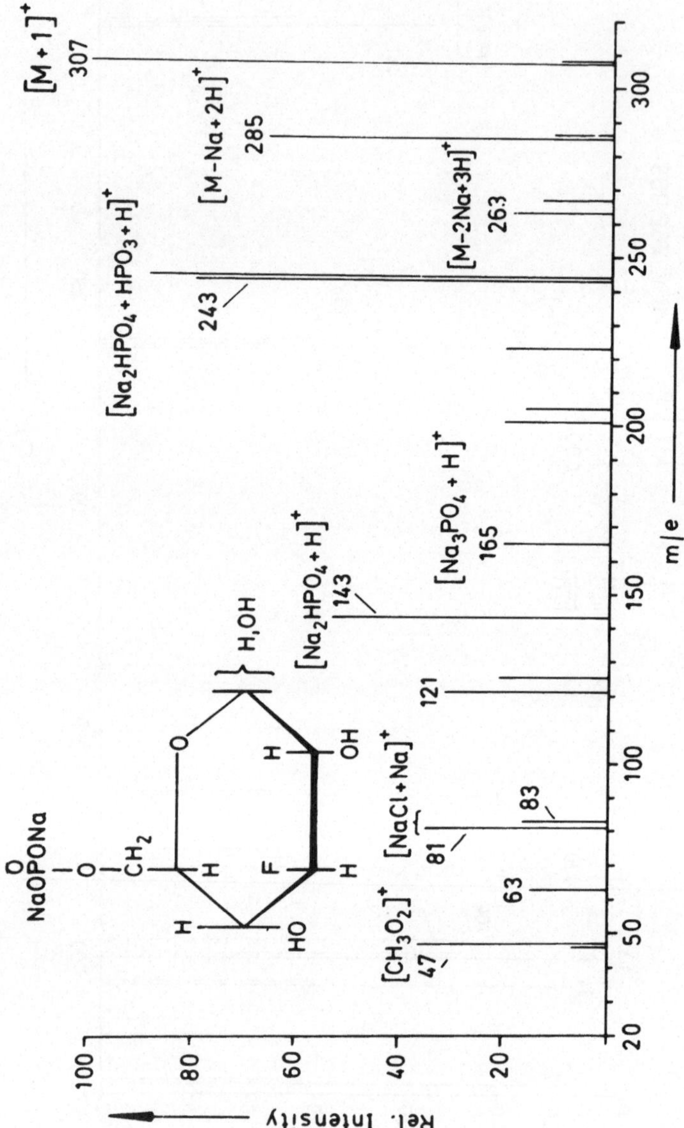

FIG. 6. FD mass spectrum of 3-deoxyfluoro-D-glucose-6-disodium phosphate.

inorganic salts, especially sulphate and phosphate salts, from the sample by a combined barium precipitation/anion exchange procedure (30). However, it has been demonstrated that an organic salt can be analysed by FDMS even in the presence of an excess of inorganic salt (31) when emission-controlled desorption and photographic detection was employed.

CONCLUSION

Inorganic as well as organic salts can be investigated by FDMS. In general the positively charged moiety of the salt is observed as the most intense ion in the FD spectra. Further, the FD spectra of salts are characterised by a number of cluster ions above the m/e value of the molecular weight of the salt from which the anionic moiety of the salt can be deduced.

The FD spectra of salts show a strong dependence on the emitter temperature program used. Relatively high anode heating currents ($\sim$20mA) and high desorption rates favour the process of ion cluster formation on the emitter surface. We obtained simple and significant FD spectra of salts by recording only the first part of a total desorption process that was run by emission-controlled desorption.

Because of the fluctuations in the ions currents observed in FDMS of salts the photoplate appears to be most reliable detection system for this class of compounds.

ACKNOWLEDGMENTS

This work was supported financially by the "Deutsche Forschungsgemeinschaft", the "Fonds der Deutschen Chemischen Industrie" and the "Ministerium für Wissenschaft und Forschung des Landes Nordrheinwestfalen".

REFERENCES

1) H.D. Beckey and H. -R. Schulten, Angew. Chem. Int. Ed. Engl., 1975, 14, 403.
2) H.D. Beckey and H. -R. Schulten, Z. Anal. Chem., 1975, 273, 345.
3) H. -R. Schulten, in "Methods of Biochemical Analysis", Vol. 24, by D. Glick, Ed., Interscience-Wiley, New York, in press, 1976.
4) H.U. Winkler and H.D. Beckey, Org. Mass. Spetrom., 1972, 6, 655.
5) H. -R. Schulten and H.D. Beckey, Org. Mass Spectrom., 1972,

6, 885.
6) H. -R. Schulten, H.D. Beckey, E.M. Bessel, A.B. Foster, M. Jarman and J.H. Westwood, Chem. Commun., 1973, 13, 416.
7) D.A. Brent, D.J. Rouse, M.C. Sammons and M.M. Bursey, Tetrahedron Lett., 1973, 4127.
8) D.A. Brent, P. deMiranda and H. -R. Schulten, J. Pharm. Sci., 1974, 63, 1370.
9) H. -R. Schulten, Biomed. Mass Spectrom., 1974, 1, 223.
10) R. Large and H. Knof, Chem. Commun., 1974, 935.
11) M.C. Sammons, M.M. Bursey and D.A. Brent, Biomed. Mass Spectrom., 1974, 1, 169.
12) D.E. Games, M.P. Games, A.H. Jackson, A.H. Olavesen, M. Rossiter and R.J. Winterburn, Tetrahedron Lett., 1974, 2377.
13) H. -R. Schulten and H.D. Beckey, in "Advances in Mass Spectrometry", **Vol**. 6, by A.R. West, Ed., Applied Science Publications, London, 1974, p. 499.
14) H. -R. Schulten and F.W. Röllgen, Angew. Chem. Int. Ed. Engl., 1975, 14, 561.
15) H. -R. Schulten and F.W. Röllgen, Org. Mass Spectrom., 1975, 10, 649.
16) M.C. Sammons, M.M. Bursey and C.K. White, Anal. Chem., 1975, 47, 1165.
17) D.E. Games, A.H. Jackson, L.A.P. Kane-Maguire and K. Taylor, J. Organomet. Chem., 1975, p. 435.
18) U. Schurath and H. -R. Schulten, Atmos. Environ., 1975, 9, 1107.
19) G.W. Wood, J.M. McIntosh and P.Y. Lau, J. Org. Chem., 1975, 40, 636.
20) M. Anbar and G.A. St. John, J. Am. Chem. Soc., 1975, 87, 544.
21) H.J. Veith, Org. Mass Spectrom., 1976, 11, 629.
22) H. -R. Schulten, J. Agric. Food Chem., 1976, 24, xxx.
23) H. -R. Schulten and W.D. Lehmann, Anal. Chim. Acta, in press, 1976.
24) H. -R. Schulten and H.D. Beckey, in "Recent Advances in Field Desorption Mass Spectrometry", Twenty-Third Annual Conference On Mass Spectrometry And Allied Topics, Houston, Texas, May 25-30, Conf. Proceedings B-1, 1975.
25) W.D. Lehmann, H.D. Beckey and H. -R. Schulten, in "Proceedings of International Symposium on Quantitative Mass Spectrometry in Life Sciences", A.P. De Leenheer and R. Roncucci, Eds., Elsevier, Amsterdam, 1976.
26) H. -R. Schulten and D. Kümmler, Z. Anal. Chem., 1976, 278, 13.
27) F.W. Röllgen and H. -R. Schulten, Org. Mass Spectrom., 1975, 10, 660.
28) H. -R. Schulten and H.M. Schiebel, Z. Anal. Chem., 1976, 280, 139.
29) G.W. Wood, P.Y. Lau and N. Mak, Biomed. Mass Spectrom., 1974, 1, 425.
30) J.A. Thompson and S. Markey, Anal. Chem., 1975, 47, 1313.
31) H. -R. Schulten and H.D. Beckey, in "Twenty-Fourth Annual

Conference on Mass Spectrometry and Allied Topics", San Diego, California, May 9-14, Conf. Proceedings K 12, 1976.
32) M. Anbar and G.A. St. John, Anal. Chem., 1976, 48, 198.
33) M. Anbar and G.A. St. John, J. Am. Chem. Soc., 1975, 97, 7195.

ROLE OF COMPUTERS IN GAS CHROMATOGRAPHY-MASS SPECTROMETRY:DETECTION, IDENTIFICATION, AND QUANTIFICATION IN DRUG METABOLISM RESEARCH

J. Roboz

Mount Sinai School of Medicine, The City University of New York, New York City, N.Y. 10029, USA

INTRODUCTION

During the last few years, rapid progress in instrumentation and techniques resulted in the commercial availability of interactive data systems with the aim to maximize the productivity of both the combined gas chromatograph-mass spectrometer-computer complex and the operator. Dedicated minicomputers, with continually increasing performance/cost ratios, are becoming the rule rather than the exception for the acquisition and handling of data. The availability of continuous repetitive scanning of mass spectra throughout the entire gas chromatographic run provides a tremendous amount of information which can be recorded, interrogated, and even interpreted in essentially real time. Alternatively, there is the option to be extremely selective and monitor only selected constituents or their fragments, disregarding everything else present. Both techniques, and various combinations, have been used for the identification and quantification of drugs and their metabolites.

Computers have 4 major areas of application in combined gas chromatography-mass spectrometry: (a) control of the operation of the mass spectrometer (and also the gas chromatograph) before and during analysis; (b) detection of drugs and metabolites, and also those endogeneous constituents influenced by the metabolism of the drug; (c) confirmation of the identity of suspected metabolites, and identification of newly detected metabolites; (d) quantification of drugs, metabolites, and endogeneous constituents of interest. The objective of this review is to describe and discuss a few selected techniques in each area of application of computers, emphasizing their advantages and limitations in the use of combined gas chromatography-mass spectrometry in drug metabolism studies.

COMPUTER CONTROLLED OPERATION OF GC/MS SYSTEM

Most mass spectrometer-computer systems have one of the two basic configurations shown in Fig. 1. Mass spectrometers with

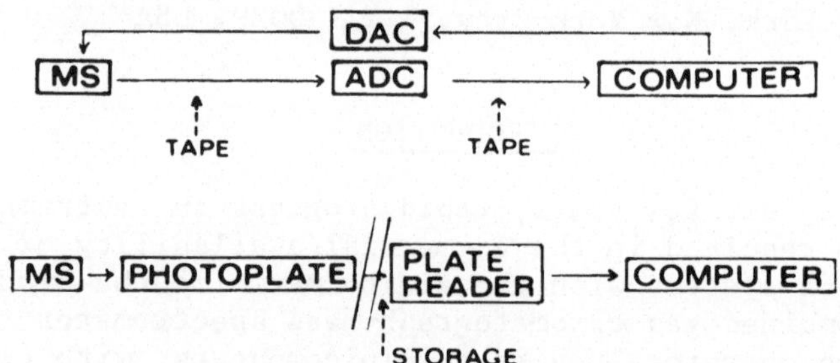

FIG. 1. Basic configurations of mass spectrometer computer system; MS=mass spectrometer; ADC=analog-to-digital converter; DAC= digital-to-analog converter.

electrical detectors, including both low- and high-resolution instruments, are usually connected to the computer *via* an analog--to-digital converter interface which enables the computer to acquire and process all ion current signals (individual and/or total ion currents) for the determination of masses and corresponding ion intensities which, in turn, are the bases of all subsequent calculations and manipulations. In interactive systems, computer control of the operation of the mass spectrometer (and possibly also of the gas chromatograph) is achieved *via* a digital-to-analog converter which should have several channels. When a non-dedicated computer is used for data processing, the ion current signals are collected on magnetic tape either in analog form directly from the final amplifier of the detector system, or in digital form after passing through an analog-to-digital converter.

The unique ion optical geometry of the Mattauch-Herzog type design permits ion detection either by electrical means or on ion--sensitive photoplates. When electrical detection is used, the conventional configuration is employed. Since photoplates must be developed manually, computers cannot be used on-line with this mode of ion detection. The photoplate reader, however, is usually interfaced with and controlled by a dedicated minicomputer (Fig. 1).

Interactive data systems are now commercially available for both quadrupole and magnetic types of mass spectrometer (Figs. 2 and 3). Dedicated minicomputers currently use a central processing

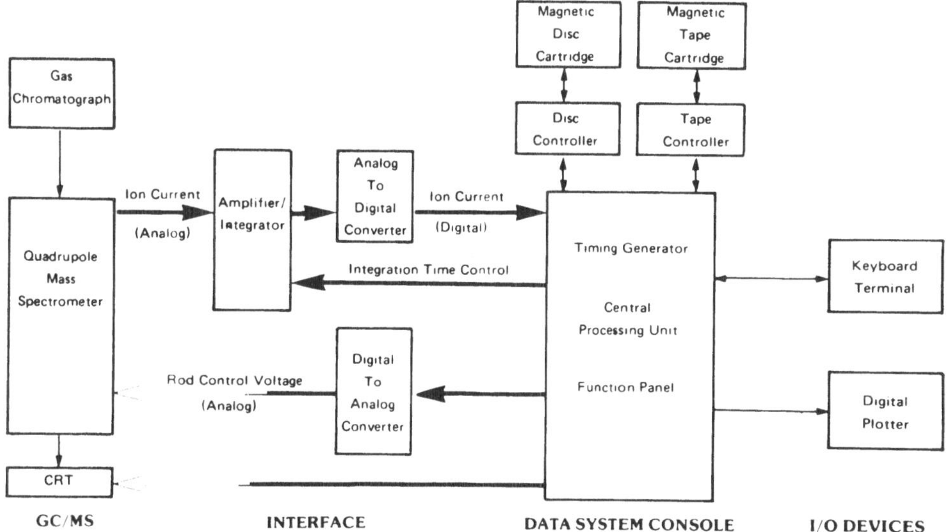

FIG. 2. Interactive data system for quadrupole mass spectrometer (Courtesy of Finnigan Corp., Sunnyvale, Calif.).

unit with 16K memory (16-bit). For real-time acquisition and manipulation of data, high capacity, rapid, and random access disks are commonly used. For inexpensive storage, data is rapidly transferred to magnetic tape cartridge. Operator interaction takes place *via* a high-resolution cathode-ray-tube (CRT) which serves several purposes: it displays data in real-time during acquisition, displays data when chromatograms or spectra are recalled from memory for examination, manipulation, or final printout on a digital plotter or hardcopy, provides a means to add alphanumeric information prior to output, and serves as a display during calibration and diagnostic routines. Instruction are entered *via*

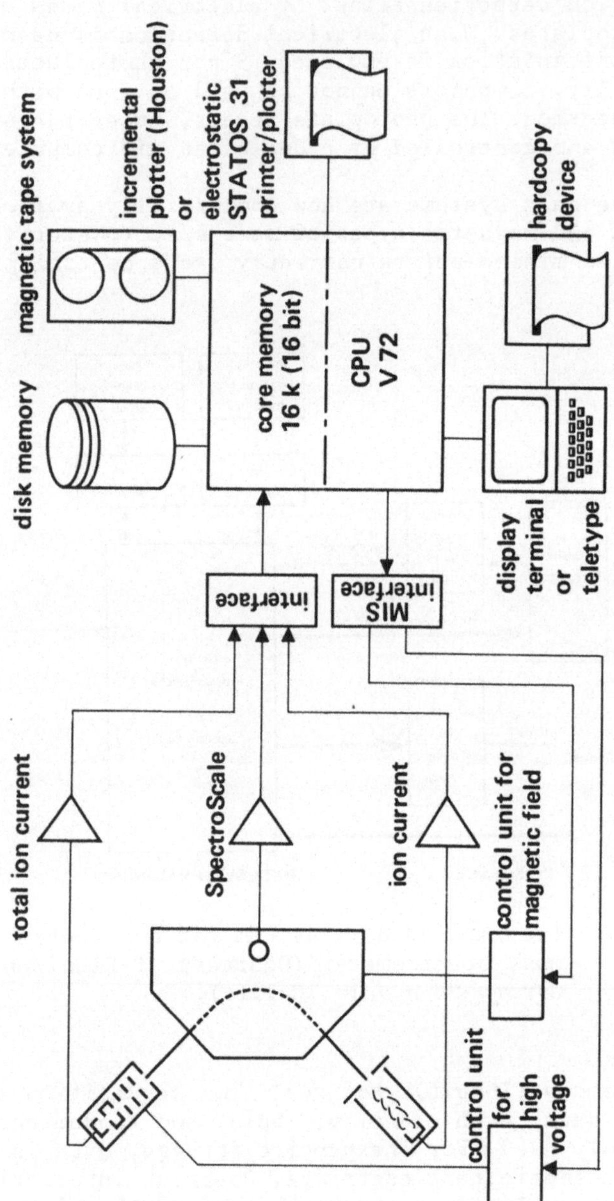

FIG. 3. Interactive data system for magnetic mass spectrometer (Courtesy of Varian/MAT, Sprinfield, New Jersey).

a more-or-less sophisticated keyboard terminal which also serves as an output device for tabulated data.

During the initial tune-up of the entire gas chromatograph--mass spectrometer system the computer performs a variety of functions, including the checking of electronic circuits, establishing proper mass calibration and corrections for mass defect, and the programming to generate appropriate control voltages (e.g., for the rods of the quadrupole) during subsequent scanning. With computer-controlled control voltages only data relevant to mass peaks are digitized and the resulting saving of time and memory capacity can be utilized to increase dwell-time on each mass peak so that a greater number of ions are collected thereby increasing sensitivity as well as dynamic intensity range.

Computer-controlled establishment of the mass scale and correction for mass defect is of great practical importance, particularly in quadrupole instruments. Mass calibration is customarily achieved by introducing a known compound whose spectrum is recognized by the computer, followed by comparison to an authentic spectrum stored in the memory. A very attractive alternate method of mass scale calibration is to utilize the compound of interest itself to establish the mass scale (assuming that the mass spectrum of the compound is known).

The mass defect of an ion is the sum of the mass defects of all elements contained in it. Mass defects of individual elements arise from the fact that only carbon-12 has an exactly integer mass (by definition); all other elements have non-integer masses. The mass defect of hydrogen is +0.00782 amu, thus hydrogen-rich compounds will have a rather large positive mass defect. Among the halogens, chlorine and bromine contribute significantly to negative mass defect, but fluorine has very little mass defect (-0.00160) making perfluoro-compounds suitable for mass scale calibration.

One technique for overcoming mass defect is illustrated in Fig. 4 (1). Assuming triangular peaks of 1.0 amu width at base (and ignoring peak intensity thresholds for simplicity), the upper part of the curve shows exact mass calibration with perfluoro-tributylamine, so that on the true mass scale m/e=314 is exactly what it is supposed to be. Next, assume that the mass scale of the instrument shifted -0.1 amu due to electronic reasons. This means that a true value od m/e=314 would now read as m/e=314.1, so that a +0.1 amu mass measurement would result. If a sample peak with +0.2 amu mass defect would have to be measured under these conditions, and no correction for mass defect were made, the measurement would be made -0.3 amu away from the maximum of the peak, and the peak abundance reading would have an error of more than 50%. A solution to the problem is to measure intensities at every 0.1 amu (shown by arrows) and also by determining peak areas

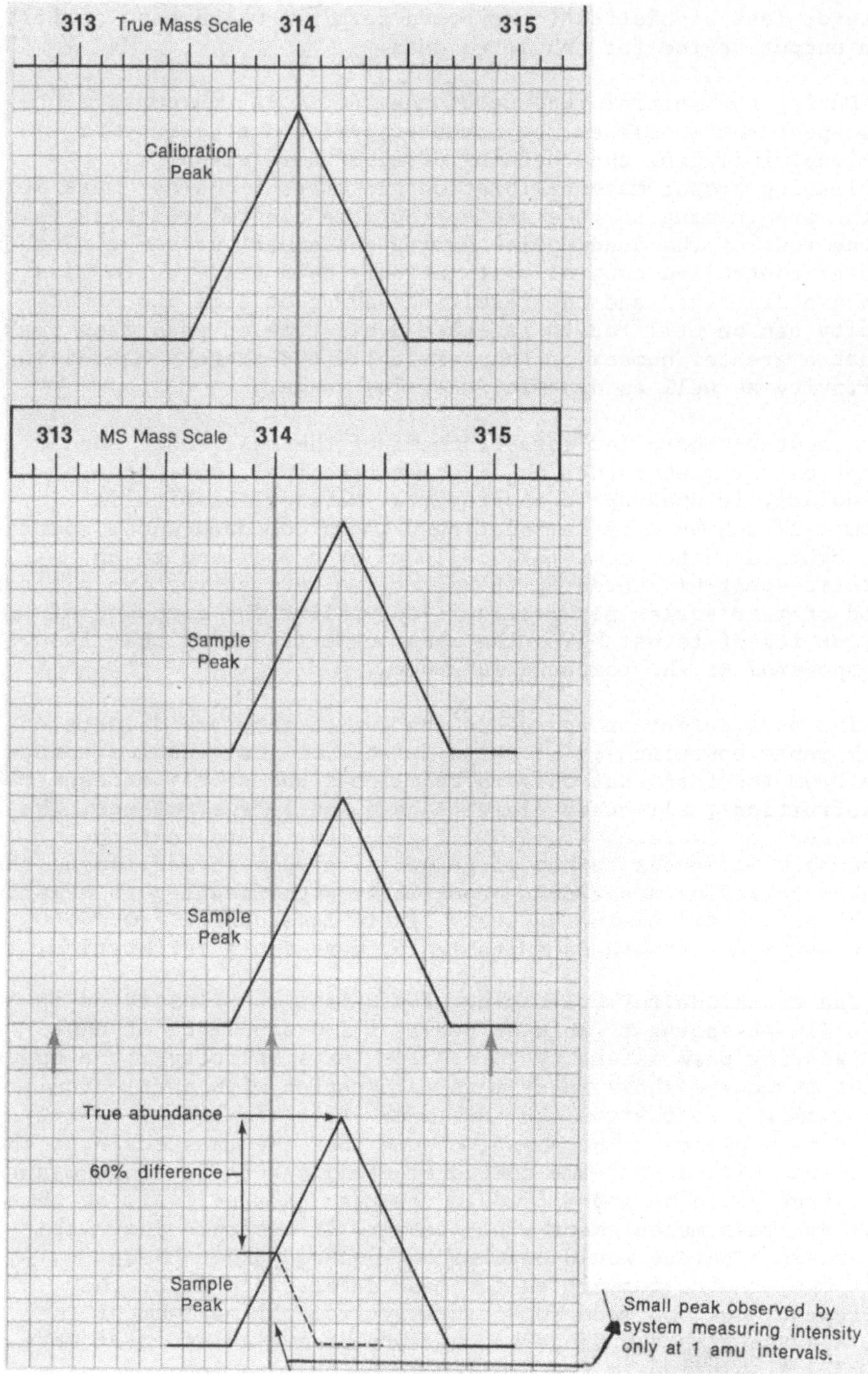
313 True Mass Scale 314 315
Calibration Peak
313 MS Mass Scale 314 315
Sample Peak
Sample Peak
True abundance
60% difference
Sample Peak
Small peak observed by system measuring intensity only at 1 amu intervals.

FIG. 4.

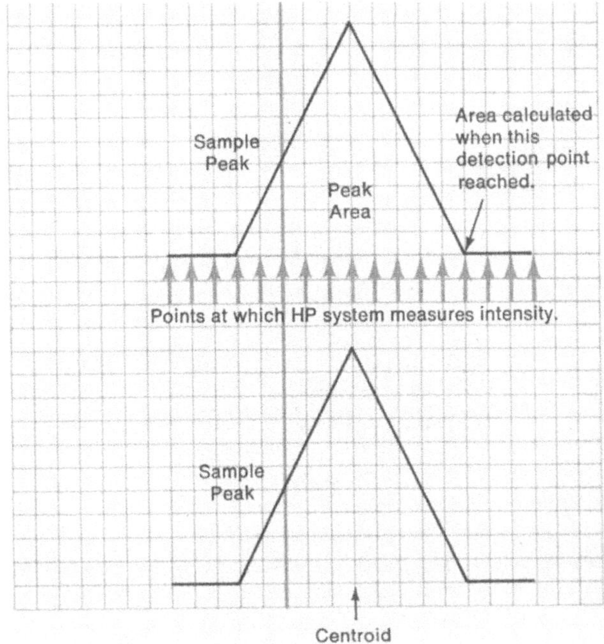

FIG. 4 (contd.) Overcoming mass defect; for detailed description, see text (Courtesy of Hewlett--Packard Corp., Palo Alto, Calif.).

instead of peak heights to estimate the abundance of peaks. Using this technique, no error is introduced by the mass defect in the determination of the mass of the peak (measure at centroid), and no error is introduced either by mass defect or mass scale shift in the determination of peak abundance (measure peak areas).

A somewhat different approach utilizes a variable "0.1 amu window" sampling to correct for mass defects (Fig. 5). For example, in the 150-299 mass range the data system may be instructed to scan masses at m/e values of 149.9, 150.0, 150.1, 150.2, and 150.3 and store the largest of these intensities as the true intensity of m/e=150. For higher masses the size of the "window" should be increased. An advantage of this approach is that it maintains the rapid scan rate necessary to acquire statistically significant data on fast eluting peaks (2).

Data acquisition in combined gas chromatograph-mass spectrometer systems (low resolution) is usually performed either in the

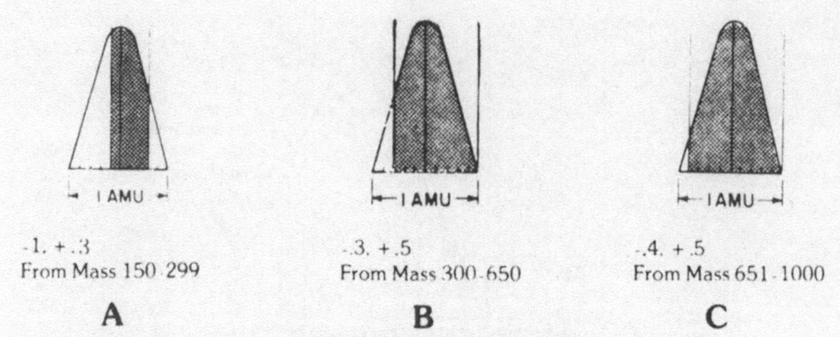

FIG. 5. Variable window, 0.1 amu sampling to accomodate mass defect (Courtesy of Finnigan Corp., Sunnyvale, Calif.).

"scanning mode", i.e., collect mass spectra continuously, or by "selected ion recording", i.e., monitor preselected masses. During the entire process of data acquisition, the computer continuously displays on the CRT selected information; for example, both total ion chromatogram and the mass spectrum of the compound being eluted may be displayed simultaneously on a split screen. This way the operator does not only follow "what is going on", but also has a chance to rapidly change experimental conditions on the bases of observed, and possibly unexpected, results displayed.

The most important advantage of the computer during data acquisition is that all data available are acquired and stored, regardless of significance and/or expectations. This aspect of computerization is indeed responsible for the enormous popularity of computerized gas chromatography-mass spectrometry today.

DETECTION OF METABOLITES

When radioactive drugs are available, the detection of metabolites is straightforward: those metabolites which contain the radioactive atom(s) can be found relatively easily by radio-scanning thin layer plates, gas chromatographic effluents, or liquid chromatographic fractions. After separation, the metabolites may be identified by standard mass spectrometric techniques.

Metabolites may be detected in humans, without the use of radioactive drugs, by utilizing specific labeling with stable isotopes. The drugs are labeled with ^{15}N, ^{13}C, or ^{2}H in such proportion that the molecular ion (M^+) and the $(M+1)^+$ ion are of approximately equal abundance. When such mixtures are administered, both the molecular and several fragment ions of the metabolites will appear as "twin ions" with characteristic intensities. For example, Knapp _et al_. (3) (1972) studied human urinary and biliary metabolites of nortriptyline by artificially creating isotopic doublets. They mixed nortriptyline and trideutero-nortriptyline (equimolar), and after administering the mixture to humans, they found both the parent drug and metabolites retaining the labeled site by searching for the presence of the M^+ and $(M+3)^+$ doublet in the trifluoroacetylated extract of basic urine. Here the search for metabolites proceeds by examining every peak in the total ion chromatogram for isotopic doublets. An added advantage is that complete gas chromatographic separation of all urinary constituents in not essential. If there is not enough material present for a full mass spectral scan for each peak, selected ion monitoring may be used in the search for the doublets. The isotopic method actually augments the specificity of selected ion monitoring. It is noted, in passing, that the same authors also used dideuterium/^{15}N-labeled nortriptyline to study the biliary metabolites.

In another study, Pohl _et al_. (4) used chemical ionization mass spectrometry to detect ion doublets in searching for the metabolic products of warfarin. The drug was labeled with ^{13}C on the benzyl carbon. In the mass spectrometer the quasimolecular ion (MH^+) was monitored for both the labeled and unlabeled drugs.

A novel computer-oriented technique for the detection of metabolites labeled with ^{13}C was developed by Sano _et al_. (5). In this method, the effluent from the gas chromatograph enters a combustion unit, and the carbon dioxide that forms is ionized in the mass spectrometer. An on-line computer monitors the peaks at m/e=44 ($^{12}CO_2^+$) and m/e=45 ($^{13}CO_2^+$) alternately at intervals of 0.5 sec, and a line graph is plotted showing the isotope ratio as a function of time. As long as the ratio stays 1.11/98.89 the line is linear. When the ratio changes, a peak is drawn, indicating that a metabolite labeled with ^{13}C appeared. From the change in the ratio one can calculate the number of carbon atoms involved. Once a metabolite is detected and its chromatographic retention time is known, a new sample is injected into the gas chromatograph and the effluent enters the mass spectrometer directly for identification.

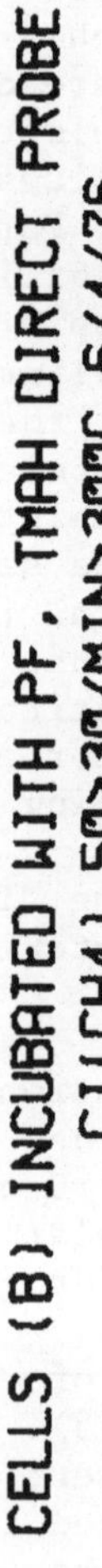

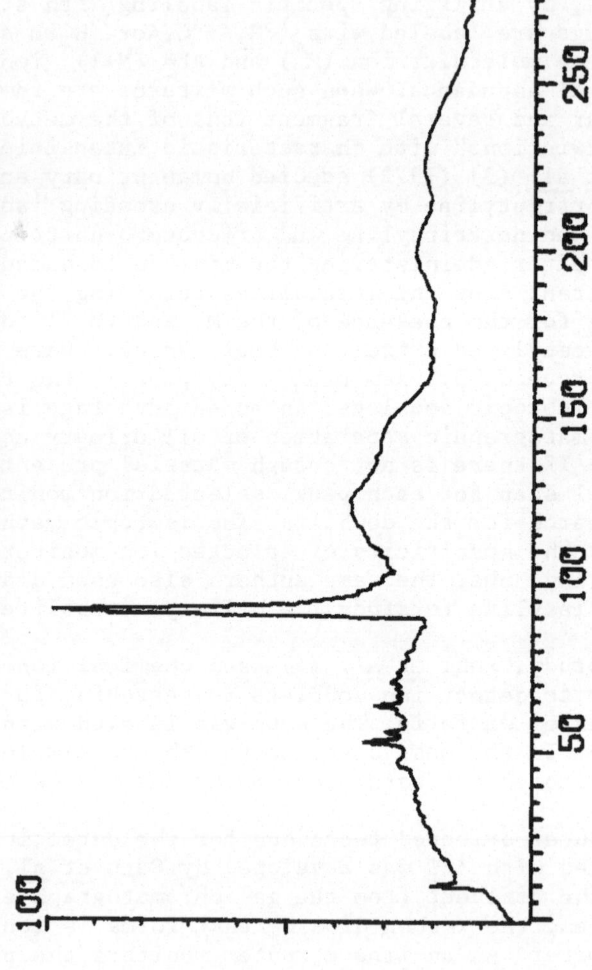

FIG. 6. Reconstructed gas chromatogram of extract from human cells incubated with pyrazofurin and methylated with trimethyl-anilinium hydroxide.

Unruh *et al.* (6) determined the limits of detection of ^{13}C--labeled drugs and metabolites in human urine. They suggest that the labeled tracer should provide at least $5x10^{-5}$g excess of carbon-13 per gram carbon in any fraction which is used for the detection of labeling.

When a pure sample of a suspected metabolite is available, computerized mass chromatography may be used for fast detection. For example, pyrazofurin (3-β-D-ribofuranosyl-4-hydroxypyrazol--5-carboxamide), an experimental antiviral/anticancer drug, is expected to metabolize by phosphorylation inside the cells (7). The chemical ionization mass spectrum of the permethylated pure sample revealed that the base peak corresponds to the molecular ion at m/e=452. A cell incubation experiment was carried out, a crude extract was prepared, and a total ion chromatogram obtained (Fig. 6). On the basis of initial experiments with the pure compound, the metabolic product was expected to appear, if at all, around 200°C, which would correspond to about scan #200. A Mass chromatogram of m/e=452 was next obtained (Fig. 7) which showed that phosphorylation did, indeed, take place inside the cells. Incidentally, an important option in the cathode-ray-tube- output of this kind of information is the possibility to amplify any part of the display so that both observation and subsequent quantification by computerized area measurement can be made more conveniently and accurately. A series of pictures by Watson *et al.* (8) in a paper on display-oriented data systems, nicely illustrates the power of this feature.

When metabolites are not available in the pure form, one might make a computerized search in the selected ion monitoring mode for masses chosen on the bases of expected structures. This was the technique used by Hammar *et al.* (9) to detect and identify several metabolites of nortriptyline. They have hypothesized the precursors and metabolites from the knowledge of the biochemistry and partially known metabolites of nortriptyline.

IDENTIFICATION OF METABOLITES

The most common method of computer-assisted identification of metabolites (or the parent drug) is, in fact, not identification but rather confirmation of suspected identity. Computerized library search for the identification of metabolites in the course of study of a new drug is of limited value. It is not likely that the metabolites will be found in existing libraries. When one is studying the metabolism of a new drug, it may be worthwhile to obtain the mass spectra of all known derivatives of the drug, and perhaps even synthesize simple derivatives for just this purpose. The objective here is to learn about the compound, to establish structure-spectra correlations which should shed some light on the

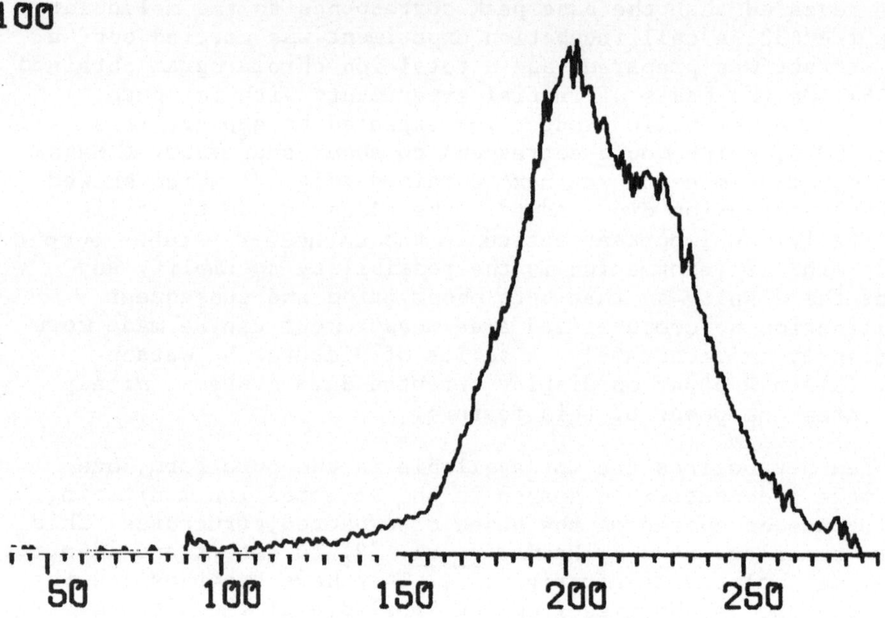

FIG. 7. Mass chromatogram of m/e=452, the quasimolecular ion of methylated pyrazofurin-5'-phosphate in crude extract from human cells. The abscissa represents scan number (also time scale). The approximate maximum at scan #200 corresponds to the appearance of the phosphate in Fig. 6.

composition and structure of unknowns.

Searching data libraries will be of value when the effect of a drug on endogeneous constituents is investigated. The mass spectra of hundreds of endogeneous constituents and known drug metabolites is now available in various libraries. In addition, computerized analysis of such "profiles" permits evaluation of the effects of the drug, both qualitative and quantitative, on dozens of constituents in a single analysis.

In file searching, the suitably coded spectrum of the "unknown" is compared against a large known file of similarly coded spectra to search for the "best fit" by some criterion; algorithms of variable sophistication have been reported for measuring the "goodness of fit" (10). Library search is usually performed either by utilizing a library provided by the manufacturer of the combined instrumentation, or via a commercially available data bank using a remote terminal connected to a time-sharing computer over a telephone line network. Of course, some users may find it advantageous to establish their own specialized libraries. Recent reviews of library search systems include those by Heller (11), Pesyna and McLafferty (12), and Karasek and Michnowicz (13). Library search is usually performed in the "forward" mode, i.e., by comparing the coded spectrum of interest to every single spectrum within the library. The method of "reverse" search, i.e. comparing each library to the unknown, is slower but allows the detection of small components (poorly resolved) in mixtures (14). A summary of the differences between forward and reverse search is given in Table 1.

When a relatively large number of gas chromatographic peaks must be identified, e.g., when the effect of a drug on endogenous constituents is investigated, several sophisticated computer techniques, all recently developed, may be utilized. Nau and Biemann (15) developed a technique utilizing both gas chromatographic retention indices and the mass spectral library search for identification in complex biological mixtures. Three standard compounds (e.g., n-tetradecane, n-docosane, and n-dotriacontane) are coinjected with the sample and the computer uses their known retention times to calculate retention indices for all eluting components. The computer will predict from the provided retention data the position (i.e., the scan number) of each compound to be searched for in the total ionization chromatogram. A certain value for the "search window" is set, and the computer will search within that window for mass spectral characteristics. Thus those components outside the "retention index window" will not interfere even if they have similar mass spectral characteristics, and also the search time is signifieantly reduced.

Essentially the same idea was further developed by Sweeley

TABLE 1.
Summary of the differences between forward and reverse search.
(Reproduced from ref. 14 with permission granted by Analytical Chemistry).

Forward search	Reverse search
Data basis of search is unknown spectrum	Data basis of search is library spectrum
Arbitrary intensities are selected from unknown	Only intensities corresponding to library compound are selected
Positive or negative deviations are approximately equal in weight	Sign and number of the deviations are diagnostic
Spectra are not adjusted for fit	Automatic renormalization
Relatively sensitive to interference or mixtures	Relatively insensitive to interference or mixtures
Qualitative data only	Qualitative and quantitative data simultaneously
Identifies complete unknowns from a large library	Identifies pre-selected compounds from a limited library
Ranked library compound suggestions as output	Yes/No responses for each library compound
Substantial operator interaction and judgment required	Automatic operation
Library size limited by peripheral storage	Library size limited by core storage
Search algorithms fixed	Search algorithms flexible

et al. (16) who developed a computer-controlled gas chromatograph--mass spectrometer system which simultaneously resolves, identifies, and quantifies hundreds of endogeneous constituents and/or metabolites of low molecular weight which might be affected by drugs or disease. The bases of this approach are gas chromatographic indices and mass chromatography. The mass spectral data library is constructed to contain the name, retention index, the m/e value of a so-called "designate" ion, and several "confirming ions" for all compounds of interest. These are established using pure compounds and is periodically updated with new compounds and/or metabolites. To search for a compound, a "search window" is determined using standard hydrocarbons for retention time measurement. Within the "window" (e.g., ± 1 min), mass chromatogram for the "designate" and "confirming" ions are next obtained. This is a unique feature since the retention index significantly restricts (by more than 90%) the size of the search to be made. When a peak is found in the "designate ion chromatogram", each chromatogram collected for the "confirmed" ion sets is searched for a peak whose apex is within one scan of that of the designate ion peak. All corresponding retention times and mass intensities are stored. Finally a series of checks and matches are made utilizing data in the library. When the specific ions which are monitored are selected with care, they will not only be characteristic to the compound in question, but also serve to differentiate from other endogenous compounds in that particular gas chromatographic elution area. Finally, when an unknown is detected, its retention index and a set of characteristic ions permit searching for the same compound in other samples even without knowing its identity.

When a compound is analyzed in a sample extracted from body fluids or tissues, the mass spectrum usually differs considerably from that of the pure compound. The reason for this is that gas chromatographic effluents are almost always mixtures including components from the column bleed, from the septum and, most importantly, from unresolved neighboring components during incomplete chromatographic separation. For identification, particularly when library matching is contemplated, spurious ion contributions must be removed. Several techniques are now available to deal with complex spectra. Dromay _et al_. (17) describe a computerized technique which results in detecting individual components by means of two histograms which statistically characterize the positions of mass fragmentogram peak modes. The basic assumption of the method is that there exist some masses for which ions occur in the mass spectrum of one component but not in the other and vice--versa (Fig. 8). By locating the resolved or singlet fragmentogram peaks at such masses tabular models of individual peak shapes can be derived. When gas chromatographic peaks are well resolved, each component will yield mass fragmentograms showing maxima at the same time; when chromatographic resolution is poor, fragmentogram maxima from neighboring components will have significant variations in terms of time. Thus, maxima must be used only from components that are unique to the eluent and are not present in neighbors. The first histogram to be obtained measures the number of singlet mass fragmentogram profiles which reach maxima in each time interval. The second histogram measures the total singlet ion intensity above background at these maxima. For more details on this elegant technique, the original reference should be consulted.

When an "unknown" metabolite is detected, the elemental composition of the molecular ion will usually be desired. It is not generally known that instruments with low resolution, such as 1000, may be used for exact mass measurement with computerized peak matching. Details and several illustrations are given by

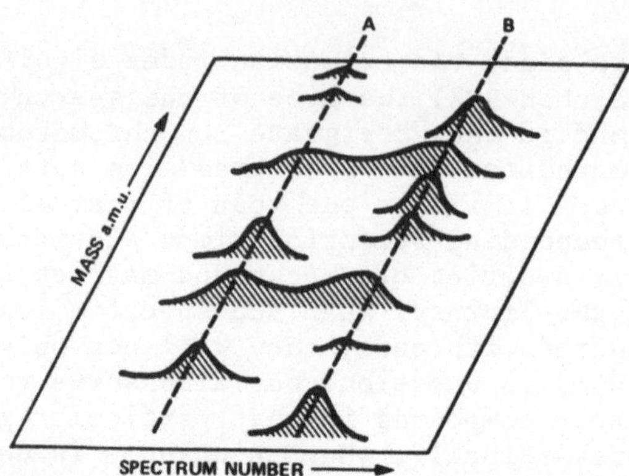

FIG. 8. Schematic representation of the set of partial mass fragmentograms for two closely spaced eluants. Components A and B have some masses in common. 'Reproduced from Dremey et al. (17) with permission granted by Analytical Chemistry).

Hammar et al. (18). Exact mass determinations are usually made with high-resolution mass spectrometers utilizing either fast scanning and electrical detection or ion-sensitive photoplate detection. In electrical detection systems data is often collected on magnetic tape (Fig. 1) to be followed by off-line computer evaluation; on-line computerized high resolution systems have been described but not too many such systems are in routine use (19). Using the photoplate detection system one can expose the photoplate during the entire duration of the elution of the gas chromatographic peak so that sensitivity is increased. Also, since all peaks are detected and integrated simultaneously and continuously, one may determine individual masses with precision and without interference from coeluted components.

During recent years some competition has developed between electric and photographic detection techniques for the determination of exact masses of gas chromatographic eluents. Habfast and Maurer (20) made comparative calculations and measurements using a Mattauch--Herzog geometry double-focusing instrument equipped with both electrical and photoplate detection systems. These authors conclude that as long as fast scanning speed (faster than 10 sec per mass decade) must be used, photographic recording appears to require less material, yields higher resolution per unit material needed, can be applied to wider mass range, and results in more accurate mass measurements. A figure of merit for comparison can be calculated by considering the ratio of L_p/L_e, where L_p represents the number of ions of one kind arriving on a photoplate during exposure divided by the minimum number of ions needed to produce a just detectable line, and L_e represents the same parameter for electrical detection as L_p does for photographic detection. When a resolution *versus* mass range diagram is constructed, operating conditions falling above the line of equal sensitivity (i.e., $L_p=L_e$) less sample is needed for photographic detection, while for the region below the line of equal sensitivity, electrical recording is more sensitive. Although these results suggest that in most cases the photographic detection is superior, most users of combined gas chromatograph-high-resolution-mass spectrometer systems prefer electrical detection.

Smith *et al.* (21) suggested that the accuracy of mass measurements in dynamic scanning is a function of the product of resolution and sensitivity. Double-beam mass spectrometry (22) separates compound and reference peaks without high resolution by electronically offsetting the separate reference and sample ion intensity outputs. When such a system is combined with an on--line computer, the accuracy of mass measurements could be easily evaluated. Aspinal *et al.* (23) made two types of measurements: In one set, 50 ng material was put on the gas chromatographic column, mass spectrometric resolution was set at 1500, and the instrument was operated in the double-beam mode. Here the product of resolution and sensitivity is about 75,000. Using these conditions masses could be measured to an accuracy of approximately 1 millimass unit, for peaks of both low and high intensity. In the second set of measurements, resolution was increased to 10,000, sample quantity was increased to 250 ng, and the instrument was operated in the single-beam mode. Here the product of resolution and sensitivity is about 3 times larger than in the first set of experiments. Accordingly, the accuracy of mass measurement became 3 times smaller. Thus, it was proven that indeed, it is the product of resolution and sensitivity that really counts in determining the accuracy of mass determinations.

High precision mass measurements of gas chromatographic effluents have been repeatedly used in drug metabolism studies both for the identification of drugs and metabolites and as a referee method where the question of several likely molecular compositions exist for a metabolic product. Ehrenthal and Pfleger (24) described in detail how computerized mass fragmentography was used for the detection of the soporific Carbomal ($C_7H_{13}N_2O_2Br$) and its metabolites in urine, and how high precision mass measurements helped to assign the proper compositions in both the fragmentation schemes and the hitherto unidentified metabolite.

High precision mass measurement may also be a useful tool in the course of developing analytical techniques for drug metabolism studies. When studying the metabolism of Virazole (1-β-D-ribofuranosyl-1,2,4-triazone-3-carboxamide), it was noted (6) that while the mass spectrum of the fully silylated drug yielded the expected molecular ion (at m/e=532) upon introduction *via* a direct probe inlet system, the molecular ion could not be detected when the compound was introduced *via* the gas chromatograph. Instead, the highest mass observed was at m/e=442. A high-resolution mass spectrum on a photoplate yielded a computer output showing a variety of possible structures corresponding to the measured mass. However, only one composition made sense on the basis of the structure of the compound: a cyano-compound formed due to dehydration in the gas chromatographic injection port, and the product contained one less trimethylsilyl than the parent compound.

QUANTIFICATION OF METABOLITES

Quantification of metabolites with the aid of computers has become common practice in recent years and dozens of publications appeared on a variety of drug and metabolites. A typical application is the quantification of tolbutamide, an oral hypoglycemic agent, and its metabolites in body fluids (25). In addition to the obvious advantage of computerized quantification in terms of convenience and speed, and accuracy of the determination of both peak heights and areas greater with computers than in manual methods. When the mass spectrometer is operated in the "scanning mode", quantification is made similarly to that used in conventional gas chromatography: internal standardization, area normalization, response factors, calculations of absolute or relative quantities. When the mass spectrometer is operated in the "selected ion recording" mode, several masses are monitored simultaneously, all areas are integrated, peak ratios calculated, and absolute or relative concentrations determined. With some additional programming capability, most computers can also immediately calculate standard calibration curves, compute least-square-fit lines, and make various statistical and curve-fitting calculations.

For proper quantification one must use an internal standard. It is still debated whether one should make an all-out effort to obtain a stable-isotope labeled version of the compound being investigated or it is adequate to use an analogue of the compound. For pharmacokinetic studies or therapeutic monitoring it appears, at least to this author, that vagaries of the biological materials and the sample preparation technique more than overshadow the question posed. An analogue, with solubility and chromatographic properties similar to those of the drug analyzed, appear adequate for most practical purposes. When stable-isotope labeling is used for internal standardization, the role of computerization increases since mass fragmentography must be used because of the problem of chromatographic retention times, and also isotope ratios must be determined precisely. According to Matthews and Hayes (26), assuming that there is no chromatographic fractionation of the isotopes, a bias of less than 1% can be obtained with unidirectional scanning for 10 or more cycles.

The role of computers in quantification is amplified with the desire to reduce sample preparation to minimum. The technique for quantification of constituents in crude mixtures is, of course, computer monitoring of selected ions.

ADVANTAGES, DISADVANTAGES, AND FUTURE TRENDS

In a recent lecture on the applications of computers in mass spectrometry, Professor Biemann (27) remarked that while on the one hand, mass spectrometry has placed computers into the chemical laboratory in a much more sophisticated manner than other instruments, such as gas chromatography it is true that now we have much more data than before and also much less time to think about them. In fact, he said, some people hardly look at mass spectra any more.

When the technique of combined gas chromatography-mass spectrometry is used in drug metabolism studies, the gas chromatograph may serve as an inlet system for the mass spectrometer, or the mass spectrometer may serve as a sophisticated (and expensive) detector for the gas chromatograph. When searching for metabolites in a complex biological mixture, the gas chromatograph is used as a tool of separation. Metabolites must be detected and identified, and the emphasis here is on mass spectrometry. The availability of computer-controlled instruments certainly provides a capability to solve very complex problems far beyond the reach of manual methods. Among the advantages one should list the continuous acquisition of data, complete mass spectral documentation of all points along the gas chromatogram, real-time availability of gas chromatographic retention indices, and almost real-time confirmation of the identity of any eluent by library search; in the case of high--precision mass measurements, computerization provides a means to speedily calculate all possible molecular compositions (with choice of heteroatoms) corresponding to a measured mass.

When drugs and metabolites are analyzed in the pharmacokinetic studies or in patient monitoring, the mass spectrometer is indeed only a fancy detector, and one does not usually study the mass spectrum in detail. Here the role of the computer is perhaps even more important: in semiroutine or routine analytical work the computer helps to detect the constituents of interest by selective ion monitoring, and performs quantification by correcting background contribution, measuring areas, making comparisons with internal standards, followed by calculations of absolute quantities

.or relative concentrations.

On the negative side, the almost complete reliance on computers certainly reduces the awareness of the operators of the functions of the instrument, and also his experience in interpretation. Often arbitrary decisions are made, e.g., on thresholding or background subtraction, followed by fast and crude comparisons.

Currently available hardware and also software is adequate, at least in the opinion of this writer, to solve most common needs in instrument control, data acquisition and processing in drug metabolism studies. In the field of computer-assisted interpretation of mass spectra, current development work in search systems will soon permit confirmation of identity,rapidly and with high probable matching factor as long as the mass spectrum of the compound is on file. This will be very useful when studying the effects of the metabolism of a drug on endogenous constituents, and also in cases where metabolic products of a drug of similar structure have already been elucidated.

As far as the computerized identification of "unknowns" are concerned, sophisticated interpretive techniques are under development. The "Probability Based Matching" system (28) utilizes a reverse searching procedure and weighting of mass and abundance data which permit matching of unknown spectra against a large data base not restricted to spectra taken under the same experimental conditions. The "Self Training Interpretive and Retrieval" system (29) selects different classes of data known to have high structural significance, such as characteristic ions, mass of neutrals lost, etc., from the unknown mass spectrum and matches these against data in the reference file to search for common structural features. These and other sophisticated interpretive techniques, now being tested on "training libraries", will eventually be of practical help in the identification of drug metabolites.

Acknowledgment. This work was supported by Grant #5P01--CA15936-03 from the National Cancer Institute, NIH, USA.

REFERENCES

1) D. Harrington, Hewlett-Packard Application note AN-176-9.
2) Finnigan Corp., Finnigan 6110 Interactive GC/MS Data System, 1976.
3) D. Knapp, T. Gaffney, R. McMahin and G. Kiplinger, J. Pharmacol. Exp. Ther., 1972, 180, 784.
4) L. Pohl, S. Nelson, W. Garland and D. Trager, Biomed. Mass Spectrom., 1975, 2, 23.
5) M. Sano, Y. Yotsui, H. Abe and S. Sasaki, Biomed. Mass Spectrom., 1976, 3, 1.
6) G. Unruh, D. Hauber, D. Schoeller and J. Hayes, Biomed. Mass

Spectrom., 1974, 1, 345.

7) J. Roboz, R. Suzuki, M. Perloff and T. Ohnuma, (Abstract), 23rd Ann. Conf. Mass. Spectrom., 1975.
8) J. Watson, D. Pelster, B. Sweetman, J. Frolich and J. Oates, Anal. Chem., 1973, 45, 2071.
9) C. Hammar, B. Alexanderson, B. Holmstedt and F. Sjoqvist, Clin. Pharmacol. Ther., 1971, 12, 496.
10) S. Grotch, Anal. Chem., 1973, 45, 2.
11) S. Heller, in "Computer Representation and Manipulation of Chemical Information", T. Wipke, S. Heller, R. Fieldmann and E. Hyde, Eds., Wiley-Intersciences, New York, 1975.
12) G. Pesyna and F. McLafferty, in "Determination of Organic Structures by Physical Methods", Vol. 6, F. Machod, J. Zuckerman and E. Randall, Eds., Academic Press, New York, 1976.
13) F. Karasek and J. Michnowicz, Res. Dev., May 1976, p. 38.
14) F. Abramson, Anal. Chem., 1975, 47, 45.
15) H. Nau and K. Biemann, Anal. Chem., 1974, 45, 426.
16) C. Sweeley, N. Young, J. Holland and S. Gates, J. Chromatogr., 1974, 99, 507.
17) R. Dromey, J. Stefik,Mark, T. Rindfleisch and A. Duffield, Anal. Chem., 1976, 48, 1369.
18) C. Hammar, G. Petterson and P. Carpenter, Biomed. Mass Spectrom., 1974, 1, 397.
19) R. Hilmer and J. Taylor, Anal. Chem., 1973, 45, 1031.
20) K. Habfast, K. Maurer, MAT 711/731. 14th Ann. Conf. Mass Spectrom. All. Topics, Dallas, 1966.
21) D. Smith, R. Olsen, F. Walls and A. Burlingame, Anal. Chem., 1971, 43, 1796.
22) W. Wolstenholme and R. Elliott, Am. Lab., November 1975.
23) M. Aspinal, J. Chapman, K. Compson, D. Hazelby and Riddoch, Proc. 21st Ann. Conf. Mass Spectrom. Allied Topics, 1973, p. 471.
24) W. Ehrenthal and K. Pfleger, Mass Spectrometry Users Meeting, Bremen, Publication No. 19, Varian, 1975.
25) S. Matin and J. Knight, Biomed. Mass Spectrom., 1974, 1, 323.
26) D. Matthews and J. Hayes, Anal. Chem., 1976, 48, 1375.
27) K. Biemann, 24th Ann. Conf. Mass Spectrom. Allied Topics, San Diego, Calif., Paper PL-1, 1976.
28) G. Pesyna, R. Venkataraghavan, H. Dayringer and F. McLafferty, Anal. Chem., 1976, 48, 1362.
29) Kwok, Kain-Sze, R. Venkataraghavan and F. McLafferty, J. Am. Clin. Soc., 1973, 95, 4185.

A REAL TIME GAS CHROMATOGRAPH-MASS SPECTROMETER MINICOMPUTER SYSTEM FOR SELECTED ION ANALYSIS

J.A. Steinborn, R.W. Silverman and D.J. Jenden

Department Pharmacology and Brain Research Institute

University California, Los Angeles, California 90024 USA

INTRODUCTION

The use of selected ion analysis (selected ion monitoring, mass fragmentography) in quantitative applications of mass spectrometry is now well established (1, 2). For many purpose the output, consisting of a few selected ion profiles representing the mass peaks of interest, can be quite satisfactorily handled without automatic data processing as from a gas chromatograph flame ionization detector (FID) system. In contrast, the more critical demands of repetitive MS scanning require a computer system for complete utilization of the data. An automatic data system becomes increasingly necessary also for selected ion analysis when large numbers of samples are to be run and when more than two isotopic variants are to be simultaneously analyzed. The latter situation arises particularly when stable isotopic labels are used as quantitative tracers, as well as in their more conventional application as internal standards (3,4). Unlike radioisotopic tracers, stable isotopic labels, in conjunction with GC-MS, allow the use of multiple tracer variants of the same compounds which can be discriminated from each other by their mass spectra and separately measured (5). However, interpretation of the relative ion abundances requires solution of a set of simultaneous linear equations for each analysis, which may be impractical without computer assistance, and a set of controls in which mass spectra of the individual variants and of a mixture of known composition are precisely measured (6). These controls allow the construction of an n-dimensional "standard curve" from which the results of subsequent individual analyses are obtained by interpolation.

The minicomputer system we describe below was developed

primarily for high throughput quantitative analyses involving two or more isotopic variants of each compound, one of which is normally used as an internal standard. The system consists of two quadrupole GC-MS units linked to a minicomputer via a specially designed interface. The computer system operates both GC-MS units independently, continuously and asynchronously for pre-selected ion monitoring. The computer system is designed to be totally transparent to the user except via instructions entered through the user console. The experimenter need only be concerned with the GC and the computer console, but can also monitor the sample output by means of a strip chart recorder. The computer and the interface are not touched by the user, and the programs are written in such a way that the user is informed throughout of actions required to continue the analysis.

The system is designed for high volume throughput of samples (∿ 15 samples/hr/GC-MS) with immediate results for each sample printed out on the console before the next sample is injected. Estimates of relative abundances of up to 12 selected ions are provided for each GC peak. For stable isotope tracer experiments, the quantities of each of the isotopic variants of the compounds are estimated (6). The raw data are saved for possible more detailed graphic display or analysis at a later time.

Fig. 1 shows the three main components of the total system and the principal data and control lines between the elements. For simplicity, only one GC-MS unit is shown. Each GC-MS is equipped with a computer terminal and a multichannel potentiometric recorder. The recorder provides ultimate backupfor monitoring and validation of the functioning of the system. The interface provides control and data linkage between the GC-MS systems and the computer, and provides hardware and calibration tests under computer control. The computer performs under control of a real time executive system that selects the specific application program required for the task to be performed. The application program, in turn, controls the GC-MS system _via_ the interface and communicates with the user _via_ the terminal concerning the experimental analysis to be performed. The disc memory is used for mass storage of application programs and experimental data. The results are printed on the user terminal. The electrostatic printer plotter permits optional graphical display of the data and results.

INTERFACE STRUCTURE AND OPERATION

The interface links an analog device - the GC-MS, with a digital device - the minicomputer. It is designed to perform the functions of timing and control of the GC-MS and to reduce the number of input-output (I/O) operations that the computer must perform. Fig. 2 illustrates the main functional components of the interface with the essential control and data lines shown.

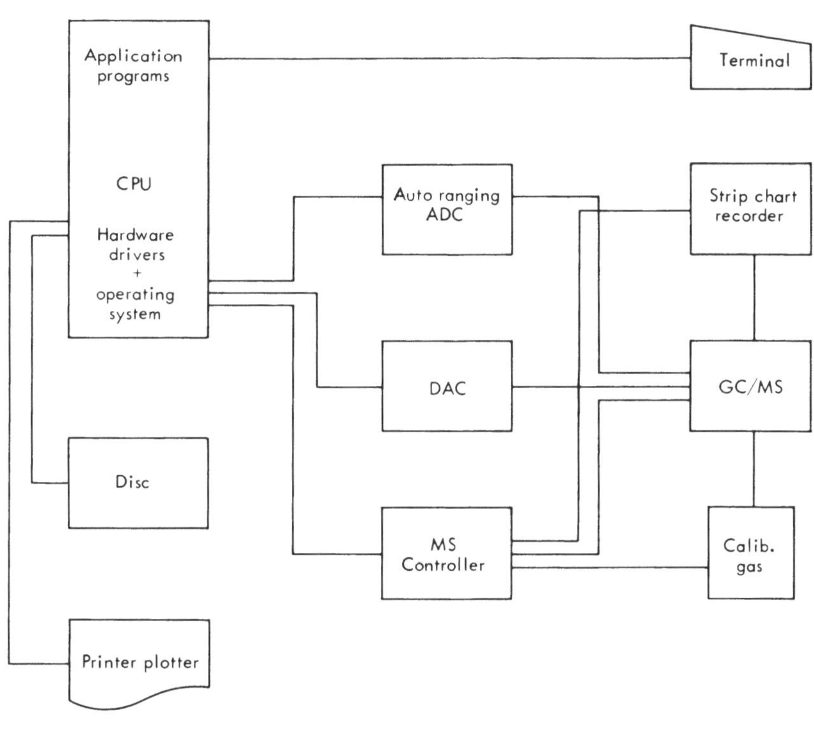

FIG. 1. Configuration of the GC-MS-computer system. The connecting lines indicate main data and control lines between the components. The central processor (CPU) is shown as a single block configured with applications programs. Each GC-MS has a terminal, recorder, GC-MS unit and automated calibration compound inlet.

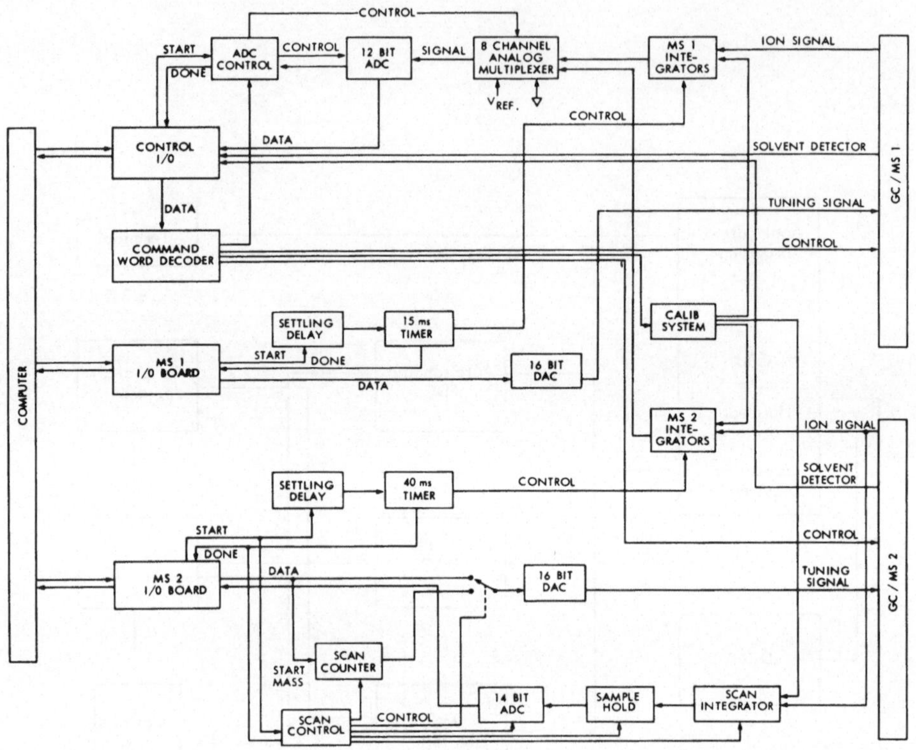

FIG. 2. Simplified schematic diagram of the interface.

The EAI QUAD 300 (GC-MS 1) spectrometer mass control signal is derived from a 16 bit digital-to-analog (DAC) converter operating under computer program control. A command is issued to focus on a specific mass and, after a 5 ms delay to allow for settling of the spectrometer, a 15 ms integration period is initiated. During this time the ion current signal is integrated simultaneously by three integrators with nominal relative sensitivities of 1, 10 and 100. Termination of the integration interval causes the interface to issue an interrupt to the computer, which in turn outputs a command to start an analog-to-digital conversion of the integrated signal. The three integrator outputs are digitized sequentially starting with the most sensitive and ending when a digital value is obtained that is less than full scale for the ADC. The number of conversions performed are counted to indicate the integrator sensitivity selected. This provides an automatic sensitivity selection function with a 110 db dynamic range and 12 bit resolution. The maximum conversion time is 60 μsec. The converted value and integrator sensitivity counter are read into the computer and the next mass control signal is output to the DAC. By issuing command words to the interface, the computer controls the initialization of a GC retention time indicator (7), admission of a mass scale calibration gas to the spectrometer, switching of the solvent vent valve (8) in the gas chromatograph, and an ion current signal demultiplexer. The ion current analog signal is demultiplexed in synchronization with the computer data collection and filtered and displayed on a multi-pen potentiometric recorder (3, 9). This provides the operator with a back-up independent of the computer data reduction system.

The interface controls the operation of the Hewlett-Packard 5981A GC-MS system (GC-MS 2) in the same manner as the EAI system with minor differences. A command is issued (via a second 16 bit DAC) to focus on an ion with a selected m/e value and after a 10 ms settling delay, a 40 ms integration period is initiated. The analog-to-digital conversion is performed by the auto-ranging system which is multiplexed to handle data from both GC-MS systems. The computer initializes a GC retention time indicator (located in the solvent detection system (7)) and controls an analog ion current demultiplexer, an integral part of the digital selected ion monitor (DSIM) (7). The DSIM provides a back-up data record for the computer data reduction system.

The interface is linked to the computer by three printed circuit assemblies (PCA) that are plugged into the computer I/O slots immediately following the system disc on the I/O priority chain. The highest priority PCA outputs a DAC word to control the mass focus of GC-MS 1. The next priority PCA outputs a DAC word to control GC-MS 2 in the selected ion monitoring mode and is the direct memory access (DMA port for digitized data in the scanning mode. The lowest priority PCA is a bidirectional 16 bit duplex register which is used to output control words to the interface for both mass spectrometers and for receiving data from the 12 bit ADC.

The entire interface system is housed in a 7 inch high rack chassis and is fabricated from over 150 integrated circuit assemblies. All analog circuitry calibration is checked automatically under computer control. The interface can be expanded to accomodate additional GC-MS systems. This requires an additional storage register for control words, a set of analog integrators and a DAC for each additional system.

THE COMPUTER PROGRAMS

The computer is a Hewlett Packard 21MX processor with 32K words of 16 bit semiconductor memory. The operating system is real time disc oriented (Hewlett Packard RTE II) which responds to usage demands of external devices on a vectored priority basis, and schedules the application programs according to task priority. The application programs that perform analysis and control of GC-MS experiments are contained in a program library stored on disc.

Under RTE II, application programs are either resident on the disc and are swapped into memory only when needed for a task or are, in selected instances, always memory resident for instant accessibility. Disc resident programs are placed into memory to perform a task, but are subject to being swapped back to disc to make way for a higher priority task.

There are approximately fifteen programs of differing priorities for each of the GC-MS systems. The following are the major application programs: the hardware input output driver routines, the dialog program to set up new problems, a data acquisition routine, calibration programs for mass scale and mass defect determination, interface checkout routines, and data analysis programs for determination of ion current ratios and estimating quantities of endogenous and tracer isotope variants of compounds. Each GC-MS system has its own copy of each of the analysis and control programs in the library, so that there is a minimal chance of interference or queueing delays. The console assigned to each GC-MS is protected from accidental running of programs for the other GC-MS system but has access to the rest of the program library.

The software for the GC-MS analysis and control is written in FORTRAN IV because of its flexibility, ease of modification and low development cost. Only the I/O drivers that directly service the three interface I/O PCA boards are written in assembly language. The drivers are configured as part of the RTE II system, and are memory resident.

The data acquisition routine (DAR) is a high priority, minimal sized program used to control the data collection throughout the entire GC run. The calibration and experiment setup routines are performed first. Then DAR sets up a memory buffer containing a control word for the control I/O board, the control numbers to the DAC via the desired GC-MS I/O Board for each selected ion, and a buffer for the data input. A call is made to the I/O driver to perform tha data collection task. When the buffer is full, DAR writes the data on disc tracks allocated to the program. After the data for a GC peak is stored, DAR schedules the analysis routine. This routine copies the data from disc tracks to a permanent file under the RTE file manager when the analysis is finished.

There is a copy of DAR resident in memory for each GC-MS system. This is to assure quick response during the data acquisition phase. The remaining analysis and control programs are disc resident and are swapped in and out of memory, as needed, on a priority basis. Time spent in DAR is dead time for data collection. When memory resident programs are not being used, the space occupied is dead space that cannot be used by other routines. By having DAR schedule tasks to the drivers and disc resident routines, dead time and memory resident space are minimized.

A utility program, GAIN, is routinely called whenever a new problem (a set of similar analyses considered as a group) is set up. It can also be called from any console by the command, "RUN, GAIN". This program measures the sensitivity and zero offsets of each of the three integrators. It also provides an overall system check since a malfunction of any of the integrators, the automatic gain ranging ADC, the control word decoder, the I/O boards or the I/O drivers would cause a failure in this program. Similarly, when the mass scale calibration program, MSCAL, is initiated, the entire system including the GC-MS is checked out, in addition to accomplishing the primary function of calibrating the mass scale with a calibration compound.

The determination of the mass defect of each of the selected ions for the compounds being analyzed is performed after the mass scale is calibrated. Fig. 3 shows the ion current traces collected during a single GC peak for fractional masses centered about a

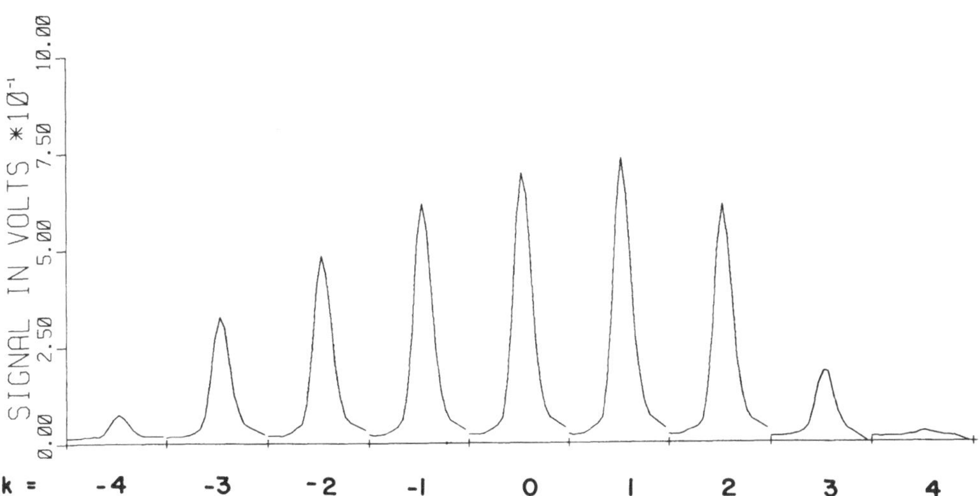

FIG. 3. Signal profiles collected during elution of a single GC peak for progressive mass functions about the nominal mass = 60 amu (m/e = 60 + k f). The data from these traces are used for mass defect determination.

precise nominal mass. From these data, the precise mass defect is determined by interpolating for the point of maximum signal. This is done simultaneously for all selected ions from the data collected for a single sample. The DAC words for each mass defect for all selected ions are stored for use in collecting data in subsequent samples.

SYSTEM OPERATION

The experimenter first turns on the user console and then types "RUN,SETUP" (Fig. 4). This program requests the user to supply the information necessary to configure the system for the specific experiments: e.g., the number and nominal mass of the

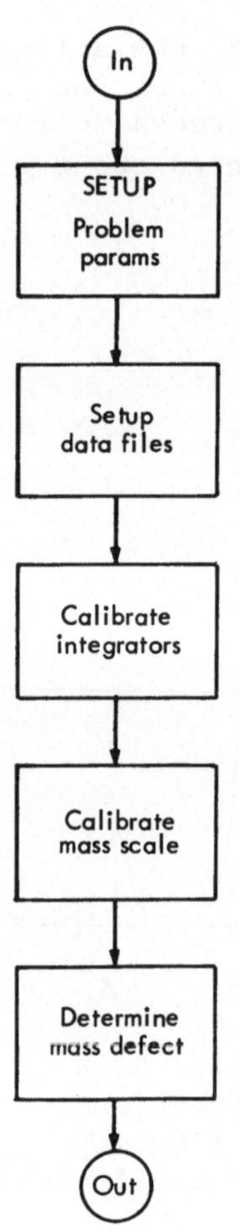

FIG. 4. Flow chart for SETUP. The initial problem setup (or re-setup) is contained in this procedure, and is initiated by the statement "RUN, SETUP" entered on the terminal. A dialog, structured by the computer, enables the user to enter information about the problem to be solved. If the experimental data is to be stored on disc files, they are configured at this stage. Calibration and mass defect programs are scheduled by this control program and the results are stored for subsequent use.

selected ions to be monitored, the number and anticipated retention times of the GC peaks, etc. SETUP can be run at any time during the experiment to respecify operational parameters. When this is completed, a summary of the experiment setup is printed on the console. At this point, the user can continue or go back and change the setup. Routine checks of the integrators and mass scale calibrations are performed by GAIN and MSCAL, respectively, and a summary printed out at that time. The user is instructed to inject a sample containing an equimolar control or standard of the compound to be analyzed. From this injection, the mass defects of each of the selected ions are measured. The program EXPR controls each GC injection (Fig. 5). The experimenter is asked to enter

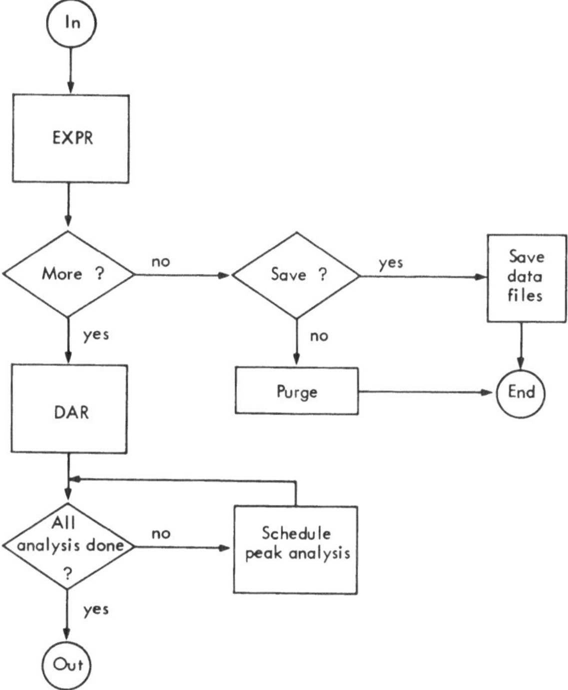

FIG. 5. The flow chart for computer controlled sample runs. The program is entered by typing "RUN,EXPR" or by scheduling from SETUP. The program asks the user if there are more samples to be run. If so, then identification information is requested from the user, and the data acquisition routine (DAR) is scheduled to begin the data acquisition procedure. After all data are required, EXPR cycles until all GC peaks have been analyzed. When there are no more samples, the choice of saving or purging the files is determined by the user. The program is then terminated.

information such as sample number, whether the injection is an unknown or a control, and the weight of the biological sample (when required). Control samples are injected first. This enables validation of the overall system performance and also establishes the basis for quantitative estimates in the subsequent samples. The data are collected using DAR. Each isotope labelled variant is given a distinct name for identification purposes. A matrix of ion current ratios for all separate isotopic variants plus a mixture of known composition of all variants for each GC peak is recorded and stored in a file (Fig. 6). When routine samples are injected, the relative ion abundances for each GC peak are computed. The quantities of the endogenous and tracer variants are then estimated (6, 9) from these relative ion abundances using the matrix of spectra from the controls and the quantity of internal standard added to the sample. If the ADC dynamic range is exceeded at any time during the data collection, a warning is presented to the operator *via* the terminal. The printout displays the ion current ratios, the abbreviated multiple mass chromatogram, and the estimated quantities of each of the isotope variants (Fig. 7).

The relative abundances of the selected ions and the estimated quantities of the variants are printed out within seconds of the completion of the GC peak. After the analysis, the data are transferred from allocated disc tracks to a permanent file. The user has the option of purging or saving the data when the experiment is finished. Data saved can be selectively retrieved for plotting on the printer plotter or for further analysis.

RESULTS

The system has been in routine operation since January 1976. A digital tape data acquisition system and off-line computer analysis, in use for several years prior to the development of the on-line computer system, provides a backup for either GC-MS in the event of a computer system failure. The results produced by the minicomputer system are consistent with the earlier system, but with increased flexibility and dynamic range. The immediate availability of results is especially valuable for discovery of problems before excessive time and samples are wasted. The potentiometric recorder output of the sample runs provides continuous, direct backup by displaying the data, as collected, in a familiar and directly accessible form. The ion current signal is displayed throughout the computer controlled run on an oscilloscope to allow the operator to assess system performance instantly.

Although the programs are written in Fortran, they are not independent of RTE II or of the computer configuration. The programs are in fact structured to operate to the best advantage of the given system. RTE II does impose some limitations of usage and requires some overhead time for its operation but this has been of

FOR PEAK # 2

PEAK WIDTH 10.4 SECONDS
PEAK NUMBER 2 OF 2
ABBREVIATED MULTIPLE MASS CHROMATOGRAM

TIME	58	60	64	71
115.70	.2703	.2415	.8128	.2318
118.70	.3690	.4502	.9652	.2493
121.70	.9624	1.1123	1.5688	.3039
124.75	1.6713	1.8372	2.0677	.3859
127.76	2.1469	1.9605	2.0964	.4445
130.77	1.6614	1.3741	1.5853	.3754
133.77	1.0993	.7928	1.2061	.3769
136.89	.6764	.5267	1.0223	.2802
139.91	.4283	.3696	.8870	.2591
142.97	.3464	.3372	.8455	.2429

M/E	PERCENT	F	FLT	CORR	T
58	100.000	1124.95	1.75181	.9908	127.98
60	98.639	1065.41	1.72798	.9903	127.01
64	73.727	997.92	1.29155	.9896	126.67
71	10.980	255.53	.19235	.9613	128.42

SUMMARY OF SPECTRA FOR DIFFERENT ISOTOPIC VARIANTS 2

	58	60	64
D0	1.00000	.00855	.01588
D4	.06355	1.00000	.00874
D9	.08379	.06743	1.00000
MIXT	1.00000	.98639	.73727

VARIANT SIGNAL/QUANTITY RATIOS
.94660 1.00000 .76844

FIG. 6. Computer printout for analysis of a derivatized GC sample of a control equimolar mixture of three stable isotope labelled variants of choline.

The abbreviated multiple mass chromatogram shows sample values for each selected ion printed every three seconds. The masses of 58, 60 and 64 are the pre-selected ions for the endogenous, undeuterated D_0 choline, the labelled tracer D_4 choline, and the internal standard D_9 choline. Mass 71, an abundant ion for unlabelled choline, is continuously sampled as a validity check. The relative abundances of the selected ions are estimated from the regression coefficients (FLT) of the data fitted to an idealized curve. The estimated retention times, T, are shown. After each control injection, a summary spectrum for all the control samples is included. The values from this matrix are used to estimate quantities in subsequent samples. The variant signal/quantity ratios are proportionality values that account for overall isotope effect manifested in unequal observed base peak signal attributable to each variant in an equimolar mixture.

M/E	PERCENT	F	FLT	CORR	T
58	100.000	973.38	2.26886	.9894	129.68
60	4.710	134.88	.10687	.9302	128.89
64	27.732	614.69	.62920	.9833	128.48
71	10.391	484.11	.23575	.9790	130.48

VARIANT	D0	D4	D9
QUANTITY	15.1537	.3100	5.0000
Q/B.W.	15.1537	.3100	5.0000
S.ACTIV	.9800	.0200	

FIG. 7. The partial computer printout for GC-MS analysis of a derivatized sample of mouse brain choline. The ion current ratios are as in Fig. 6. The estimated quantities (nanomoles) of D_0 (endogenous) choline and D_4 (tracer) are shown. The Q/BW are the quantities per unit of brain weight (the weight was not entered and defaulted to unity). The specific activities are $D_0/(D_0 + D_4)$ and $D_4/(D_0 + D_4)$.

negligible consequence particularly when considering its advantage for multitask real time operation: both GC-MS systems are able to collect data and perform analyses while the system console is being used for program development - without noticeable degradation of performance of any of the operations. Moreover, one of the GC-MS systems (GC-MS 2) can be operated in a scanning mode of operation while the other is operated in the selected ion mode.

The principal advantage of developing an in-house system over a commercial system is flexibility rather than economy. Proprietary systems are not readily modified for the specific needs of a research group because complete program listings and documentation are not available to the user and the routines are often written in assembler rather than Fortran. The system that we describe is an optimal hardware and software combination for selected ion monitoring with the capacity to be modified and expanded as changing needs of the research group require.

ACKNOWLEDGMENTS

The authors express their appreciation of Fred Lee for the construction and testing of the interface and Ann Chang for preparation of the computer programs. This work was supported by MH 17691 and DA 1057.

REFERENCES

1) F.C. Falkner, B.J. Sweetman and J.T. Watson, Appl. Spectrosc. Rev., 1975, 10,51.
2) D.J. Jenden and A.K. Cho, Annu. Rev. Pharmacol., 1973, 13, 371.
3) D.J. Jenden, A.K. Cho and R.W. Silverman, in "First Int. Conf. on Stable Isotopes in Chem., Biol., and Med., Conf. no. 730525", Argonne National Laboratory, May 1973, p. 291.
4) D.J. Jenden, M. Roch and R.A. Booth, Anal. Biochem., 1973, 55,438.
5) D.J. Jenden, V.G. Carson and R.A. Booth, Proc. Second Int. Conf. on Stable Isotopes, in press, 1976.
6) D.J. Jenden, Proc. Second Int. Conf. on Stable Isotopes, in press, 1976.
7) R.W. Silverman and D.J. Jenden, in Proc. Second Int. Conf. on Stable Isotopes , in press, 1976.
8) R.W. Silverman and D.J. Jenden, J. Chromatogr. Sci., 1974, 12,505.
9) D.J. Jenden and R.W. Silverman, J. Chromatogr. Sci., 1973, 11,601.

CAN GAS CHROMATOGRAPHY - MASS SPECTROMETRY INFLUENCE TRENDS IN DRUGS ABUSE?

H. Brandenberger

Institute of Legal Medicine, University of Zurich

Zurich, Switzerland

THE DRUG-USE PHENOMENON

Let me begin with a quotation, recently made by Robert L. DuPont, director of the National Institute of Drug Abuse: "What happened in the 16th century with the introduction of tobacco into the world and what happened with the introduction of distilled spirits in the 17th century is the kind of global drug-use phenomenon that we're seeing now with other drugs". Does that mean that we are drifting toward a society which tolerates drug abuse and its commercialization with all the dangerous side effects, toward a society in which heroin and its substitutes can play the same role as alcohol and tobacco today?

You will say that I am comparing dissimilar things, that alcohol and tobacco are legally accepted in spite of their damaging and unhealthy side-effects, while real drugs such as narcotics are presently under strict international and national control, by the Division of Narcotic Drugs of the United Nations, by Interpol and by a large number of national authorities and organizations. You will say that all these organizations are trying to restrict and control the growth of drug-containing plants as well as the synthetic drug manufacture, that they are trying to abolish the illicit drug market by imposing heavy penalties on drug dealers, and that the possession and use of drugs is considered illegal and even immoral in most countries.

I agree with you that all these organizations are working hard in order to diminish the dangers connected with the illicit drug market and drug abuse. But will they succeed? Let's look back into the past! How successful were the rulers of Constantinople in the 16th century? One of them, in 1524, prohibited coffee drinking. Violators, he decreed, were to be drowned. But did you ever drink a stronger cup of coffee than Turkish coffee? Another sultan - I believe it was Murad the 4th - inflicted the death penalty upon tobacco smokers. And the long range effect? "Fumer comme un Turc" is the French description for very heavy smoking. And a more recent example: How did prohibition affect the drinking habits of our American colleagues? From observations made at some international congresses, I have concluded that they are no drier than we Europeans.

LATEST TRENDS IN DRUG ABUSE

Let's turn to the present. What are the latest trends in the field of illicit drugs and drug abuse?

Opium and Opiates. There is probably less illegal traffic of the resin itself. However, the extraction of morphine and its conversion to the potent heroin by acetylation have moved to opium--producing areas which cannot be controlled, areas such as Burma, Thailand, Malaysia. Thus, the illicit traffic can handle a more potent and less bulky material. It is more difficult to curb.

Cannabis. The contraband of marijuana (Cannabis leaves) and haschisch (Cannabis resin) is steadily being substituted by the smuggling of an oily extract of Cannabis containing up to 80% of the active principle THC (tetrahydrocannabinol). Compared with the few percents in the untreated natural product, this is a very high concentration. It facilitates the illicit traffic and puts the Cannabis threat into a new dimension.

Cocaine. Contraband from the South to the North American continent is increasing. The first larger shipments have reached southern Europe, we expect to find cocaine on the black markets in central Europe very soon.

Synthetic drugs. European black markets are selling all sorts of illegal drugs, whether they are under narcotic law such as the large number of total synthetic opiates or LSD, or just pharmaceutical specialities with narcotic or hypnotic action, legally only obtainable under a medical prescription. Not only are the regulations circumvented, quite often the buyers are also cheated, since the products do not always contain what they are being sold as. Compound drugs and all kinds of mixtures of very small amounts of LSD or heroin with pharmaceuticals are becoming more and more common.

Until now, all the combined efforts to fight the abuse of ancient and modern drugs do not seem to be very successful. However, we do not know where we would stand without these endeavours. The situation might be much worse.

THE ROLE OF CHEMISTRY

In the fight against drugs, chemistry plays an important part, although, unluckily, chemists are working on both sides. On one side, extraction procedures are adapted to the simple equipment available in clandestine laboratories. Active components are converted to more potent drugs by simple chemical modifications. Synthetic routes leading to substances with narcotic, hypnotic or hallucinogenic action are copied and misused. On the other side, analytical methods are being developed for the detection of drugs in products of plant origin, in so-called "street samples" (confiscated powders, tablets or solutions), and in the human body. Some of the methods applied are dedicated to the large-scale screening of human urines or street samples for specific types of common drugs, others deal with the detection and identification of unknown or unexpected components, a last group of procedures centers on the ultra-trace analysis of drugs of abuse in body fluids.

Even synthetic chemists are involved in the war against illegal drug production, as shown by a recent example. As you probably know, the restrictions imposed on the cultivation of morphine-producing *Papaver somniferum* has led to a shortage of codeine for medical purposes. Since codeine can also be made from thebaine, an alkaloid biologically synthesized by the species *Papaver bracteatum*, and since illicit heroin production from thebaine seems unlikely (Table I), it has been suggested (1) to substitute the cultivation of *P. somniferum* for *P. bracteatum*. In order to evaluate possible risks involved in such a change, the United Nations Narcotics Laboratory has recently convened a working group of experts in Geneva to consider the feasibility of an illicit conversion of thebaine into drugs of abuse. The experts came to the conclusion that the available methods are not likely to be carried out in clandestine laboratories. Should the usual source for heroin production disappear, it would be much more probable that the illicit manufacture would shift to certain entirely synthetic materials of high physiological potency which are fairly easy to prepare.

It became clear at that meeting that trends in drug abuse are largely influenced by the availability of the illegal compounds. For us analysts, it may be more interesting to know whether they also depend upon the effectiveness of our analytical control and whether the development of GC-MS has had a practical impact. In order to answer these questions, I am going to give some examples taken from our own work in Zurich.

TABLE I. Possibilities for synthetical heroin manufacture from Papaver species

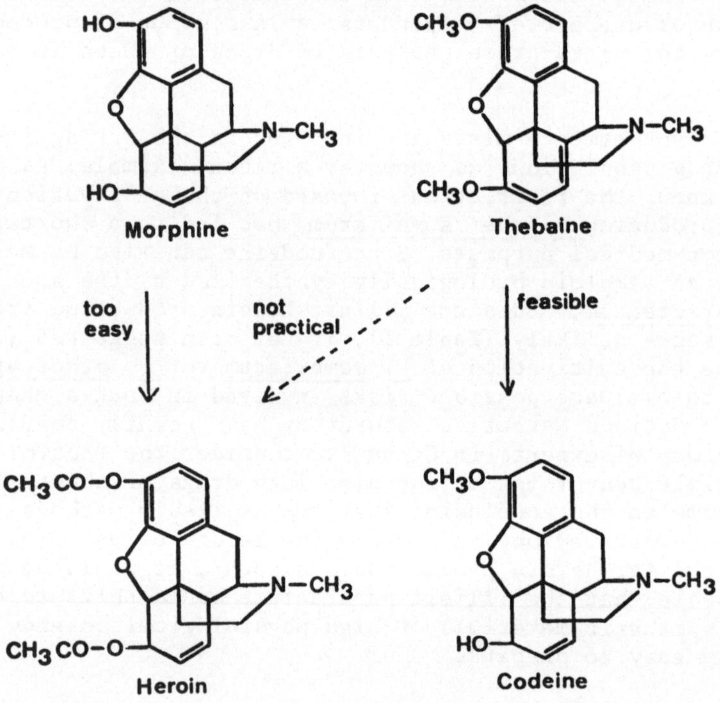

HO
O
N–CH3
HO
Morphine

CH3O
O
N–CH3
CH3O
Thebaine

too easy

not practical

feasible

CH3 CO–O
O
N–CH3
CH3 CO–O
Heroin

CH3O
O
N–CH3
HO
Codeine

DRUG SCREENING (AMPHETAMINES) (2)

Ten years ago, we were asked to develop a quick screening test for the control of doping by amphetamines. Fig. 1 illustrates our answer to this request. The body fluid is injected into a sealed serum vial containing solid carbonate. Amphetamines are set free from their salts and - in part - displaced into the gas phase. Aliquots of the head space gas are removed through the membrane and analyzed by GC.

The method possesses only a modest sensitivity, if flame ionization GC is used. However, if a mass spectrometer focussed on the appearance of the ion with mass 91 is available as a GC monitor, amphetamine concentrations of 100 ng per ml and methyl amphetamine concentrations as low as 10 ng per ml of body fluid can still be detected. Fig. 2 gives an example of an application, the analysis of the urine of a doped competitive cyclist.Polyethylene imine was used as a partition phase, the appearance of ion 91 recorded. The intensities of the 3 single mass recordings differ by factors of 10. The retention time corresponds to amphetamine. The method is fast and easy to carry out, but - seen in retrospect - its development was probably a loss of time. A laboratory which can block its mass spectrometer with such screening work is hard to find. Screening of large sample numbers should be done with less sophisticated equipment, with methods which can be carried out by unskilled workers.

Methods which satisfy these requirements are the immuno tests. An enzyme immunoassay has been developed for amphetamines. Its specificity - illustrated in Table II - is rather poor, but the method is so simple and fast that it will probably become the screening test of choice for amphetamines. Of course, it should be used only with due consideration of its fallibility; positive results must always be verified by more reliable procedures.

IDENTIFICATION OF STREET SAMPLES

As I have pointed out before, street samples do not always contain what is claimed. They are often mixtures, and they may hold components which are unknown to us. In such cases, GC-MS is of great help. A product which was submitted to our laboratory as Mescaline did not contain a trace of 3,4,5-trimethoxyphenyl-ethylamine but instead 2 different organic bases could be isolated and identified by GC-MS. Fig. 3 gives the mass spectrum of the major component which led to its identification as 1-(1-phenyl)--cyclohexylpiperidine or phencyclidine, the active ingredient of Sernyl, the so-called peace drug.

The molecular ion is seen at m/e 243 with loss of 43 (m/e 200), 57 (186) and 77 (166). Other important ions are due to the tropylium ion (m/e 91) and phenyl residue (m/e 77).

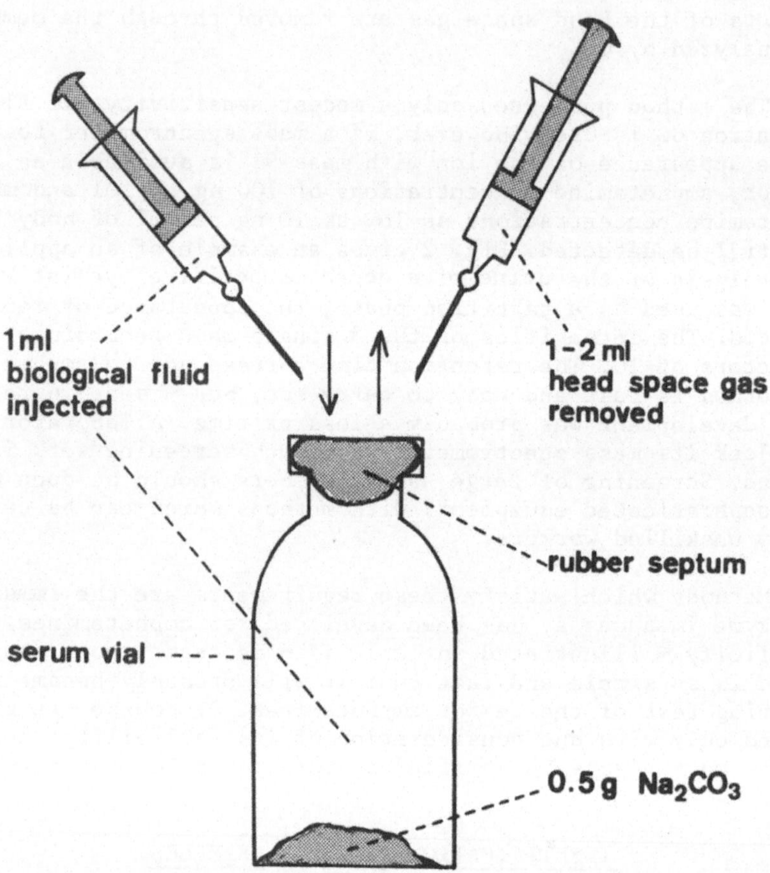

FIG. 1. Head space gas sampling of amphetamines.

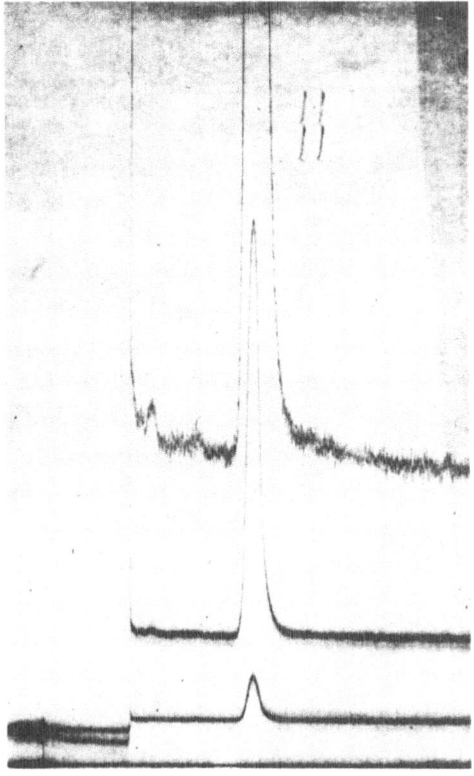

FIG. 2. Detection of amphetamine in the urine of a cycle racer by GC with single mass detection (m/e 91).
Column: Carbowax 20M on potassium hydroxide coated Chrom. W.

TABLE II.
Specificity of enzyme immunoassay for amphetamine

Drug	Concentrations in Urine giving a Signal equivalent to 1.0 μg/ml Amphetamine	Relative Reponse
Amphetamine	1.0 μg/ml	1.0000
Methamphetamine	1.5 μg/ml	0.6800
Mephentermine	1.9 μg/ml	0.5400
Phentermine	3.3 μg/ml	0.3000
Benzphetamine	11.7 μg/ml	0.0850
Cyclopentamine	15.0 μg/ml	0.0670
Phenmetrazine	16.7 μg/ml	0.0600
Ephedrine	23.3 μg/ml	0.0430
Phenylpropanolamine	25.0 μg/ml	0.0400

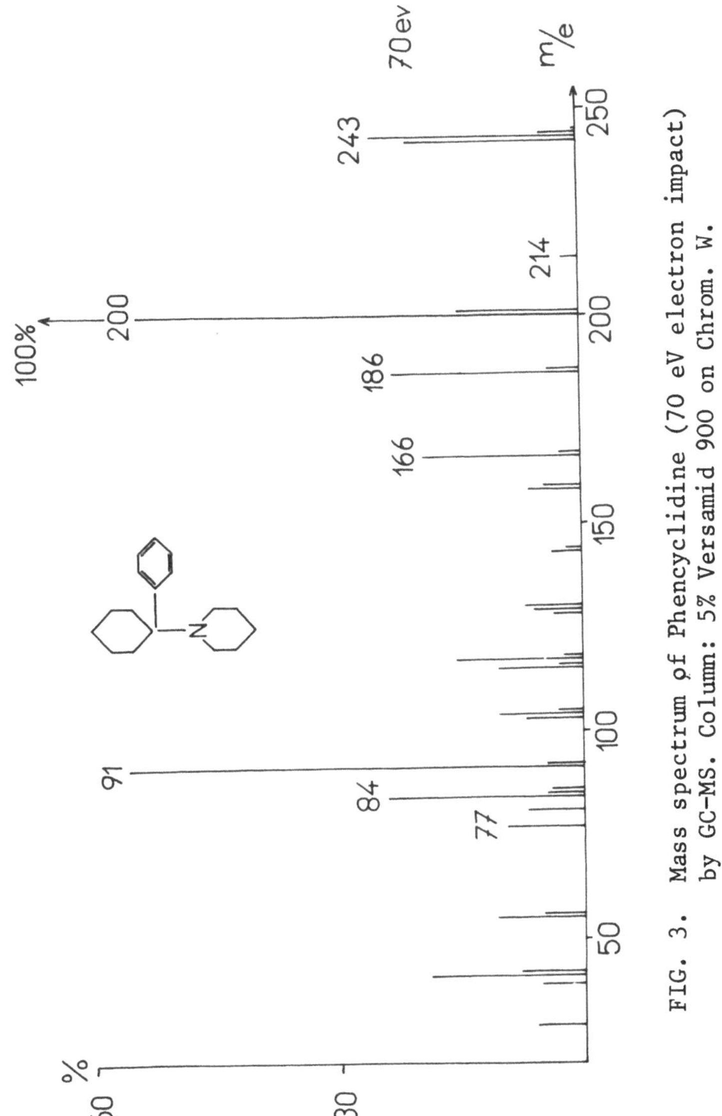

FIG. 3. Mass spectrum of Phencyclidine (70 eV electron impact) by GC-MS. Column: 5% Versamid 900 on Chrom. W.

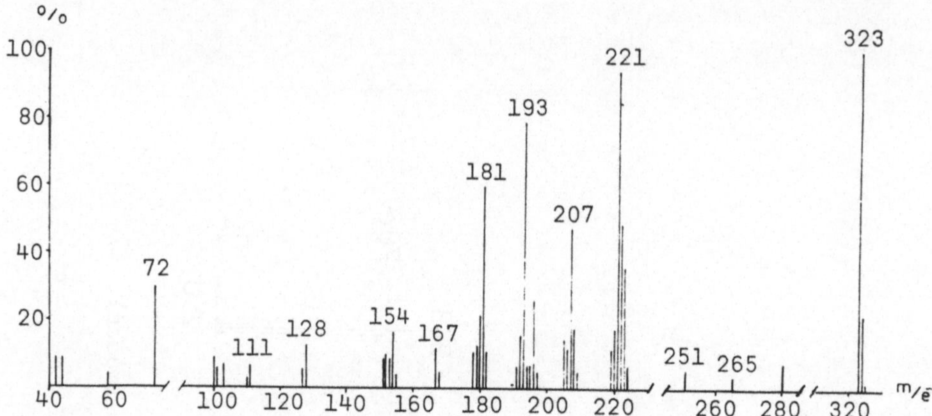

FIG. 4. Mass spectrum of LSD (70 eV electron impact) by GC-MS. Column: 1% SE-30 on Chrom. W.

Fig. 4 gives the mass spectrum of the minor component, present in a very small concentration only. You will recognize LSD. (M^+ at m/e 323 with largest fragment ions resulting from the loss of the side chain with ring dehydrogenation; m/e 72 points to the split of the peptide bond).

A brownish looking powder confiscated in Milano, where a young Swiss had been trying to sell it, proved to be more difficult to identify. It contained a base which - after purification - gave the mass spectrum plotted in Fig. 5. With the additional help of UV-, IR- and NMR-spectrophotometry, the base could be identified as 1-dimethylaminoethyl-2-p-ethoxybenzyl-5-nitro-benzimidazole or etonitazene, probably the most potent analgesic ever synthesized (Formula A) (3). The M^+ at m/e 396 indicates an even number of nitrogens. The ion at m/e 135 represents the side chain in the 2-position, it loses ethylene and yields the ion at m/e 107. The base peak (m/e 86) represents the side chain in the 1-position (cleavage between the 2 nitrogens) which also undergoes a McLafferty rearrangement to give ion at m/e 58. The ion at m/e 30 indicates the presence of an aromatic nitro group. All too often, analytical identification methods are used by specialists, each method alone. With mass spectrometry, this approach is quite common. But we should not forget, that a mass spectrometer is not only a solo instrument; it can also contribute to the right sound as a partner in an instrumental band.

DRUGS IN BODY FLUIDS

In the search for unknown drugs and/or pharmaceuticals in body fluids, a combined instrumental approach is especially helpful. Table III shows our extraction scheme. For drug analysis, steam distillation is usually omitted. Different extraction steps isolate the strongly acidic, weakly acidic, neutral, basic and amphoteric compounds in separate fractions. All are analyzed by UV-spectrophotometry, in organic solution and (with the exception of the neutral extract) also in water at 3 different pH-values. Thin-layer chromatography, color tests, IR- and fluorescence-spectrophotometry may yield additional information. Fractions of interest are analyzed by GC-MS.

We still work with an 8 year old LKB-9000 and without a computer. Data acquisition and presentation are done by a multichannel analyzer. The mass spectra are printed out as modulo-14 tables. This can be done on-line (without normalization) or off-line (with normalization and listing of the total ion current). Table IV shows a typical example, a barbiturate which had already been indicated by UV-spectrophotometry. The mass spectrum identifies it as the not very common 5-allyl-5-n-butyl-barbituric acid. The molecular ion (m/e 224) is not visible. The base peak results from the loss of the butyl residue (M^+-57 = 167). M^+-29 and M^+-43 are also visible and so are the ions at m/e 15, 29, 43 and 57 in the vertical column +9. The butyl residue is therefore not branched. Two dominant masses are 41 (allyl residue) and 124 (by ring contraction of the base peak fragment (m/e 167) with elimination of CONH). We are using such modulo-14 tables in place of plots.

We also did not build our own MS-library. Instead, we are using Cyphernetics MSSS. If the work load in a laboratory is heavy and the manpower short, this is certainly the faster approach. We do hope that MSSS will be further perfected by incorporation of more spectra of biologically important substances and by omitting unneeded duplications. Table V shows a dialogue which led to identification of an unexpected drug with a mass spectrum composed of only a few quite unspecific ions, a dominant base ion with mass 58 and some small fragments (5% intensity or less) with the masses 91, 208, 115, 105 and 77. Only 3 questions were needed to obtain the identification. 368 spectra in the MSSS library have 58 as base ion. Fortyfive of them contain a tropylium

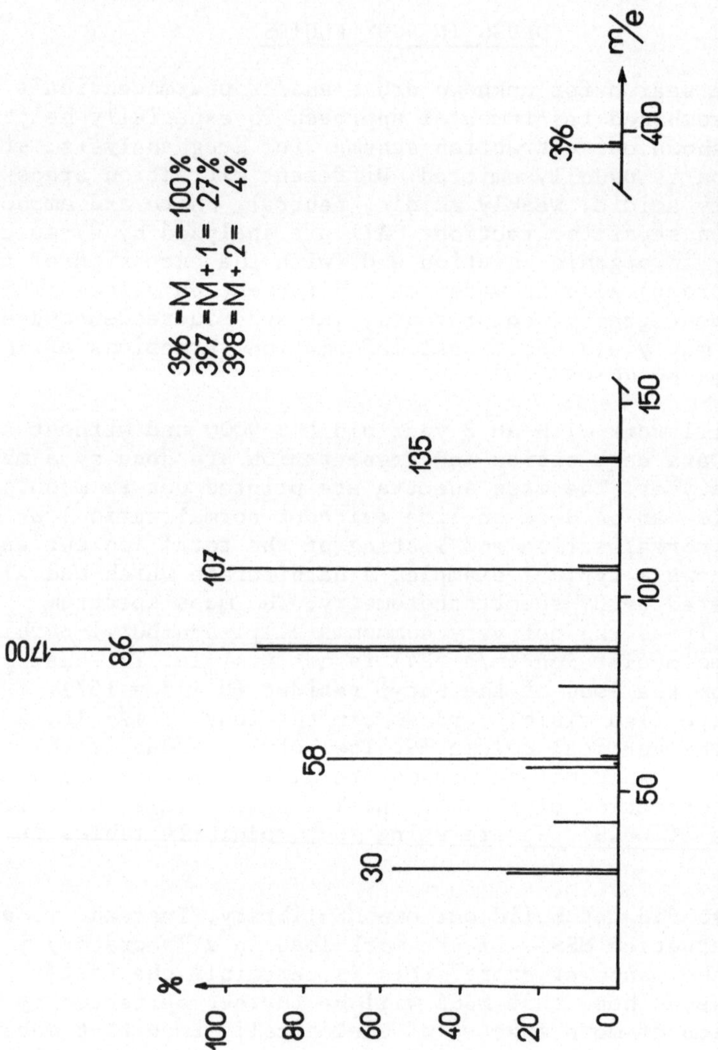

FIG. 5. Mass spectrum of 1-dimethylaminoethyl-2-p-ethoxy-benzyl-5-nitro--benzimidazole (70 eV electron impact, solid inlet).

TABLE III.
Scheme for the extraction of pharmaceuticals and drugs from body fluids

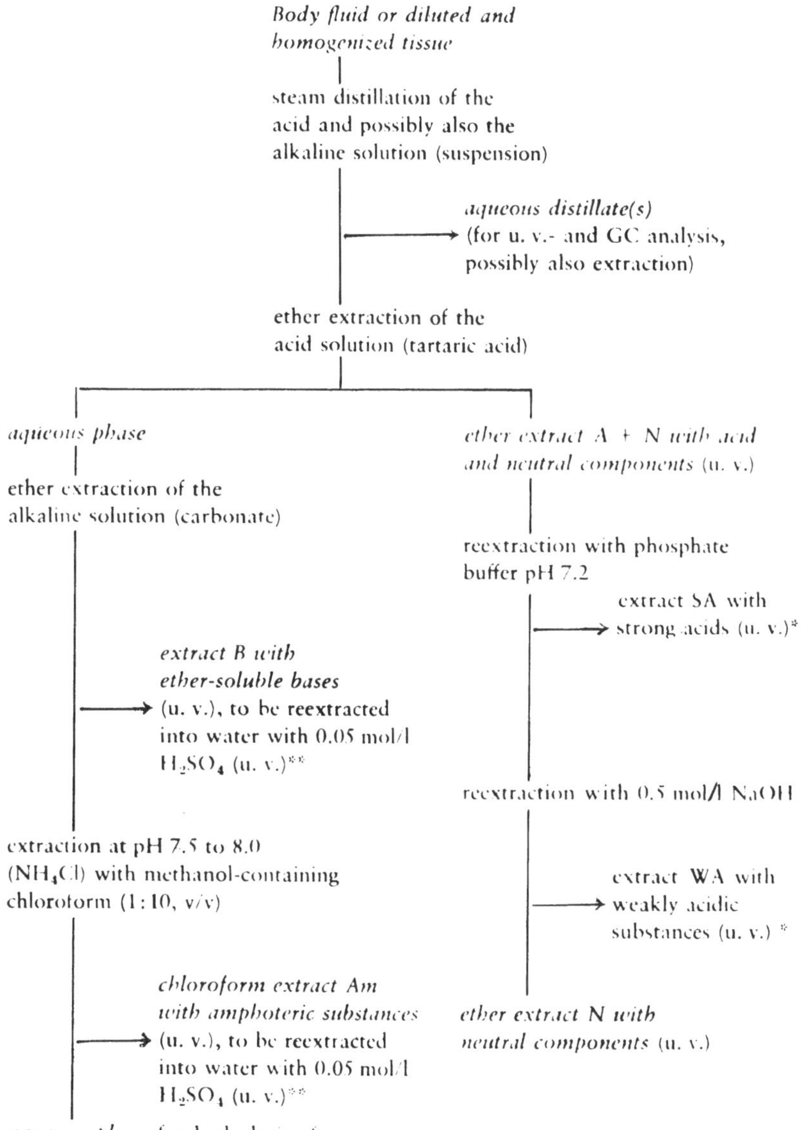

* after recording of u. v.-spectra (at 3 pH-values) acid extraction into ether (u. v.).
** after recording of u. v.-spectra (at 3 pH-values) alkaline extraction into organic phase.
*** with new extractions after hydrolysis.

FALL 115/76
ELUT EXTRAKT WS

CHR.2 2.MS

TABLE IV. Presentation of the mass spectrum of 5-allyl-5-n--butyl-barbituric acid as modulo-14 table. On-line output by way of a multi-channel analyzer data system.

		+1	+2	+3	+4	+5	+6	+7	+8	+9	+10	+11	+12	+13
	12													
	12													
6										81		12	9	
2.							96	792	215	647	51	187	47	
34				13	41	9.4	194	2553	298	3.9	359	82		
48			44	93	2.1	499	252	354	152	54	7	86	16	
62		11		141	156	491	194	156	143	59		58	55	
76		55	153	149	251	248	115	1.		45				
9.		12	11	42	49	2.5	4.7	746	547	18				
1.4			281	7	2.9	88	81	87	46					
118					84	17	1459	2.7			47			
132					44		56	1.6	49	685	2.3			
146						14.	51	412	1.	4				
16.							113	5651	35.4	4.6	53			
174								7.6	193	149				
188								59						
2.2														
216														
23.														
244														
258														
272														
286														

TABLE V.
Identification of the drug Darvon from its mass spectrum by consulting Cyphernetics MSSS

```
OPTION? PEAK

PEAK AND INTENSITY SEARCH


TYPE PEAK, MIN INT, MAX INT
CR TO EXIT, 1 FOR ID,MW,MF AND NAME

USER:58,99,100

  # REFS     M/E PEAKS

     368          58

NEXT REQUEST: 91,2,8

  # REFS     M/E PEAKS

      45          58  91

NEXT REQUEST: 208,2,8

  # REFS     M/E PEAKS

       4          58  91 208

NEXT REQUEST: 1

  ID#      MW      MF                  NAME

15632     339 C22.H29.N.02          4-DIMETHYLAMINO-3-METHYL-
                                   1,2-DIPHENYL-2-BUTYLPROPIONAT
E
                                    (PROPOXYPHENE,DARVON)
23411     339 C22.H29.N.02          4-DIMETHYLAMINO-3-METHYL-
                                   1,2-DIPHENYL-2-BUTYL PROPIONA
T
                                   E/PROPOXYPHENE
23412     339 C22.H29.N.02          4-DIMETHYLAMINO-3-METHYL-
                                   1,2-DIPHENYL-2-BUTANOL PROPIO
N
                                   ATE               (PROPOXYPHEN
E
35592     339 C22.H29.N.02          4-DIMETHYLAMINO-3-METHYL-
                                   1,2-DIPHENYL-2-BUTANOL PROPIO
N
                                   ATE(PROPOXYPHENE)
```

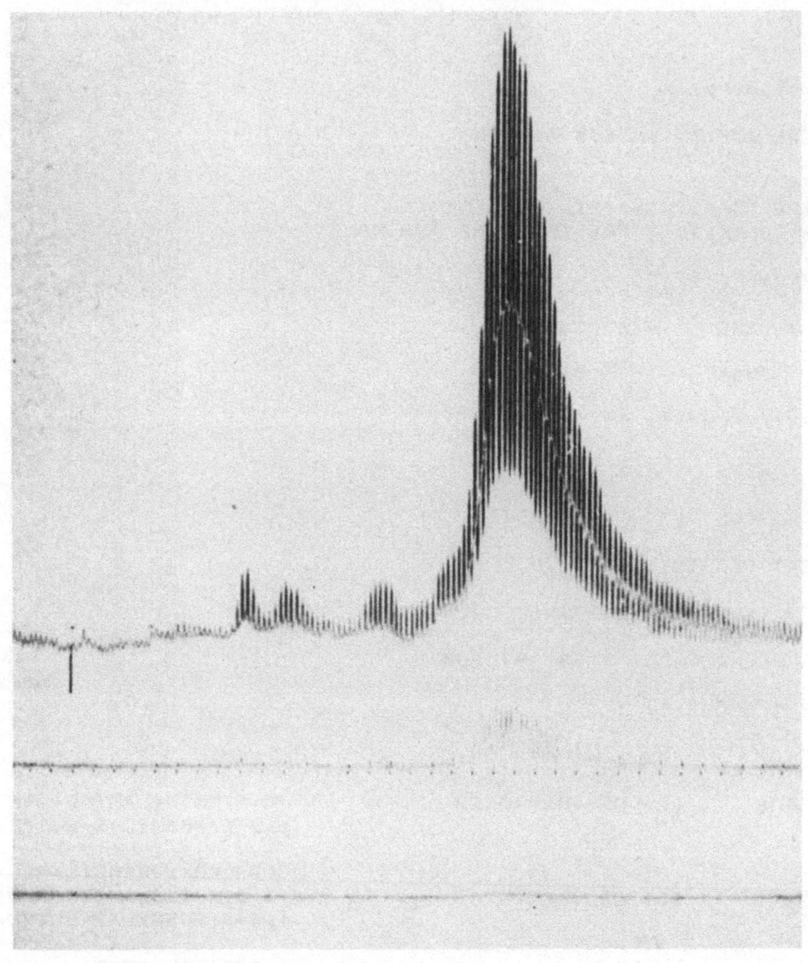

FIG. 6. Chromatogram of THC, recorded by mass specific detection of M^{+} (m/e 314, upper trace), M^{+}-15 (middle trace) and M^{+}+1 (lower trace) (20 eV electron impact). Column: 1% SE-30 on Gas Chrom. Q.

ion, 4 also the ion 208. The 4 entries are identical: propoxyphene, the active ingredient of the drug Darvon.

TRACE ANALYSES BY GC-MD (THC) (4,5)

Another possibility of using a mass spectrometer in drug control is given by its suitability as a mass specific detector for GC. Mass fragmentography or GC with mass specific detection (GC-MD), as we call it, has opened a new dimension in trace analysis. It permits detection of drugs in the low ng- and pg-range, often with a specificity which is much superior to that of conventional GC.

Fig. 6 shows a chromatogram of THC, the active component of cannabis (Formula B), recorded by mass specific detection of the ions 314 (M^+), its isotope 315 (lower trace) and 299 (M^+-15). The identity of the drug is given by the retention time and the intensity ratios between the recorded ions. They must correspond to the respective peak ratios of the mass spectrum taken under the same conditions, in this case at 20 eV. The spectrum is shown in Fig. 7. Fragment 299 corresponds to almost 60% of the molecular ion. M^+-1 is 23% as intense as M^+, as can be expected from the 20 carbons in the molecule. With this approach, THC and other cannabinoids can be detected in a tiny trace of dust. Fig. 8 shows the chromatogram of a hexane extract of such a powder. The mass recording 314 and 299 indicate the presence of cannabidiol and THC, the mass recording 295 is used for the detection of cannabinol which forms the big last peak. Differentiation between cannabidiol and THC is possible with the peak intensity ratio which is low for cannabidiol (first peak) and 60% in favor of 314 for THC.

Another application we have reported already in 1972 is the detection of THC in saliva after ingestion of 2 mg of THC on sugar. A few ml of saliva, collected after 30 min., were extracted with hexane, the extract concentrated to 100 µl and 2 µl injected. Fig. 9 shows the chromatogram obtained. The large peak recorded by the two masses 314 and 299 is THC.

Agurell and others have reported the detection of THC in blood and the identification of THC-metabolites in urine, also by mass fragmentography (6). We shall hear more about that in the following paper (7).

For sensitive and specific mass fragmentography, a careful selection of the tracer ions should be made. The simpler the fragmentation, the higher the abundance of the potential tracer ions, the more promising is the approach for ultra-trace analysis. It might be advantageous to convert the drug in question into a

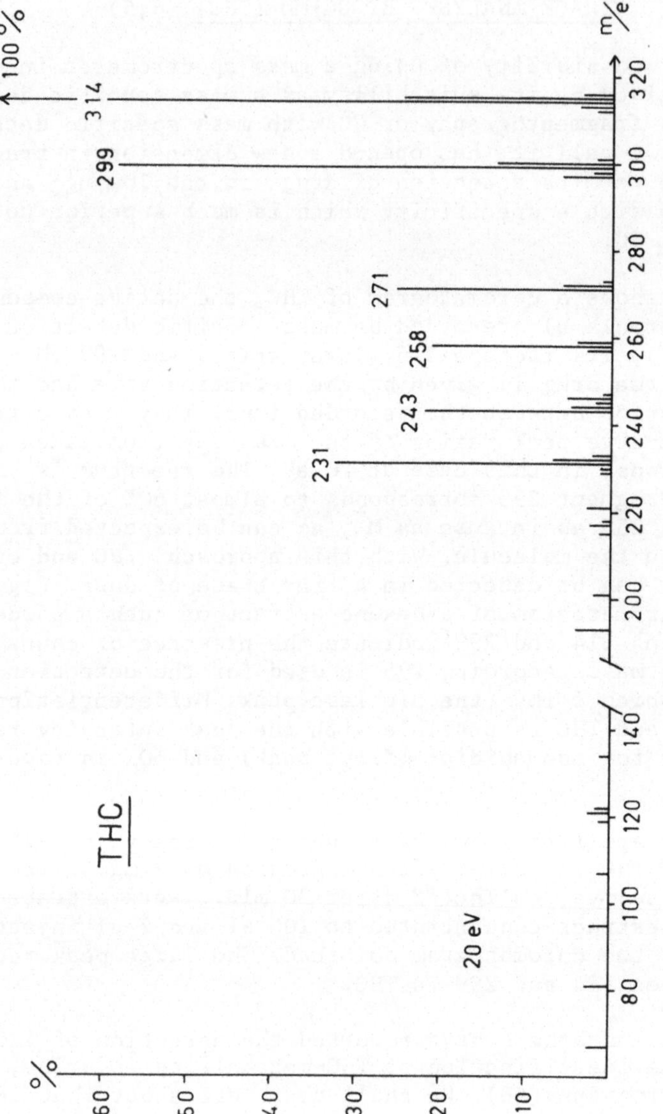

FIG. 7. Mass spectrum of THC (20 eV electron impact) by GC-MS.

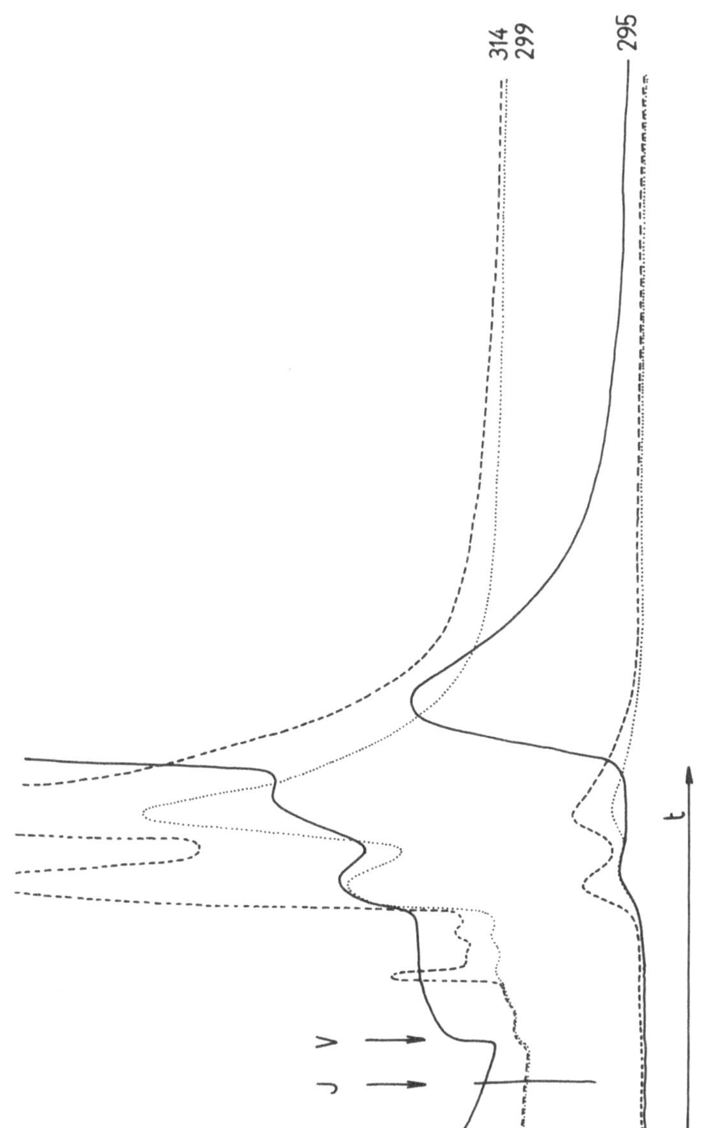

FIG. 8. Identification of Cannabidiol, THC and Cannabinol in a trace of dust by GC with mass specific detection of m/e 314, m/e 299 and m/e 295 (20 eV electron impact). Column: 1% SE-30 on Gas Chrom.Q.

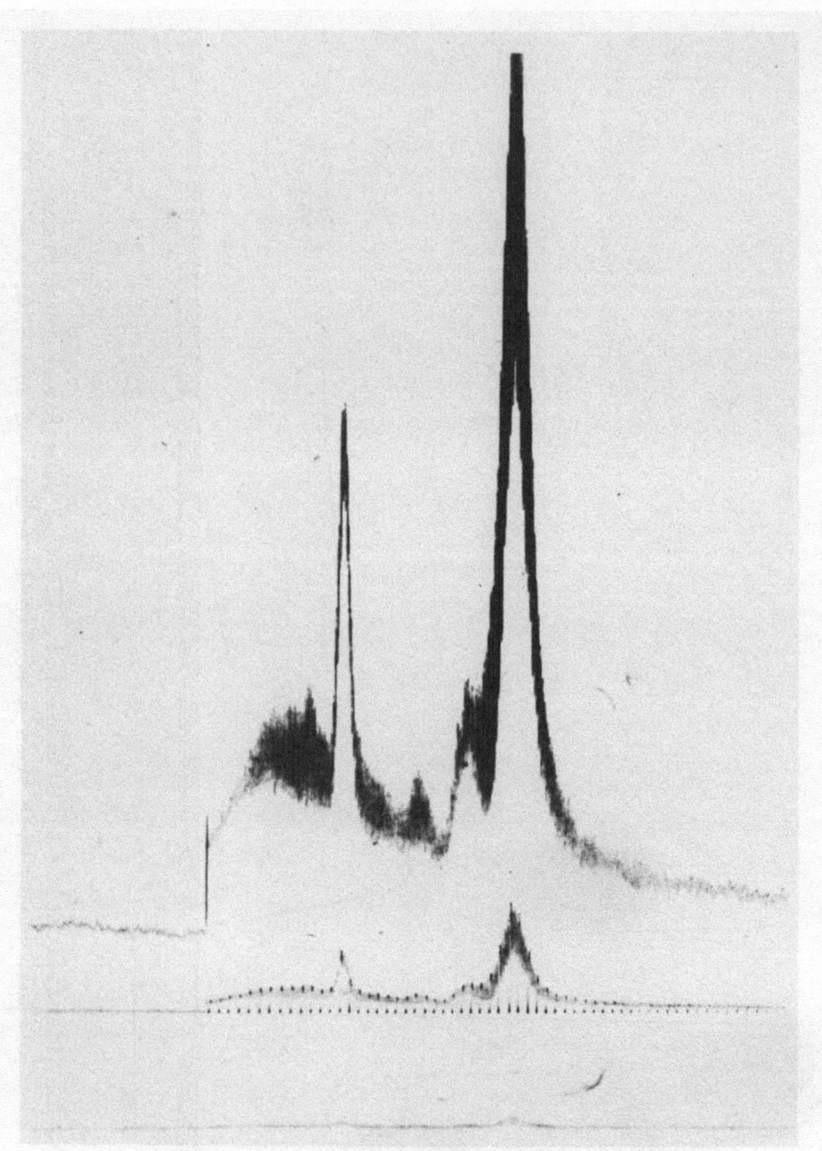

FIG. 9. Detection of THC in saliva by GC with mass specific detection of m/e 314 and m/e 299 (20 eV electron impact). Column: 1% SE-30 on Gas Chrom.Q. The last peak corresponds to THC.

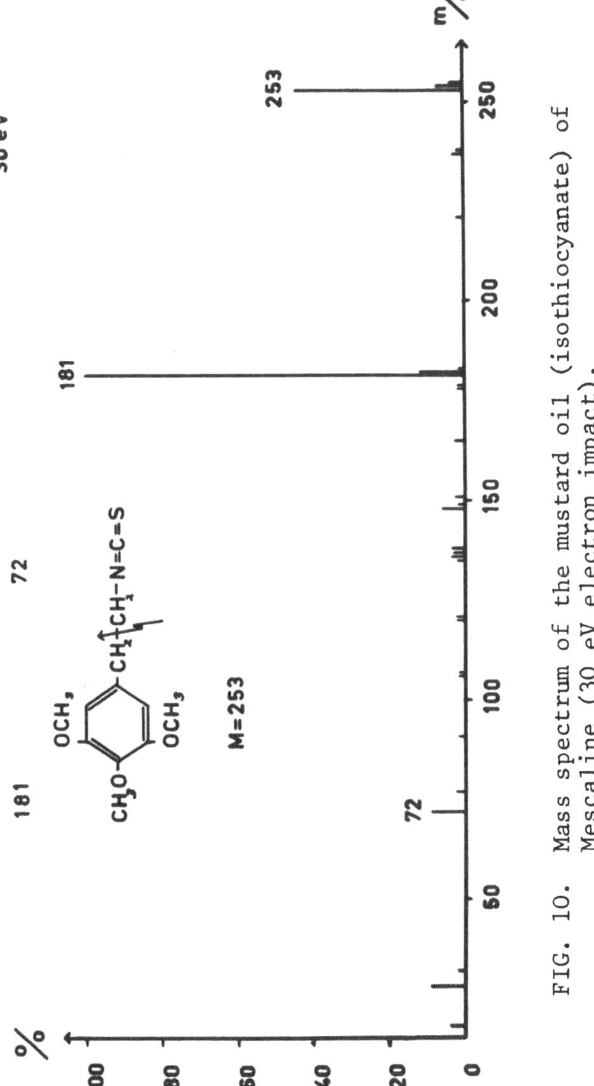

FIG. 10. Mass spectrum of the mustard oil (isothiocyanate) of Mescaline (30 eV electron impact).

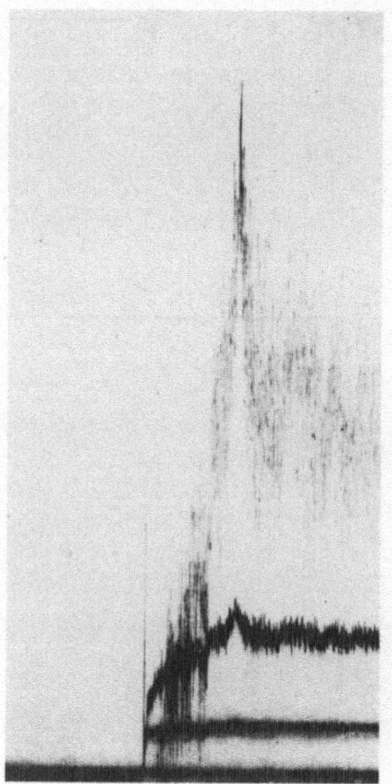

FIG. 11. Gas chromatographic detection of 13 pg of Mescaline after conversion to the mustard oil by single ion detection of M^{+} (m/e 253). Column: 1% SE-30 on Chrom.W.

derivative better suited for chromatographic separation and for mass specific detection. We usually do that with amphetamines and hallucinogens with related structures such as mescaline and STP. We convert them to their isothiocyanates, just by addition of carbon disulfide to the extract (8). Fig. 10 shows the mass spectrum of the derivative of mescaline. The fragmentation is very simple. The molecular ion at m/e 253 and the base peak at m/e 181 contain practically all the spectrometric information. Fig. 11 shows a gas chromatogram of 13 pg of mescaline after conversion to the mustard oil, recorded with mass specific detection of the molecular ion. This is a good example of the sensitivity of the method.

THE ROLE OF GC-MS IN THE FIGHT AGAINST DRUG ABUSE

What can GC-MS do in the fight against drug-abuse?

1. It will certainly not become a tool for the screening of large sample numbers. The equipment is too sophisticated; it needs too skilled an operator.
2. It is, however, a good back-up method for the verification of the positive findings of the screen.
3. It is an ideal tool for the identification of the unsuspected or the totally unknown in street samples and body fluids. But we should take care to use it in conjunction with other instrumental methods.
4. It is a tool for the ultra-trace analysis of drugs in body fluids.
5. It will help to draw light on the metabolic pathways of the drugs.

Will GC-MS help to curb the illicit use of drugs or to change the trends in drug abuse? This is hard to answer. It is to be expected that in the future, uncommon drugs will have less chance to pass unnoticed. The time between the appearance of a new drug on the black market and the onset of counter-measures will be shortened. GC-MS, I think, has had already an impact in the field of doping, especially as a back-up method. It has helped to turn the battle against amphetamines into a success. There is reason to hope that this will happen also with other groups of drugs, such as opiates, LSD or the cannabinoids.

REFERENCES

1) United Nations Narcotics Laboratory, Report of an Expert Group on "the Feasibility of the Conversion of Thebaine into Drugs of Abuse and of Potential Abuse", January 1976, Geneva.
2) H. Brandenberger, GC-MS Symposium, March 1969, Zurich.
3) H. Brandenberger, Dtsche Lebensm. Rundsch., 1974, 70, 31.

4) H. Brandenberger, in "International Meeting of Forensic Sciences", Edinburgh, September 1972.
5) H. Brandenberger and D. Frangi-Schnyder, in "International Symposium on Microchemical Techniques", Pennsylvania State University, August 1973.
6) S. Agurell, B. Gustafsson, B. Holmstedt, K. Leander, J.-E. Lindgreen, I. Nilsson, F. Sandberg and M. Asberg, J. Pharm. Pharmacol., 1973, 25, 554.
7) M. Nordqvist, J.-E. Lindgren and S. Agurell, in "International Symposium on Mass Spectrometry in Drug Metabolism", Milan, June 1976.
8) H. Brandenberger and D. Schnyder, Z. Anal. Chem., 1972, 261, 297.

CHARACTERIZATION OF IN VIVO LIVER METABOLITES OF Δ^1-TETRAHYDRO-CANNABINOL BY EI AND CI MASS SPECTROMETRY

D.J. Harvey, B.R. Martin and W.D. Paton

University Department of Pharmacology, South Parks Road, Oxford OX1 3QT, U.K.

SUMMARY

The *in vivo* liver metabolism of Δ^1-tetrahydrocannabinol (Δ^1-THC) was examined in mouse, guinea-pig and rat by electron impact and chemical ionization mass spectrometry. Twenty-four metabolites were identified and considerable species variation was observed. The predominant biotransformation steps in the mouse involved hydroxylation in the 6α- and 7- positions followed by oxidation to the 6-oxo and 7-carboxylic acid metabolites respectively. Mono- and dihydroxy derivatives of these compounds were also produced. All of these contained hydroxy groups in the pentyl side-chain. In the guinea-pig, 6β-hydroxylation occurred in preference to hydroxylation at the 6α-position and oxidation of the pentyl side-chain at predominantly $C_{3''}$ produced major metabolites. The metabolism of Δ^1-THC in the rat was similar to that in the mouse but fewer trisubstituted metabolites were observed.

INTRODUCTION

The metabolism of Δ^1-tetrahydrocannabinol (Δ^1-THC, I) has been studied extensively in recent years and it is evident that a large number of metabolites are formed (1, 2). Three regions of primary biotransformation have been described; hydroxylation of the 6 and 7 positions allylic to the Δ^1-double bond, oxidation of the double bond to an epoxide (3, 5) and hydroxylation of the pentyl side-chain in various positions (6). Each initial reaction may be followed by the introduction of a second hydroxyl group and, in addition, oxidation of one or more of the hydroxyl groups can produce a variety of polar metabolites including aldehydes

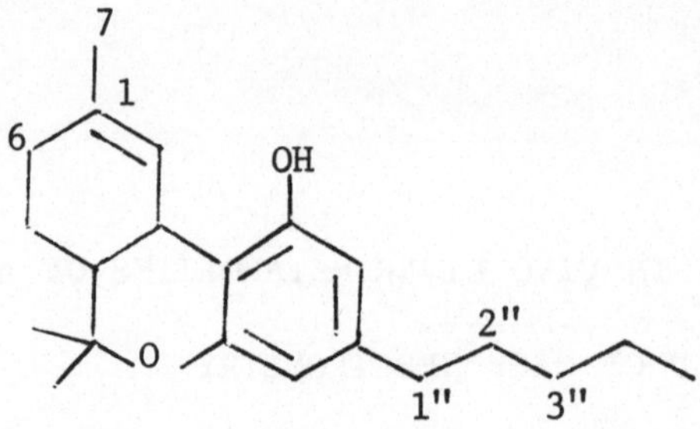

I

(7), ketones (8), acids (9, 11); and hydroxy-acids (12, 14). It is also apparent that considerable species variation exists (6, 10). Administration of Δ^1-THC to animals leads to a considerable accumulation in various tissues, notably liver, lung, spleen, kidney and adrenal gland (1, 15, 16) and it has been shown that both Δ^1-THC and a number of metabolites are present (13, 14). There is also evidence to suggest that there are metabolic differences between tissues (5, 17).

We are currently investigating the nature of these tissue metabolites using mass fragmentography, but before this can be done, the compounds have to be identified and suitable ions chosen for monitoring purposes. Initially large doses of Δ^1-THC were administered in order to obtain strong spectra, but subsequent experiments using lower doses (10 mg/kg) showed a similar metabolic profile. This paper describes the identification of a number of new metabolites present in the livers of three species following intraperitoneal administration of Δ^1-THC.

EXPERIMENTAL

Dosage of animals. Δ^1-THC was suspended in Tween 80 and isotonic saline and was administered intraperitoneally to male mice (Charles River CD1, 25 g), rats (Sprague Dawley, 250 g) and guinea-pigs (Duncan Hartley, 400 g). Each animal received a dose of 100 mg/kg at 26 h and at 2 h before sacrifice.

Extraction and separation of metabolites. Livers were removed from the animals and homogenized in 5 volumes of isotonic saline. Unchanged Δ^1-THC and its metabolites, together with much lipid material, were extracted with three portions of redistilled ethyl acetate (homogenate volume) and the solution was dried over magnesium sulphate. After removal of the solvent, the residue from approximately 1.5 to 2.0 g of rat or guinea-pig liver or pooled mouse livers (n=3) was chromatographed on Sephadex LH-20 in redistilled chloroform. Fractions were taken as shown in

Scheme 1. Aliquots of these fractions, equivalent to 1 g of liver, were converted into derivatives for GC-MS as described below.

Preparation of derivatives. (a) Trimethylsilyl (TMS) derivatives. 5 μl of a solution of N,O-bistrimethylsilyltrifluoroacetamide (BSTFA, 2 parts), trimethylchlorosilane (TMCS, 1 part) and acetonitrile (2 parts) was added to the dried (N_2 stream) metabolite sample which was then heated at 60° for 10 min. The total sample was injected into the gas chromatograph. (b) d_9-TMS derivatives. These were prepared by the addition of d_9--bistrimethylsilylacetamide (d_9-BSA, 3 μl), acetonitrile (2 μl) and TMCS (trace) to the sample which was then treated as above. (c) Methyl ester-TMS derivatives. A solution of diazomethane in ether (prepared from N-methyl-N-nitroso -N'-nitro-guanidine as described by Fales *et al.* (18)) was added to a solution of the metabolites in methanol (20 μl). After 3 min the excess diazomethane and solvents were removed (N_2 stream) and the residue was converted into TMS derivatives as described above. (d) Methyloxime-TMS derivatives. The metabolites were dissolved in pyridine (20 μl) and an excess of methoxyamine hydrochloride was added. The mixture was then heated at 60° for 1 h, the solvents were removed, and the residue was converted into TMS derivatives. (e) Lithium aluminium deuteride reduction. The metabolites, dissolved in dry ether, were refluxed with lithium aluminium deuteride for 1 h and worked up using a standard procedure. The reduced metabolites were then converted into TMS derivatives as described above.

Mass spectrometry. Low resolution mass spectra were recorded at 25 eV with a V. G. Micromass 12B mass spectrometer. This was interfaced *via* a glass jet separator to a Varian 2400 gas chromatograph fitted with a single 2 m x 2 mm glass column packed with 3% SE-30 on 100-120 mesh Gas Chrom Q (Applied Science Laboratories, State College Pa., U.S.A.). The column was temperature programmed from 170° to 280° at 2°/min with the injector block and transfer line temperatures maintained at 270° and 230° respectively. Helium at 30 ml/min was used as the carrier gas. The mass spectrometer was operated at a resolution of 1000 with an accelerating voltage of 2.5 kV, and an ion source temperature of 260°. Spectra were recorded with a V.G. Data System type 2040 which was set to acquire spectra repetitively throughout the elution of the chromatogram. A scan speed of 3 sec/decade with an inter-scan delay of 2 sec was employed, and data acquisition was started when the gas chromatographic column reached 190°.

Chemical ionization mass spectra were recorded with a V. G. Micromass 70/70 mass spectrometer interfaced to a similar chromatographic system to that described above. Isobutane was used as the reagent gas, and the mass spectrometer was operated with an accelerating voltage of 4 kV and an ion source temperature of 220°. Spectra were recorded at 3 sec/decade with the above data system.

SCHEME 1.

LIVER

homogenate

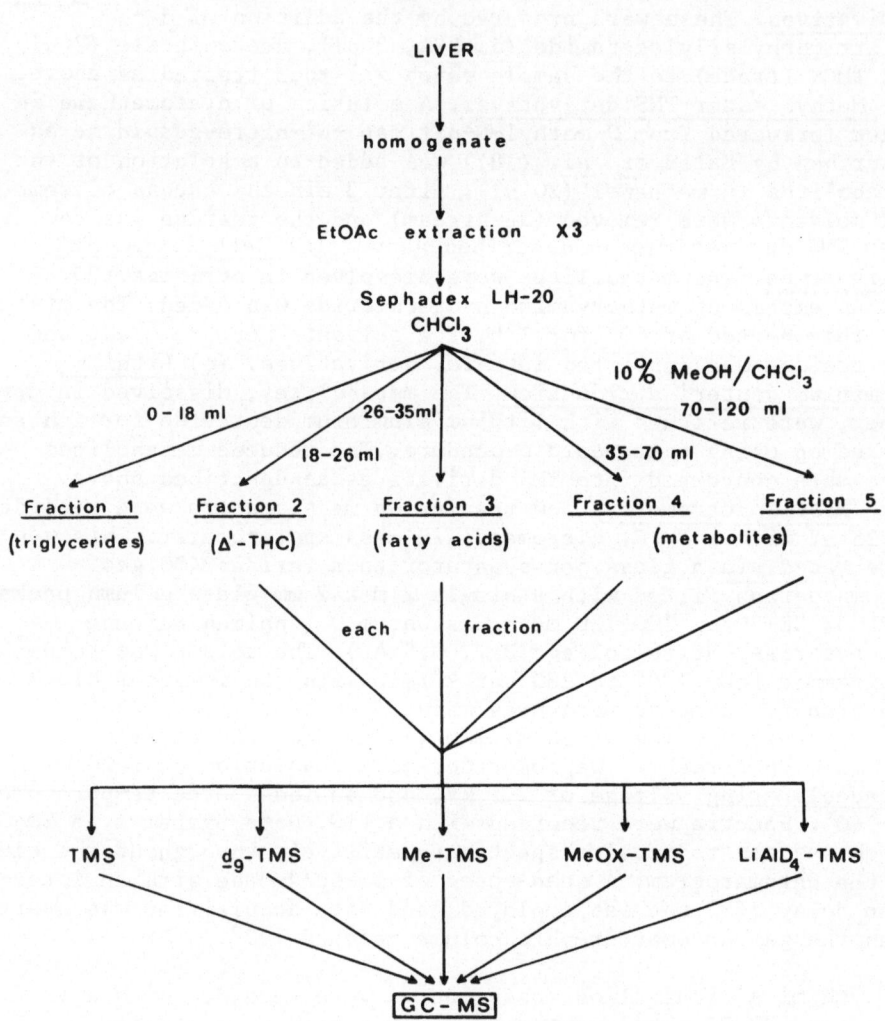

EtOAc extraction X3

Sephadex LH-20

$CHCl_3$

0–18 ml

18–26 ml

26–35ml

35–70 ml

10% $MeOH/CHCl_3$ 70–120 ml

Fraction 1 (triglycerides)

Fraction 2 (Δ^1-THC)

Fraction 3 (fatty acids)

Fraction 4 (metabolites)

Fraction 5

each fraction

TMS

d_9-TMS

Me-TMS

MeOX-TMS

$LiAlD_4$-TMS

GC-MS

RESULTS AND DISCUSSION

In order to reduce the possibility of degradation and loss of metabolites during extraction and isolation, these procedures were reduced to a minimum and full use was made of the gas chromatographic separation combined with computer processing of the mass spectra. The metabolites were extracted from the homogenized livers with ethyl acetate, but as this also resulted in the extraction of much lipid material, the extract was fractionated on Sephadex LH-20 in order to isolate the metabolites. Elution of the column with chloroform and chloroform-methanol mixtures as outlined in Scheme 1 gave two metabolite fractions, the first (Fraction 4) consisting mainly of monohydroxy compounds and the second (Fraction 5) of more polar metabolites. Initially the second metabolite fraction was subdivided using a graded elution from the column, but as the separation achieved was similar to that obtained by gas chromatography, there was little to be gained, and in subsequent experiments the fraction was examined intact.

The metabolites were then converted into a number of derivatives for examination by GC-MS. TMS derivatives were used to record general metabolic profiles, and the shifts in the ions observed in the spectra of the $\underline{d}_9$-TMS derivatives indicated the number of TMS groups and hence hydroxyl or acid groups present (19). Ketones were examined as methyloxime-TMS derivatives and carboxylic acids were converted into methyl esters or reduced to primary alcohols with lithium aluminium deuteride. The latter procedure also reduced ketones to secondary alcohols.

Mass spectra were recorded by computer on a magnetic disc repetitively throughout the elution of the gas chromatographic run. In this way, recordings could be made of the spectra of minor metabolites which did not give distinct peaks in the total ion current trace or whose peaks were obscured by those produced by major components or residual tissue constituents. Once acquired, these spectra could be processed in a number of ways. Known metabolites were identified by comparison of their mass spectra with those of authentic samples, either directly or with the computer library search programme. Minor metabolites and components of mixed GLC peaks were examined using specific ions. Single ion chromatograms traced through the chromatographic run facilitated the identification of metabolites containing common functional groups. For example the ion at m/e 145 ($TMS\overset{+}{O}=CH.CH_2.CH_2.CH_3$, $\underline{a}$) was specific for metabolites containing a 2"-hydroxyl group (9, 20, 21). By noting the scan in which a number of ions maximized, it was possible to build up the spectra of many of the less abundant compounds. This was best done with the data system "MassMax" program which rejected all ions except those which maximized in a given scan. Using this it was possible to extract the spectra of individual components from multi-component GLC

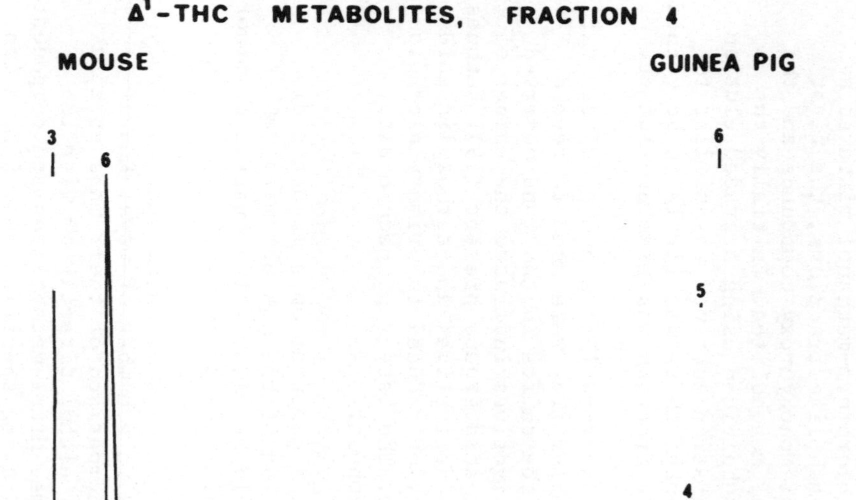

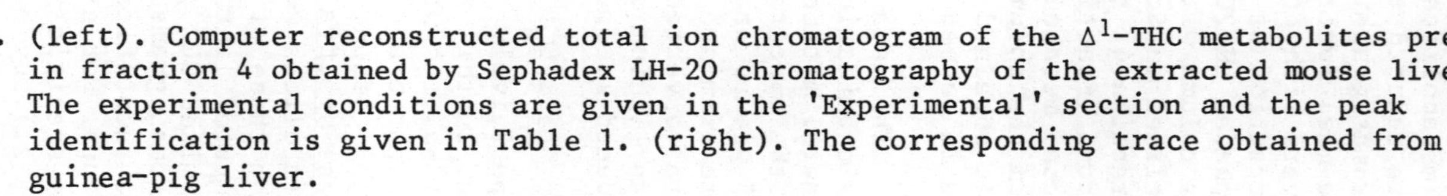

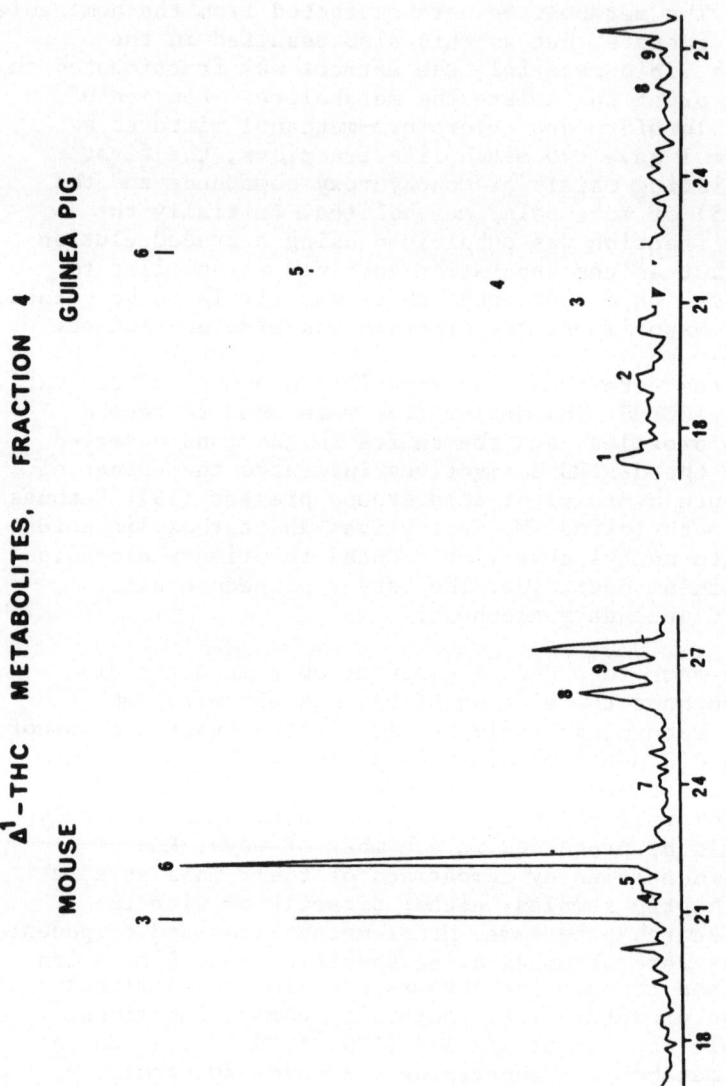

FIG. 1. (left). Computer reconstructed total ion chromatogram of the Δ^1-THC metabolites present in fraction 4 obtained by Sephadex LH-20 chromatography of the extracted mouse livers. The experimental conditions are given in the 'Experimental' section and the peak identification is given in Table 1. (right). The corresponding trace obtained from guinea-pig liver.

peaks, providing that the separation of individual compounds was such that their ions maximized in different scans. By plotting reconstructed total ion chromatograms using only the ions extracted by the computer, it was possible to enhance considerably the resolution of the chromatograms. Further enhancement of the resolution was achieved by using only the ions above mass 300 as, with the exception of m/e 145 and 117, the majority of diagnostic and abundant fragment ions from the THC metabolites were in the upper mass range. Many ions of low mass produced by endogenous tissue material and contaminants were then filtered out. The metabolic profiles shown in Figs.1-4 were produced by the data system in this way. Chemical ionization spectra were used to identify molecular ions.

Seven compounds were identified in fraction 4 from mouse liver (Fig. 1a). Four of them, including the two major metabolites 6α- (3, Table 1) and 7-hydroxy-Δ^1-THC (6) have been reported previously. The mass spectrum of the TMS derivative of 6α-hydroxy-Δ^1-THC (3) was characterized by ions at m/e 459 ($[M-15]^+$, Table 2) and a base peak at m/e 384 ($[M-90]^+$) whereas the spectrum of the corresponding derivative of 7-hydroxy-Δ^1-THC (6) exhibited a molecular ion at m/e 474 and a base peak at m/e 371 ($[M-103]^+$). The third metabolite, produced in small quantities, was identified as 6β-hydroxy-Δ^1-THC (5) by its characteristic base peak at m/e 343 ($[M-131]^+$) (9). Trace amounts of 3"-hydroxy-Δ^1-THC (4) were also detected and characterized by an abundant ion at m/e 330 ($[M-144]^+$, b, Scheme 2) (9, 20, 21).

The other three metabolites in fraction 4 gave molecular ions at m/e 488 (TMS derivatives) which showed 18 atomic mass unit (amu) shifts in the spectra of the d_9-TMS derivatives, thus indicating the presence of two TMS groups. The mass increment of 14amu from the molecular weight of the monohydroxy-Δ^1-THC metabolites was indicative of oxidation of a methylene group to a carbonyl. These compounds did not react with diazomethane and were thus not carboxylic acids ($CH_2OH \longrightarrow COOH$). The presence of ketone groups was subsequently confirmed by the preparation of methyloximes (Table 3). The base peak of metabolite 8 at m/e 385 (TMS derivative) was produced by elimination of a CH_2-OTMS group as shown by deuterium labeling (d_9-TMS) and indicated the presence of a primary hydroxyl group probably at C_7 by analogy with the fragmentation shown by 7-hydroxy-Δ^1-THC. In the spectra of the TMS derivatives of the other two hydroxy-ketones the base peaks at m/e 145 (a) and m/e 344 ($[M-144]^+$) respectively indicated hydroxyl substitution in the 2"- and 3"-positions of the side-chain (9, 20, 21). None of these spectra gave definitive information on the position of the ketone group. Location of this group at C_6 in all three compounds was established by reduction with $LiAlD_4$ to the corresponding deuterated dihydroxy-Δ^1-THC's. Metabolite 8 was reduced to a diol showing GC-MS properties identical to those

TABLE 1. Structures and relative proportions of the Δ^1-THC metabolites in the three species.

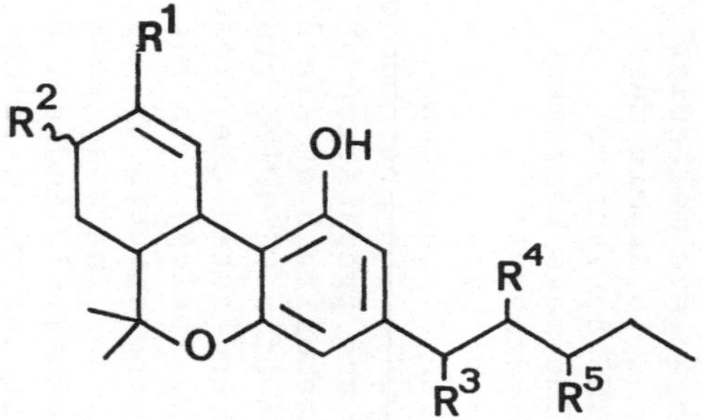

No.[a]	Compound	R^1	R^2	R^3	R^4	R^5	Relative abundance Mouse	G-pig	Rat	Table[c]
1	1"-OH-Δ^1-THC	CH_3	H	OH	H	H	–	+	–	2
2	2"-OH-Δ^1-THC	CH_3	H	H	OH	H	–	+	–	2
3	6α-OH-Δ^1-THC	CH_3	α-OH	H	H	H	+++	++	++	2
4	3"-OH-Δ^1-THC	CH_3	H	H	H	OH	+	++	+	2
5	6β-OH-Δ^1-THC	CH_3	β-OH	H	H	H	+	+++	+	2
6	7-OH-Δ^1-THC	CH_2OH	H	H	H	H	+++	+++	+++	2
7	2"-OH-Δ^1-THC-6-one	CH_3	=O	H	OH	H	+	–	–	3
8	7-OH-Δ^1-THC-6-one	CH_2OH	=O	H	H	H	++	+	+	3
9	3"-OH-Δ^1-THC-6-one	CH_3	=O	H	H	OH	++	+	++	3
13[d]	6α,7-di-OH-Δ^1-THC	CH_2OH	α-OH	H	H	H	+++	++	+++	6
14	Δ^1-THC-7-oic acid	COOH	H	H	H	H	+++	++	+++	4
15	2",7-di-OH-Δ^1-THC	CH_2OH	H	H	OH	H	++	+	–	6
16	6β,7-di-OH-Δ^1-THC	CH_2OH	β-OH	H	H	H	++	+++	++	6
17	6α-OH-Δ^1-THC-7-oic acid	COOH	α-OH	H	H	H	+++	+	++	4
18	3",7-di-OH-Δ^1-THC	CH_2OH	H	H	H	OH	++	+	+	6
19	2",6α,7-tri-OH-Δ^1-THC	CH_2OH	α-OH	H	OH	H	++	+	–	6
20	2"-OH-Δ^1-THC-7-oic acid	COOH	H	H	OH	H	+++	++	++	4

./.

TABLE 1 (contd.).

No.[a]	Compound	R^1	R^2	R^3	R^4	R^5	Relative abundance Mouse	G-pig	Rat	Table[c]
21	3",6α,7-tri-OH-Δ^1-THC	CH_2OH	α-OH	H	H	OH	++	++	-	6
22	2",6α-di-OH-Δ^1-THC-7-oic acid	COOH	α-OH	H	OH	H	+++	-	+	4
23	3"-OH-Δ^1-THC-7-oic acid	COOH	H	H	H	OH	+++	++	++	4
24	3",6α-di-OH-Δ^1-THC-7-oic acid	COOH	α-OH	H	H	OH	++	-	+	4

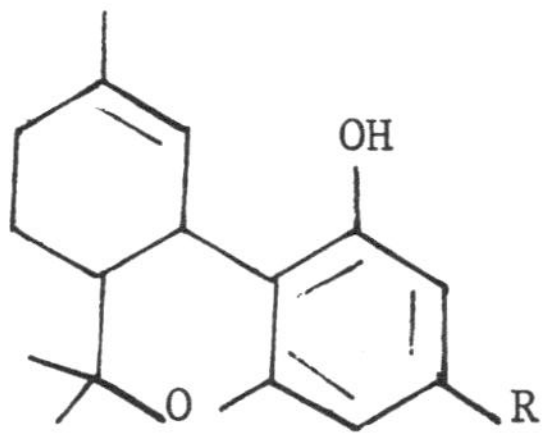

No.	Compound	R	Relative abundance in liver Mouse	Guinea-pig	Rat	Table
10	Δ^1-THC-2"-oic acid	CH_2-COOH	-	+	-	5
11	Δ^1-THC-3"-oic acid	CH_2-CH_2-COOH	+	+++	+	5
12	Δ^1-THC-4"-oic acid	CH_2-CH_2-CH_2-COOH	-	+	-	5

a) Compounds in order of elution: 1-9, Fraction 4; 10-24, Fraction 5.
b) +++, major metabolite; ++, minor metabolite; +, trace.
c) Table containing GC-MS data.
d) Compounds 10-12 at end of table.

TABLE 2. GC-MS data for the monohydroxy-Δ^1-THC metabolites (TMS derivatives).

Compound	No.	R. T. (min)	M^+	Base	Other diagnostic ions
1"-OH-Δ^1-THC	1	18.2	474[a] (38) $\underline{18}$	417[b] $\underline{18}$	-
2"-OH-Δ^1-THC	2	19.5	474 (16) $\underline{18}$	145[c] $\underline{9}$	330 (11) $\underline{9}$
6α-OH-Δ^1-THC	3	21.1	474 (0)	384[d] $\underline{9}$	459[e] (6) $\underline{18}$
3"-OH-Δ^1-THC	4	21.3	474 (16) $\underline{18}$	330[f] $\underline{9}$	459[e] (6) $\underline{18}$
6β-OH-Δ^1-THC	5	21.5	474 (60) $\underline{18}$	343 $\underline{9}$	384 (21) $\underline{9}$
7-OH-Δ^1-THC	6	22.1	474 (9) $\underline{18}$	371[g] $\underline{9}$	-

a) Relative intensity in parenthesis, shift in $\underline{d}_9$-TMS derivative is underlined.
b) $[M-C_4H_9]^{+\cdot}$.
c) Ion $\underline{a}$ ($TMS\overset{+}{O}=CH-CH_2-CH_2-CH_3$).
d) $[M-TMSOH]^{+\cdot}$.
e) $[M-CH_3]^{+}$.
f) $[M-144]^{+\cdot}$.
g) $[M-CH_2-OTMS]^{+}$.

SCHEME 2.

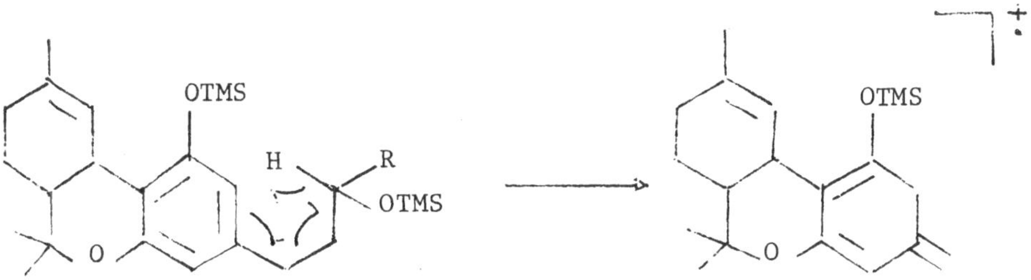

OTMS
H
R
OTMS
O

OTMS
O
b

$R = C_2H_5$ (4)

$R = H$

TABLE 3. GC-MS data for the ketol metabolites of Δ^1-THC.

Compound	No.	Derivative	R.T. (min)	M^+	Base
2"-OH-Δ^1-THC-6-one	7	TMS	23.8	488[a] (19) 18	145[b] 9
		MeOX-TMS	25.3	517 (16) 18	145[b] 9
7-OH-Δ^1-THC-6-one	8	TMS	26.1	488 (57) 18	385[c] 9
		MeOX-TMS	26.9	517 (5) 18	414[c] 9
3"-OH-Δ^1-THC-6-one	9	TMS	26.6	488 (49) 18	344[d] 9
		MeOX-TMS	27.8	517 (27) 18	373[d] 9

a) Relative intensity in parenthesis, shift in d_9-TMS derivative is underlined.

b) Ion a.

c) $[M-CH_2-OTMS]^+$.

d) $[M-144]^+$.

of 6α,7-dihydroxy-Δ^1-THC (13) except that it contained deuterium at C_6 (base peak at m/e 473 produced by elimination of TMSOH and $[M-103]^+$ at m/e 460) thus establishing its structure as 7-hydroxy-Δ^1-THC-6-one (8). The diols from the other two ketols were identified as 2",6α-dihydroxy-Δ^1-THC and 3",6α-dihydroxy-Δ^1-THC again with deuterium incorporation at C_6, confirming the structures of the ketols as 2"- (7) and 3"-hydroxy-Δ^1-THC-6-one (9) respectively.

The metabolic profile of fraction 4 from the guinea-pig (Fig. 1b) was considerably different from that of the mouse. Six monohydroxy metabolites were identified. The major metabolite was again 7-hydroxy-Δ^1-THC (6) but much less 6α-OH-Δ^1-THC was produced. Instead, 6β-OH-Δ^1-THC (5) was present in a high concentration. Three side-chain hydroxy metabolites were also detected. 3"-Hydroxy-Δ^1-THC, which was present in trace amounts in mouse liver, was fairly abundant (peak 4). Small quantities of 1"-hydroxy-Δ^1-THC (peak 1) and 2"-hydroxy-Δ^1-THC (peak 2) were also found. Their spectra were characterized by abundant ions at $[M-57]^+$ (loss of $C_4H_9$from the side-chain) and m/e 145 (a) respectively (9, 20, 21). The two ketols, 3"- and 7-hydroxy-Δ^1-THC-6-one, were also detected (peaks 9 and 8). Table 1 summarized the amounts of the metabolites found in each species.

The metabolic profile produced by fraction 4 from the rat was similar to that from the mouse. 7-Hydroxy-Δ^1-THC was the major metabolite and this was accompanied by smaller amounts of 6α-hydroxy-Δ^1-THC and 7-hydroxy-Δ^1-THC-6-one and trace amounts of 3"- and 6β-hydroxy-Δ^1-THC and 3"-hydroxy-Δ^1-THC-6-one. 1"- and 2"-hydroxy-Δ^1-THC and the 2"-hydroxy-ketone were not found.

Fraction 5 contained mainly polyfunctional metabolites and was much more complex than fraction 4. However the fragmentations observed to be characteristic of the position of hydroxyl substitution in the monohydroxy Δ^1-THC metabolites were also present in high abundance in most of the poly-substituted compounds. By using single-ion chromatograms of these ions from a number of derivatives in combination with retention time measurements it was possible to identify most of the metabolites.

The major metabolites in fraction 5 (Fig. 2) from the mouse all formed methyl esters with diazomethane and were thus carboxylic acids (Fig. 3). In all cases the carboxylic acid group was located at C_7,Δ^1-THC-7-oic acid (14) itself being the most abundant of these compounds. The other 5 compounds were hydroxylated derivatives of this acid containing a single hydroxyl group at the 2"- (20), 3"- (23), or 6α- (17) positions or two hydroxyl groups in the 2",6α- (22) or 3",6α- (24) positions. Detailed mass spectra of these compounds have been reported previously (13,14). Table 4 summarizes the major ions and retention times. Δ^1-THC-7-oic acid was also found as a major metabolite in the rat but was

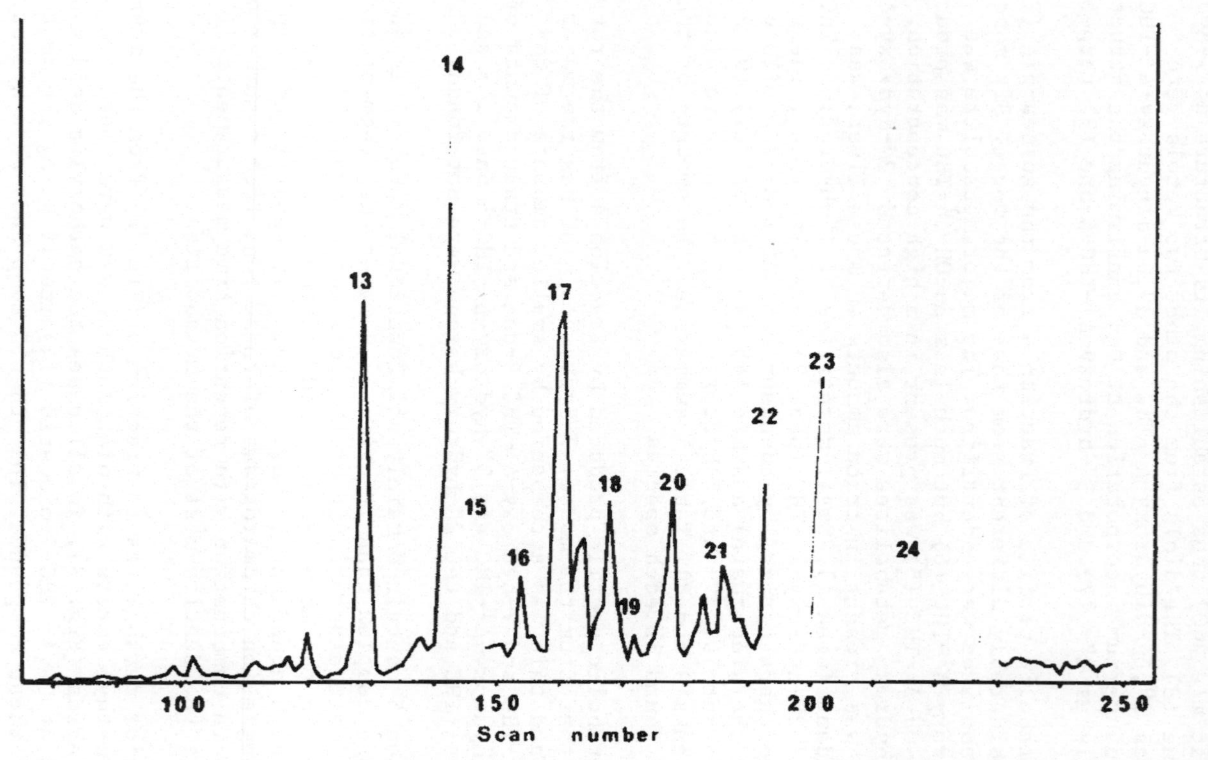

FIG. 2. Computer reconstructed chromatogram of ions m/e 300–680 of the metabolites present in fraction 5 from mouse liver. The experimental conditions are given in the 'Experimental' section and the peak identification is given in Table 1.

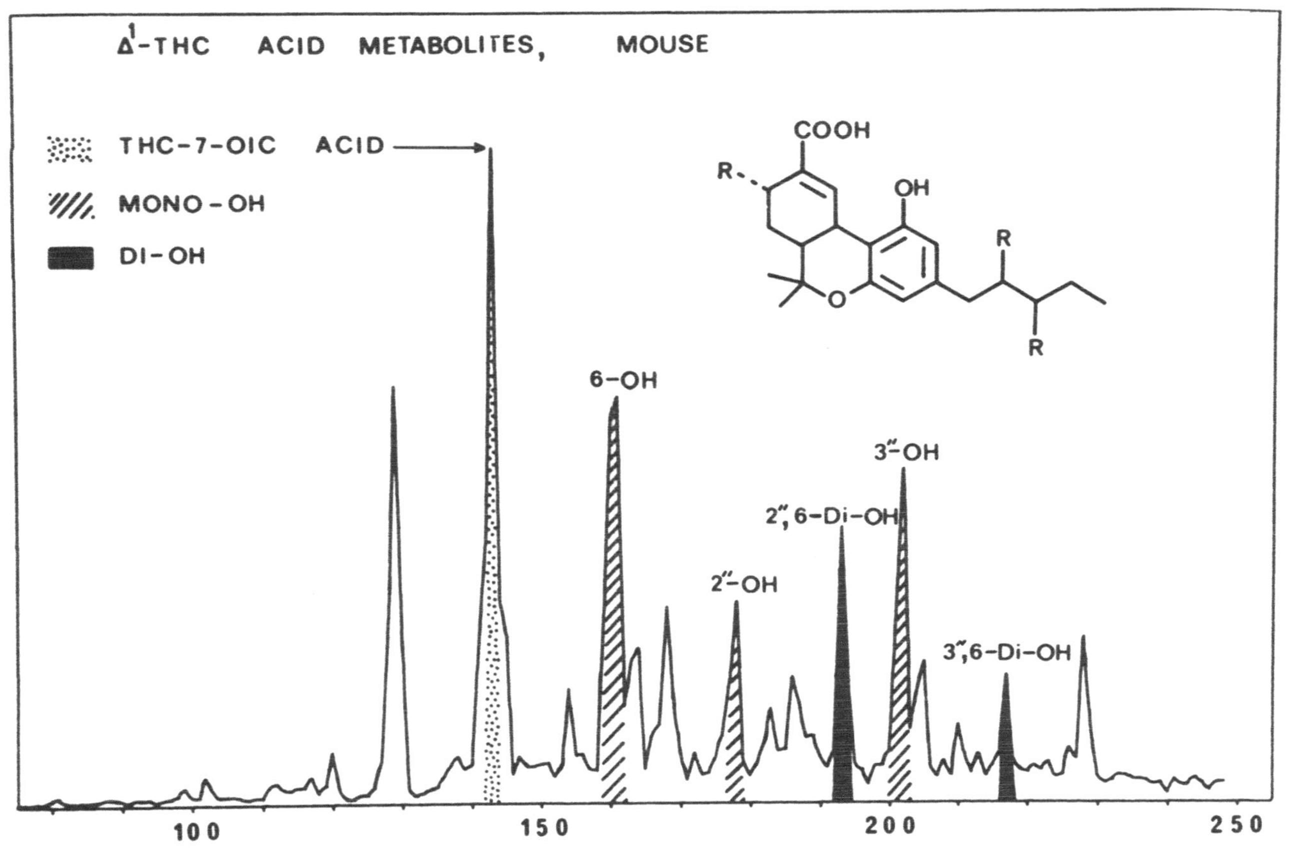

FIG. 3. Acid metabolites present in fraction 5 from mouse liver (Fig. 2). The R groups in the formula show the possible positions of hydroxylation. Mono- and di-hydroxy metabolites are identified as shown in the key.

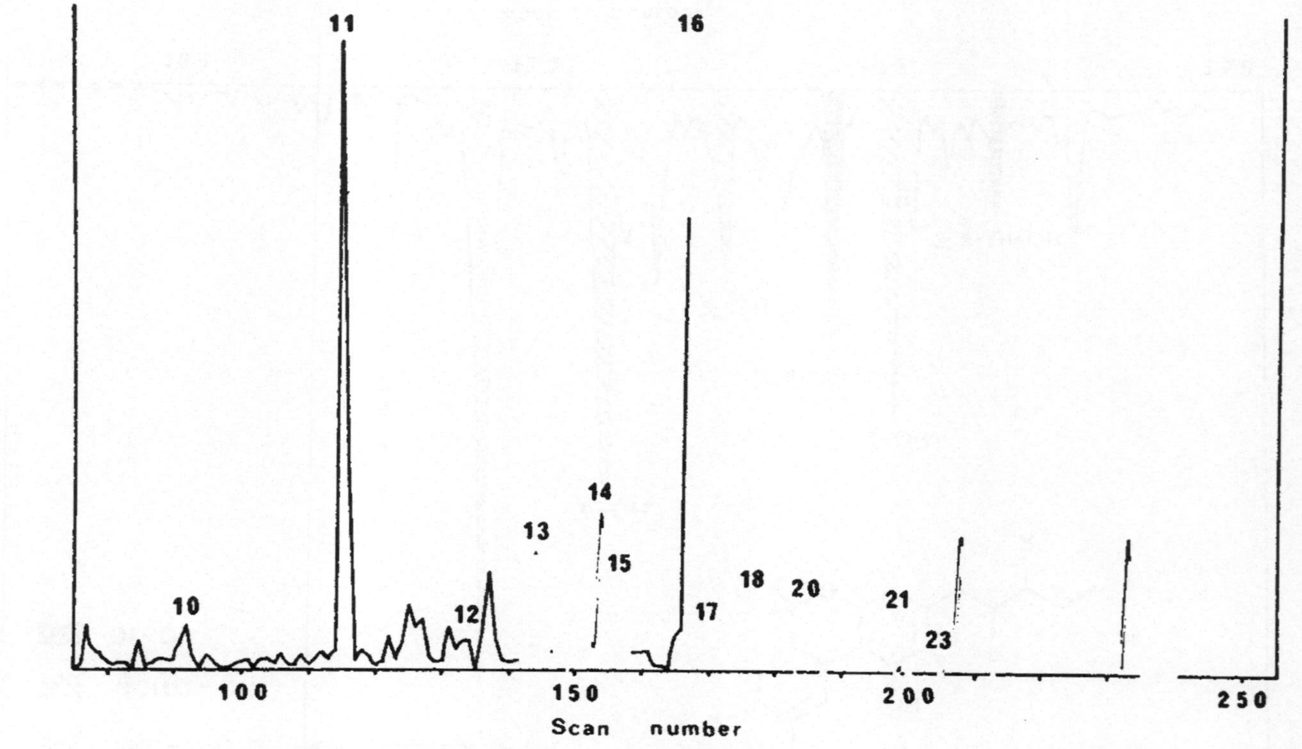

FIG. 4. Computer reconstructed chromatograms of ions m/e 300-680 of the metabolites present in fraction 5 from guinea-pig liver. The separation conditions are as for Fig. 2 and the peak identification is given in Table 1.

less abundant in the guinea-pig (Fig. 4). The two side-chain hydroxy acids (2"- and 3"-hydroxy-Δ^1-THC-7-oic acids, 20 and 23) were also present in substantial amounts in each species, but the 6α-hydroxy-Δ^1-THC-7-oic acid (17), although of comparable abundance in the rat, was a minor metabolite in the guinea-pig. Trace amounts of the two di-hydroxy acids were detected in the extracts of rat liver but not in those from the guinea-pig.

A second series of acids was produced by the guinea-pig (Fig. 4), one of these being the major metabolite (11). Its retention time was shorter than that of Δ^1-THC-7-oic acid and the shifts in its ions observed in the methyl ester-TMS and d_9-TMS derivatives indicated that it contained one acid group (Table 5). The molecular ion at m/e 460 (TMS derivative, Fig. 5) was 28 amu lower than that of Δ^1-THC-7-oic acid, indicating loss of C_2H_4. That this loss had occurred from the side-chain and that this compound was 4",5"-bisnor-Δ^1-THC-3"-oic acid (11) was supported by the absence of the characteristic ion at m/e 303 (a) produced by compounds containing an intact pentyl side-chain, and its replacement by m/e 377 (d). Loss of the side-chain gave ion g, m/e 315 (Scheme 3).

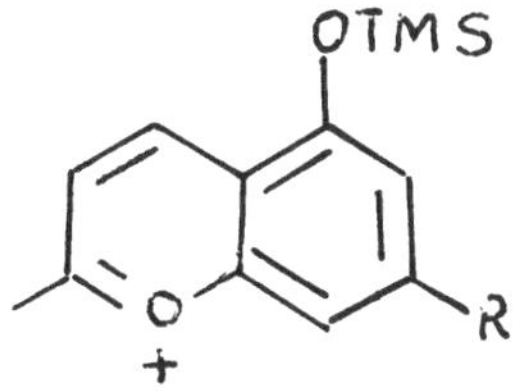

OTMS

O
+

R

c	m/e 303	$R = C_5H_{11}$
d	m/e 377	$R = (CH_2)_2COOTMS$

Finally reduction of this acid with lithium aluminium deuteride gave the corresponding primary alcohol with the incorporation of two deuterium atoms. The spectrum of this compound was compared with that of the synthetic reference compound 3"-hydroxy-4",5"-bisnor-Δ^6-THC (22). The spectra of the TMS derivatives of both compounds were characterized by a molecular ion, $[M-15]^+$ and the base peak which resulted from loss of the side-chain as shown in Scheme 2 (R = H).

Two other homologous side-chain acids were also characterized in this fraction. Their spectra were similar to that of the 3"-acid (Table 5) with the appropriate shifts in the ions corresponding to the structures 3",4",5"-trisnor-Δ^1-THC-2"-oic acid (10) and 5"-nor-Δ^1-THC-4"-oic acid (12). Only the 3"- side-chain acid was detected in the other two species and then only in trace amounts.

TABLE 4. GC-MS data for the Δ^1-THC-7-oic acid metabolites.

Compound	No.	Derivative	R.T. (min)	M^+	Base	Diagnostic ions $[M-90]^+$	Diagnostic ions Side-chain ion
Δ^1-THC-7-oic acid	14	TMS	25.7	488 (43)18[a]	371 9	-	-
		Me-TMS	23.0	430 (92)9	371 9	-	-
6α-OH-Δ^1-THC-7-oic acid	17	TMS	27.7	576 (0)	486 18	486	-
		Me-TMS	27.0	518 (1)18	428 9	428	-
2"-OH-Δ^1-THC-7-oic acid	20	TMS	29.7	576 (10)27	145 9	-	145 (100)9
		Me-TMS	29.0	518 (23)18	145 9	-	145 (100)9
3"-OH-Δ^1-THC-7-oic acid	23	TMS	32.4	576 (19)27	432 18	-	432 (100)18
		Me-TMS	32.6	518 b	374 9	-	374 (100)9
2",6α-di-OH-Δ^1-THC-7-oic acid	22	TMS	31.6	664 (0)	145 9	574 (6)27	145 (100)9
		Me-TMS	32.0	606 (4)27	145 9	516 (50)18	145 (100)9

./.

TABLE 4 (contd.).

Compound	No.	Derivative	R.T. (min)	M^+	Base	Diagnostic ions $[M-90]^+$	Diagnostic ions Side-chain ion
3",6α-di-OH-Δ^1-THC-7-oic acid	24	TMS	34.2	664 (0)	574 <u>27</u>	574 (100)<u>27</u>	430[c] (32)<u>18</u>
		Me-TMS	36.6	606 (0)	516 <u>18</u>	516 (100)<u>18</u>	372 (50)<u>9</u>[c]

a) Relative intensity in parenthesis, shift in $\underline{d}_9$-TMS derivatives is underlined.

b) Impure spectrum.

c) $[M-90-144]^+$.

SCHEME 3.

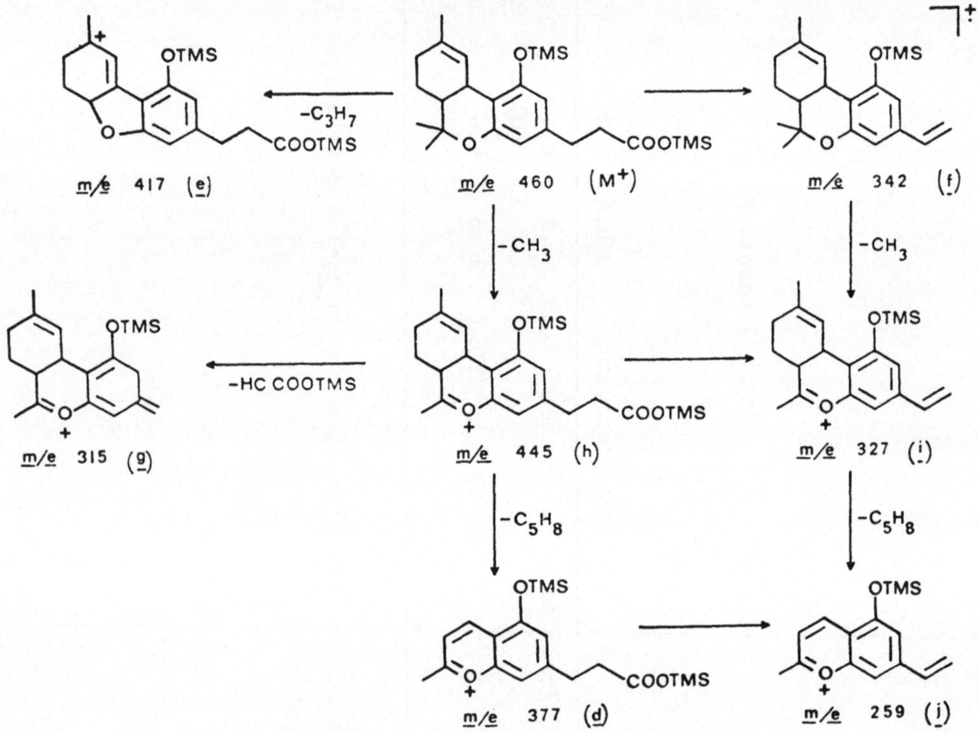

OTMS
$-C_3H_7$
COOTMS
m/e 417 (e)

OTMS
COOTMS
m/e 460 (M^+)

OTMS
m/e 342 (f)

$-CH_3$

$-CH_3$

OTMS
-HC COOTMS
m/e 315 (g)

OTMS
COOTMS
m/e 445 (h)

OTMS
m/e 327 (i)

$-C_5H_8$

$-C_5H_8$

OTMS
COOTMS
m/e 377 (d)

OTMS
m/e 259 (j)

Several di- and trihydroxy metabolites were present in fraction 5 from the livers of all three species. Peak 13 in the mouse profile (Fig. 2) was identified by its retention time and mass spectrum as 6α,7-dihydroxy-Δ^1-THC, a known metabolite of Δ^1-THC in the mouse (8). Its mass spectrum (Table 6) had a base peak at m/e 472 (loss of TMSOH from the 6α-OTMS group) and a weak ion at m/e 459 produced by loss of the 7-CH_2-OTMS group. The molecular ion was absent but a weak ion was observed at $[M-15]^+$ (m/e 547). This compound was also a major metabolite in the rat, but not as abundant in the guinea-pig livers.

The major diol produced by the guinea-pig (peak 16) was the isomeric 6β,7-dihydroxy-Δ^1-THC characterized by its base peak at m/e 369 (loss of CH_2OTMS (103 amu) and TMSOH (90 amu)) and ions at M^+ (m/e 562), $[M-90]^+$ (m/e 472) and $[M-103]^+$ (m/e 459) (9). This compound was less abundant in the mouse and rat livers.

Two other diols produced in moderate abundance by the mouse were identified as 2",7- (15) and 3",7-dihydroxy-Δ^1-THC (18) by their molecular ions at m/e 562 (27 amu shift in the d_9-TMS spectra) and the presence of the characteristic side-chain fragment ions at m/e 145 (a) and $[M-144]^+$ (m/e 418) in the spectra of peaks 15 and 18 respectively. Neither compound reacted with diazomethane, methoxylamine hydrochloride or lithium aluminium hydride. The retention times of the TMS derivatives of the 6α,7-, 2",7- and 3",7- diols showed the same relationship to 7-hydroxy-Δ^1-THC (6) as the corresponding 6α-, 2"- and 3"- hydroxy-Δ^1-THC-7-oic acids had to Δ^1-THC-7-oic acid (14), giving further support to the structural assignments. Only trace amounts of the side-chain diols were detected in the other two species.

Two trihydroxy metabolites were also characterized in the extracts from mouse and guinea-pig livers. These were substituted in the same positions as were the dihydroxy acids (22 and 24), namely 6α,7- and either the 2"- or 3"- positions of the side-chain. Their retention times also bore the same relationship to the mono- and dihydroxy metabolites as the dihydroxy acids did to the 7-acid and its mono-hydroxylated derivatives. Both compounds exhibited abundant ions at m/e 560 (loss of TMSOH) (Table 6) and the side-chain fragment ions at m/e 145 (peak 19) and m/e 416 (loss of 144 amu from $[M-90]^+$, peak 21) characterized these triols as 2,6α,7- and 3",6α,7-trihydroxy-Δ^1-THC respectively.

Chemical ionization mass spectra. Several of the metabolites described above, particularly the di- and trihydroxy compounds did not produce molecular ions. Chemical ionization mass spectra using isobutane as the reagent gas were therefore used in an attempt to confirm the molecular weight of these compounds. The spectra of a number of reference cannabinoids were recorded and in

TABLE 5. GC-MS data for the side-chain acid metabolites of Δ^1-THC.

Compound	No.	Derivative	R.T.(min)	M^+	$[M-CH_3]^+$	$[M-C_3H_7]^+$	$[M-CH_3-C_5H_8]^+$	$[M-HCOOR]^+$	m/e 315
Δ^1-THC-2"-oic acid	.10	TMS	16.4	446 (100)$\underline{18}$[a]	431 (78)$\underline{18}$	403 (19)$\underline{18}$	363 (26)$\underline{18}$	328 (48)$\underline{9}$	315 (32)$\underline{9}$
		Me-TMS		388 (100)	373 (73)	345 (31)	305 (67)	270 -	315 (86)
Δ^1-THC-3"-oic acid	11	TMS	20.2	460 (100)$\underline{18}$	445 (64)$\underline{18}$	417 (14)$\underline{18}$	377 (19)$\underline{18}$	342 (22)$\underline{9}$	315 (22)$\underline{9}$
		Me-TMS		402 (100)	387 (72)	359 (23)	319 (31)	284 (5)	315 (39)
Δ^1-THC-4"-oic acid	12	TMS	23.1	474 (100)$\underline{18}$	459 >100[b] $\underline{18}$	431 (26)$\underline{18}$	391 (14)$\underline{18}$	356 (10)$\underline{9}$	315 (56)$\underline{9}$
		Me-TMS		416 (100)	401 (76)	373 (32)	333 (72)	298 -	315 (76)

a) Relative intensity in parenthesis, shift in $\underline{d}_9$-TMS spectrum is underlined.

b) Mixed spectrum.

TABLE 6. GC-MS data for the diol and triol metabolites of Δ^1-THC (TMS derivatives).

Compound	No.	R.T. (min)	Diagnostic ions				
			M^+	$[M-90]^+$	$[M-103]^+$	Side-chain	Other
6α,7-di-OH-Δ^1-THC	13	24.2	562 (0)	472[a] (100)18	459 (8)18	-	-
2",7-di-OH-Δ^1-THC	15	26.3	562 (4)27	-	459 (76)18	145 (100)9	-
6β,7-di-OH-Δ^1-THC	16	27.3	562 (35)27	472 (28)18	459 (80)18	-	369[b] (100)9
3",7-di-OH-Δ^1-THC	18	28.7	562 (-)[c] 27	-	459 (100)18	418 (60)18	-
2",6α,7-tri-OH-Δ^1-THC	19	28.4	650 (0)	560 (57)27	547 (8)27	145 (100)9	635[d] (3)
3",6α,7-tri-OH-Δ^1-THC	21	30.8	650 (0)	560 (100)27	547 (42)27	416[e] (39)27	-

a) Relative intesity in parenthesis, shift in d_9-TMS spectra is underlined.

b) $[M-90-103]^+$.

c) Mixed spectrum.

d) $[M-15]^+$.

e) $[M-90-144]^+$.

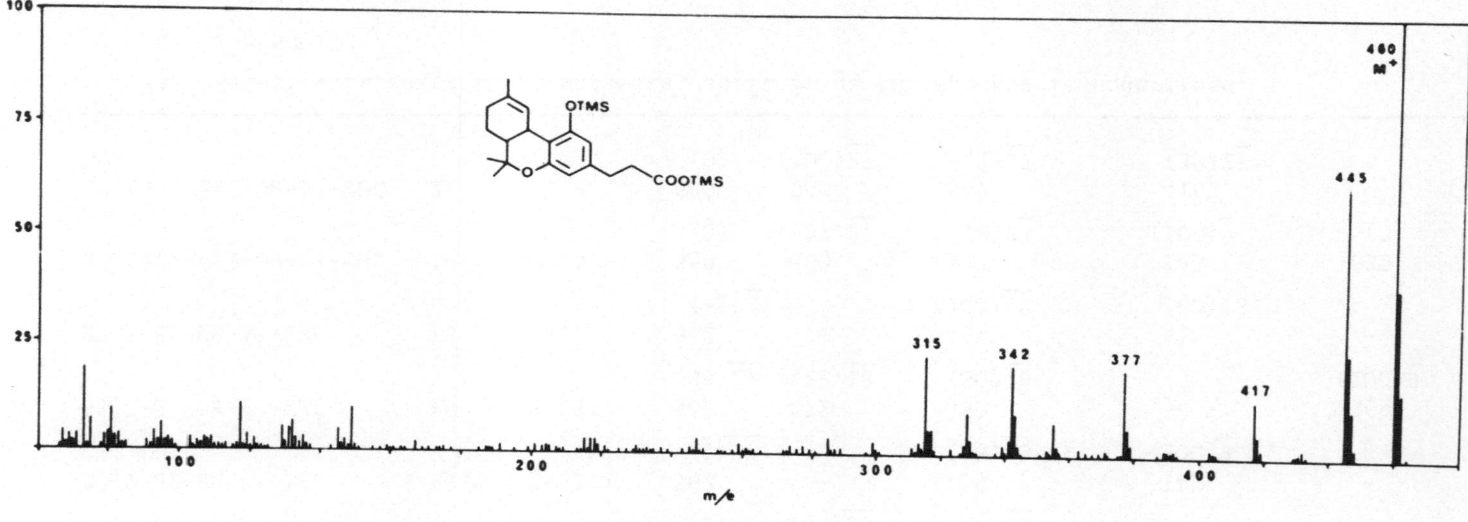

FIG. 5. 25 eV mass spectrum of the TMS derivative of 4",5"-bisnor-Δ^1-THC-3"-oic acid (<u>11</u>).

all cases, the basic cannabinoids (Δ^1-THC, Δ^6-THC, the *n*-heptyl homologue of Δ^1-THC and cannabinol) gave spectra dominated by the quasi-molecular ion $[MH]^+$ with negligible fragmentation. The base peaks in the spectra of the monohydroxy metabolites, 6α- and 7-hydroxy-Δ^1-THC and 7-hydroxy-Δ^6-THC produced $[M + 1]^+$ ions of about 80% relative intensity but in all cases the base peak resulted from loss of TMSOH (90 amu) from the quasi-molecular ions. The ions characteristic of the position of substitution which were abundant under electron-impact conditions were absent from the chemical ionization spectra. Nevertheless the high abundance of $[M + 1]^+$ was advantageous, especially in the case of 6α-hydroxy-Δ^1-THC which did not give a molecular ion under electron impact conditions. By using mass fragmentography the molecular ions of the two triols (19, 21) and the di-hydroxy acids (22 and 24) were observed in low abundance.

Acknowledgments. We thank Professor R. Mechoulam for a gift of 6α-hydroxy-Δ^1-THC, Dr. M.C. Braude of the National Institute for Mental Health for supplies of Δ^1-THC and 7-hydroxy-Δ^1-THC through the Medical Research Council, and Dr. E.W. Gill for 3"-hydroxy-4",5"-bisnor-Δ^6-THC. We are indebted for support by a Programme Research Grant of the Medical Research Council and a grant from the Wellcome Trust.

REFERENCES

1) S. Agurell, J. Dahmen, B. Gustafsson, U.-B. Johansson, K. Leander, I. Nilsson, J.L.G. Nilsson, M. Nordqvist, C.H. Ramsay, A. Ryrfeldt, F. Sandberg and M. Widman, in "Cannabis and its Derivatives", W.D.M. Paton and J. Crown, Eds., 1972, Oxford University Press, London, p. 16.
2) R. Mechoulam, N.K. McCallum and S. Burstein, Chem. Rev., 1976, 76, 75.
3) O. Gurney, D.E. Maynard, R.G. Pitcher and R.W. Kierstead, J. Am. Chem. Soc., 1972, 94, 7928.
4) Z. Ben-Zvi and S. Burstein, Biochem. Pharmacol., 1975, 24, 1130.
5) M. Widman, M. Nordqvist, C.T. Dollery and R.H. Briant, J. Pharm. Pharmacol., 1975, 27, 842.
6) S. Agurell, M. Binder, K. Fonseka, J.-E. Lindgren, K. Leander, B. Martin, I.M. Nilsson, M. Nordqvist, A. Ohlsson and M. Widman, in "Marihuana. Chemistry, Biochemistry and Cellular Effects", G.G. Nahas, Ed., 1976, Springer-Verlag, New York, p. 141.
7) Z. Ben-Zvi and S. Burstein, Res. Commun. Chem. Pathol. Pharmacol., 1974, 8, 223.
8) G. Jones, M. Widman, S. Agurell and J.-E. Lindgren, Acta Pharm. Suec., 1974, 11, 283.
9) M.E. Wall and D.R. Brine, in "Marihuana. Chemistry, Biochemistry and Cellular Effects", G.G. Nahas, Ed., 1976, Springer-Verlag,

New York, p. 51.
10) B.R. Martin, D.J. Harvey and W.D.M. Paton, J. Pharm. Pharmacol., in press, 1976.
11) M. Nordqvist, S. Agurell, M. Binder and I.M. Nilsson, J. Pharm. Pharmacol., 1974, 26, 471.
12) S.H. Burstein, J. Rosenfeld and T. Wittstruck, Science, 1972, 176, 422.
13) D.J. Harvey and W.D.M. Paton, in "Marihuana. Chemistry, Biochemistry and Cellular Effects", G.G. Nahas, Ed., 1976, Springer-Verlag, New York, p. 93.
14) D.J. Harvey and W.D.M. Paton, Res. Commun. Chem. Pathol. Pharmacol., 1976, 13, 585.
15) A. Ryrfeldt, C.H. Ramsay, I.M. Nilsson, M. Widman and S. Agurell, Acta Pharm. Suec., 1973, 10, 13.
16) E.B. Jr. Truitt and M. Braude, in "Research Advances in Alcohol and Drug Metabolism", Vol. 1, R.J. Gibbins, Y. Israel, H. Kalant, R.E. Popham, W. Schmidt and R.G. Smart, Eds., 1974, Wiley, New York, p. 199.
17) K. Nakagawa and E. Costa, Nature, (London), 1971, 234, 48.
18) H.M. Fales, T.M. Jaouni and J.F. Babashak, Anal. Chem., 1973, 45, 2301.
19) J.A. McCloskey, R.N. Stillwell and A.M. Lawson, Anal. Chem., 1968, 40, 233.
20) M. Binder, in "Marihuana. Chemistry, Biochemistry and Cellular Effects", G.G. Nahas, Ed., 1976, Springer-Verlag, New York, p. 159.
21) M.S. Binder, S. Agurell, K. Leander and J.-E. Lindgren, Helv. Chim. Acta, 1974, 57, 1626.
22) D.K. Lawrence, D. Phil. thesis, University of Oxford, 1974.

DETECTION AND QUANTIFICATION OF CANNABINOIDS IN HUMAN BLOOD PLASMA

A.Ohlsson[1],J.-E.Lindgren[2,3],K.Leander[2,4],S.Agurell[1,3].

[1]University of Uppsala,Uppsala;[2]Karolinska Institutet,

Stockholm,[3]Astra,Södertälje,[4]Univ.Stockholm,Stockholm,S.

INTRODUCTION

Techniques for the quantitative determination of Δ^1-tetrahydrocannabinol (Δ^1-THC) and related cannabinoids have long been desired. Recent developments in this area has been summarized e.g. in a review by Grlić (1).

It should be emphasized that what is a "satisfactory method" will vary with the aims of the investigator. For epidemiological studies of Cannabis use, simple but somewhat ambiguous screening procedures will probably be satisfactory. In forensic toxicology, the requirement for unequivocal identification hardly allows the use of procedures yielding ambiguous results. Finally, in pharmacokinetic studies it may be necessary to have a specific as well as accurate quantitative procedure. These considerations would suggest that eventually different procedures will be used for the determination of cannabinoids in body fluids.

Gas chromatography in combination with sensitive detection systems has been widely investigated for this purpose (cf. ref. 2). In general, the problems have been to remove lipid materials interfering with the gas chromatographic separation of e.g. THC and to increase the sensitivity in the detection step.

We have published (3, 4) methods for the quantitative determinations of Δ^1-THC and Δ^6-THC in the blood of Cannabis smokers. These methods are based upon the purification of THC extracted from blood plasma by liquid chromatography on Sephadex LH-20 followed by mass fragmentographic assay using the deuterated THC-analogue as internal standard. Rosenfeld *et al*. (5) also used mass fragmentography for detection of Δ^1-THC *in humans* after smoking.

The latter authors used trimethylanilinium hydroxide for on column methylation of the phenolic group and relied on the tri-deuteromethyl ether as internal standard in the last step. The purpose of the present paper is to present information on the mass fragmentographic procedures for THC and other cannabinoids developed in our laboratory including a silylation procedure. For a more detailed account and literature survey we refer to Ohlsson *et al*. (2).

METHODS

Synthesis of deuterated cannabinoids. A detailed description of the synthetic procedures is given in Ohlsson *et al*. (2).

Olivetol-d_7. The synthetic procedure of Pitt *et al*. (6) was used with some modifications to increase the incorporation of deuterium. Methyl 5-(3,5-dimethoxyphenyl)penta-2,4-dienoate was reacted with $LiAlD_4$ followed by hydrogenation with D_2 using the catalyst tris(triphenylphosphine)rhodium chloride. The 5-(3,5--dimethoxyphenyl)pentan-1-ol-1,1,2,3,4,5,-d_6 was then reacted with PBr_3, $LiAlD_4$, and BBr_3 according to Pitt *et al*. (6) to yield olivetol-d_7.

Δ^1-THC-d_7, CBD-d_7 and CBN-d_7. Δ^1-THC-d_7 and cannabidiol-d_7 (CBD-d_7) were prepared from olivetol-d_7 according to standard methods (7). Δ^6-THC-d_7 was aromatized to cannabinol-d_7 (CBN-d_7) with chloranil and a trace of *p*-toluenesulfonic acid (cf. ref. 8). The usual dehydrogenation method (7) using sulphur causes extensive deuterium scrambling. The deuterium content in these cannabinoids was (m/e 314-321): d_0 2%, d_1 2%, d_2 2%, d_3 3%, d_4 3%, d_5 9%, d_6 30%, and d_7 100%. The purity of these cannabinoids by gas liquid chromatography (GLC) was over 95%.

Olivetol-d_3, Δ^1-THC-d_3 and CBN-d_3. Olivetol-d_3 was prepared as described by Pitt *et al*. (6). Δ^1-THC-d_3 and CBN-d_3 were synthesized as described for the d_7-analogues. Δ^1-THC-d_3 with deuterium content (m/e 314-317): d_0 2%, d_1 6%, d_2 12%, and d_3 100%; CBN-d_3 with a deuterium content of (m/e 310-313): d_0 2%, d_1 5%, d_2 15%, d_3 100% was obtained. The purity of Δ^1-THC-d_3 and CBN-d_3 by GLC was over 95%.

Olivetol-d_2, Δ^1-THC-d_2, CBD-d_2, and CBN-d_2. The methyl ester of 3,5-dimethoxybenzoic acid was reduced with $LiAlD_4$ and the obtained alcohol was reacted with PBr_3. The bromide was mixed with Cu(I)I and butyllithium to yield O,O-dimethylolivetol-1',1'--d_2, which showed the deuterium content (m/e 180 - 182): d_0 0.3%, d_1 4%, and d_2 100%. For the demethylation hydroiodic acid was used (9).

Δ^1-THC-d_2 was prepared as Δ^1-THC-d_3 and more than 95% pure by GLC. Deuterium content (m/e 314-316): d_0 10%, d_1 20%, and d_2 100%. CBD-d_2 was synthesized as described for the d_7-analogue. The CBD-d_2 was 97% pure by GLC and showed a deuterium content (m/e 314-316): $\underline{d}_0$ 8%, $\underline{d}_1$ 12%, and $\underline{d}_2$ 100%. CBN-$\underline{d}_2$ was prepared from Δ^6-THC-$\underline{d}_2$ by hydrogenation with sulphur (7). The purity by GLC was over 97% and the deuterium content (m/e 310-312): $\underline{d}_0$ 12%, $\underline{d}_1$ 46%, and $\underline{d}_2$ 100%.

Non-labelled cannabinoids. Δ^1-THC, Δ^6-THC, CBD, and CBN were synthesized according to standard procedures and carefully purified and dried before use. Stock solutions were maintained in ethanol (1-5 mg/ml, 4°C, stored in the dark).
GLC was carried out on a Varian 2100 FID gas chromatograph using 2 mm (i.d.)x180 cm glass columns with 2% SE-30 ultraphase on Gas Chrom Q (100-120 mesh) at 250°C.

Analysis of THC in blood plasma. The procedure of Agurell *et al.* (4) was used for Δ^6-THC. Analogous procedures utilizing the proper deuterated internal standards and mass numbers are applicable for Δ^1-THC, CBN, and CBD. The procedure for Δ^1-THC has been published (3).

Blood samples. Blood samples (10 ml) were collected as desired in heparinized tubes. Plasma is obtained by centrifugation and stored in silanized glass tubes at -20°C until analysed.

Extraction. To a 5.0 ml plasma sample is added 100 ng (THC levels above 5 ng/ml) or 10-20 ng (below 5 ng/ml) of deuterated internal standard dissolved in 50 µl ethanol. The plasma sample is extracted three times with an equal volume of light petroleum containing 1.5% isopentanol in a glass stoppered centrifuge tube. After centrifugation the light petroleum is drawn off and the combined organic extracts evaporated under nitrogen at 50°C almost to dryness. This extract is quantitatively transferred to the Sephadex LH-20 column using three 0.2 ml portions of the elution solvent.

Liquid chromatographic purification. Mantled silanized glass columns (1 x 40 cm, void volume ca. 15 ml) operated at 12°C (const. temp. water cooling system) containing Sephadex LH-20 and eluted with light petroleum-chloroform-ethanol (10:10:1) were used for purification of plasma extracts. Each column was provided with a 200 ml solvent reservoir and after each purification the column was washed with 40 ml solvent. If not used, the columns were washed twice weekly with fresh solvent to maintain constant elution volumes.

The elution volume for Δ^6-THC was determined by calibration with ng amounts of Δ^6-THC-^{3}H. The calibration can also be carried out by GLC analysis of the concentrated fractions after application of 10 µg quantities to the column. The column was run at 0.2 ml/min and the pertinent 7-8 ml fraction was collected and evaporated under nitrogen. The residue was dissolved in ethanol and transferred to a 50 µl conic vial (Reactivial, Pierce) dried, and finally dissolved in 10-15 µl of ethanol and stored at 4°C in the dark until analysis. This solution was subjected to mass fragmentography directly or after silylation.

Silylation procedure. The dried sample was dissolved in 25 µl dry acetonitrile and mixed with 10 µl of the silylating reagent BSA [N,O-bis(trimethylsilylacetamide)] and kept at 50-60°C for 10 min. The solution was evaporated to dryness under a stream of nitrogen and dissolved in 10 µl acetonitrile and 2µl was subjected to mass fragmentography.

Mass fragmentography. Mass fragmentography was carried out using an LKB 9000 GC-MS instrument. The column was a 1.4 m x 2 mm i.d. silanized glass column containing 3% OV-17 on Gas Chrom CLP 100/120 mesh. Temperatures were in the column 180-210°, flash heater 250°, and source 290°. Helium was carrier gas (25 ml/min) and typical retention times for non-derivatized cannabinoids are: CBD 3.4 min, Δ^6-THC 4.0 min, Δ^1-THC 4.5 min, CBN 6.1 min. For mass fragmentography a multiple ion detector was added (10) and used at 50 eV. For Δ^6-THC, the mass spectrometer was set to continuously record the intensities of m/e 314 (molecular ion of non-labelled Δ^6-THC) and m/e 318 (internal standard, Δ^6-THC-d_4) as well as m/e 299, 303. For TMS-Δ^1-THC-d_0 (TMS=trimethylsilyl) the intensity of m/e 386 (molecular ion) and for TMS-Δ^1-THC-d_3 the intensity of m/e 389 (molecular ion) were recorded. The standard curves were prepared by adding known amounts of Δ^6-THC (0.5-100 ng/ml) or Δ^1-THC (0.5-20 ng/ml) to blank plasma samples and carrying out the described procedure. The Δ^1-THC samples were in these experiments all silylated before being subjected to mass fragmentography.

RESULTS AND DISCUSSION

The method described for the determination of Δ^6-THC (cf. ref. 4) in blood is basically identical to the one published (3) for Δ^1-THC. The method is based upon the addition of the proper deuterated internal standard due to the blood plasma sample. After extraction with light petroleum, the THC containing fraction is purified by liquid chromatography on a Sephadex LH-20 column. The pertinent fraction containing THC (absorbed by smoking a spiked Cannabis sample) and deuterated THC (added as internal standard) is collected. The relative amounts of the two compounds are determined by mass fragmentography monitoring the molecular peaks of the silylated or non-silylated cannabinoids.

As subsequently discussed, this method can also be used for the analysis of other cannabinoids, such as THC-analogues of different molecular weight, CBD, and CBN. However, at present less information is available with regards to these latter compounds. There are generally only ng-amounts of cannabinoids present in blood plasma and although the deuterated internal standards serve as carrier, the method requires scrupulously clean, silanized glass ware and redistilled solvents.

Δ^6-THC-d_4. The synthesis of this internal standard was described (4) previously. The label is located in the 1"- and 2"-positions of the pentyl side chain.

Other deuterated cannabinoids. Preparations of Δ^1-THC-d_4 as well as olivetol-d_7, olivetol-d_3, and olivetol-d_2 were published earlier (2, 3). From these intermediates correspondingly labelled Δ^1-THC, CBD, and CBN can be prepared.

The sensitivity in the final mass fragmentographic assay may be limited by the amount of non-labelled (d_0) compound present in the deuterated internal standard. This interferes with the non--labelled cannabinoid present in the plasma. Thus, we have tried to minimize this interference by preventing exchange reactions, increasing the number of hydrogens substituted with deuterium, and by limiting the amount of internal standard in samples containing low amounts of cannabinoids. Thus, the present limit of sensitivity (ca. 0.3 ng THC/ml plasma and after silylation ca. 0.1 ng THC/ml plasma) is partly due to the amount of THC-d_0 in the internal standard and not due to chromatographic, derivatization procedure or mass spectrometric problems per se. As expected, d_7-containing Δ^1-THC, CBD, and CBN showed, together with Δ^1-THC-d_3, the least contamination with d_0-analogue (2%). This is in the same range as found (4) for Δ^6-THC-d_4 (1.5%).

Extraction and purification. Δ^6-THC and its Δ^1-THC isomer appear to be stable in human plasma for months if stored at -20°C in silanized glass tubes. The extraction and purification procedures for Δ^1-THC, CBD, or CBN are analogous.

The extraction procedure for Δ^6-THC, as revealed by experiments with [^{3}H]-Δ^6-THC, is quite efficient and the recovery after both extraction and liquid chromatography is usually over 80%. Also, early studies on Δ^1-THC showed recoveries of 70 ± 6% (s.d.) after the column purification (3). The Sephadex LH-20 separation is essential in removing interfering lipids and metabolites before mass fragmentography. The elution volume for e.g. Δ^6-THC is stable for months if the column is operated at low const. temp. (12°C) and, when not in use, washed with fresh solvent regularly. Our present setup contains five columns which can be handled by a technician (10 samples per day). However, such systems can be automated (11). Over 90% of the Δ^6-THC peak is eluted in a 5 ml volume but a 7-8 ml fraction is collected. As a precaution ca. 3-4 ml on each side of the peak is collected and stored.

Mass fragmentography. The sensitivity achieved in the quantification of Δ^6-THC and other cannabinoids is partly due to the liquid chromatography clean-up of blood plasma extract. Mainly, however, the sensitivity and specificity is dependent upon the mass fragmentographic analysis.

We have used the deuterium labelled analogue Δ^6-THC-d_4 for the analysis of Δ^6-THC in blood plasma from male, casual Cannabis smokers who had smoked 8 mg of Δ^6-THC.

Two standard curves were prepared (0-10 and 10-100 ng THC/ml)

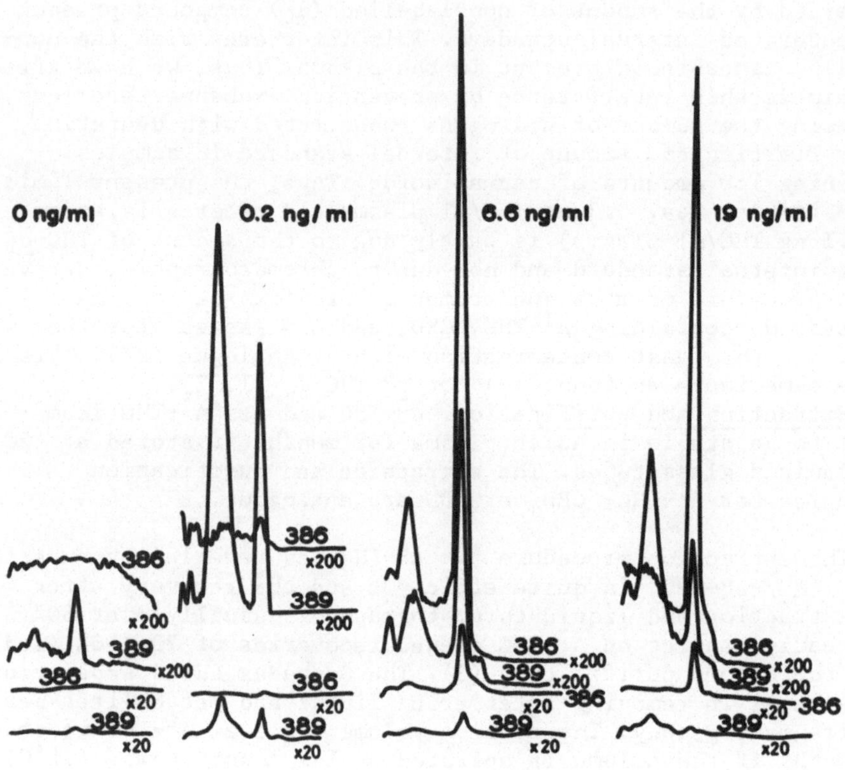

FIG. 1. Mass fragmentograms of trimethylsilyl-Δ^1-tetrahydrocannabinol (TMS-Δ^1-THC) (m/e 386) with TMS-Δ^1-THC-$\underline{d}_3$ as internal standard (m/e 389) from purified plasma extracts. Plasma levels of 0, 0.2, 6.6, and 19 ng/ml.

by adding known amounts of Δ^6-THC to blank plasma samples and carrying out the described procedure. Peak height ratio (m/e 314-318) was plotted against known amounts of Δ^6-THC. Since the same ratios were obtained simply by mixing known amounts of Δ^6-THC and Δ^6-THC-$\underline{d}_4$, usually such standard curves were used. For the analysis of Δ^1-THC in blood plasma from another group of Cannabis users, we used the deuterium labelled analogue Δ^1-THC-$\underline{d}_3$ and the silylation technique. Such mass fragmentograms are shown in Fig. 1.

A standard curve of (0.5-20 ng/ml) TMS-Δ^1-THC with TMS-Δ^1--THC-$\underline{d}_3$ as internal standard was also prepared according to the procedure described for the Δ^6-THC standard curve. These Δ^1-THC samples were silylated before mass fragmentography.
Peak height ratio (m/e 386-389) was plotted against known amounts of Δ^1-THC (Fig. 2).

There are advantages as well as disadvantages in the use of deuterium labelled analogues as internal standards. They are carriers for the minute amounts of cannabinoids present in biological samples and, being almost identical to the analyzed compound they can be added to the original plasma sample to compensate for variations in extraction and purification recoveries. A disadvantage is that one or more deuterium labelled standards has to be synthesized but hopefully these might be provided by NIDA.

The present method for Δ^6-THC without silylation can be used down to 0.3 ng/ml. As pointed out (see discussion on "Reference cannabinoids") the previous sensitivity is partly limited by the small amount of Δ^6-THC-$\underline{d}_0$ present in the deuterated internal standard. Thus, we have tried to eliminate $\underline{d}_0$-contamination in the internal standards for Δ^6-THC, Δ^1-THC, CBD, and CBN - with best results in the $\underline{d}_7$-analogues. The sensitivity increases at least twice if the THC samples are silylated before mass fragmentographic analysis.

Δ^1-THC and CBD can both be determined in the same fragmentogram using the same channels - m/e 314 and e.g. 316 and 317 for internal standards - whereas CBN is simultaneously assayed using m/e 310 and 313 (Fig. 3).

So far we have not developed assays for any of the possibly important Δ^1-THC metabolites, e.g. 7-hydroxy-Δ^1-THC. However, it is likely that if quantitative metabolic studies in man warrant, this metabolite could also be quantitated after elution and derivatization using fragmentography.

Plasma levels. The plasma levels of Δ^6-THC in man after smoking 8 mg Δ^6-THC - of which about half is absorbed in the lungs - is shown in Fig. 4. Immediately after smoking high values (>100 ng/ml) of Δ^6-THC are recorded but drop rapidly to 10-20 ng/ml at 0.5 hour and are about 1 ng/ml at 4 hours.

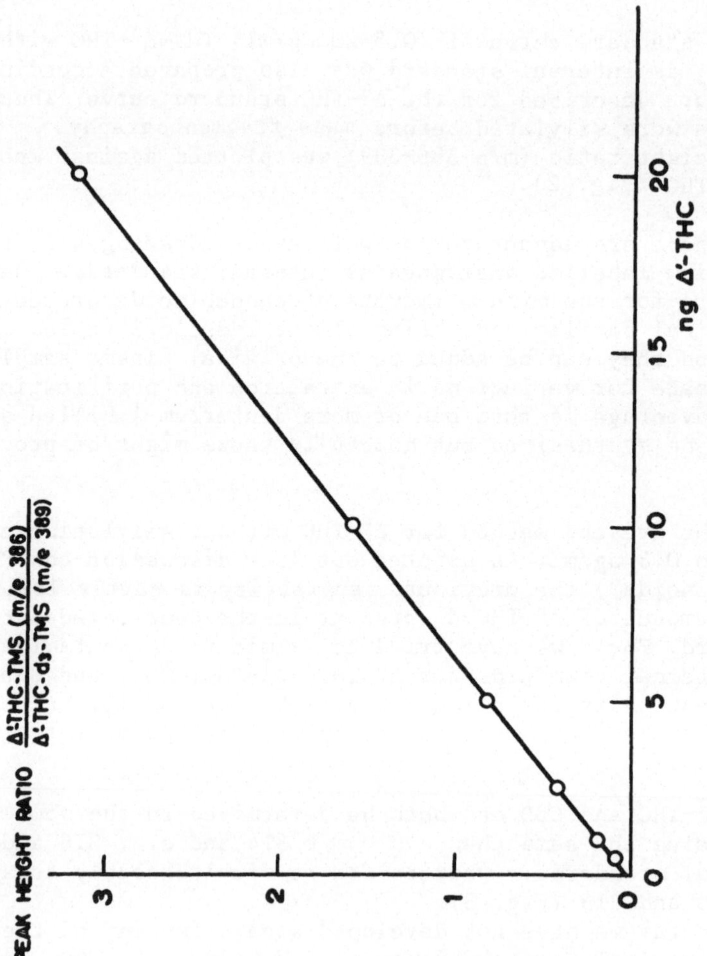

FIG. 2. Standard curve for trimethylsilyl-Δ^1-tetrahydrocannabinol (TMS-Δ^1-THC) (0–20 ng/ml) in plasma.

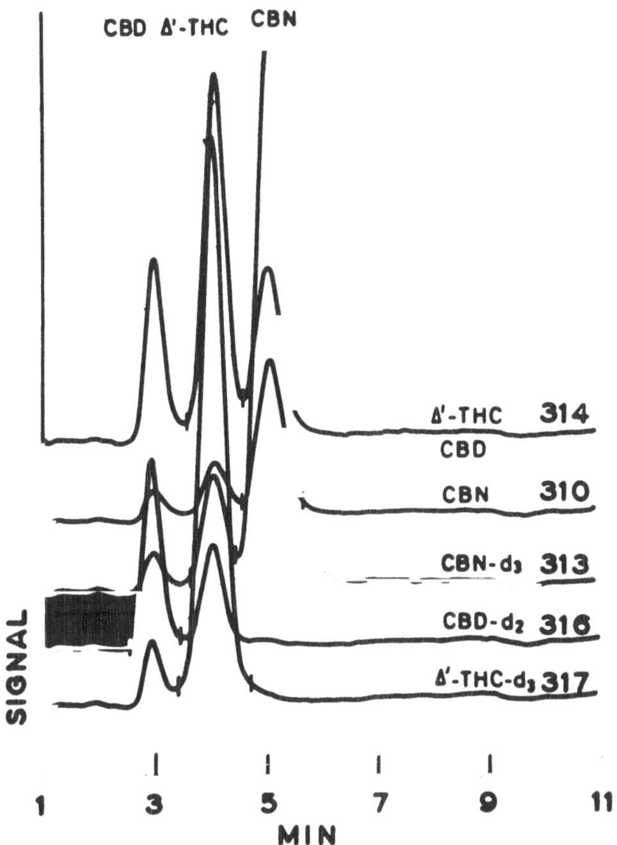

FIG. 3. Mass fragmentograms of CBD (m/e 314) and internal standard CBD-$\underline{d}_2$ (m/e 316, retention time 3.0); Δ^1-THC (m/e 314) and internal standard Δ^1-THC-$\underline{d}_3$ (m/e 317, retention time 4.0), and CBN (m/e 310) and internal standard CBN-$\underline{d}_3$ (m/e 313, retention time 5.0). Column temperature 220°C. (Δ^1-THC=tetrahydrocannabinol, CBD=cannabidiol, CBN=cannabinol).

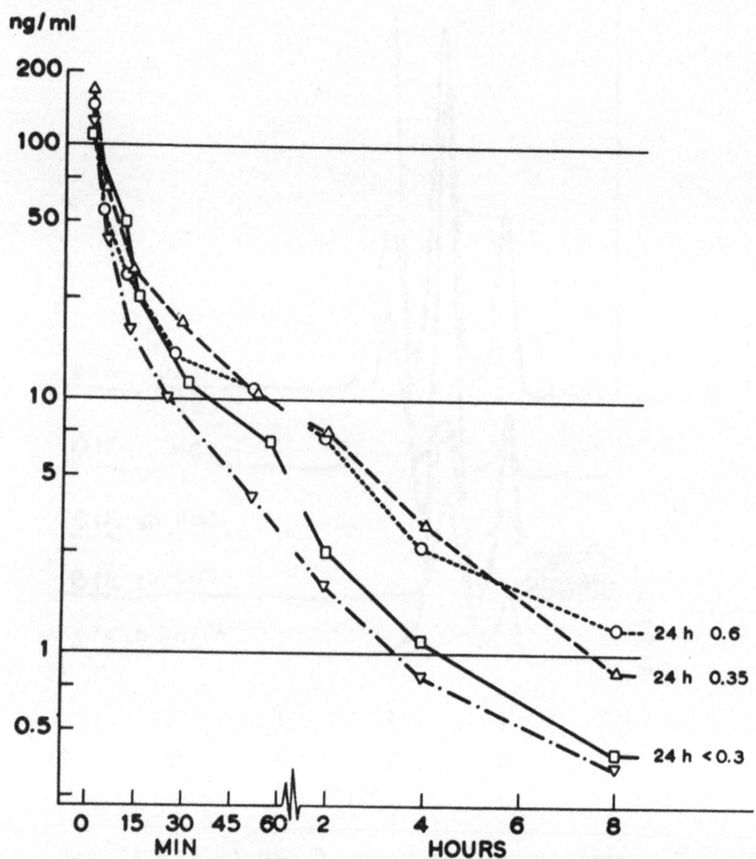

FIG. 4. Plasma levels of Δ^6-tetrahydrocannabinol (Δ^6-THC) after smoking 8.3 mg Δ^6-THC.

Garrett and Hunt (12) have shown that the terminal half-life of Δ^1-THC in the dog is reached only slowly. They also found that the return of Δ^1-THC from the tissues is the rate determining step of the drug elimination process after the initial distribution and metabolism.

Similar plasma levels of Δ^1-THC as for Δ^6-THC (Fig. 4) have earlier been found in man by us (3) and Rosenfeld _et al_.(5) using mass fragmentography, and by Galanter _et al_. (13) using Δ^1-THC-^{14}C. Wall _et al_. (14) have published plasma levels of both Δ^1-THC, CBD, and CBN and certain metabolites as well as subjective psychological effects after i.v. administration of the labelled drug.

Hence, the sensitivity requirements for the determination of Δ^1-THC are indicated in Fig. 4. Plasma levels will obviously be modified by the amount of THC absorbed and by the rate of absorption but if levels are to be measured later than 4 hours after administration a sensitivity of 1 ng/ml is required. Such as sensitivity combined with specificity is perhaps limited to the mass fragmentographic techniques. We have so far encountered little interference in the mass fragmentographic determination of Δ^1-THC provided redistilled solvents, particularly ethanol, and all silanized glassware are used. If interference does occur the silylation technique provides an alternative.

The capacity of the mass fragmentographic technique is limited by the main time requiring step being the liquid chromatography purification. With the non-automated system now in use, a technician can process about ten plasma samples per day. This might be improved by automation, by high pressure liquid chromatography or by using the double extraction technique of Rosenfeld _et al_. (5).

Acknowledgments. The support of the Swedish Medical Research Council is appreciated. The mass spectrometric work was supported by the grants to the Department of Toxicology.

REFERENCES

1) L. Grlić, Acta Pharm. Jugoslav., 1974, _24_, 63.
2) A. Ohlsson, J.-E. Lindgren, K. Leander and S. Agurell, Research Monograph Series, NIDA, 1976.
3) S. Agurell, B. Gustafsson, B. Holmstedt, K. Leander, J.-E. Lindgren, I. Nilsson, F. Sandberg and M. Asberg, J. Pharm. Pharmacol., 1973, _25_, 554.
4) S. Agurell, S. Levander, M. Binder, A. Bader-Bartfai, B. Gustafsson, K. Leander, J.-E. Lindgren, A. Ohlsson and B. Tobisson, in "Pharmacology of Cannabis", S. Szara and M. Braude, Eds., Raven Press, New York, USA, in press, 1976.

5) J.J. Rosenfeld, B. Bowins, J. Roberts, J. Perkins and A.S. McPherson , Anal. Chem., 1974, 46, 2232.
6) C.G. Pitt, D.T. Hobbs, H. Schran, C.E. Twine Jr. and D.L. Williams, J. Label. Comp., 1975, 11, 551.
7) T. Petrzilka, W. Haefliger and C. Sikemeier, Helv. Chim. Acta, 1969, 52, 1102.
8) R. Mechoulam, B. Yagnitinsky and Y. Gaoni, J. Am. Chem. Soc., 1968, 90, 2418.
9) P.K. Bäckström and G. Sundström, Acta Chem. Scand., 1970, 24, 716.
10) K. Elkin, L. Pierrou, U.G. Ahlborg, B. Holmstedt and J.-E. Lindgren, J. Chromatogr., 1973, 81, 47.
11) W.G. Sippell, P. Lehman and G. Hellman, J. Chromatogr., 1975, 108, 305.
12) E.R. Garrett and C.A. Hunt, to be published, 1976.
13) M. Galanter, R.J. Wyatt, L. Lemberger, H. Weingartner, T.B. Waughan and W.T. Roth, Science, 1972, 176, 934.
14) M.E. Wall, D.E. Brine and M. Pereze-Reyes, in "Pharmacology of Cannabis", S. Szara and M. Braude, Eds., Raven Press, New York, in press, 1976.

IDENTIFICATION OF A MAJOR METABOLITE OF TETRAHYDROCANNABINOL IN PLASMA AND URINE FROM CANNABIS SMOKERS

M. Nordqvist[1], J.-E.Lindgren[2,3],S. Agurell[1,3]

[1]University of Uppsala, Uppsala; [2]Karolinska Institutet, Stockholm; [3]Astra, Södertälje, Sweden

INTRODUCTION

Intake of cannabis can not be confirmed by analysis of urine samples based upon detection of Δ^1-tetrahydrocannabinol (Δ^1-THC), the psychoactive constituent of the drug. It has been shown that the drug is almost completely metabolized in human after i.v. administration of radiolabelled Δ^1-THC (1-3). The major excretion route was via faeces. An active metabolite, 7-hydroxy-Δ^1-THC, constituted 20% of the faecal radioactivity, but only trace amounts were present in the urine in which acidic metabolites accounted for the major part. Less than 0.02% of the administered dose of Δ^1-THC was excreted unchanged via urine.

The low amounts of Δ^1-THC and 7-hydroxy-Δ^1-THC excreted in the urine seem to be a result of further metabolism yielding more polar compounds with the 7-methyl group being further oxidized to a carboxyl group via the aldehyde (4) with or without introduction of additional hydroxyl group(s). The Δ^1-THC-7-oic acid has been identified as a major metabolite in urine, faeces, and plasma from humans given Δ^1-THC or 7-hydroxy-Δ^1-THC i.v. (3).

The plasma level of unchanged drug 12 h after the inhalation of a normal occasional dose of Δ^1-THC is about 0.3 ng/ml, the earlier limit of our method for detection of THC (5). The newly introduced silylation method (6) results in a higher sensitivity but for detection of Δ^1-THC or 7-hydroxy-Δ^1-THC in urine or in plasma samples taken after 24 h perhaps a greater sensitivity is needed. The concentrations of 7-hydroxy-Δ^1-THC and other known neutral metabolites in plasma are always much lower than that of the parent compound (3). On the other hand, the amount of

Δ^1-THC-7-oic acid in plasma is almost equal to that of Δ^1-THC within 50 min and then remains slightly higher for more than 24 h.

Our proposed method for the identification of Cannabis intoxications is based upon the finding that Δ^1-THC-7-oic acid is a major metabolite of Δ^1-THC in man. For forensic purposes, when the identification but not accurate quantitation is required, it should be a practical method if gas chromatography-mass spectrometry (GC-MS) equipment is available. It is related to our method for detection of THC in blood plasma (7) and can, as described here, be used for qualitative identification and semi-quantitative assay of a major acidic metabolite in urine and plasma from Cannabis users. This method with slight modifications probably is also applicable to related acidic metabolites. A large part of the metabolites in human plasma, urine, and faeces was classified as more polar acids by Wall et al. (3). They are probably dioic acids or dihydroxylated monooic acids. The dioic acids 2",3",4",5"-tetranor-Δ^1-THC-7,1"-dioic acid, 4",5"-bisnor-Δ^1-THC-7,3"-dioic acid, and Δ^1-THC-7,5"-dioic acid are major metabolites in rabbit urine (8, 9) together with monohydroxylated monoic acids (9, 10). Dihydroxylated THC-monooic acids have been identified by Harvey and Paton (11) in mice.

EXPERIMENTAL

Analysis of THC-7-oic acid in blood plasma. All glassware was carefully washed and silanized before use. The solvents were of analytical grade and the chloroform and ethanol distilled twice. 50 ng Δ^6-THC-7-oic-acid (12) dissolved in 25 µl ethanol was equilibrated with 2.5 ml human plasma in a glass stoppered centrifuge tube; 2.5 ml 1 M citrate-HCl buffer (pH 4.1) was added before extraction with 10 ml diethyl ether. After centrifugation for 10 min the ether layer was transferred into a conic tube and evaporated to dryness under a stream of nitrogen at 40°. The residue was dissolved in 100 µl methanol, an ethereal solution of diazomethane was added in excess, and the solution was allowed to react for 10 min at room temperature before evaporation of reagent and solvent under nitrogen. The resulting residue was transferred to a jacketed Sephadex LH-20 column (1 x 55 cm, V_0= 17 ml) using three consecutive 200 µl portions of the elution solvent hexane-chloroform-ethanol(10:10:1). At the elution rate 0.2 ml/min and the temperature 12° the methyl ester of Δ^6-THC-7-oic acid was eluted between 37 and 47 ml. This fraction was collected in a 10 ml conic tube. The solvent was evaporated under nitrogen, the residue dissolved in ethanol and transferred to a 1.0 ml conic vial, where it was stored at 4° in the dark. On the day of analysis the methylated extract was dried under nitrogen, dissolved in 25 µl dry acetonitrile, mixed with 10 µl silylating agent (N,O-bis-(trimethylsilyl)acetamide) and kept at 50-60° for

10 min. It was then dried again under nitrogen and redissolved in 25 µl acetonitrile, of which 5 µl was subjected to mass fragmentography (3% SE-30 Gas Chrom Q, 100/200 mesh, 200°). The mass spectrometer (LKB Model 9000) was adjusted to record the intensity of m/e 430 (M^+), 415 (M^+-CH_3), and 374 ($M^+-C_4H_8$) at 70 eV on three different channels. For further details, see previous paper in this volume (Fig. 3).

Analysis of THC-7-oic acid in urine. The procedure described above for analysis of human plasma was also applied to urine. 100 ng [^{3}H]-Δ^6-THC-7-oic acid was equilibrated for 20 min with 5.0 ml human urine. Addition of 2.5 ml 1 M citrate-HCl buffer kept the pH at 4.1 during the extraction with equal volume of diethyl ether (7.5 mlx1). The mass spectrometer was set to record the intensity of m/e 430 (M^+), 415 (M^+-CH_3), and 371 ($M^+-COOCH_3$) in the final silylated purified ether extract.

RESULTS AND DISCUSSION

The partition of [^{3}H]-Δ^6-THC-7-oic acid (prepared by acid catalyzed exchange) between buffer solutions of different pH and organic phase (diethyl ether, light petroleum with 1.5% isopentanol, and toluene, respectively) was studied to determine the combination of pH and extraction solvent giving the best recovery. Diethyl ether proved to be the most efficient solvent (log D=2.2, pH 4.1). The recovery from plasma of added [^{3}H]-Δ^6--THC-7-oic acid (in the range of 50-200 ng) was over 85% after extraction with a double volume of ether. The recovery from urine was over 90% using equal volumes. The in vivo metabolite of Δ^1-THC, Δ^1-THC-7-oic acid is expected to be extracted identically. The diazomethane was effective in esterifying the Δ^6-THC-7-oic acid but the corresponding acid metabolite of Δ^1-THC was more difficult to esterify. However, the procedure described using a 10 min incubation with excess of reagent was applicable to both acids.

The elution volume on Sephadex LH-20 of the methyl ester of Δ^6-THC-7-oic acid was determined by calibration with ng amounts of tritiated compound. The ester of Δ^1-THC-7-oic acid had a slightly higher elution volume but was collected quantitatively within the wide fraction selected. Total recovery of [^{3}H]-Δ^6--THC-7-oic acid in plasma and urine after extraction, methylation, and column purification was around 70%.

The silylated methyl ester of Δ^6-THC-7-oic acid is not fully separable from the Δ^1-isomer on the conventional GLC columns tested (SE-30, SE-52, OV-17, JXR, XE-60). When not silylated, the methyl esters were resolved but that of the Δ^1-isomer partly decomposed on the column. Thus, the silylated methyl ester of Δ^6-THC-7-oic acid was used as an external standard for the Δ^1-

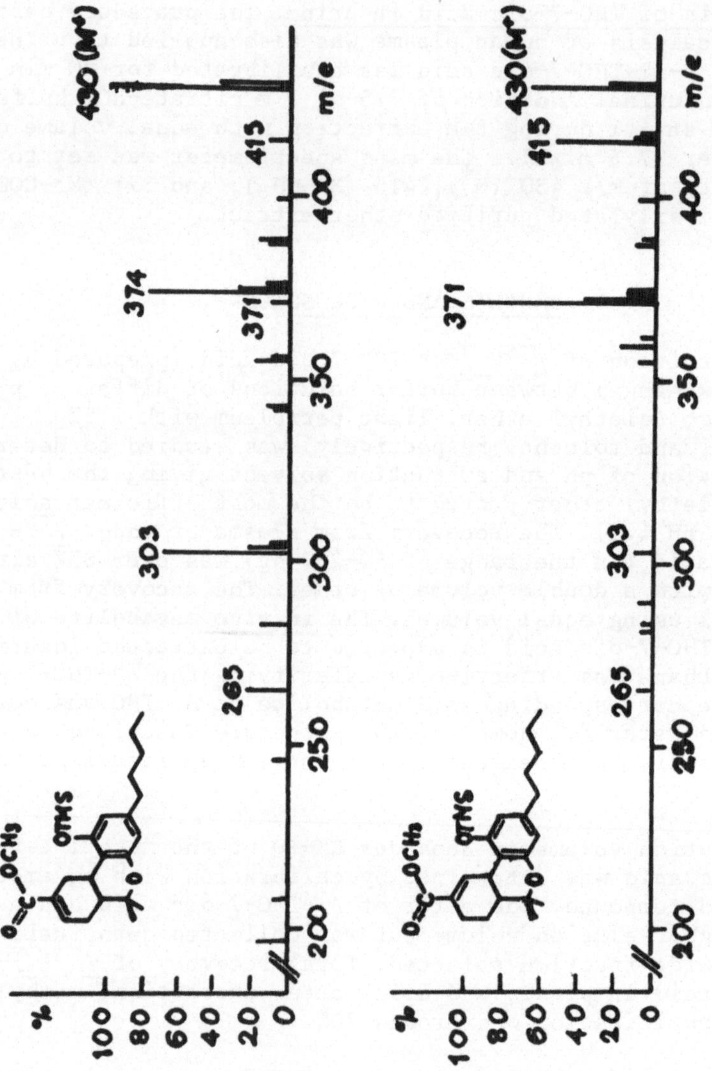

FIG. 1. Mass spectra of the silyl ethers of Δ^6-(upper panel) and Δ^1-THC-7-oic acid methyl ester (lower panel).

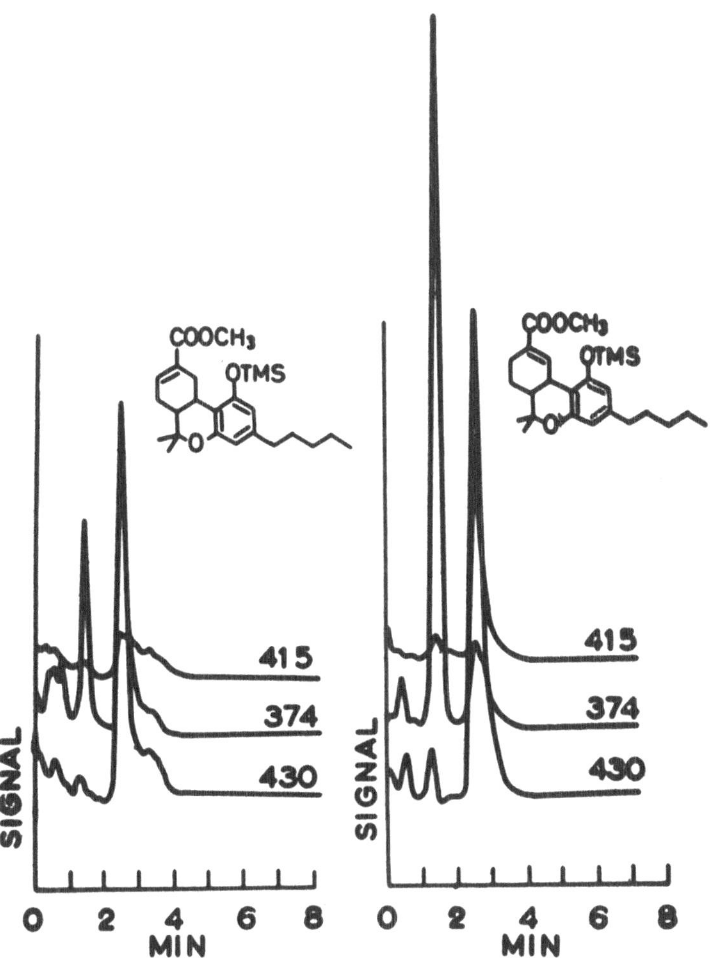

FIG. 2. Mass fragmentograms of the silyl ethers of the isomeric THC-7-oic acid methyl esters from purified plasma extracts.

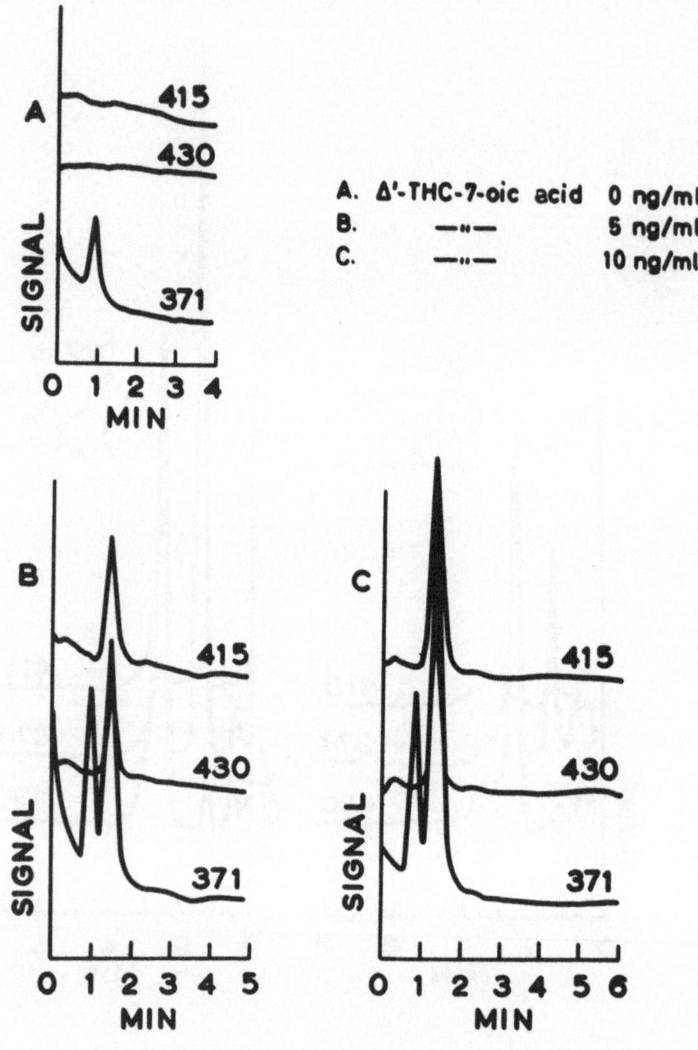

FIG. 3. Mass fragmentograms of the silyl ethers of methyl esters from purified urine extracts
A. 5 ml urine (control).
B. 5 ml urine and 25 ng Δ^1-THC-7-oic acid.
C. 5 ml urine and 50 ng Δ^1-THC-7-oic acid.

-isomer since both compounds have very similar retention times but different intensities in the mass fragmentograms (Fig. 1,2). The Δ^6-isomer can also serve as a reference for semiquantitative estimations of Δ^1-THC-7-oic acid in plasma or urine from smokers. The possible concomitant occurrence of Δ^6-THC-7-oic acid should be of little significance since Δ^6-THC at most occurs in 2% of the amount of Δ^1-THC in Cannabis (13). There is only a limited need for an accurate quantitation method for Δ^1-THC-7-oic acid since pharmacological tests with Rhesus monkeys (14) have shown that the isomeric Δ^6-THC-7-oic acid is inactive in doses up to 10 mg/kg i.v. If necessary, the deuterium labelled Δ^1-THC-7-oic acid can be synthesized as an internal standard according to methods described by Pitt and Wall (15) and Pitt _et al_. (16). There are no specific studies on the urinary excretion rates of Δ^1-THC-7-oic acid in humans after Cannabis smoking, but preliminary experiments indicate a prolonged excretion. Identification of this acid may be a practical method to identify Cannabis users. The sensitivity of the method described here is at least in the range of a few ng/ml. Further studies are in progress.

Acknowledgments. The support of the Medical Research Council was appreciated. Mass spectrometry was supported by grants to the Department of Toxicology. Reference acids were kindly provided by Drs. R. Mechoulam and M. Wall.

REFERENCES

1) L. Lemberger, S.D. Silberstein, J. Axelrod and I.J. Kopin, Science, 1970, _170_, 1320.
2) L. Lemberger, J. Axelrod and I.J. Kopin, Ann. N.Y. Acad. Sci., 1971, _191_, 142.
3) M.E. Wall, D.E. Brine and M. Perez-Reyes, in "Pharmacology of Cannabis", S. Szara and M. Braude, Eds., Raven Press, New York, USA, 1974.
4) Z. Ben-Zvi and S. Burstein, Res. Commun. Chem. Pathol. Pharmacol., 1974, _8_, 223.
5) A. Ohlsson, J.-E. Lindgren, K. Leander and S. Agurell, in "Research Monograph Series", NIDA, 1976.
6) A. Ohlsson, J.-E. Lindgren, K. Leander and S. Agurell, This volume, 1976.
7) S. Agurell, B. Gustafsson, B. Holmstedt, K. Leander, J.-E. Lindgren, I. Nilsson, F. Sandberg and M. Asberg, J. Pharm. Pharmacol., 1973, _25_, 554.
8) M. Nordqvist, S. Agurell, M. Binder and I.M. Nilsson, J. Pharm. Pharmacol., 1974, _6_, 471.
9) M. Nordqvist, S. Agurell, M. Binder, I.M. Nilsson, unpublished results.
10) S. Burstein, J. Rosenfeld and T. Wittstruck, Science, 1972, _176_, 422.
11) D.J. Harvey and W.D.M. Paton, in "Marihuana Chemistry,

Biochemistry and Cellular Effects", G.G. Nahas, Ed., Springer-Verlag, New York, 1976, p. 93.

12) R. Mechoulam, Z. Ben-Zvi, S. Agurell, I.M. Nilsson, J.L.G. Nilsson, H. Edery and Y. Grunfeld, Experientia, 1973, 29, 1193.

13) A. Ohlsson, C.I. Abou-Chaar, S. Agurell, I.M. Nilsson, K. Olofsson and F. Sandberg, Bulletin Narc., 1971, 23, 29.

14) R. Mechoulam, Z. Ben-Zvi, A. Shani, H. Zemler, S. Levy, H. Edery and Y. Grunfeld, in "Cannabis and its derivatives", W.D.M. Paton and J. Crown, Eds.,O.U.P., 1972.

15) C.G. Pitt and M.E. Wall, in "Annual report from Research Triangle Institute", North Carolina, USA, to National Institute on Drug Abuse, 1974.

16) C.G. Pitt, M.S. Fowler, S. Sathe, S.C. Srivastava and D.L. Williams, J. Am. Chem. Soc., 1975, 97, 3798.

ANALYSIS OF IMPURITIES IN ILLICIT HEROIN BY MASS SPECTROMETRY

M. Klein

Research Laboratory, 7704 Old Springhouse Rd.

McLean, Virginia 22101, USA

INTRODUCTION

This paper presents a mass spectral study of impurities in illicit heroin. Identification of these impurities is of importance because of their potential contribution to pharmacological effects. The major impurities, 6-acetyl-morphine (I-c) and 6-acetylcodeine (I-k), are predictable, owing to their origination from the naturally-occurring alkaloids-morphine (I-a) and codeine (I-j)(1) after reaction with acetic anhydride. In addition, the presence of a trace impurity (m/e 397, $C_{22}H_{23}NO_6$, at less than 0.4%) was observed by gas chromatography-mass spectrometry (Fig. 2).

The identical compound (confirmed by MS fragmentation and GC retention time) was synthesized from morphine N-oxide and acetic anhydride (obtained in 45% yield). Its isolation in a moderately pure form (ca. 95%) allowed for its further study. Based on high resolution mass spectrometry, infrared and nuclear magnetic resonance examination, it is believed to be 3,6,17-triacetyl-*nor*morphine, also referred to as N-acetyl-*nor*heroin (I-l). Its synthesis from morphine N-oxide and acetic anhydride was consistent with the Polonovski reaction (2) proceeding by a free radical mechanism, reported by Craig *et al.* (3). Synthesis of 3,6,17-triacetyl-d_9-*nor*morphine (I-m) aided in the structural assignment (Fig. 3). Important features of its mass spectrum involved fragmentations dramatically different from heroin. These included acetyl fragmentations and splitting the ethanamine bridge. The acetyl groups of heroin fragment: $[M-CH_2CO\ (42)]^{+\cdot}$, $[M-CH_3CO_2\ (84)]^{+\cdot}$ and $[M-(CH_2CO+CH_3CO_2)\ (101)]^{+\cdot}$. The trace constituent provided different prominent ions: $[M-CH_2CO\ (42)]^{+\cdot}$, $[M-2\ CH_2CO\ (84)]^{+\cdot}$, $[M-(CH_2CO+CH_3CO_2H)\ (102)]^{+\cdot}$ and $[M-(2\ CH_2CO+CH_3CO)\ (127)]^{+\cdot}$.

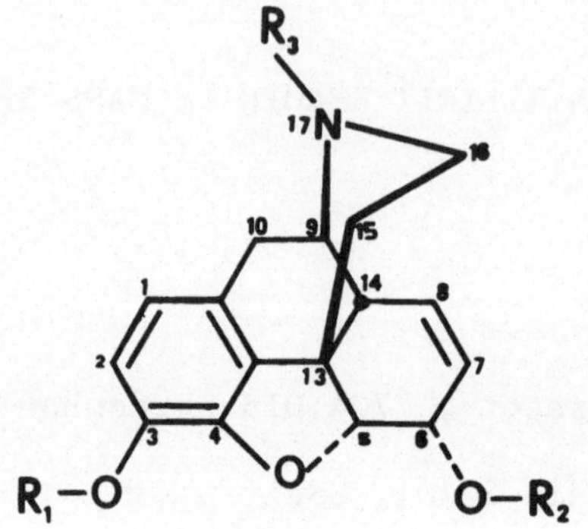

(I)

(I) a) $R_1=R_2=H$; $R_3=CH_3$

b) $R_1=COCH_3$; $R_2=H$; $R_3=CH_3$

c) $R_1=H$; $R_2=COCH_3$; $R_3=CH_3$

d) $R_1=R_2=COCH_3$; $R_3=CH_3$

e) $R_1=COCD_3$; $R_2=H$; $R_3=CH_3$

f) $R_1=COCD_3$; $R_2=COCH_3$; $R_3=CH_3$

g) $R_1=H$; $R_2=COCD_3$; $R_3=CH_3$

h) $R_1=COCH_3$; $R_2=COCD_3$; $R_3=CH_3$

i) $R_1=R_2=COCD_3$; $R_3=CH_3$

j) $R_1=CH_3$; $R_2=H$; $R_3=CH_3$

k) $R_1=CH_3$; $R_2=COCH_3$; $R_3=CH_3$

l) $R_1=R_2=R_3=COCH_3$

m) $R_1=R_2=R_3=COCD_3$

FIG. 1. Acetylated derivatives of morphine ($R_1=R_2=H$; $R_3=CH_3$) and codeine ($R_1=$ $=R_3=CH_3$; $R_2=H$).

In addition, fragment ions, occurring at m/e 87 (C_4H_9NO) and m/e 209 ($C_{14}H_9O_2$), were prominent. This paper, describing the characterization of its structure, includes a discussion of: 1) Reaction of morphine N-oxide and acetic anhydride (Ac_2O). 2) IR and NMR spectra of N-acetyl-*nor*heroin. 3) Mass spectral fragmentation of N-acetyl-*nor*heroin; a) Fragmentation of ester moieties and synthesis of deuteroacetylated derivatives of morphine, and b) fragmentation of the morphinan ring system of N-acetyl-*nor*heroin.

DISCUSSION AND RESULTS

Reaction of morphine N-oxide and Ac_2O. The Polonovski reaction (2, 4, 5) has been described, whereby *tert*-amine oxides containing at least one N-methyl group are converted by acetic anhydride into *sec*-amines and formaldehyde. The reaction mechanism by J.C. Craig *et al*. (3) proposed that it proceeded by a free radical sequence. As applied to the substrate morphine N-oxide (II) (Scheme 1), deoxygenation occurs through the N-acetoxyl intermediate (III). Two potential free radical initiators are thus generated: •OAc and the "heroin free radical" (IV), which after reaction with acetate ion gives (V). Exchange of •OAc between (V) and (III) generates additional "heroin free radical" (V) and the intermediate, 17-acetoxylmethylene-heroin (VI).Rearrangement of (VI), with loss of formaldehyde, acetylates the amine to give N-acetyl-*nor*heroin (I-1), thus terminating free radical propagation. An alternate pathway involves generation of additional •OAc and heroin (I-d) by reaction of (V) and acetic acid.

IR and NMR spectra of N-acetyl-norheroin. The infrared spectrum of N-acetyl-*nor*heroin indicated that it contained a tertiary amide and two ester groups. The carbonyl stretching bands of the phenolic acetate (1760 cm^{-1}) and the allylic acetate (1735 cm^{-1}) were observed in heroin and N-acetyl-*nor*heroin. In addition, the IR of (I-1) indicated a strong amide I band at 1632 cm^{-1} which was not sensitive to solution concentration or change of phase. Since the presence of free N-H stretching bands was not observed in the regions 3500-3400 cm^{-1} (in dilute solution) and 3340-3060 cm^{-1} (at higher concentrations), an N,N-disubstituted amide was indicated (6).

The nuclear magnetic resonance spectra of alkaloids in the morphine series have been studied in detail, in which chemical shifts and coupling constants have been documented (7-9). The two aromatic protons H_1 and H_2 of heroin appeared as doublets centered at δ 6.54 and 6.82, respectively, (J_{12} 8.4 Hz). The strong singlets, due to isolated methyl groups, dominated the spectrum: N-CH_3 (δ 2.39), 3-$OCOCH_3$ (δ 2.23) and 6 -$OCOCH_3$ (δ 2.10). The proton at C-5 (adjacent the oxygen in the furan ring) formed a doublet at δ 5.1 (J_{56} 0.9 Hz). The ethylenic protons occurred as doublets (H_8 δ 5.36; H_7 δ 5.63), which were somewhat resolved and mutually coupled (J_{78}

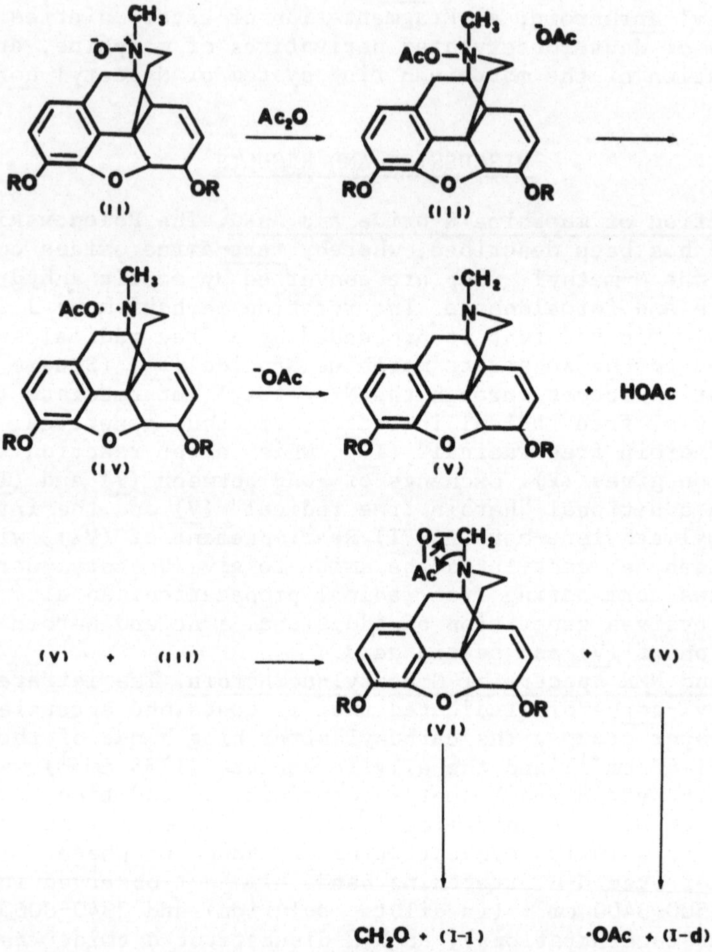

SCHEME I. Free radical mechanism of the Polonovski reaction of morphine N-oxide (II, R=H) and acetic anhydride yielding 3,6,17-triacetylnormorphine (I-l, R=Ac), and heroin (I-d, R=Ac).

10.2 Hz); each resonance was then further split by interaction with the 6β-, and 14β- protons of the allylic system. The 6β-proton (which is on the carbon atom containing the acetyl ester) was not resolved, probably resulting from its chemical shift into the region of resonances of H_5, H_7 and H_8. The NMR spectrum of non-acetylated compounds, as in the case of codeine, includes a multiplet for H_6 centered at δ 4.20. Acetylation shifts generally observed for six-numbered ring protons fall downfield between 1.0 and 1.1 p.p.m. (10). The doublets for the protons at the 10α and 10β positions (δ 2.60 and 3.25) and the multiplet for H_9 (δ 3.30) were not resolved, due to coupling from H_{14} and overlapping resonances of $N-CH_3$ and the H_{14} multiplet itself (δ ca. 2.71). Protons at C_{15} and C_{16} in the ethanamine ring also were not resolved, occurring in the region of methyl resonances.

The NMR spectrum of N-acetyl-<u>nor</u>heroin did not contain the $N-CH_3$ singlet at δ 2.39. The 3 $-OCOCH_3$ singlet (δ 2.25) was observed. An unresolved singlet (which integrated as 6 protons) resonated at δ 2.11 and was attributed to the 6 $-OCOCH_3$ and $-NCO-CH_3$ groups. Various peaks indicated that the remainder of the morphine carbon skeleton appeared unaltered: 1) aromatic protons H_1 and H_2 were doublets centered at δ 6.57 and 6.82, respectively, (J_{12} 8.4 Hz); 2) H_5 (adjacent the oxygen in the furan ring) formed a doublet at δ 5.11 (J_{56} 0.9 Hz); 3) the ethylenic protons occurred as doublets (H_8 δ 5.43; H_7 δ 5.72; J_{78} 10.2 Hz).

The multiplet resulting from the H_9methine was not resolved, although it appeared to shift downfield from its location in the NMR of heroin to the region δ 3.4-4.0. Similarly complex, the 10α and 10β protons were not resolved. The resonance resulting from the 10β proton occurred in the region δ 3.0-3.6. Two peaks (δ 2.84 and 2.75) appeared to contribute to the doublet from the 10α proton, although this was not fully ascertained due to partial overlap from the H_{14} multiplet (δ 2.70).

<u>Mass spectral fragmentation of N-acetyl-norheroin</u>. a) <u>Ester fragmentation</u>. The major ions in the mass spectrum of N-acetyl-<u>nor</u>heroin resulting from ester and amide fragmentation of the molecular ion (m/e 397) occurred at m/e 355 $[M-CH_2CO]^{+\cdot}$ (<u>VII</u>), m/e 313 $[M-2CH_2CO]^{+\cdot}$ (<u>VIII</u>), m/e 295 $[M-(CH_3COOH+CH_2CO)]^{+\cdot}$ (<u>IX</u>) and m/e 270 $[M-(2CH_2CO+CH_3CO)]^{+\cdot}$ (<u>X</u>), (shown in Scheme 2). Analogous fragments resulted from the molecular ion (m/e 406) of the deuterated derivative, 3,6,17-triacetyl-d_9-<u>nor</u>morphine: m/e 362 $[M-CD_2CO]^{+\cdot}$, m/e 318 $[M-2CD_2CO]^{+\cdot}$, m/e 299 $[M-(CD_3COOH+CD_2CO)]^{+\cdot}$ and m/e 272 $[M-(2CD_2CO+CD_3CO)]^{+\cdot}$. High resolution mass spectrometric studies gave some accurate mass measurements which supported these processes: $C_{22}H_{23}NO_6$ (calculated 397.15253; obtained 397.15200); $C_{20}H_{21}NO_5$ (calculated 355.14196; obtained 355.14450);$C_{18}H_{19}NO_4$ (calculated 313.13140; obtained 313.12900); $C_{18}H_{17}NO_3$ (calculated 295.12084; obtained 295.12800). "Metastables" at 317.5 (m/e 397→m/e 355) and at 276.1 (m/e 355→m/e 313) were

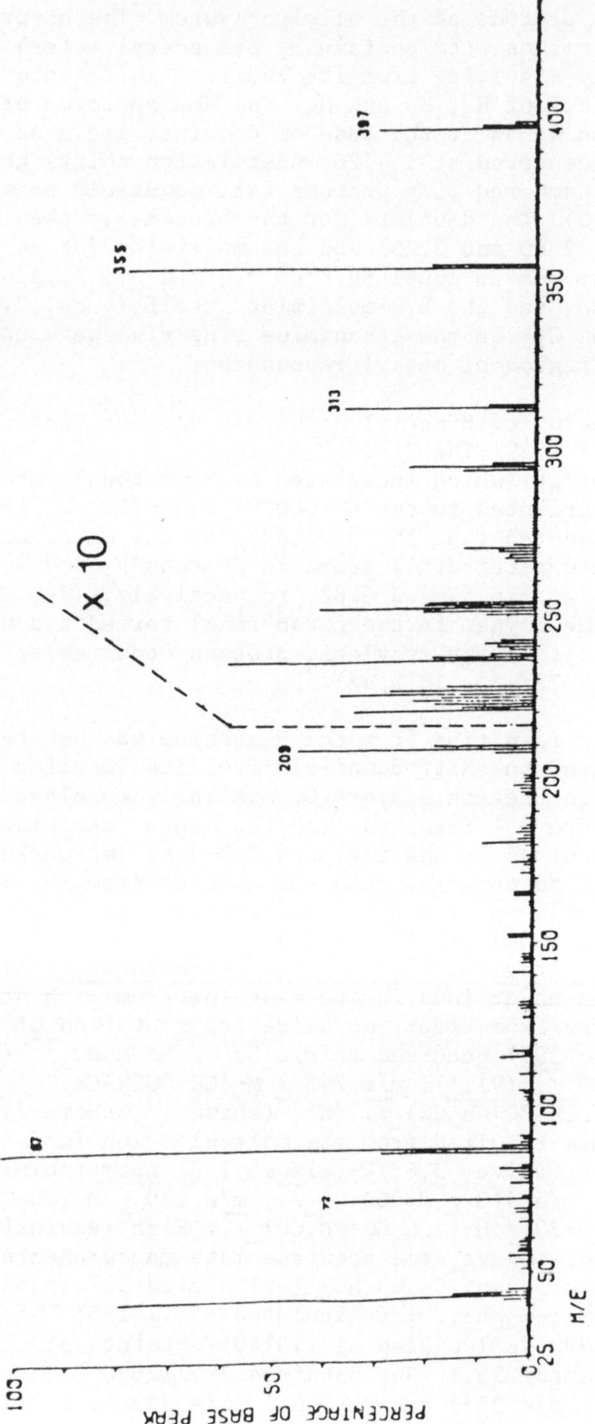

FIG. 2. Mass spectrum of 3,6,17-triacetylnormorphine (I-1).

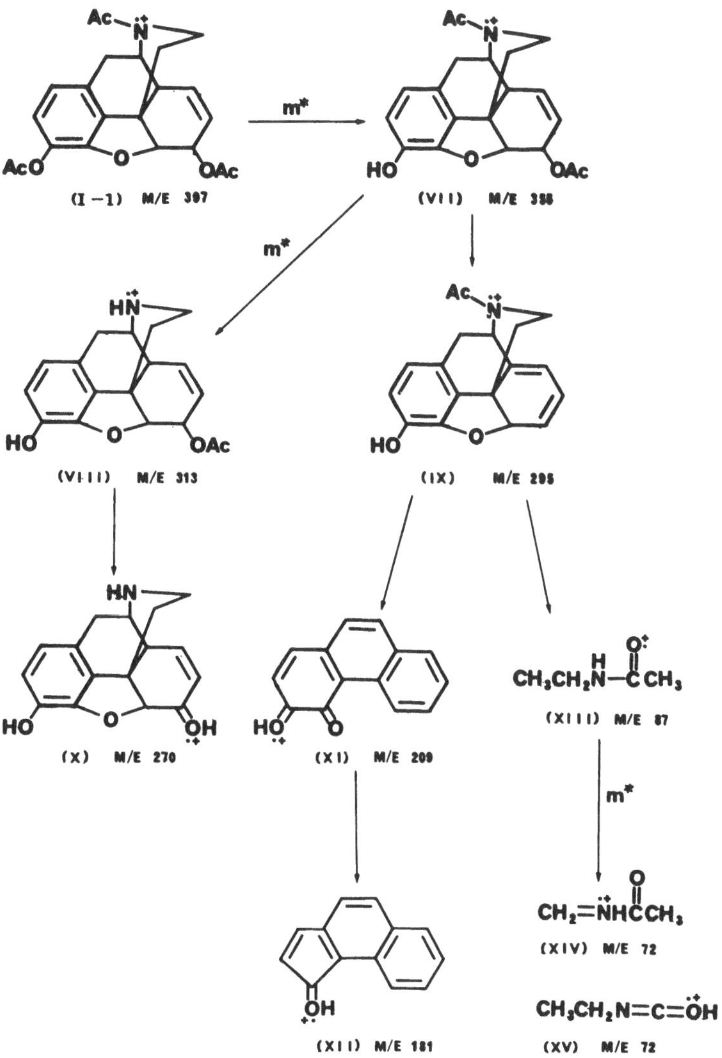

SCHEME II. Mass fragmentation of 3,6,17-triacetyl-normorphine (I-1) with formation of major ions.

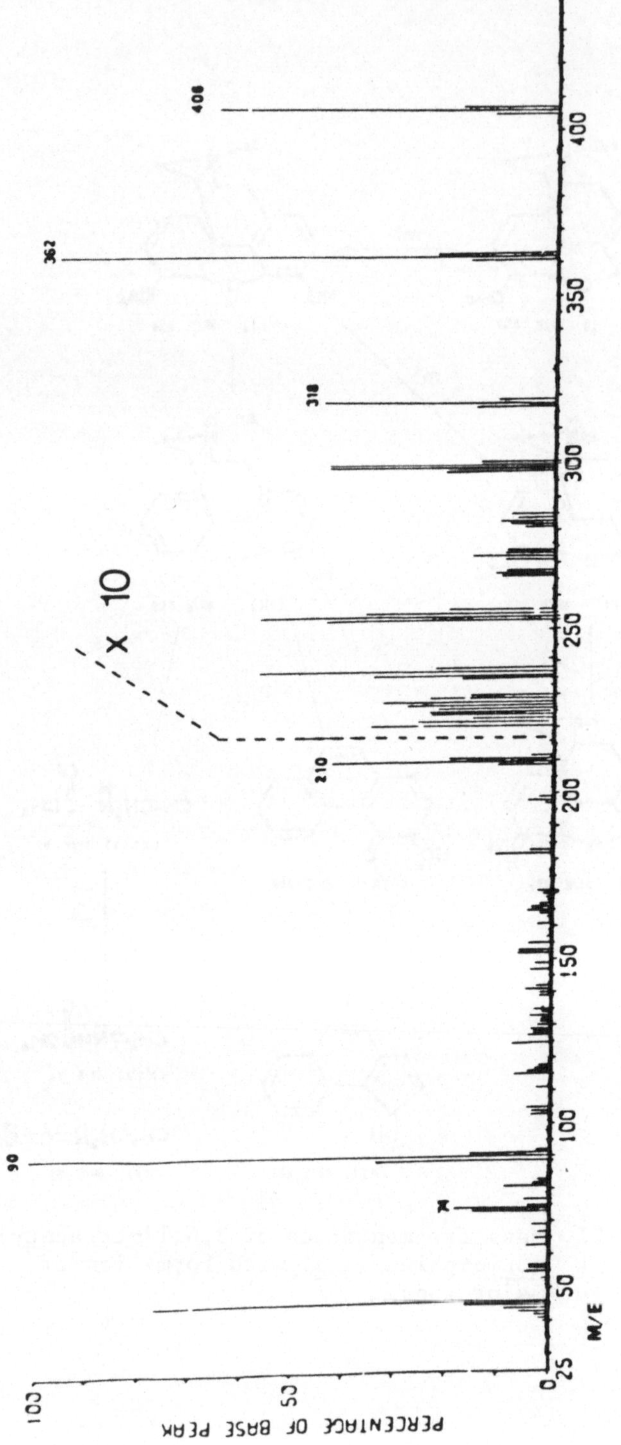

FIG. 3. Mass spectrum of 3,6,17-triacetyl-d_9-normorphine (I-m).

observed. The more facile loss of the 3-O-phenolic acetate was primarily attributed to the fragment at m/e 355. The fragment at m/e 313 was attributed to fragmentation of the acetyl group from the amide. The ion m/e 295 resulted from loss of the 6-O-allylic acetate (CH_3COOH) and (CH_2CO) from the phenolic acetate or the acetylamide. These conclusions were based upon comparisons of the mass fragmentation processes of the ester groups of heroin (M^+ at m/e 369) (I-d), deuteroacetylated heroin derivatives (I-e to I-i) and 6-acetyl-codeine (M^+ at m/e 341) (I-k). The deuterated analogs were derived from morphine (I-a) by the different reactivities of its phenolic and alcoholic groups with esterifying reagents (11). They included 3-acetyl-d_3-morphine (M^+ at m/e 330) (I-e), 6-acetyl-d_3-morphine (M^+ at m/e 330) (I-g), 3-acetyl-d_3-acetyl-morphine (M^+ at m/e 372) (I-f), 3-acetyl-6-acetyl-d_3-morphine (M^+ at m/e 372) (I-h) and 3,6-diacetyl-d_6-morphine (heroin-d_6) (M^+ at m/e 375) (I-i). (I-e) was synthesized by treatment of morphine with acetic anhydride -d_6 at room temperature. (I-g) was synthesized via formation of heroin-d_6 by heating a solution of morphine and acetic anhydride-d_6, followed by selective acid hydrolysis of the 3-phenolic ester. Mixed deuterated esters were synthesized by treatment of 6-acetylmorphine and 6-acetyl-d_3-morphine with acetic anhydride-d_6 and acetic anhydride, respectively, at room temperature.

The heroin derivatives displayed prominent molecular ions. Base peaks were derived from the facile fragmentation of the phenolic acetates: [M-CH_2CO]$^{+\cdot}$ in heroin (m/e 327) and 3-acetyl-6-acetyl-d_3-morphine (m/e 330); [M-CD_2CO]$^{+\cdot}$ in 3-acetyl-d_3-morphine (m/e 285), 3-acetyl-d_3-6-acetyl-morphine (m/e 328) and 3,6-diacetyl-d_6-morphine (m/e 331). Fragmentation of the 6-allylic acetate was represented by: [M-CH_3CO_2 (59)]$^{+\cdot}$ in the mass spectrum of heroin (m/e 310), 3-acetyl-d_3-6-acetylmorphine (m/e 313) and 6-acetylcodeine (m/e 282); [M-CD_3CO_2 (62)]$^{+\cdot}$ in 6-acetyl-d_3-morphine (m/e 268), 3-acetyl-6-acetyl-d_3-morphine (m/e 310) and 3,6-diacetyl-d_6-morphine (m/e 313). A prominent ion resulting from fragmentation of both esters [M-(CH_2CO+CH_3CO_2) (101)]$^{+\cdot}$ was observed to occur from heroin at m/e 268. Similarly, 3-acetyl-d_3-acetylmorphine produced m/e 269 [M-(CD_2CO+CH_3CO_2) (103)]$^{+\cdot}$, 3-acetyl-6-acetyl-d_3-morphine produced m/e 268 [M-(CH_2CO+CD_3CO_2) (104)]$^{+\cdot}$ and 3,6-diacetyl-d_6-morphine produced m/e 269 [M-(CD_2CO+CD_3CO_2) (106)]$^{+\cdot}$. b) Fragmentation of morphinan ring system. The major ions resulting from fragmentation of the morphinan ring system of (I-1) were observed at m/e 209 (XI), 181 (XII), 87 (XIII), 86 and 72 (XIV, XV). Related fragments resulting from 3,6,17-triacetyl-d_9-normorphine (I-m) occurred at m/e 210, 182, 90, 89, 75, 74 and 73. Accurate mass measurements from high resolution studies provided the following formulae: $C_{14}H_9O_2$ (calculated 209.06025; obtained 209.06250); C_4H_9NO (calculated 87.06841; obtained 87.06540); C_4H_8NO (calculated 86.06059; obtained 86.06340); C_3H_6NO (calculated 72.04494; obtained 72.04870). The "metastable" at m/e 59.6 indicated the process: m/e 87 $\rightarrow$ m/e 72. M/e 209 was attributed to the dioxyphenanthrene moiety, which resulted from elimination of the 3,6-ester groups and the ethanamine ring, although its contribution in heroin, morphine and codeine was less pronounced. Elimination of CO from m/e 209 gave m/e 181. N-methyl morphine derivatives have been reported to exhibit a rather abundant peak at m/e 59 ($CH_3\overset{+\cdot}{N}HCH_2CH_3$), resulting from fragmentation of the bridge (12-14). Similarly, dihydro-norcodeine (NH instead of NCH_3) produced an intense ion at m/e 45 and a minor fragment at m/e 59 (11, 12). The presence of other N-substituents has been reported to produce analogous ions: m/e 161 from N-(γ-phenylallyl)- -and m/e 99 from N-(cyclopropylmethyl)-(13). Similarly, m/e 87 was attributed to $[CH_3CH_2NHCOCH_3]^{+\cdot}$ from N-acetyl-norheroin. Formation of m/e 72 resulted from loss of methyl from m/e 87. Fragmentation of $[CH_3CH_2NHCOCD_3]^{+\cdot}$ from (I-m) produced ions m/e 75, 74, 73, indicating loss of methyl from its ethyl or acetyl portion, followed by rearrangement: $[CH_2{=}\overset{+\cdot}{N}HCOCD_3]$ or $[CH_3CH_2{-}N{=}C{=}\overset{+\cdot}{O}D]$.

EXPERIMENTAL

Instrumentation. Gas chromatography-mass spectrometric studies were performed on Finnigan GC/MS 3100D quadrupole instrument with jet separator (column: 3 ft. x 1/8 in.); temperature: initial 150°C (3 min.), final 250°C, program rate 10°C/min. Electron beam energy was 70 eV. Heroin hydrochloride concentrations were 40 mg/ml in methanol and 2-4 μl were injected.

High resolution mass analyses were performed on Hitachi Perkin-Elmer RMU-6L single focusing magnetic mass spectrometer by direct probe. Accurate mass measurements were determined using perfluoro-tri-n-butylamine as standard. Electron beam energy was 60 eV. Temperature of the heated inlet was gradually increased from 120° to 200°C in analysis. Mass spectra were normalized to the base peak. Infrared spectra were obtained by Perkin-Elmer 621 Grating Infrared Spectrophotometer. N-Acetyl-norheroin was examined by KBr disc and in anhydrous chloroform solutions. Nuclear magnetic resonance spectra were performed on JEOL C-60HL High Resolution NMR Instrument (60 MHz) and recorded downfield from reference standard, tetramethylsilane. Samples were run in $CDCl_3$.

Quantitation of N-acetylnorheroin in heroin hydrochloride samples was by gas chromatography on a Perkin Elmer 900 Gas Chromatograph (column: 6 ft. x 1/4 in.); temperature: initial 180°C, final 270°C, program rate 4°C/min.; Flame Ionization Detector, interfaced with Spectra-Physics Autolab System I Computing Integrator, measured peak areas in μ volts; Carrier gas: nitrogen 60 ml/min. Authentic heroin hydrochloride samples were prepared by dissolving ca. 60 mg/ml in methanol. Standard N-acetylnorheroin (prepared by procedure described below) was 5.0 mg/ml. Sample injections were 8 μ liters for heroin hydrochloride and 2 μl for N-acetyl-norheroin. Retention times were 12.5 min. (6-acetylcodeine), 13.0 min. (6-acetylmorphine), 14.6 min. (heroin) and 19.4 min. (N-acetylnorheroin).

Materials. Standards of morphine, 6-acetylmorphine, 6-acetylcodeine, heroin and morphine N-oxide were supplied by the Drug Enforcement Administration, Special Testing and Research Laboratory, McLean, Virginia. Acetic anhydride (Analytical Reagent) was by Mallinckrodt Chemical Works, St. Louis, Missouri (Lot ZGB-B). Acetic anhydride-d_6 was by Merck and Company, Inc., Rahway, New Jersey (Lot E-1086). Gas chromatographic column material was 3% OV-1 Gas Chrom Q, 100/120 mesh, by Supelco, Inc., Bellefonte, Pennsylvania (Lot D2066). Acid Washed Celite 545 for partition columns was by Johns Manville, Englewood Cliffs, New Jersey (Lot 50AW). Ether (anhydrous) was by J.T. Baker Chemical Company, Phillipsburg, New Jersey (Lot 415 4809). Chloroform was by Fisher Scientific Company, Fair Lawn, New Jersey (Lot 761081). Perchloric acid (70%) was by Mallinckrodt Chemical Works, St. Louis, Missouri (Lot AKE).

Synthesis of morphine esters. a) 3-Acetyl-d_3-morphine (I-e). Morphine base (50 mg) was dissolved in 2 ml acetic anhydride-d_6 and remained at room temperature for 20 min. Water (5 ml) was added, and after the solution became homogenous, solid sodium carbonate was added to pH 8.5. The mixture was extracted with 3-20ml ether portions. The ether extracts were combined and evaporated to dryness. The product was eluted with 2% ethanol in benzene from 4.0 gram silica (adsorbosil-5) to give (I-e). b) 3-Acetylmorphine (I-b). Morphine base (50 mg) was treated with 2 ml acetic anhydride at room temperature for 20 minutes. 3-Acetylmorphine was isolated for mass analysis by the same procedure as (I-e). c) 6-Acetyl-d_3--morphine (I-g). 3,6-Diacetyl-d_6-morphine (50 mg) was converted to its hydrochloride with HCl (g) in anhydrous ether. The white precipitate was collected on a Buchner funnel. The product was then dissolved in 20 ml water and 0.1 gram hydroxylamine hydrochloride was added. After 2 minutes, solid sodium carbonate was added to pH 8.5 and the solution was extracted with 3-20 ml ether portions. The ether extracts were combined and evaporated to dryness. The product remaining was dried in a vacuum desiccator. d) 3,6-Diacetyl--d_6-morphine (I-i). Morphine base (50 mg) was dissolved at room temperature in acetic anhydride-d_6 (2 ml) and the resulting solution was heated at reflux temperature for 2 hours. The solution was cooled to room temperature and diluted with 5 ml water. Solid sodium carbonate was added to pH 8.5 and the white precipitate of (I-i) formed, was collected on a Buchner funnel and dried in a desiccator. e) 3-Acetyl-6-acetyl-d_3-morphine (I-h). 6-Acetyl-d_3--morphine (30 mg) was dissolved in acetic anhydride (1.0 ml) at room temperature and allowed to stand for 5 minutes. The solution was diluted with 3 ml water and after homogeneous, solid sodium carbonate was added to pH 8.5. The precipitate formed and was collected on a Buchner funnel and dried in a desiccator. f) 3-

-Acetyl-d_3-6-acetylmorphine (I-f). Heroin hydrochloride (50 mg) was dissolved in 10 ml water and 0.1 g hydroxylamine hydrochloride was added. The solution remained at room temperature for three minutes, followed by the addition of solid sodium carbonate to pH 8.5. The product was extracted with 3-20 ml portions of ether. The ether extracts were combined and evaporated to dryness to give 6-acetylmorphine (I-c), confirmed by mass analysis. (I-c) was then dissolved in two ml acetic anhydride-d_6 and remained at room temperature for 45 minutes. Water (10 ml) was added, and, after solution, solid sodium carbonate was added to pH 8.5. The white precipitate formed, was collected on a Buchner funnel, washed with water and dried in a vacuum desiccator.

Isolation of (I-1) and (I-m). a) 3,6,17-Triacetylnormorphine (I-1). Morphine N-oxide (10 mg) was heated at reflux temperature in 5 ml acetic anhydride-d_6 for one hour. In the first several minutes of the reaction, the solution appeared exothermic with simultaneous color changes (yellow to red to light brown). The solution was then cooled to room temperature, and water (10 ml) was added. When the mixture became homogeneous, solid sodium carbonate was added to pH 9.0. The mixture was extracted with 3x125 ml portions of ether.The extracts were combined and evaporated to dryness. The ether extract was mixed till homogeneous with 8g celite, 10 ml 1% perchloric acid and 10 ml ether to comprise the top layer of a partition column (2.5 in. diam.). The bottom layer of the column consisted of a homogeneous mixture of celite (5 g) and 10 ml 1% perchloric acid. The column was eluted with 300 ml anhydrous ether. The ether eluate was evaporated to dryness. GC/MS analysis indicated the presence of (I-i) (M^+ at m/e 397). GC quantitation established its purity as 94.7%. Elution of the celite column with 300 ml water-washed chloroform followed. The chloroform eluate was evaporated to dryness and analyzed by GC/MS. The major constituent was heroin (M^+ at m/e 369), as well as several other minor products. b) 3,6,17-Triacetyl-d_9-normorphine (I-m). Morphine N-oxide (10 mg) was dissolved in 5 ml acetic anhydride-d_6 and heated at reflux temperature for one hour. The reaction mixture was worked up by the procedure described above for 3,6,17-triacetylnormorphine (I-1). 3,6,17-Triacetyl-d_9-normorphine (M^+ at m/e 406) was obtained in the ether eluate from a 1% perchloric acid-celite column and was analyzed by GC/MS. The major constituent from the chloroform eluate was heroin-d_6 (M^+ at m/e 375).

APPENDIX

Major ions in mass spectra.

Morphine (I-a): 285 (100), 284 (18), 268 (14), 216 (10), 215 (31), 174 (16), 162 (38), 124 (19), 115 (11), 81 (10), 70 (12), 59 (8), 44 (13), 42 (21).

3-Acetylmorphine (I-b): 327 (75), 310 (9), 285 (82), 268 (19), 267 (21), 215 (34), 162 (100), 124 (33), 115 (26), 81 (32), 77 (12), 70 (20).

6-Acetylmorphine (I-c): 327 (100), 284 (12), 268 (60), 215 (25), 211 (10), 210 (8), 204 (19), 181 (6), 174 (7), 162 (14), 146 (15), 124 (9), 115 (8), 94 (13), 81 (15), 70 (16), 59 (21).

Heroin (I-d): 369 (80), 328 (21), 327 (100), 326 (13), 310 (47), 284 (9), 269 (10), 268 (54), 267 (14) 215 (30), 204 (25), 162 (12), 146 (13), 124 (13), 115 (9), 94 (10), 81 (18), 70 (12), 59 (10), 44 (16), 43 (49), 42 (28).

3-Acetyl-d_3-morphine (I-e): 330 (100), 313 (14), 286 (75), 269 (16), 268 (20), 257 (5), 243 (7), 241 (6), 239 (6), 238 (6), 229 (9), 226 (26), 220 (11), 216 (28), 211 (36), 205 (26), 183 (15), 175 (16), 165 (21), 162 (75), 152 (18).

3-Acetyl-d_3-acetylmorphine (I-f): 372 (82), 328 (100), 313 (57), 284 (12), 269 (57), 256 (20), 254 (16), 241 (27), 239 (25), 226 (29), 216 (48), 211 (45), 183 (25), 178 (50), 175 (28), 162 (31), 146 (27), 141 (35), 128 (27), 124 (29), 115 (33).

6-Acetyl-d_3-morphine (I-g): 330 (100), 286 (11), 285 (9), 268 (60), 252 (6), 239 (1), 227 (13), 215 (54), 211 (22), 207 (37), 182 (16), 178 (15), 174 (22), 165 (22), 163 (26), 152 (22), 146 (51), 144 (19), 141 (24), 128 (25), 125 (28), 115 (41).

3-Acetyl-6-acetyl-d_3-morphine (I-h): 372 (85), 330 (100), 310 (54), 285 (9), 268 (59), 257 (3), 253 (6), 239 (7), 227 (10), 226 (9), 215 (43), 207 (62), 181 (11), 176 (13), 174 (14), 165 (14), 163 (26), 153 (26), 152 (16), 146 (35), 144 (13), 141 (11), 128 (13), 125 (20), 115 (31);

Heroin-d_6 (I-i): 375 (80), 331 (100), 313 (57), 285 (9), 269 (56), 256 (16), 254 (11), 241 (16), 239 (19), 226 (44), 216 (47), 211 (63), 207 (86), 183 (32), 178 (32), 165 (47), 163 (48), 152 (36), 146 (49), 141 (46), 139 (38).

Codeine (I-j): 299 (100), 298 (15), 284 (9), 282 (13) 280 (10), 270 (7), 256 (9), 242 (15), 230 (12), 229 (40), 225 (11), 214 (25), 188 (20), 181 (9), 175 (9), 162 (51), 146 (10), 128 (9), 124 (21), 115 (17), 81 (11), 70 (9), 59 (15), 44 (18), 42 (34).

6-Acetylcodeine (I-k): 341 (100), 298 (7), 282 (55), 240 (7), 29 (31), 225 (10), 214 (10), 204 (30), 188 (8), 162 (14), 149 (21), 124 (18), 115 (10), 81 (29), 77 (9), 70 (24).

3,6,17-Triacetylnormorphine (I-l): 397 (3), 355 (8), 313 (4), 295 (2), 270 (1), 252 (4), 236 (6), 226 (4), 210 (19), 209 (44),

181 (10), 153 (4), 152 (5), 141 (3), 115 (5), 91 (3), 87 (100), 86 (29), 81 (6), 72 (37), 55 (4), 44 (15), 43 (79).

3,6,17-Triacetyl-d_9-normorphine (I-m): 406 (6), 362 (9), 318 (5), 299 (4), 272 (1), 254 (6), 237 (6), 228 (3.2), 211 (19), 210 (41), 209 (10), 182 (10), 153 (6), 152 (5), 141 (3), 115 (6), 91 (16), 90 (100), 89 (33), 81 (10), 75 (11), 74 (18), 73 (16), 57 (4), 55 (5), 46 (79), 45 (15).

Acnowledgments. The author wishes to express his gratitude to Stanley P. Sobol, Laboratory Director, Special Testing and Research Laboratory, and Martin H. Studier, Senior Chemist, Argonne National Laboratory, Argonne, Illinois, for their recommendations and advice in this reseacrh project; to James M. Moore, Special Testing and Research Laboratory, and Professor Charles Hammer, Georgetown University, Washington, D.C., for their technical assistance; to Roger F. Canaff, Special Testing and Research Laboratory, for his review of this manuscript; and to Mrs. Lani Hidalgo for typing the final manuscript. The NMR spectra were obtained from Theodore C. Kram.

REFERENCES

1) S.P. Sobol and A.R. Sperling, ACS Symposium Series, 1975, 13, 170.
2) M. Polonovski and M. Polonovski, Bull. Soc. Chim. Fr., 1927, 41, 1190.
3) J.C. Craig, F.P. Dwyer, A.N. Glazer and E.C. Horning, J. Am. Chem. Soc. , 1961, 83, 1871.
4) M. Polonovski and M. Polonovski, Compt. rend., 1927, 184, 331.
5) M. Polonovski and M. Polonovski, Bull. Soc. Chim. Fr., 1927, 41, 1186.
6) C.N.R. Rao, Ed., Chemical Applications of Infrared Spectroscopy, Academic Press Inc., London, 1963, p. 259.
7) W. Fulmor, J.E. Lancaster, G.O. Morton, J.J. Brown, C.F. Howell, C.T. Nora and R.A. Hardy, J. Am. Chem. Soc., 1967, 89, 3322.
8) T. Rull, Bull. Soc. Chim. Fr., 1962, p. 586.
9) T. Rull and D. Gagnaire, Bull. Soc. Chim. Fr., 1963, p. 2189.
10) N.S. Bhacca and D.H. Williams, Eds., in "Applications of NMR Spectroscopy in Organic Chemistry-Illustrations from the Steroid Field", Holden-Day Inc., San Francisco, p. 77.
11) M.H. Studier, M. Klein and L.P. Moore, in "Chemistry of Morphine and Related Compounds", Drug Enforcement Administration, U.S. Department of Justice, Washington, D.C., 1974.
12) H. Audier, M. Fetizon, D. Ginsburg, A. Mandelbaum and R. Rull, Tetrahedron Lett., 1965, 13.
13) A. Mandelbaum and D. Ginsburg, Tetrahedron Lett., 1965, 2479.

14) D.M.S. Wheeler, T.H. Kinstle and K.L. Rinehart, J. Am. Chem. Soc., 1967, 89, 4494.

SCREENING BY GAS CHROMATOGRAPHY-MASS SPECTROMETRY FOR METABOLITES OF FIVE COMMONLY USED ANABOLIC STEROID DRUGS

R.J. Ward, A.M. Lawson and C.H.L. Shackleton

Division of Clinical Chemistry, Clinical Research Centre, Watford Road, Harrow, Middlesex HA1 3UJ, England

SUMMARY

A gas chromatographic-mass spectrometry (GC-MS) technique is described for the detection and identification of anabolic steroid drugs in urine samples of athletes competing in major international events.

The metabolism of some commonly used drugs, in particular, Dianabol (17α-methyl-1,4-androstadien-17β-ol-3-one), Stanazolol (17α-methyl-5α-androstan-17β-ol-[3,2-c]-pyrazole), Nilevar (17α-ethyl-4-estren-17β-ol-3-one), Orabolin (17α-ethyl-4-estren-17β-ol), Durabolin (4-estren-17β-ol-3-one phenylpropionate) have been studied. The approach to processing the data from the GC-MS analysis does not preclude the detection of other additional anabolic steroid drugs.

INTRODUCTION

Self-administered anabolic steroid drugs have been used increasingly by athletes over the last ten years in attempts to improve performance. Reports about the side effects of such drugs and concern of unfair advantage to the athlete taking them resulted in their use being banned in 1974 by various sport federations and presently by the International Olympic Committee. Unlike the stimulant drugs, these anabolic steroids are taken in the months prior to major competition and not necessarily the day of the event. In order to effectively implement the rules governing their misuse, it has been necessary to develop methods which could initially screen urine samples from athletes for these drugs and

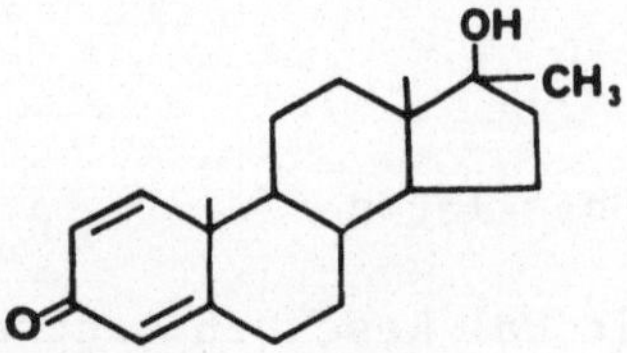

17α-methyl-1, 4-androstadien 17β-ol-3-one (Dianabol)

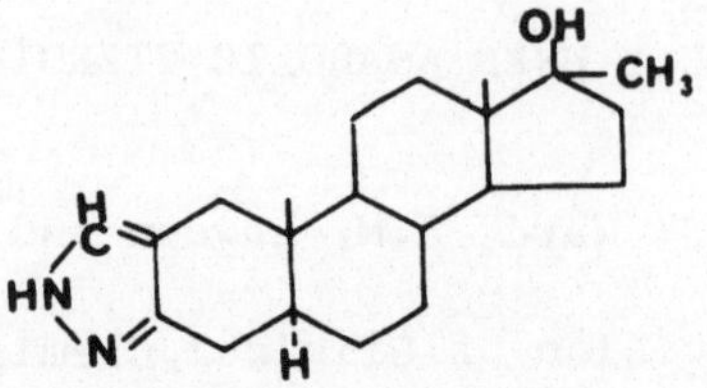

17α-methyl-5α-androstan-17β ol-(3., 2-C)-pyrazole (Stromba)

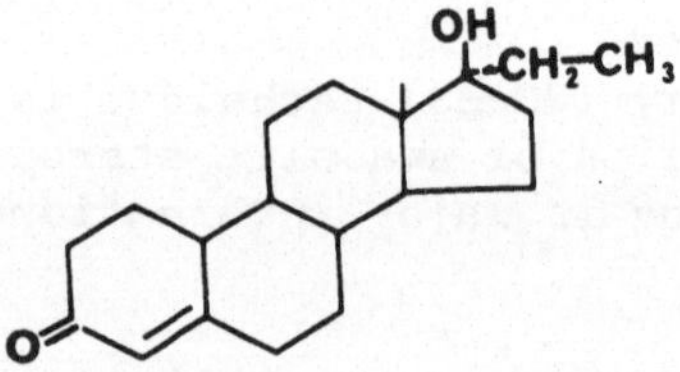

17α-Ethyl-4-estren-17β-ol-one (Nilevar)

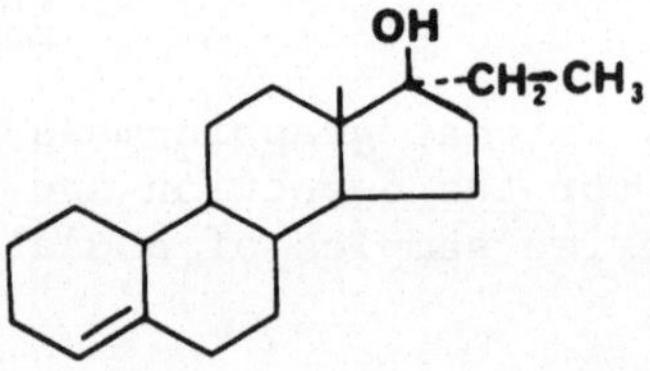

17α-Ethyl-4-estren-17β-ol (Orabolin)

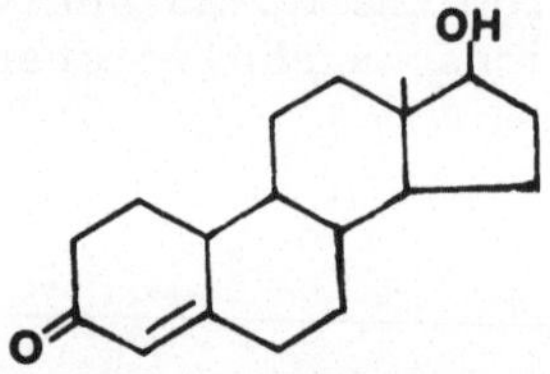

19-nortestosterone

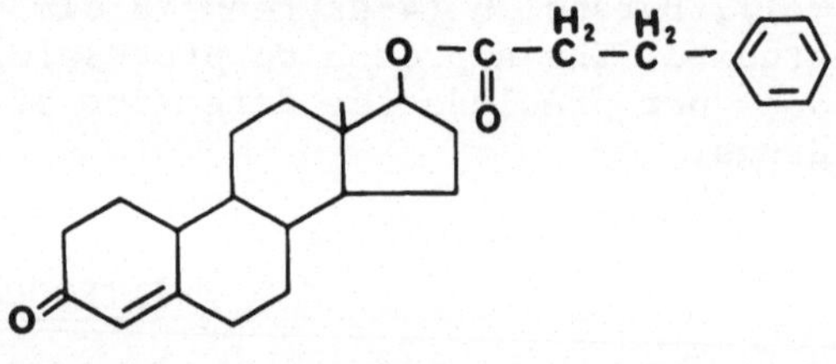

Nandrolone phenylpropionate (Durabolin)

FIG. 1. Structure and systematic names of the anabolic steroid drugs screened.

then in any positive case, unequivocally identify the drug by gas chromatographic-mass spectrometry (GC-MS) analysis.

Only a small amount of anabolic steroid drug is excreted in the urine unchanged, the major portion being metabolised and conjugated with glucuronic or sulphuric acid before excretion in the urine.

Initial studies with Dianabol resulted in the development by Prof. R.V. Brooks of a radioimmunoassay which was suitable for screening for the 17α-methyl and 17α-ethyl groups of anabolic steroids in the urine of athletes (1). Antiserum has recently been raised which is specific for the 19-*nor* compounds.

In the present study the technique of GC-MS has been used as a means of identifying the anabolic drug in samples which have given a positive response to the radioimmunoassay test. These studies have necessitated (2) an investigation into the metabolism of the anabolic steroids most likely to be used in sport, and to establish the most specific and sensitive means of applying GC-MS methods. It has been important to retain in the method the capability of identifying metabolites of previously unencountered drugs.

Urine samples from athletes, competing in European and World championship sporting events have been analysed. Initially urine was screened for Dianabol, Stanazolol, Nilevar, Orabolin and Durabolin (Fig. 1). In the 19-*nor*-testosterone ester class (e.g. Durabolin) there are several drugs which are metabolised in similar manner.

MATERIALS AND METHODS

Urine samples. The urine samples analysed were from athletes competing in international fixtures in 1975 for which positive results for the 17α-methyl or 17α-ethyl containing anabolic steroids had been given by the radioimmunoassay screen. An investigation of the anabolic drugs was made following their administration to male volunteers (permission for these investigations to be carried out was obtained from the Northwick Park Hospital Ethical Committee). Urine samples were collected for several days after administration.

Extraction and derivative formation. The analytical procedures used for the recovery and purification of the steroids from urine have been described previously (2). Trimethylsilyl (TMS) ethers were prepared of the steroids prior to gas chromatographic analysis.

Gas chromatography (GC). Preliminary GC analysis was performed on packed glass columns (6' 3% OV-1) in a Pye 104 instrument, using temperature programming (200-270°C at 2°C/min) with an injector temperature of 250°C and 40 ml/min argon carrier flow rate.

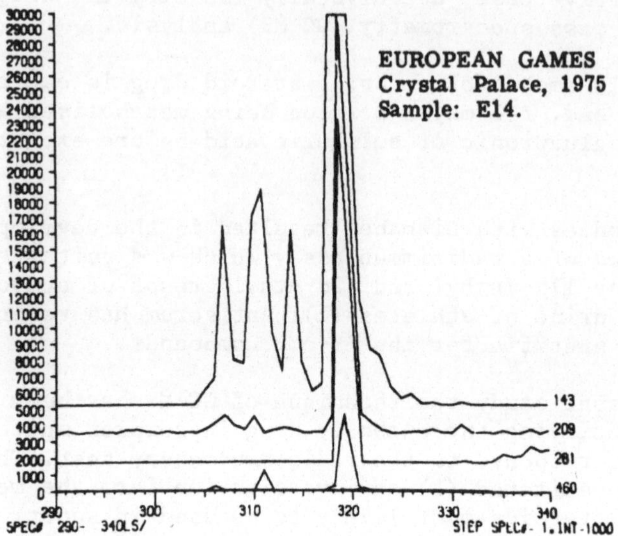

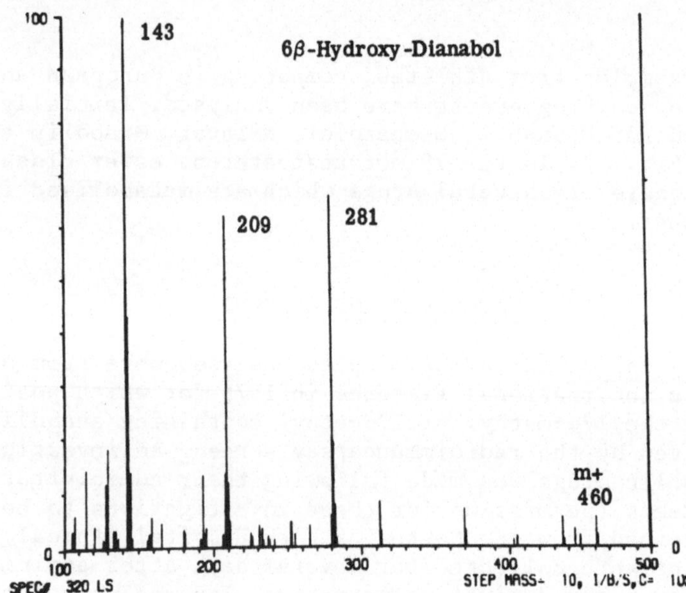

FIG. 2. Data retrieval following repetitive scanning. Selected ion plot (top figure) for ions characteristic of the major urinary metabolite of Dianabol, 6β-hydroxy-Dianabol. Complete spectrum (lower figure) of scan 320 in which all selected ions are peaking.

Gas chromatography-mass spectrometry. The two GC-MS systems used were : (a) A Varian MAT-731 high resolution mass spectrometer operated at 1000 RP, 8 KV, 70eV and source temperature of 250°C. It was connected to a Varian 2700 series GC by a dual stage Watson--Biemann separator (250°C). The GC contained a 6' 3% OV-1 packed glass column with an injection temperature of 250°C, and a helium carrier gas flow of 20 ml/min. (b) A Varian MAT-112 employing 0.8 KV, 70 eV, a 250°C source temperature and coupled by a slit separator (275°C) to a Varian 1400 series GC. A packed glass 6' 3% OV-1 column and helium carrier gas at a flow rate of 20 ml/min was used.

Selected ion monitoring of up to six ions covering the complete mass range was possible using the multiple ion selection (MIS) facility of the MAT-112 in the magnetic field switching mode.

Both of these mass spectrometers were connected on-line to a Varian SpectroSystem 100 MS. Data was processed by this system and off-line by an ICL 1903 computer. Several programs have been written for use with the latter system, including programs for the selection of intense and maximising ions (3). These will be described in detail elsewhere.

The derivatised extracts of all urine samples were initially analysed by respective scanning techniques using the MAT-731 instrument. Scans were made at 8 second intervals throughout the elution period of the sample. As the metabolism of Dianabol, Nilevar, Stanazolol, Orabolin and Durabolin has been previously studied (2) and the major metabolites identified, characteristic ions for the detection of a drug or its metabolite were known. The repetitive scan data was then searched for these characteristic ion patterns and full spectra printed of the scan number of any positive response.

Thin layer chromatography (TLC). In cases where greater sensitivity was required, TLC on Kieselgel HF 254 plates (20 cm x 20 cm x 0.25 mm, $CHCl_3$:MeOH, 90:10) was used to purify and concentrate extracts. Compounds of interest were eluted from the Kieselgel by methanol and sonication. Eluates were blown to dryness (N_2, 60°C) and the residue derivatised (HMDS/TMCS).

RESULTS AND DISCUSSION

Dianabol. 6β-Hydroxy-Dianabol, the major metabolite of Dianabol, was detected in the urine of several athletes. Fig. 2 shows the response from the repetitive scan data for the selected characteristic ions of 6β-hydroxy-Dianabol m/e 143, 209, 281 and 460. Conclusive evidence of the presence of the metabolite was then obtained by requesting a complete spectrum of scan number 320. The unchanged drug and other minor metabolites of Dianabol

(4,5) have also been observed. In the present study the 6β-hydroxy metabolite was always excreted in greater amounts than other metabolites and hence its detection provided the most sensitive means of determining possible drug ingestion.

Until now, positive results for anabolic steroid drugs in the urine of athletes have been reported to the particular Sports Federation in the form of a mass spectrum of the drug or its metabolite obtained after repetitive scanning techniques. However, if selected ion monitoring was considered acceptable as evidence for the identification of the steroid drug, much greater sensitivity can be obtained. Some initial preselection of the ions is possible from a knowledge of the specificity of the antisera with which the positive sample has reacted. The ions for each particular drug containing that grouping can then be selected in turn (complete mass range) on the Varian MAT-112, and the sample is monitored isothermally. No processing of data is necessary and each analysis takes only a few minutes. The disadvantage in this approach is that to obtain a full mass spectrum the sample has to be re-run with the data processing system.

Although greater sensitivity is possible using the Varian MAT-731, only 4 ions within 20% of the lowest mass can be monitored simultaneously with the present peak switching unit. Although m/e 143 is the base peak in the spectrum, this is a relatively common ion in the spectra of trimethylsilyl derivatives of endogenous steroids which makes it unsuitable for use in the absence of considerable purification of the extract. By comparison m/e 209 gives much greater selectivity and although the relative intensity is only 30%, it is the ion of choice for single ion monitoring (2). The ultimate detection limit has not been measured, but samples containing 350 pg give a readily detectable response (signal: noise ratio, 100:1).

Nilevar and Orabolin. Nilevar and Orabolin were principally metabolised by A-ring reduction, to tetrahydro-Nilevar (m/e 450 [M^+], 157 [base], 360, 270 and 144), by hydroxylation of the ethyl side-chain and reduction to form a 19-*nor*-pregnanetriol (19-*nor*- 5ξ,17α-pregnane-3ξ,17β,21-triol). The mass spectrum of this compound has been reported previously (2) and has characteristic ions at m/e 538 (M^+), 245 (base), 448, 421, 331, 241, 217 and 103. Unfortunately, spectra of this compound obtained following repetitive scanning included ions at m/e 157 and 144, ions characteristic of an unhydroxylated ethyl side-chain. However, the employment of "maximising ion" criteria (6) resolved this dilemma since summed ion intensity plots separated the 19-*nor*--pregnanetriol from an endogenous steroid 5-pregnene-3β,16α,20α--triol, the source of the m/e 157 and 144 peaks. A selected ion plot of m/e 245 from the maximising ion data showed a similar increase in resolution from selected ion plots from unprocessed data (Fig. 3). Metabolites of Nilevar and Orabolin have not yet

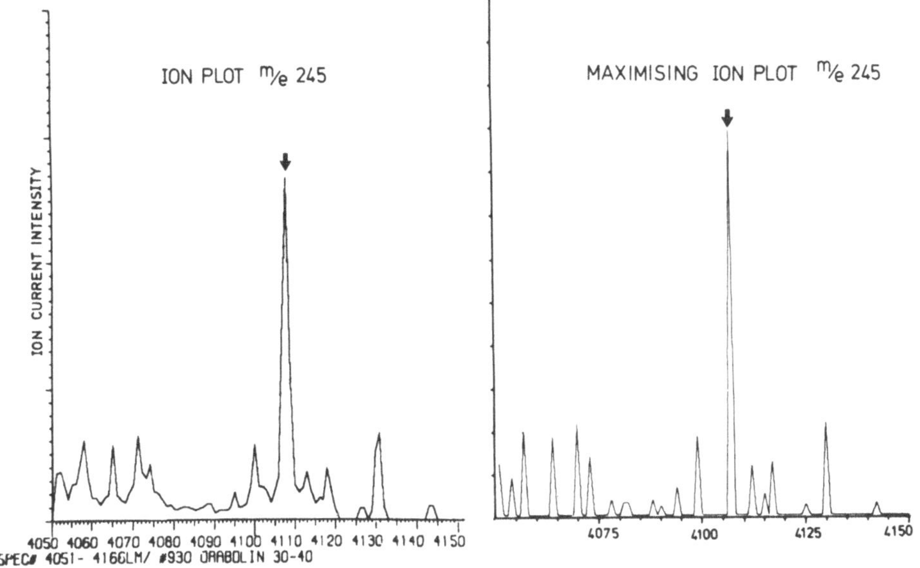

FIG. 3. Ion plot of m/e 245 obtained from repetitive scan data of a urine extract containing Orabolin metabolites. The plot on the right hand side shows the improved apparent resolution by applying 'maximising ion criteria" to the data. The arrow indicates the location of the major metabolite of the drug.

been detected in urine of athletes.

Stanazolol. The gas chromatographic properties of the trimethylsilyl derivative of this drug are poor due to its absorption on the column. It also gives an unfavourable mass spectrum with a base peak at m/e 143 and no intense high mass ions. These facts coupled with its low level of excretion in urine make it difficult to detect. Nevertheless, it has been successfully identified in a urine sample from an athlete competing in a major event. This was achieved using both repetitive scanning and selected ion monitoring after subjecting the sample to prior concentration on TLC.

Further investigation of the metabolism of Stanazolol has been made by administering a randomly labelled radioactive form of the

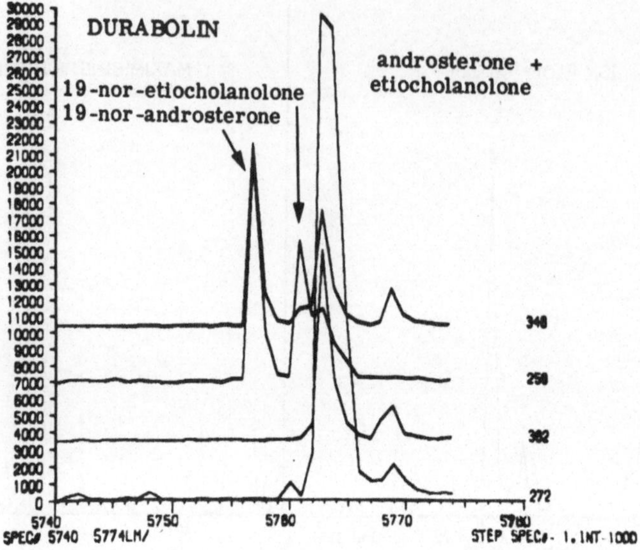

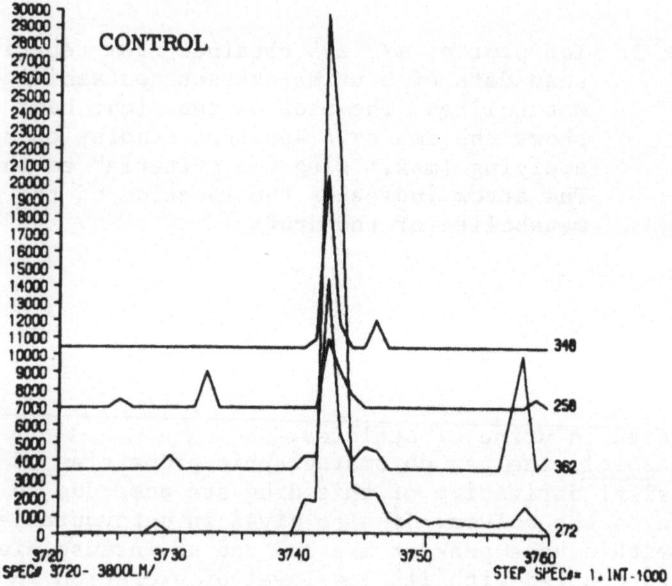

FIG. 4. Data retrieval following repetitive scanning. Selected ion plots (top figure) to differentiate endogenous androsterone and etiocholanolone (m/e 362 and 272) from major Durabolin metabolites 19-nor-androsterone and 19-nor-etiocholanolone (m/e 348 and 258). Lower figure shows ion plots from undrugged control sample.

drug to a marmoset monkey. Only 16% of the radioactivity was excreted in the urine in the first day while 40-60% was excreted in the faeces. On the 2nd and 3rd subsequent days, 0.9% and 0.4% were detected respectively in the urine. Similar results have been obtained (7) in man. This observation supported the contention that the low level of urinary excretion in man may be due to the drug's elimination via the hepatic portal system.

Durabolin. Unlike the drugs previously mentioned, Durabolin is administered intramuscularly and is readily detected in urine many days after ingestion, possibly due to more complete adsorption. Currently, it has not been detected in athlete samples.

However, the two major metabolites were 19-nor-etiocholanolone and 19-nor-androsterone. The spectra of these compounds are very similar but ion intensity differences and their retention time permits their distinction. The ions used for their detection are m/e 129, 239, 257, 258, 333, 348.

The metabolites of Durabolin can be distinguished from the endogenous steroids, androsterone and etiocholanolone by selecting the two ions m/e 348 and 258 characteristic for 19-nor-androsterone, 19-nor-etiocholanolone and the m/e 362 and 272 androsterone and etiocholanolone. If this sample is compared with a normal control the peaks produced by the drugs metabolites are clearly observed (Fig. 4).

CONCLUSIONS

The metabolism of the five drugs, Stanazolol, Dianabol, Nilevar, Orabolin and Durabolin has been studied and the metabolites identified.

Using repetitive scanning techniques we have obtained unequivocal proof of identity of the administered drug in urine samples received from athletes competing in major international competition. The alternative method of selected ion monitoring as a means of identifying drug metabolites has been investigated and despite its limited selectivity has superior sensitivity and speed of analysis. Nevertheless, with the repetitive scanning method, all data is retained which allows a fully comprehensive analysis and permits other steroid drugs, as yet uninvestigated, to be identified at a later stage.

Acknowledgments. The work was supported by a grant from the British Sports Council. We are also indebted to the Medical Research Council for allowing their facilities at the Clinical Research Centre to be used for this study.

The authors wish to thank Mr. M. Chu and Mr. M. Madigan for technical assistance with mass spectrometry, also Mr. A.C.S. Thomas and Mr. M. Bhasin for computer programming.

REFERENCES

1) N.A. Summer, J. Steroid Biochem., 1975, 5, 307.
2) R.J. Ward, C.H.L. Shackleton and A.M. Lawson, Brit. J. Sports Med., 1975, 9, 93.
3) A.C.S. Thomas, M. Bhasin and A.M. Lawson, unpublished results.
4) A.M. Lawson and C.J.W. Brooks, Biochem. J., 1971, 4, 25.
5) M. Donike, Sportarzt und Sportmed., 1975, 1, 1.
6) J.E. Biller and K. Biemann, Anal. Lett., 1974, 7, 515.
7) R.V. Brooks, personal communication.

MASS SPECTROMETRY IN ACUTE INTOXICATION DIAGNOSTICS: ASPECTS AND EXPERIENCES

G. Brante, S. Jönsson and A. Andersson

Centrallaboratoriet, Sit Lars Stukhus, Fack, 220 06

Lund 6, Sweden

THE ADVANTAGES OF USING MASS SPECTROMETRY IN EMERGENCY TOXICOLOGY

In the last 5 years or so the usefulness of mass spectrometers in emergency toxicology has repeatedly been stressed. It this justified? This question should be subdivided into:

1. Are chemical analyses of toxic agents of considerable value in emergency toxicology?
2. Are methods employing mass spectrometry (MS) of any advantage in such analyses?

1. Opinions regarding the value of the analyses are controversial as illustrated by the published statements given in Table 1, which represent two current schools, A with a positive, B with a negative attitude towards emergency drug screening.

Support for the A school are to be found in several articles and letters (3-9). The experiences of the first four groups are much the same as those of McCarron's *et al.* (1) and are based on frequent meetings or on enquetes with emergency room physician: "As stated by others (1), and verified by our experience, there are two schools ...... In our experience, the first group (=B) appears to be diminishing as they find our results entirely accurate and timely" (3); "... many physicians find an emergency drug screen service extremely valuable" (4); "The attitude of the physicians with whom we have worked parallels much more closely the sentiments expressed by McCarron and her coworkers (1)" (5); "The requested analyses we do are generally appreciated and occasionally are very important in determining the patient's care" (6). Green *et al.* (8) make the statement: "In emergency toxicology unequivocal identification and quantification of toxic agent(s)

TABLE 1.
Excerpts from two answers to the same inquiry: Are emergency toxicology measurements really used?

A. McCarron *et al*. (1)	One group (of clinicians): Therapy is not altered by toxicological data. The other group: correct diagnosis and management is aided greatly ... We have observed ... transitions of hundreds of clinicians from the former to the latter group owing ... to ... availability of clinical toxicology laboratory testing of high quality.
B. Wiltbank *et al*. (2)	In our experience the *emergency* toxicology screen appears to be providing little information that would justify its 24-h availability.

and speed are essential for the institution of rational therapeutic measures"; Teitelbaum *et al*. (9): "The result of this study support the need for twentyfour hour a day, on-line, analytical toxicology services, and refute the claims of physician impression as a satisfactory diagnostic tool in clinical toxicology".

In keeping with opinion B there is the discussion by the poisoning treatment expert Newton (10): "What advantage would it be to know what the drugs are?" In the same book, Ciba Foundation Symposium 26 (new series) (11), a lot of similar attitudes are exhibited, to be sure intermixed with attitudes of type A. Very recently results by Shipe and Savory (58) were thought to "indicate that use of GC-MS for routine toxicology does not significantly increase the quality of the service."

It seems rather obvious that the advocates of type A largely belong to the laboratory profession and those of type B to the treatment clinicians. Thus we can suspect a certain professional subjectivity but there is certainly also experience behind the positions: many clinicians have been disappointed by incomplete, misleading or delayed answers from the toxicological laboratory and certainly may have reason to be wary even to-day (12). They also point to the paucity of specific treatments available. On the other hand, the advocates referred to as type A obviously are leaders of, or have access to, modern highest class laboratories, most of them utilizing MS and impressed by their stimulating experiences.

As to the magnitude of the problem it can be calculated that in the western world 1-2 per thousand inhabitants will try suicide by drugs per year. If alive on arrival to the hospital some 98-99% of the victims will survive. There is some variation depending on the drug involved and the statistics consulted (Table 2). The stay

TABLE 2.
Death frequency in hospital of acute overdose cases:

	%	
Barbiturates, short-acting	0.5	Average ≤2.5
Barbiturates, long-acting	2.0	
Methaqualone	∿ 3.0	
Tricyclic antidepressants	0.7	
Neuroleptics	∿ 1.0	
Paracetamol	3-5	
Propoxyphene	(30?)	
Other drugs	≤ 1.0	

(Data mostly obtained from references 11, 13, 14).

in hospital is said, in most cases, to be only of one or of a few days, the average being about 3 days. This, of course, may be much longer in individual, serious cases.

The frequency and severity of intoxications now seem to be decreasing considerably, at least in Sweden (13). This is said to be due to a more restrictive prescription policy, as regards dangerous drugs, and improved patient information . A similar tendency is believed to occur or be near in other countries.

In a new case suspected of intoxication the emergency room physician may reasonably need or want to know: if there are poisons in the patient, what they are,whether they are present at a toxic level, and sometimes their more exact concentrations. Primarily he tries to support the patient's life disregarding the cause of his state, but he also wants to remove any poison present and to counteract its negative effects in the body or, alternatively, in intoxication-mimicking non-poisoned cases, to remit the patient to other proper treatments, e.g. surgery.

The clinician's instruments for decisions in poisoning cases can be briefly summarized as follows: a. Direct informations (by patient, relatives etc), b. Confiscated matters (pills, drug containers etc), c. Clinical symptoms (coma, miosis, convulsions, arrhytmias. Comatose patients less than 50 years of age are considered with high probability to be intoxication cases), d. Laboratory examinations (physiological, such as blood pressure, ECG, chemical etc).

A majority of the cases is apparently easily diagnosed and/or managed without chemical analyses, or with only simple, suspicion--directed ones, although occasional mistakes, sometimes harmful to the patients, are claimed then to occur (9). In the remaining more complicated and/or serious cases of unknown cause many clinicians claim that chemical analyses will influence their management only in those cases in which a more or less specific therapy exists and is needed. Such cases include: Intoxications by salicylates, long acting barbiturates, tricyclic antidepressants, which may require haemodialysis (15) or charcoal hemoperfusion (16), i.e. more laborious measures; by paracetamol, propoxyphene and others, for which "specific" antidotes are tried (although the gain might partly be doubted, cf 17); by amphetamine, phencyclidine a.o., which may require pH adjustments in the body for enhancement of excretion; by iron and methanol, which may require ferredoxamine and ethanol, respectively, in their treatment. And fascinating therapeutic opportunities with specific antibodies are forthcoming (Smith, et al.,(59)). In such cases, the chemical analyses could be said to be able to save lives but such cases are less than 1% of all cases admitted alive to hospital. Therefore, from this viewpoint the opinion of the restrictive clinicians may be understandable. However, other gains may be possible, as summarized in Tables 3 and 4.

TABLE 3.
Cases when comprehensive-exact-rapid-repeatable systems for toxic substance analysis in (suspected) overdose cases are urgent for decision making:

1. Effective specific antidote is available.
2. Special, although unspecific, treatment is advantageous (e.g. dialysis).
3. New kinds of treatment are to be tried.
4. Specific treatments are forthcoming.
5. Specific treatment is overweighted but unnecessary.
6. There is no intoxication a) of a certain kind or b) at all.
7. Prognostic information is urgent and possible to obtain.
8. Intoxication pattern in society is changing, e.g. new intoxication types appear.

TABLE 4.
Aims when employing analyses in overdose cases:

For the patients:	Increased chance of survival Shortened disease time Fewer complications Less severe complications Fewer after-effects
For the doctor and coworkers:	Safer handling Better utilization of resources Increased knowledge
For the society:	Saving of resources New or better prophylactic possibilities

A common and very important problem is (point 6 in Table 3) to be able to rule out intoxication in cases where another pathological process mimics it, not the least in the case of children (18). This requires especially comprehensive and accurate analytical systems to achieve diagnostic certainty. It is also of great advantage to be prepared for new management possibilities (point 3, 4 and 8 in Table 3) by having adequate diagnostic facilities and experience available.

Table 4 summarizes the benefits to be derived for the various people concerned in intoxication and, in principle, obtainable in some situations by chemical analyses. Finally, as pointed out by Bickel (7): "There can be little doubt, however, that a treatment of intoxications based on pharmacokinetic knowledge is superior to a treatment guided by vague clinical impressions. Furthermore the success of measures aimed at favourably altering pharmacokinetics can be objectively assessed by means of drug analyses which then can serve as a feed-back for the improvement of treatment." It may, indeed, be considered remarkable if, contrary to those in other fields, the doctors involved in intensive care of poisoning should not desire to use all relevant data obtainable about what they are trying to cure.

To sum up, the answer to subquestion 1 is that there are clinical needs for emergency toxicological chemical analyses although in some cases these needs are not currently acknowledged by all. If new specific therapeutic measures are developed then the needs may be increased. Ideally the analytical system should be rapid, comprehensive, accurate and precise enough and available on a 24-hour, 7-day-week basis. (In non-emergency situations the importance of chemical analyses for clinical, forensic, epidemiological and research purposes is well recognized).

2. The answer to subquestion 2 about the preferability of MS, depends on the existence and effectiveness of nonMS-alternatives for the analyses concerned. Table 5 shows such alternatives. They

TABLE 5.
Conventional (non-MS-) systems for analyses in emergency toxicology.

1. Group separation by extractions
2. "Spot tests" - simple test tube, largely colour reactions
3. Photometric methods: visible and UV light, fluorimetry
4. Thin layer chromatography methods
5. Gas chromatography methods using FID-, EC-, and N-detectors
6. Immunoassay methods
7. Combinations of 1-6

Prerequisites: existence of simple, rapid, reasonably specific versions.

(References: 19-22).

have been used and continuously improved for a long time. In high class laboratories they may certainly fulfill their function (cf 1, Table 1). However, they are not of a generally and positively diagnostic nature as are the MS-alternatives. Apart from the possibility of overlooking some metabolites - especially fatal when the original drug is not present in the sample - these methods suffer from the drawbacks listed in Table 6. Toxicologists using conventional methods often feel a need to confirm their findings by MS and an increasing number are achieving this. On the other hand some of the non-MS methods may even be preferable to MS-systems and be used in a complementary fashion. If the nature of the toxic substance is strongly suspected these methods are highly sensitive, and specific. Examples may be certain colorimetric tests for salicylate, paracetamol a.o., direct thin-layer application of serum for carbamazepine quantitation and some other TS and GC methods (e.g. for alcohols and some narcotics). These are very rapid, simple to perform and may be reliable enough even in untrained hands. The future of the immunoassay methods is hard to evaluate for the present - they may develop to be of great utility but most probably only in those cases where there are strong suspicions of the chemicals involved and even so for the detection of a single substance or some group of drugs, but not when general screening is desired.

Table 7 illustrates some current trends in intoxication

TABLE 6.
Shortcomings of conventional systems.

1. Ambiguous results and risk of misjudgement e.g. when two substances do not separate and specific reagents are not available
2. Appropriate rapid methods are still lacking for several toxic substances and metabolites
3. The time and personel requirements are often inappropriately high due to a primary need for application of several discrete methods, and a secondary need for confirmation of the primary results by other methods

TABLE 7.
Trends in intoxication patterns and advantageous qualities of MS-systems.

Drug effect increases = the drug concentration in the samples decreases, the drug number increases - i.e. analytical systems must become more sensitive, specific and comprehensive

MS-systems can afford:

High sensitivity	for identification, quantification
High specificity	due to scans, exact R_t and ideal internal standards
Comprehensiveness	still uniform and simple and progressive (one extract, one column and no derivatization is usually needed)
Rapidity	due to temperature programming and use of computers

patterns together with some impressive advantages of the MS-systems. For the latter the possibility of including a great number (hundreds) of common toxic agents into one single rapid procedure which is sensitive and specific enough and capable of giving a rough quantification should be emphasized. This has been expressed in the way that the MS-system is generally diagnostic and this on a positive identification level for chemical individuals (23).

Some of the weaknesses inherent in present MS-systems are

listed in Table 8. By capacity limitations is meant the fact that

TABLE 8.
Difficulties with current MS-systems.

Costs (purchase, running)
Management (training, service requirements, down-time)
Capacity limitations
Quantitation problems
Contamination problems
Gaps of knowledge and experience
Time schedule problems

there are some kinds of toxic substances, which are not easily analysed by MS methods, e.g. very volatile or non-volatile or column-adhering substances. Other problems arise due to pyrolysis, further adsorption phenomena, extraordinary fragmentation, uncharacteristic fragments of some substances and so on. Among the "difficult" substances the following have been mentioned: alcohol, ethchlorvynol, chloral hydrate, polar drugs of high molecular weight (e.g. cardiac glycosides, many antibiotics), quaternary salts, inorganic poisons (e.g. As, CN), LSD, mushroom and botulinus poisons, and drugs not easily dissolved in organic solvents such as diphenoxylate, furosemide, phenylpropanolamine chlorazepate, benzoylecgonine (3, 4, 18, 24). By use of special measures most of them can however be analysed by MS methods. The other pitfall also can usually be circumvented by various instrumental and methodological measures some of which will be mentioned in section II.

By quantitation problems is meant that the recoveries of the many different drugs are highly variable in the compact extraction procedure used, and it is impossible to include a sufficient number of appropriate internal standards in the initial analysis. Sometimes for more reliable quantification to be obtained, the sample has to be run repeatedly, with appropriate internal standards included. Methods other than MS may also have to be employed. Many endo- and exogenous compounds occurring in biological samples and detected by MS are still unidentified and many MS scans present interpretation difficulties. These problems reflect the fact that much remains to be learnt before the full potential of MS systems can be exploited. The running costs of MS may ultimately not be higher than those of the current non-MS systems. Laboratory technicians can be trained to carry out these analytical procedures. Finally, the necessity to interrupt other projects which require

MS for the occasional but urgent acute intoxication analysis may be disturbing and demand strict rules of order to be established. To sum up the points regarding question 2, despite some present weaknesses, the MS-systems have properties which make them most attractive as central analytical units in emergency toxicology. They could advantageously replace most, but not all, "conventional" non-MS methods currently used for similar purposes.

MS-SYSTEMS USED IN EMERGENCY TOXICOLOGY

General information on the use of MS in chemical analyses is available from several recent articles (25-28). Of the many reviews available on MS, a number deal with its application to problems in toxicology (3, 4, 24, 29, 30).

In Table 9 are included the places and names of members of

TABLE 9.
Locations of MS groups active in the toxicology field.

Country	Number	Location
USA	10	Bethesda (31) Salt Lake City (29) Baltimore (32) Boston (24) San Francisco (33) Palo Alto (8) Washington (34) Columbus (4) Gary (35) Rhode Island (60)
England	1	Middlesex (27)
Canada	1	Toronto (36)
West-Germany	2	Saarland (37) Bremen (38)
Norway	1	Oslo (39)
Sweden	∿ 2	Stockholm (30 , 40) Lund (41)
Total number	17	

laboratory groups active in the field, which I have been able to collect from the literature. I realize that the list cannot be complete, and possibly it is incorrect in some details. The dominance of USA in the activity hitherto is obvious, however. It is of interest to note that most of laboratory leaders included do not belong to the medical profession, although often they are more or less intimately connected with clinicians. This fact, beside the economic one, might have contributed to the resistance to the development in the field.

The MS-instrumentations and -strategies used have been highly variable. I will give a brief presentation of these emphasizing some recent approaches.

1. Table 10 contains a concise list of the instrumentations

TABLE 10.
Mass spectrometry instrumentation.

MS type	magnetic sector, single or double focusing quadrupole, dodecapole tandem
Inlet system	direct introduction through gas chromatograph (GC),i.e. GC-MS through liquid chromatograph (LC)
Interface	frit, jet or membrane separator
Ion source	electron impact ionization (EI) chemical ionization (CI) field ionization (FI) field desorption (FD) atmospheric pressure ionization (API)
Ion collector	ion current measurement or ion counting
Computer (COM)	dedicated mini-computer with accessories and programs

which has been utilized, proposed or discussed. Most workers have used low resolution instruments. Interesting possibilities are, however, inherent in high resolution MS (38). Polypole instruments because of their more rapid scanning power might have a certain advantage if capillary tube GC is to be used. Use of tandem MS has been proposed by Szabo (42) because of the possibility of selecting the chemical ionization source (see also 43). It has, however, not yet been tested on real life intoxication samples, and is somewhat more expansive than the simpler MS instrument.

Direct introduction of sample was employed by Milne, Fales and Axenrod (44) (the early Bethesda group, Table 9), Billets *et al.*(32) and Boerner *et al.* (33). The second group still preferred it in 1975 (6). Milne et al. (44) used CI while Fenselau (6) used EI. The Boerner group (33) applied a unique "chemical vapor analysis" system with a membrane separator as a direct sampling inlet allowing the analyses to be done very rapidly (2 mins).

The other groups in Table 9 have used GC-MS with conventional packed columns. Capillary columns are presently being discussed and tried and may have some advantages and further possibilities to offer (45) but perhaps also disadvantages due to unnecessary elaborateness. LC might have a future in the field, possibly in combination with API. However, it has not been tried in toxicological screening yet, as far as I know. The time for the GC-MS run is normally between 15-30 mins.

In the type of ion source used the trend seems to be turning over from EI to CI (4) or, better, to use both. FI or FD or API (46) are to my knowledge still untried in the field. An ion counting technique has been used by Abramson (34), other workers have used ion current measurement.

A number of different computers with various capabilities have been used with a preference for the on-line variety. The common fundamental features required are: 1. Collection and storing of the data. 2. Calculations such as background subtraction and normalization, and of the concentration level. 3. Control of operation. 4. Presentation of data. (See also Table 12).

The hardware in our laboratory in Lund is listed in Table 11.

TABLE 11.
Hardware S:t Lars Lund.

GC	Pye 104 Temp programmer, packed columns
MS	LKB 2091 Mass marker LKB 2091/110
COM	Alpha LSI-2 24K (16 bits)
DEC writer LA-36	
Disc	Pertec D 3341 2.5 M words (16 bits)
Display	Tectronix model 603
Plotter	Houston DP 1
MID, 4 channels, Altema	
Magnetic tape recorder	

It combines the LKB 2091 GC-MS instrument with a computer system purchased from a Swedish firm, Altema. The various accessories to the CPU represent the capacity desirable in a well-developed system which allows, among other things, an efficient application of MS methods to problems in toxicology (see Table 12). As an

TABLE 12.
MS procedures.

Running alternatives:

DI-MS or GC-MS:
Scans on selected TIC peaks or
Repetitive scans
Library search
Mass chromatography (MC)
Mass fragmentography (=SIM, selected ion monitoring,(57))

Computer utilization:

Data collection, intensity summation, retention time (R_t) or index (R_i) registration
Display and selection
Plotting of TIC, MS, MC, SIM
Library search: "Forward"
"Reverse"

option - mostly for drug monitoring in therapy - we have a hardware multiple ion detector (MID), which of course is useful also in toxicology MS. Altogether our equipment, purchased between 1974-1975, has cost $ 150,000, about half of which was spent for the computer + programs. The service has almost entirely been performed by our electronics engineer (Anders Andersson). The down-times have been very short and infrequent (in average less tha 1 day/month) during our 2½ years at work.

2. The procedures generally used are summarized in Table 12. I have already mentioned a little about DI-MS. I only want to add that it might be a possible in the future to use a well-controlled heating system in DI-MS together with a double-focusing instrument and computer to base the analyses on exact mass determinations together with the volatilization temperature (cf. 47).

As yet most workers have used GC-MS and packed columns,

often using OV-17, 1 or 3%. Capillary columns can give cleaner MS making the interpretation easier. On the other hand, it can be foreseen that if used with CI, repetitive scans and high resolution instruments, the time factor will be more critical. On such occasions it might be necessary to rerun the sample one or more times in order to get the desired data.

Initially GC-MS analysis was carried out by manually taking scans on peaks as they successively appear on the total ion current (TIC)-curve during a GC-run. Background scans had to be collected simultaneously, as close to the respective peak as possible. This was sometimes difficult to do in a single run and rather often the GC-MS run had to be repeated.

Manual calculations and presentation of spectra can be very time-consuming. Nevertheless one should be prepared to do it since even if a computer is included this may fail. Furthermore there is also an important educational aspect involved in the task.

With a computer the most effective mode of running is by suitably frequent repetitive scanning over the whole or certain preselected parts of the TIC-curve. The computer collects all the data, and during and after the run it is possible to display or recall those mass spectra, which seem of interest in relation to the TIC-curve. After the GC-run it is also possible to produce mass chromatograms (MCs) for selected fragments, a highly effective procedure to get quantitation as well as further identification (cf also Fig. 5).

Most of the computer features shown in the figure have already been touched upon. I only want to add the importance of using as exact as possible R_t:s or, better, R_i:s for a) further identification and b) managing the computer in quantitation work based on fragments which occur repeatedly in the chromatogram, in MC as well as with SIM.

Finally in this table, the library search is divided into two kinds: The so called forward variant which has been mostly used and the reverse one which seems a most interesting development. The latter has recently been discussed in papers by McLafferty *et al*. (48) and by Abramson (34), and was recently used by Green *et al*. (8), apparently with great success.

There are attractive features in the reverse search which, as I can understand it, ought to be utilized as the primary computerized search in situations such as toxicological MS, where a library limited to 100 compounds often will suffice. In cases where this primary search does not result a forward search is still possible, "manually" or by computer. In the latter case the comprehensive library in a time sharing computer may be used (cf. 29). A rerun on a double focusing instrument, if available, after the use of a low resolution instruments may be used as an ultimate resource.

A manual library search may take a lot of time. Convenient aids are available, however (3, 4, 24, 29), and with experience it may be preferable sometimes to use time-consuming time-sharing search (3, 49).

3. The types of drugs etc, usually found in the intoxication samples are shown in Table 13. It should be pointed out 1. that

TABLE 13.
Drugs and other compounds frequently occurring in samples from intoxication cases.

Hypnotics	
Barbiturates	Amo-, Pheno-, Seco-, and other barbitals
Nonbarbiturates	Methaqualone, Glutethimide, Ethchlorvynol
Analgesics	Acetylsalicylic acid, Paracetamol, Propoxyphene
Anticonvulsants	Diphenylhydantoin
Tranquillizers	Benzodiazepines, Meprobamate, Tricyclic antidepressives, Phenothiazines, Antihistamines
Miscellaneous	Alcohols, Narcotics, Chloral hydrate, Hezapropymate, Fe, Li
Artifacts, Non-Drugs	Phthalates and Adipate, and further plasticizers and vacutainer contents, Fatty acids and -esters, Cholesterol, Squalene (in saliva)

the list is all but complete - in rare cases further types of compounds such as LSD, mushroom-poisons, potassium cyanide, etc. can occur. 2. That each group contains many more compounds than the ones exemplified in the figure. 3. That the poisons often are extensively metabolized and may after a time be present preferentially or only in metabolite form in a sample . Therefore it may indeed be a formidable task to clarify a very complex case of poisoning. Fortunately the common cases are much simpler, more than one toxic substance occurring in only some 40% of the cases. Furthermore as little as a dozen of the substance types may be responsible for more than 90% of the intoxications - which dozen depends on the region: in the USA barbiturates dominate the picture (3), in Lund (Sweden) the benzodiazepines (50).

Altogether some 150 drugs, metabolites and artifacts are encountered in such work (3). The number of original intoxication drugs indicated during several years has been 26 (Billets _et al._ (32)) - 66 (Foltz _et al._ (4)) - 72 (Costello _et al._ (24)) - 73 (Law (3)). Therefore a library of less than 100 selected drugs, metabolites and artefacts is comprehensive enough to cover most of the requirements. However, at least some of the groups (24, 29) have libraries of more than 400 compounds. A large library is an advantage when the problem is to rule out any intoxication at all, certainly an important case. If the case is only to indicate the positive occurrence of one or a few of the more common dangerous drugs it may be of some inconvenience to have a very large library. To be consistent, however, in principle the goal should be to individuate _all_ important toxic substances in _all_ cases. This may perhaps be a possible approach in the future with the availability of extended libraries and refined retrieval possibilities.

4. The _sample materials_ obtained are commonly stomach-contents (particularly when using DI methods, (32)), blood (the most common), urine. Remnants of pills, cerebrospinal fluid and saliva are also examined. The latter is being more widely used in drug therapy monitoring. Examination of saliva may also be important in cases of acute toxicology where, at an early stage, concentrations in saliva may be higher than in other body fluids (51). We have examined saliva in a case of amitriptyline intoxication and found very high concentrations of ami - and nortriptyline, higher than that in serum (52). (Boerner _et al._ (33) could not find toxic substances in saliva in cases of massive overdose but this might have been due to insensitive methods or improper sampling time). Saliva may be collected also in comatose cases, from d. gland. parotis under stimulation. In Sweden we usually do not get samples from stomach contents because of the reluctance of using aspiration in comatose cases.

5. As regards _preparation of the samples_ for MS, Boerner _et al._ (33) successfully tried direct injection of unextracted original samples in the mass spectrometer-inlet. The separator membranes, however, became contaminated so this otherwise attractive mode had to be abandoned in favour of extracts. The extraction procedure can vary from a simple chloroform extraction without pH-adjustment to extracts performed at acidic as well as basic pH, with or without preceeding hydrolysis. Thus, Finkle _et al._ (29) seem to prepare at least four extracts for a sample, a tedious and time-consuming process, probably impractical in acute situations although desirable in non-acute cases and forensic work. A rapid procedure is used by Law (3) and involves mixing the different samples

such as gastric contents, blood and urine in known volumes and extracting with 10% ethanol-methylene chloride at acid as well as basic pH. The extracts are combined, evaporated and the concentrates are injected in the MS inlet. We have used this extraction method but avoid combining the samples. Foltz et al. (4) used similar extractions. Other extraction possibilities including absorption on charcoal or XAD2 resin and "salting" out procedures have been tried with less satisfactory results, however, due to losses of important sample constituents or increased "contamination" with endogenous components etc. (3,4). In our hands the Law extraction procedure takes at least 30 min to perform whereas the method of Boerner et al. (33) requires only 8 min.

Usually in intoxication cases there is enough of the original poison in a free form in the sample to be detectable (cf 4). In the few cases where this is not so it is necessary to include a hydrolysis procedure for its release and detection. This may be done rapidly if the purpose is preferentially qualitative. We found this situation in a case of intoxication from several substances one of which was suspected to be diazepam. This was detectable only after hydrolysis with 6 N HCl for 5 mins. As hydrolysis may destroy other compounds of interest, e.g. tricyclic antidepressants, it should be used only in a second run, after a first one without hydrolysis. Similarly derivatization of the sample is usually unnecessary but may be called for if the preliminary findings are considered incomplete or require confirmation.

We have studied recoveries for our extraction method from urine and serum, and found highly varying recoveries for different substances. This precludes accurate quantitation. The problem can be solved by inclusion of suitable internal standards in a second analysis when we know what drugs are present in the sample. With such a strategy the initial analysis provides a qualitative identification of as many as possible of toxic substances present. A second analysis then can indicate the intoxication level obtained directly from the TIC-curve or from the MC done immediately after the GC-run. If more precise quantitations are requested these may be done subsequently and with separate methods (MS or non-MS, with IS included). Usually it is sufficient to determine if the concentration is of a toxic or only of a therapeutic level (53, 54). A lot of the intoxication cases have taken therapeutic doses of drugs besides those overdosed.

6. Green et al. (8) in Palo Alto have recently published an enthusiastic abstract of their current MS system for emergency toxicology. They describe it as an entirely automated GC-MS--instrument, employing McLafferty's probability based matching (PBM) algorithm, (48), which was found to be eminently suited for the purpose. PBM panels for acidic or basic toxic agents have

been evaluated for their efficiency in screening blood or urine extracts containing unknown toxic substances by comparing the PBM findings with the results obtained by conventional methods. An MS reverse search technique which reduces interferences arising from co-eluting GC peaks and allows more accurate quantitation (by MS) than does GC alone, is coupled with GC retention time screening in addition to the sensing of mass spectral patterns. This system seems to have been composed with care using the latest developments in the field. It will be very interesting to compare their results with those obtained by conventional methods. The authors say that their system readily allowed the detection of additional toxic substances which sometimes were missed by conventional methodology, in cases of multiple drug intake. In Oslo there a similar comparison is under investigation (39).

Miscellaneous additional data. The mass range for the MS scans have usually been 40-450 for EI and 90-450 for CI. The time necessary to deliver the result for a sample with GC-MS has been reported to be in the average of 1-2 hours (3, 4), which has been considered adequate by the clinicians.

The population covered by the different MS groups have been of the order of 0.5-1 million people (5, 18). The frequency of analyses requested has been 2-8/week. It is to be observed that the MS possibility has been utilized only or preferentially in cases of difficult diagnosis, perhaps some 10-20% of all cases in the region. Perhaps the populations covered could without disadvantage be extended to 2 millions if the service of the MS laboratory is adequately restricted. The time remaining on the MS equipment ought very advantageously to be used in other drug analyses, e.g. in therapy monitoring and narcotic addiction diagnostics, as well as in drug research work.

To sum up this section: There exists today an abundance of equipment and methodological possibilities and proposals for toxicology-MS. Several of them are already at the developmental stage and some are used routinely. Without doubt further developments in this direction are under way.

MS APPLICATION RESULTS

In this section we only want to show a few of our own examples which illustrate certain characteristic advantages of the MS methods. Similar or other illustrative examples have been presented (3, 4, 24, 55; and other references quoted below).

Fig. 1 shows how we could identify glutethimide with enough certainty in spite of the presence of a phthalate which co-eluted on GC. Fragments 149 and 104 characteristic of the phthalate are "contaminants" in the glutethimide mass spectrum, whose fragment

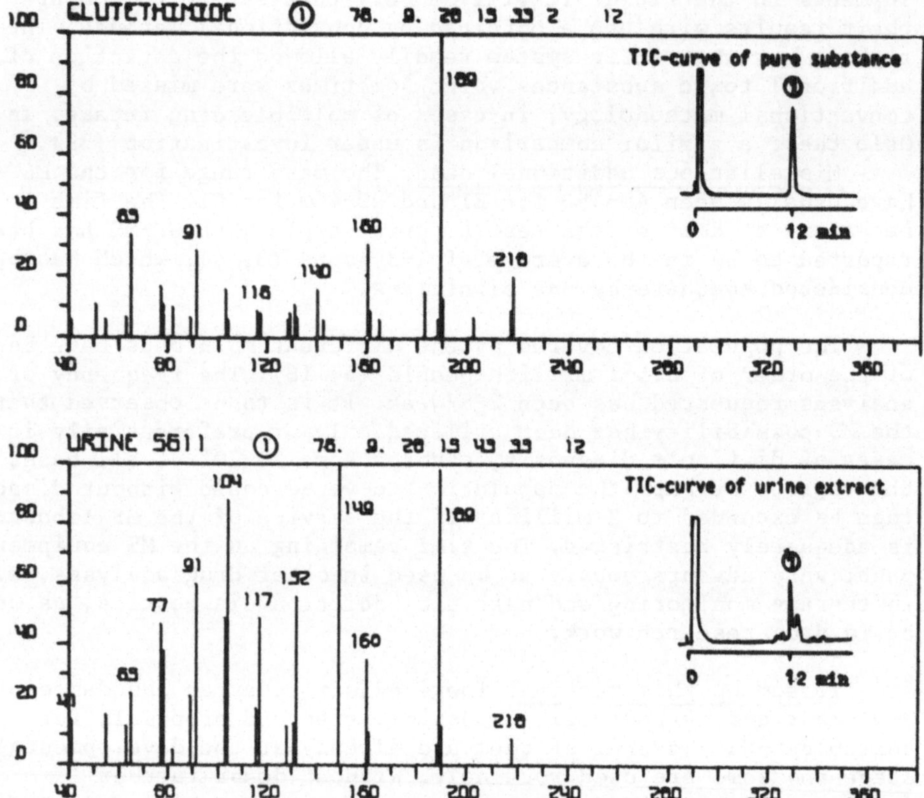

FIG. 1. Detection by MS of glutethimide in the presence of dibutyl phthalate in the sample (in the inlet, TIC curves).

pattern, however, is characteristic enough for diagnosis to be made. If only flame ionization detection had been used the levels of glutethimide would have been overestimated.

An example of the identification of more than one compound in the same spectrum was published by Billets et al. (32). Chlorpromazine, glutethimide and caffeine could be identified in the same mass spectrum which was obtained with DI-MS and EI. CI may facilitate such interpretations (4) because of the simpler MS and greater occurrence of molecular ions observed.

Fig. 2 shows how in a case of hexapropymate intoxication (a poisoning of serious nature, a total of 41 deaths registered in Sweden up to 1971) the urine did not contain any detectable amount of the original compound. Instead 4 of its metabolites were observed (at least one of them was a hydroxy-derivative (52)). In the serum (Fig. 3) hexapropymate itself, was found. If only a urine sample had been examined with non-MS-methods, the nature of the intoxication might have been missed or have taken a long time to individuate, particularly since at the time when this case was studied only little was known of the metabolism of hexapropymate. (From interpretations of the fragmentation patterns evidence for the presence of the metabolites in urine has been obtained in 3 further hexapropymate intoxication cases. They all had more or less of the same components while normal and patients with intoxications from other drugs did not have detectable amounts of them (52)).

The 2 examples presented above were examined before we had obtained our computer. In the following the merits of the latter component should be apparent. Fig. 4 shows a case with a composite multiple intoxication picture. The dots just under the curve line indicate repetitive scans taken automatically and collected by the computer. During the GC the apparently important scans were plotted out in a background-subtracted normalized form and exposed to interpretation including a library search. It is hard to see how non-MS-systems could have been so informative in an hour's time. Possibly with non-MS methods one had become satisfied to indicate one or two of the drugs, disregarding the presence of the other ones, with potential consequences for the treatment and for the interpretation of the case.

Without computer assistance the organisation of the scan material in Fig. 4 had indeed been very time-consuming. A further important advantage of the computer is its ability to present on a plotter or display in a few minutes MC of any fragment ions contained in the scan range. By selection of those ions characteristic of drugs suspected to be present in a case it is possible to "search and scan", i.e. to explore if a fragment occurs in the chromatogram, determine its R_t and roughly its concentration

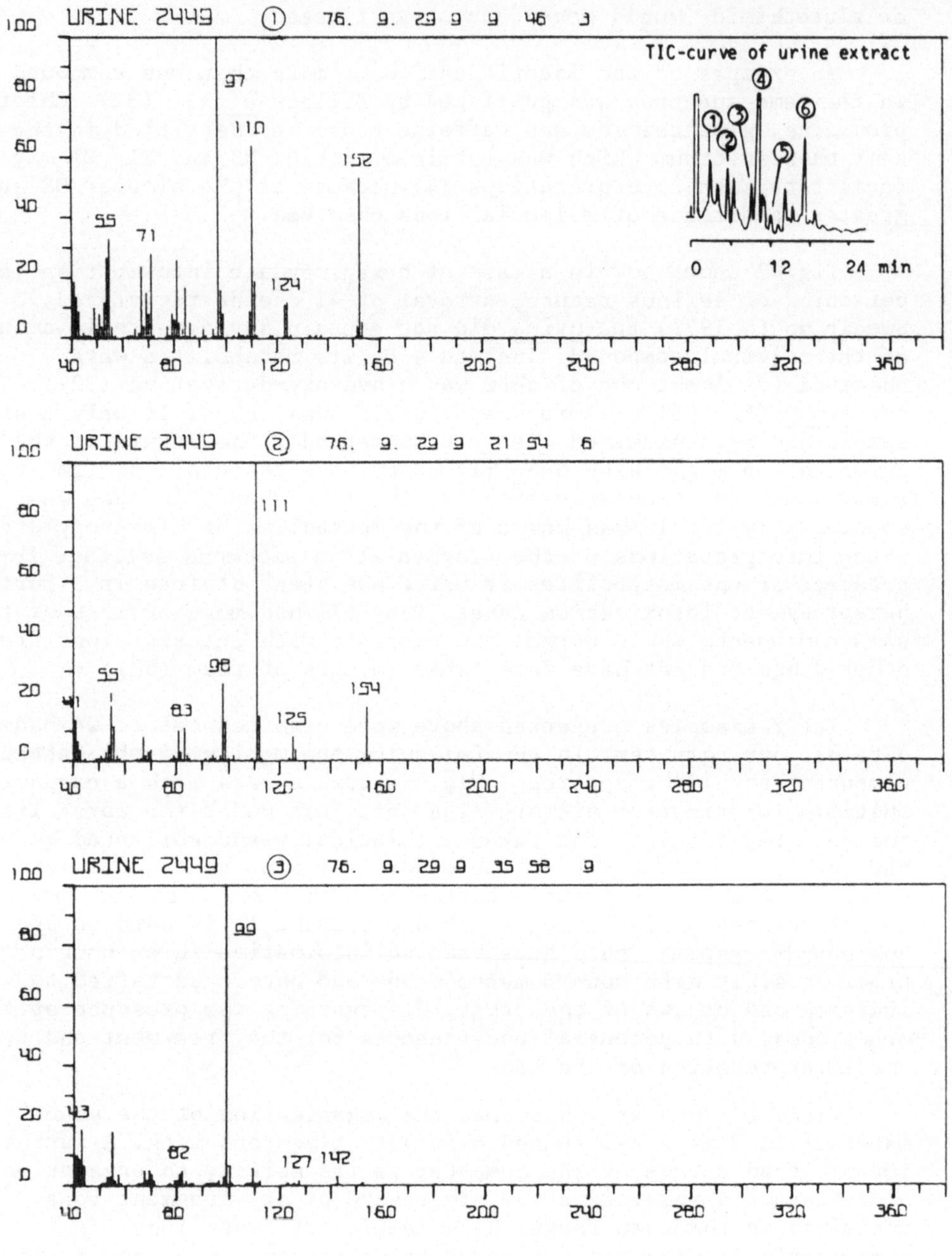
URINE 2449 ① 76. 9. 29 9 9 46 3
TIC-curve of urine extract
0 12 24 min
97
110
152
55
71
124
URINE 2449 ② 76. 9. 29 9 21 54 6
111
98
55
154
83
125
41
URINE 2449 ③ 76. 9. 29 9 35 58 9
99
43
82
127 142
40 80 120 160 200 240 280 320 360

FIG. 2.

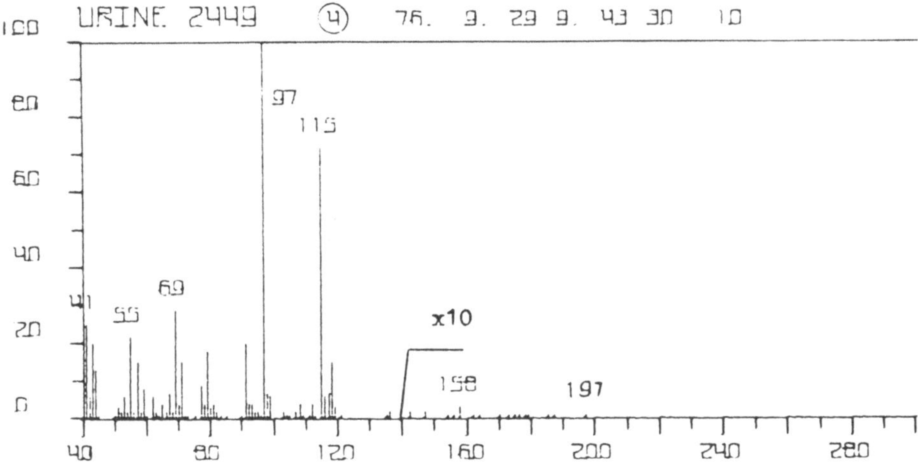

FIG. 2. (Contd.)
Detection by GC-MS of probable hexapropymate metabolites in the urine of a case of hexapropymate intoxication (in the inlet, TIC curves).
1. Hexapropymate metabolite I (R_t 2.9 min); 2. Metabolite II (R_t 5.7 min); 3. Metabolite III (9.5 min); 4. Metabolite IV (Hydroxy) (R_t 10.3 min); 5. Dibutyl phthalate (R_t 12.4 min); 6. Dioctyl phthalate (R_t 17.5 min).

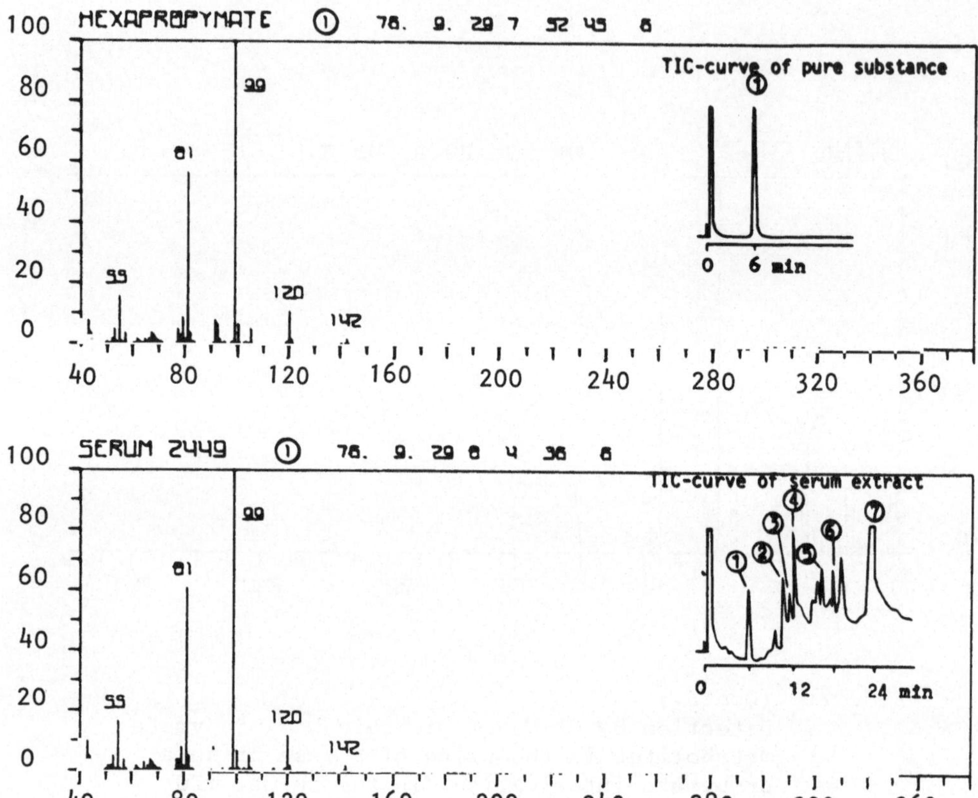

FIG. 3. Detection by GC-MS of hexapropymate in serum from the same cases as in Fig. 2. 1. Hexapropymate (R_t 6.1 min); 2. Methyl palmitate (R_t 11.1 min); 3. Dibutyl phthalate (R_t 12.3 min); 4. Methyl stearate (R_t 13.0 min); 5. Dioctyl phthalate (R_t 17.2 min); 6. Frequent, unidentified, contaminant (R_t 18.8 min); 7. Cholesterol (R_t 25.5 min).

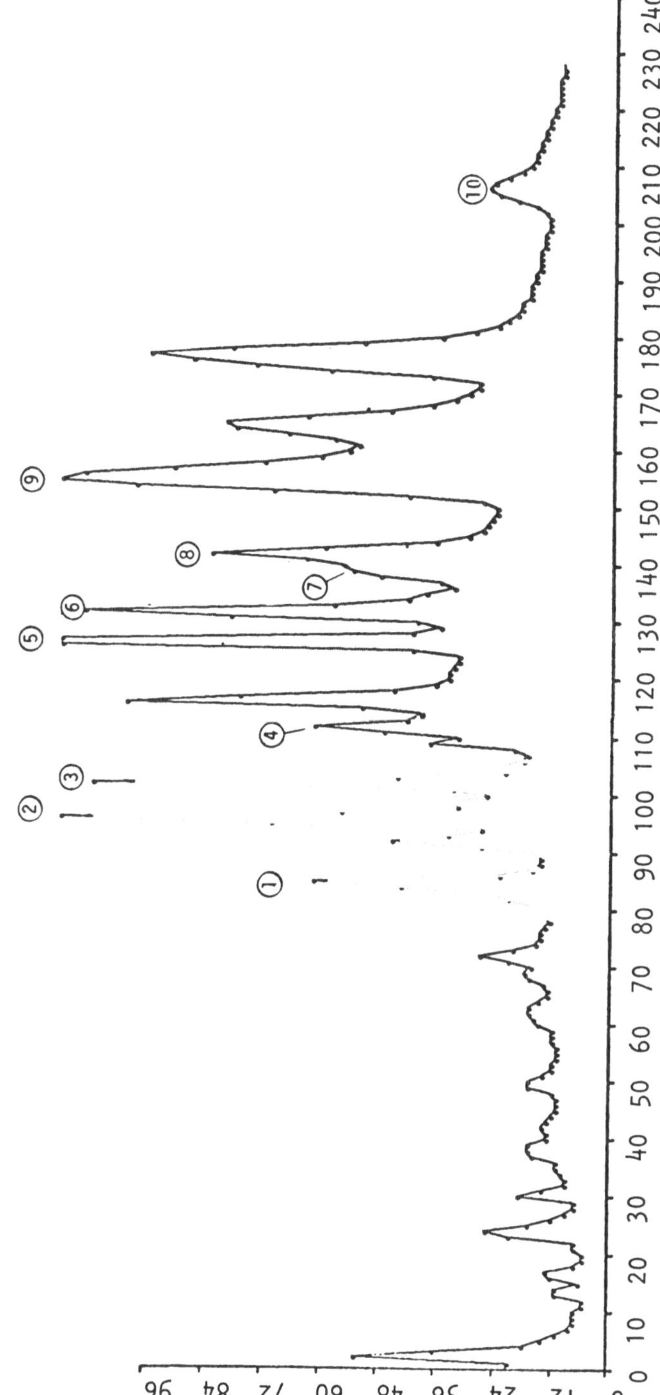

FIG. 4. TIC curve of urine extract in a case of multiple overdosage studied by repetitive scanning mass spectrometry. Drugs, metabolites and "artifacts" identified are: 1. Pentobarbital; 2. Diphenhydramine; 3. Dibutyl phthalate; 4. Methaqualone; 5. Dioctyl phthalate; 6. Codeine; 7. Methaqualone metabolite; 8. Frequent, unidentified, contaminant; 9. Thioridazine metabolite; 10. Cholesterol.

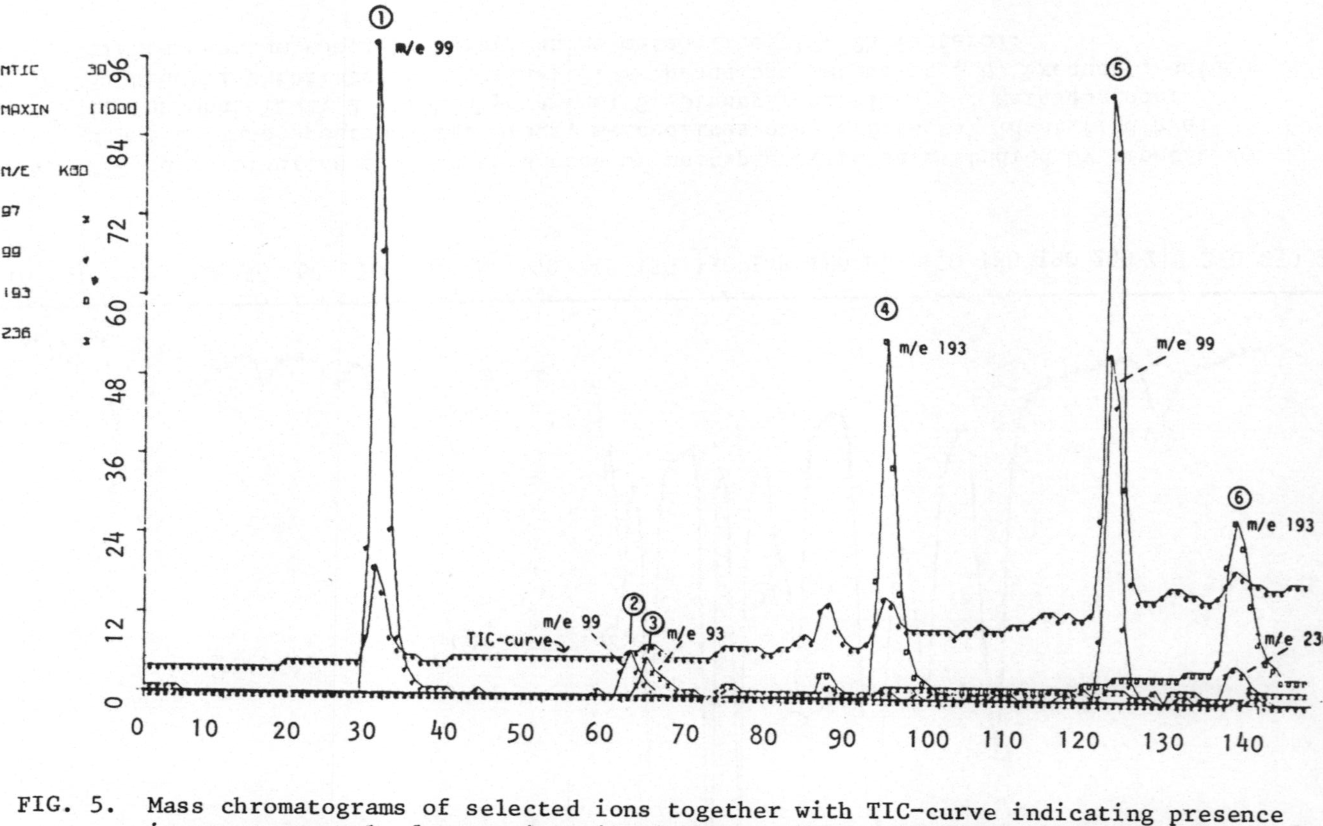

FIG. 5. Mass chromatograms of selected ions together with TIC-curve indicating presence in a serum sample from an intoxication case of hexapropymate and -metabolite and carbamazepine. 1. Hexapropymate; 2. Hexapropymate metabolite; 3. Hexapropymate metabolite; 4. Iminostilbene (pyrolysis product of carbamazepine); 5. Tri-2--butozyethyl phosphate (plasticizer in vacutainer); 6. Carbamazepine.

level, if then indicated, to order out a full mass spectrum at its R_t to be presented for closer identification. This may give peaks even in those regions of the TIC curve where no peaks are apparent. Enlightening examples have been given by Ehrenthal and Pfleger (37) and Ehrenthal et al. (56). We want to add an own experience in Fig. 5, where a combination of four selected characteristic ions (97, 99, 193 and 236) produced mass chromatograms of intoxication drugs partly much more expressive than the TIC chromatogram.

FINAL COMMENTS AND CONCLUSIONS

In this survey and analysis evidence has been given to show the need of MS analyses in clinical toxicology. These methods have been used for a long time in other fields and to a certain extent also in acute emergency cases. In scattered laboratories in the western world this fact is accepted and explorative studies and routine services have been developed over several years.

If 1) the treatment and the prognosis required in future in cases of acute intoxication should become more dependent on exact diagnosis of the substances involved, and 2) the MS-equipment should develop to a level where it is suitably automated and simple to manage so that personnel not specifically trained in MS can use it to provide a 24 hour high-quality comprehensive toxicological service, then it is likely that the clinicians, today a little cold-hearted, will welcome the support. And if the cost will be exorbitant the politicians may give their economic support, when they realize the gain in efficiency, without necessarily increasing the personnel, and the usefulness of the available spare time on the MS for therapy drug monitoring in routine, drug treatment, research work, etc. The trend in MS development today is such that the second requirement seems possible to fulfil in 5-10 years. Whether new antidotes or other specific treatments are discovered remains to be seen although immunological developments in such directions have been suggested. In the meantime it is important that existing units in the field continue their efforts. The increased knowledge obtained will aid the development of instrumentation and methods.

ACKNOWLEDGMENT

Arne Gustafson, MD, head of the medical intensive care unit of Lasarettet (the University Hospital of Lund), put the samples from the poisoning cases at our disposal.

REFERENCES

1) M.M. McCarron, C.B. Walberg and G.D. Lundberg, Clin. Chem., 1974, 20, 118.
2) T.B. Wiltbank, H.E. Sine and B.B. Brody, Clin. Chem., 1974, 20, 116.
3) N.C. Law, Final Report, Suburban Hospital Ass. (Stencil copy), 1974.
4) R.L. Foltz, P.A. Clarke, D.A. Knowlton and J.R. Hoyland, Research report, Batelle, Columbus Laboratories (Stencil copy), 1974.
5) C. Costello, Personal communication, 1975.
6) C. Fenselau, Personal communication, 1975.
7) M.H. Bickel, Int. J. Clin. Pharmacol. Ther. Toxicol., 1975, 11, 145.
8) D.E. Green, F.-C. Chao, K.O. Loeffler and R.A. Lemmon, Fed. Proc. Fed. Am. Soc. Exp. Biol., 1976, 35, 212.
9) D.T. Teitelbaum, J. Morgan and G. Gray, Clin. Toxicol., 1976, 9, 29.
10) R.W. Newton, in "The Poisoned Patient: the role of the laboratory", Ciba Foundation Symposium 26 (new series), ASP, Amsterdam, 1974, p. 14.
11) Ciba Foundation Symposium 26 (new series), "The Poisoned Patient: the role of the laboratory", Elsevier-Excerpta Medica - North Holland, Amsterdam, 1974.
12) E.C. Dinovo and L.A. Gottschalk, Clin. Chem., 1976, 22, 843.
13) H.-O. Malmlund, Socialstyrelsens kommitté för läkemedelsinformation, Symp. mars 1976, nr 1, Läkemedlens plats inom psykiatrin, 1976.
14) A.B. Baker, Med. J. Aust., 1969, 497.
15) H.A. Lee and A.C. Ames, Brit. Med. J., 1965, 1, 1217.
16) J.A. Vale, A.J. Rees, B. Widdop and R. Goulding, Brit. Med. J., 1975, 1, 5.
17) A.P. Douglas, A.N. Hamlyn and O. James, Lancet, 1976, i, 111.
18) N.C. Law, Personal communications, 1974, 1975, 1976.
19) L. Abrahamsson and A.M. Wassén, in "Methods in Clinical Chemistry", M. Roth, Ed., S. Karjer, Basel, 1970.
20) B.S. Finkle, D.M. Taylor and E.J. Cherry, J. Chromatogr. Sci., 1971, 9, 383.
21) L. Abramhamsson and A.M. Wassén, Scand. J. Clin. Lab. Invest., 1972, 29, Suppl. 126, abstract No. 13.7.
22) B. Widdop, in "The Poisoned Patient: the role of the laboratory", Ciba Foundation Symposium 26 (new series), Elsevier, Amsterdam, 1974.
23) B. Holmstedt, Läkartidningen, 1971, 68, 57.
24) C.E. Costello, H.S. Hertz, I. Sakai and K. Biemann, Clin. Chem., 1974, 20, 255.
25) C. Fenselau, Appl. Spectros., 1974, 28, 305.
26) J. Roboz, Adv. Clin. Chem., 1975, 17, 109.
27) A.M. Lawson, Clin. Chem., 1975, 21, 803.

28) A.M. Lawson and G.H. Draffan, Progr. Med. Chem., 1975, 12, 1.
29) B.S. Finkle, R.L. Foltz and D.M. Taylor, J. Chromatogr. Sci., 1974, 12, 304.
30) B. Holmstedt and J.E. Lindgren, in "The Poisoned Patient: the role of the laboratory", Ciba Foundation Symposium 26 (new series), ASP, Amsterdam, 1974, p. 105.
31) N.C. Law, H.M. Fales and G.W.A. Milne, Clin. Toxicol., 1972, 5, 17.
32) S. Billets, J. Carruth, N. Einolf, R. Ward and C. Fenselau, Johns Hopkins Med. J., 1973, 133, 148.
33) U. Boerner, S. Abbott, J.C. Eidson, C.E. Becker, H.T. Horio and K. Loeffler, Clin. Chim. Acta, 1973, 49, 445.
34) F.P. Abramson, Anal. Chem., 1975, 47, 45.
35) M.E. Caplis and L.S. Eichmeyer, Clin. Toxicol., 1976, 9, 15.
36) E.M. Sellers, S.M. MacLeod, B. Kapur, J. Marshman and C. Stapleton, Clin. Chem., 1975, 21, 973.
37) W. Ehrenthal and K. Pfleger, Varian Mat Publication, 1975, Nr 19.
38) A. Zune, P. Dobberstein, K.H. Maurer and U. Rapp, Varian Mat Publication 20, 1975.
39) J.E. Pettersen, Klin-Kem avd, Rikshospitalet, Oslo 1975, Personal communication.
40) R. Bonnichsen, C.G. Fri, B. Hedfjäll and B. Ryhage, Z. Rechtsmed., 1972, 70, 150.
41) G. Brante, Läkartidningen, 1976, 73, 631.
42) I. Szabo, To be presented at the 7th Int. Mass Spec. Conf., Florence, Italy, 1976-08-30 to 1976-09-03, Manuscript, 1976.
43) J. Sunner and I. Szabo, To be presented at the 7th In. Mass Spectrometry Conference, Florence, Italy, 1976-08-30 to 1976-09-03, Manuscript, 1976.
44) G.W.A. Milne, H.M. Fales and T . Axenrod, Anal. Chem., 1971, 43, 1815.
45) LKB Application sheet 197, 1975.
46) D.I. Caroll, L. Dzidic, R.N. Stillwell, K.D. Haegele and E.C. Horning, Anal. Chem., 1975, 47, 2369.
47) D. Schuetzle, Biomed. Mass Spectrom., 1975, 2, 288.
48) F.W. McLafferty, R.H. Hertel and R.D. Villwook, Org. Mass Spectrom., 1974, 9, 690.
49) W.H. Mc Fadden, in "Techniques of combined gas chromatography/ mass spectrometry: applications in organic analysis", John Wiley & Sons, New York, 1973.
50) A. Gustafson, Intoxikationer vårdade å MIVA 1973 och 1974. (Intoxications treated in the intensive care medical unit, University Hospital of Lund, 1973 and 1974). Personal communication, 1975.
51) Varian Mat, Product-Information Application no. 5, 1975.
52) G. Brante, S. Jönsson and A. Andersson, Unpublished results.
53) R.C. Baselt, J.A. Wright and R.H. Cravery, Clin. Chem., 1975, 21, 44.

54) C.L. Winek, Clin. Chem., 1976, 22, 832.
55) R.L. Foltz, in "Advances in mass spectrometry", Vol. 6, Appl. Science Publ., England, 1974.
56) W. Ehrenthal, K. Pfleger and M. Möller, Varian Mat Publication, No 22, 1975.
57) J.Th. Watson, F.C. Falkner and B.J. Sweetman, Biomed. Mass Spectrom., 1974, 1, 156.
58) Shipe, J.R., and Savory, J. (1976): A critical evaluation of the contributions of gas chromatography-mass spectrometry to emergency toxicology service. Clin. chem. 22:1198.
59) Smith, Th.W., Haber, E., Yetman, L., and Butler, V.P. (1976): Reversal of advanced digoxin intoxication with F_{AB} fragments of digoxin-specific antibodies. New Engl. J. Med. 294:797-800.
60) Ullucci, P.A., Cadoret, R., Stasiowski, P.D., and Martin, H.F. (1976): A GC/MS drug screening procedure in current use. Clin. Chem. 22:1197.

PARTICIPANTS

- A. ASSANDRI - Lepetit S.p.A., Laboratorio Farmacocinetica, Via Durando 38, Milano, Italy.
- P. BACCHIN - Snam Progetti, Monterotondo (Roma), Italy.
- M.K. BALDWIN - Shell Research Limited, Tunstall Laboratory Sittingbourne Research Centre, Sittingbourne, Kent, U.K.
- A.S. BELOUSOV - Director Botkina Hospital, Moscow, USSR.
- G. BELVEDERE - Istituto Ricerche Farmacologiche "Mario Negri", Via Eritrea 62, Milano, Italy.
- A. BENAKIS - Chief of Laboratory of Drug Metabolism, Dept. of Pharmacology, Univ. of Geneva, 20 rue de l'Ecole de Médicine, 1211 Geneva, Switzerland.
- L. BERGSTEDT - LKB Produkter AB, Box 76, S-161 25 Bromma 1, Sweden.
- E. BERETTA - Gruppo Lepetit S.p.A., Via Durando 38, Milano, Italy.
- P.A. BIONDI - Istituto Fisiologia Veterinaria e Biochimica, Via Celoria 10, Milano, Italy.
- N.O. BODIN - Astra Pharmaceuticals AB, Toxicology Laboratories, S-151 85 Södertälje, Sweden.
- G. BONOMI - Industrie Pirelli S.p.A., Viale Sarca 202, Milano, Italy.
- K.O. BORG - AB Haessle, Fack, S-431 20 Moelndal, Sweden.
- H. BRANDENBERGER - Institute of Forensic Medicine, Univ. of Zürich, Zürichbergstrasse 8, CH-8032 Zürich, Switzerland.
- G. BRANTE - Clin. Chem. Laboratory, S:t Lars sjokhus, 22006 Lund, Sweden.
- L. BRESCHIGLIARO - Clinica Medica I, Università di Padova, Policlinico, Padova, Italy.
- P. BRUEGGER - F.Hoffmann-La Roche & Co. A.G., Grenzacherstrasse 124, CH-4052 Basel, Switzerland.
- V. BOZZATTA - Alfa Farmaceutici, Via Ragazzi del '99, n.5, Bologna, Italy.
- M. CAGNASSO - Istituto Fisiologia Veterinaria e Biochimica, Via Celoria 10, Milano, Italy.
- A. CANUTI - Laboratorio Chimico Provinciale di Igiene e Profilassi, Via S.Maria in Betlem V, Cremona, Italy.
- F. CANEPARI - V.le Cirene 7, Milano, Italy.
- L. CAPPELLINI - Istituto Ricerche Farmacologiche "Mario Negri", Via Eritrea 62, Milano, Italy.

- M. CASTEGNARO - International Agency for Research on Cancer, 150 Cours Albert Thomas, 69008 Lyon, France.
- F. COLUCCI D'AMATO - Società Italiana di Neurochimica, Via S. Pasquale 55, Napoli, Italy.
- R.A. COOMBS - Sandoz Inc., Rt.10, East Hanover, N.J. 07936, USA.
- E.L. CRAMPTON - The Boots Co.Ltd., Research Department, Pennyfoot St., Nottingham, U.K.
- C. CREMONESI - Istituto Ricerche Farmacologiche "Mario Negri", Via Eritrea 62, Milano, Italy.
- A. CRUYL - Lab. Medische Biochemie, Klinische Analyse, Rijksuniversiteit, Akademisch Ziekenhuis, De Pintelaan 135, 9000 Gent, Belgium.
- F. CUGURRA - Istituto di Farmacologia e Farmacognosia,Università di Genova, Via al Capo di S.Chiara 5, Genova, Italy.
- A. DE LEENHEER - Lab. Medische Biochemie, Rijksuniversiteit Gent, De Pintelaan 135, 9000 Gent, Belgium.
- A.M. DE MASSARI - Istituto di Medicina Legale e delle Assicurazioni, Policlinico di Borgo Roma, Verona, Italy.
- G. DENTI - Università della Calabria, Dipartimento di Chimica, Arcavacata di Rende (Cosenza), Italy.
- A. DE PASCALE - Istituto Ricerche Farmacologiche " Mario Negri", Via Eritrea 62, Milano, Italy.
- G. DIJKSTRA - State University of Utrecht, Analytical Chemical Laboratory, Croesestr. 77A, Utrecht, The Netherlands.
- H. DOLEZALOVA - University of Connecticut, U-154 Storrs, Connecticut 06268, USA.
- B. DONOVAN - St.Mary's Hospital, Medical School, Dept. of Biochemistry, London W2 1PG, U.K.
- L.G. DRING - Department of Biochemistry, St. Mary's Hospital, Medical School, London W2 1PG, U.K.
- S. DURU - University of Hacettepe, Faculty of Pharmacy, Sihniye, Ankara, Turkey.
- W. EHRENTHAL - Institute of Pharmacology and Toxicology, Univ. of the Saarland, D-665 Homburgh/Saar, West Germany.
- H. EHRSSON - Karolinska Institute, Fack, 10401 Stockholm 60, Sweden.
- L.F. ELSOM - Dept. of Metabolism, Huntingdon Research Centre, Huntingdon, Cambs, U.K.
- Y.A. ELTEKOV - Institute of Physical Chemistry, Leninskii prospect 31, Moscow B71, USSR.
- R. FANELLI - Istituto Ricerche Farmacologiche "Mario Negri", Via Eritrea 62, Milano, Italy.
- C. FARINA - ISF, Italseber S.p.A., Via L. Da Vinci 1, Trezzano Sul Naviglio, Milano, Italy.
- A. FORGIONE - Istituto Ricerche C. Erba, Via Imbonati 24, Milano, Italy.
- A. FRIGERIO - Istituto Ricerche Farmacologiche "Mario Negri", Via Eritrea 62, Milano, Italy.
- J. GAL - Dept. of Pharmacology, School of Medicine, Center for Health Sciences, University of California, Los Angeles, California 90024, USA.
- S. GARATTINI - Istituto Ricerche Farmacologiche "Mario Negri", Via Eritrea 62, Milano, Italy.

- E. GHISALBERTI - Department of Organic Chemistry, University of W.A., Nedlands, 6009, Western Australia.
- B. GIOIA - Farmitalia, Via dei Gracchi 35, Milano, Italy.
- P. GIULIDORI - Zambon Research Laboratories, Via L. del Duca 10, Bresso (Milano), Italy.
- A.G. GIUMANINI - Centro di Spettrometria di Massa, Università Via S. Giacomo 5, Bologna, Italy.
- E. GRASSI - Zambeletti S.p.A., Via Zambeletti, Baranzate di Bollate (Milano), Italy.
- H. GREENWOOD - Department of Chemical Pathology, St. Bartholomew's Hospital, 51-53 Bartholomew Close, London EC1, U.K.
- B. GRIBAUDI - Eulo, Facoltà di Medicina, Via Valsabbina 19, Brescia, Italy.
- L.A. GRIFFITHS - University of Birmingham, Dept. of Biochemistry, Birmingham 13, U.K.
- P. HADJIEVA - Department of Chemistry, University of Sofia, Bul.Anton Ivanov 1, 1126 Sofia, Bulgaria.
- K.D. HAEGELE - University of Texas,Health Science Center, Dept. of Pharmacology, San Antonio, Texas 78284, USA.
- R. HAENNI - F.Hoffmann-La Roche & Co., A.G., Grenzacherstrasse 124, CH-4052 Basel, Switzerland.
- H. HAMBÖCK - Ciba-Geigy SA, R - 1061.417 Basel, Switzerland.
- D.J. HARVEY - University of Oxford, Department of Pharmacology, South Parks Road, Oxford, OX1 3Q7, U.K.
- C. HIGNITE - V.A. Hospital, 4801 Linwood Blvd., Kansas City, Missouri 64128, USA.
- K.-J. HOFFMANN - AB Haessle, Fack, S-431 20 Moelndal, Sweden.
- E.C. HORNING - Institute for Lipid Research, Baylor College of Medicine, Houston, Texas 77030, USA.
- M.G. HORNING - Institute for Lipid Research, Baylor College of Medicine, Houston, Texas 77030, USA.
- M. ISHIBASHI - Analytical Instrument Div., Geboow 105, Schiphol-Oost, The Netherlands.
- M. KLEIN - Research Laboratory, 7704 Old Springhouse Rd., McLean, Virginia 22101, USA.
- T. KUHARA - Kurume University School of Medicine, 67 Asahi-machi, Kurume-shi, Fukuoka-ken, Japan.
- M. JARMAN - Institute of Cancer Research, Block F, Clifton Avenue, Sutton, Surrey SM2 5PX, U.K.
- G. LANDI - Via Pezzotti 63, Milano, Italy.
- Y. LANGOURIEUX - Cen-Saclay, B.P. n° 2, Service des Molécules Marquées, 911900 Gif-sur-Yvette, France.
- J. LANZONI - Istituto Ricerche Farmacologiche "Mario Negri", Via Eritrea 62, Milano, Italy.
- R.M. LEE - Research Institute Smith, Kline & French Labs.Ltd., Welwyn Garden City, Herts, U.K.
- B. LEGG - Beecham Research Laboratories, Fourth Avenue, Harlow, Essex, U.K.
- W.D. LEHMANN - Institute of Physical Chemistry, University of Bonn, 5300 Bonn, Wegelerstr. 12, West Germany.
- A. LEONARDI - Istituto Ricerche Farmacologiche "Mario Negri", Via Eritrea 62, Milano, Italy.
- O.R.W. LEWELLEN - Drug Metabolism Department, Research Institute May & Baker Ltd., Dagenham, Essex, U.K.

- G. LHOEST - Université Catholique de Louvain, Ecole de Pharmacie, 73 Av. Emanuel Mounier, 1200 Bruxelles, Belgium.
- J. LIEHR - Ciba-Geigy AG, R 1062.204, CH-4002 Basel, Switzerland.
- D.C.K. LIN - Battelle Memorial Institute, 505 King Ave. Columbus, Ohio 43201, USA.
- C. LINDBERG - Organic Pharmaceutical Chemistry, University of Uppsala, Box 574, 752 63 Uppsala, Sweden.
- J.-E. LINDGREN - Astra Läkemedel AB, 15185 Södertälje, Sweden.
- A. LONGO - Tecnofarmaci S.p.A., Pomezia (Roma), Italy.
- J.W. LOWN - University of Alberta, Department of Chemistry, Edmonton, Alberta T6 G2 G2, Canada.
- R. MAFFEI FACINO - Chimica Farmaceutica e Tossicologica, Università di Milano, Viale Abruzzi 42, Milano, Italy.
- G.H. MAHRAN - University of Cairo, Egypt, Faculty of Pharmacy, Kasr.El-Aini, Cairo, Egypt.
- P. MANITTO - Istituto Chimica Organica, Università di Milano, Milano, Italy.
- P. MARTELLI - Istituto Biochimico Italiano, Via Lorenzini 2, Milano, Italy.
- I. MATSUMOTO - Kurume University, School of Medicine, 67 Asahi-machi,Kurume-shi, Fukuoka-ken, Japan.
- K.H. MAURER - Varian Mat GmbH, D-2800 Bremen, West Germany.
- E. MELILLO - 75 Young Street, Welland, Canada.
- G. MELLERIO - Centro Spettrometria di Massa, Università degli Studi di Pavia, Via Taramelli 10, Pavia, Italy.
- F. MENEZ - Service Scientifique, Ambassade de France, Piazza Farnese, Roma, Italy.
- T. MEYER - National Institute of Forensic Toxicology, Sognasvannsveien 28, Oslo 3, Norway.
- F. MIKES - Pharmacological Institute, University of Zürich, Gloriastr. 32, 8006 Zürich, Switzerland.
- B.J. NAHRUNG - Box 54 Post Office, Mossman, Queensland 4873, Australia.
- M.G. NEALE - Fisons Ltd., Pharmaceutical Division, Bakewell Road, Loughborough, Leicestershire, LE11 0QY, U.K.
- P. NEGRINI - Istituto Ricerche Farmacologiche "Mario Negri", Via Eritrea 62, Milano, Italy.
- S. NELSON - Laboratory of Chemical Pharmacology, NIH, Bethesda, Md. 20014, USA.
- V. NICOLELLA - Farmitalia, Via dei Gracchi 35, Milano, Italy.
- C. NICOTRA - C.E.A.-CEN Grenoble, Scag-Eapc, BP 85X, 38041 Grenoble, Cedex, France.
- M. NORDQVIST - Dept. of Pharmacognosy, Faculty of Pharmacy, Box 579, S-75123 Uppsala, Sweden.
- A. OHLSSON - Dept. of Pharmacognosy, Faculty of Pharmacy, Box 579, S-75123 Uppsala, Sweden.
- R.A. OKERHOLM - Parke, Davis & Company, 2800 Plymouth Road, Ann Arbor, Michigan 48106, USA.
- J.S. OLIVER - Glasgow University, Dept. Forensic Medicine, Glasgow G12 8QQ, U.K.
- C. PANTAROTTO - Istituto Ricerche Farmacologiche "Mario Negri", Via Eritrea 62, Milano, Italy.
- G.L. PASSETTI - Zambeletti S.p.A., Via Zambeletti, Baranzate di Bollate (Milano), Italy.

- D.W. PAYLING - Fisons Ltd., Pharmaceutical Division, Bakewell Road, Loughborough, Leicestershire, LE11 0QY, U.K.
- E. PELLA - Istituto Carlo Erba per Ricerche Terapeutiche, Via Imbonati 24, Milano, Italy.
- G. PIFFERI - ISF - Italseber S.p.A., Via L. da Vinci 1, Trezzano Sul Naviglio (Milano), Italy.
- D. PITRÈ - Bracco Industria Chimica, Via E. Folli 50, Milano, Italy.
- A. PREMOLI - Gruppo Lepetit, Via per Rescalda 34, Gerenzano (Varese), Italy.
- P. PROFUMO - LKB Strumenti S.p.A., Via G.B. Morgagni 30/E, Roma, Italy.
- J.-C. PROME' - Centre de Recherche de Biochimie et Génetique du CNRS, 118 route de Narbonne, 31400 Toulouse, France.
- A. PROX - Dr. Karl Thomae GmbH, Birkendorfer Strasse 65, 7950 Biberach an der Riss, Germany.
- M. PRZYBYLSKI - Johannes Gutenberg-Universität, Institut für Organische Chemie, D-65 Mainz, Johann-Joachim-Becher-Weg 18-20, Germany.
- I. RACZ - Pharmaceutical Institute of the Med.University of Budapest, 1165.Budapest, XXXIV.u.3., Hungary.
- RAYNAUD - Rhone-Poulenc Industries, 13 quai Jules Guesde, (94) Vitry S/Seine, France.
- H. RINGSDORF - Johannes Gutenberg-Universität, Institute für Organische Chemie, D-65 Mainz, Johann-Joachim-Becher-Weg 18-20, Germany.
- J. ROBOZ - Mount Sinai School of Medicine, 11 East 100th St., New York City, N.Y. 10029, USA.
- R. RONCUCCI - Continental Pharma Research Laboratories, 30, st.op Haacht, B-1830 Machelen (Bt), Belgium.
- E. ROSSI - Istituto Ricerche Farmacologiche "Mario Negri", Via Eritrea 62, Milano, Italy.
- L. ROSSI - Istituto Ricerche Farmacologiche "Mario Negri", Via Eritrea 62, Milano, Italy.
- J.J. RYAN - Health Protection Branch, Tunney's Pasture, Ottawa, Canada KIA 0L2.
- E. SACCHETTI - Clinica Psichiatrica dell'Università, Via F. Sforza, Milano, Italy.
- H.-R. SCHULTEN - Institute of Physical Chemistry, University of Bonn, 5300 Bonn, Wegelerstr. 12, West Germany.
- M. SENN - Boehringer Mannheim, 68 Mannheim 31, Sandhoferstrasse 112, Germany.
- C. SIGNORINI - E.U.L.O., Facoltà di Medicina, Via Valsabbina 19, Brescia, Italy.
- J.A. STEINBORN - UCLA, Pharmacology Department, Los Angeles, California 90066, USA.
- M. STEPITA-KLAUCO - University of Connecticut, U-154, Storrs, Connecticut 06268, USA.
- F. SUGNAUX - Laboratory of Drug Metabolism, Department of Pharmacology, University of Geneva, 20 Bd. d'Yvoy, 1211 Geneva, Switzerland.
- H.R. SULLIVAN - The Lilly Research Laboratories, P.O.Box 618 Indianapolis, Indiana 46206, USA.
- J. TASKINEN - Orion Pharmaceutical Co., Box 10019, SF-00101 Helsinki 10, Finland.

- H. TAUSCH - Österr. Studienges. F. Atomenergie, A 1081 Wien, Lenaugasse 10, Austria.
- G. TEAL - I.C.I. Plant Protection Division, Jealotts Hill Research Station, Bracknell, Berks, U.K.
- J.-P. THENOT - Baylor College of Medicine, Institute for Lipid Research, 1200 Moursund, Houston, Texas 77030, USA.
- A. TREBBI - Zambeletti S.p.A., Via Zambeletti, Baranzate di Bollate (Milano), Italy.
- D. TRIPIER - Hoechst AG., Pharma-Synthese, Frankfurt (M) 80, Postfach 80-03-20, Germany.
- TUONG CHI CUONG - Parcor, 195 Route d'Espagne, 31024 Toulouse, France.
- U. VALCAVI - Istituto Biochimico Italiano, Via Lorenzini 2, Milano, Italy.
- G. VAN LEAR - Department of Pharmacodynamics and Analytical Research, Lederle Labs., Pearl River, N.Y. 10965, USA.
- D.P. VENTER - Department of Pharmacology, Potchefstroom University, Potchefstroom 2520, South Africa.
- P. VENTURA - ISF - Italseber, Via Leonardo da Vinci 1, Trezzano Sul Naviglio (Milano), Italy.
- N.P.E. VERMEULEN - Department of Pharmacology, Gorleus Laboratories, State University, P.O.Box 75, Leiden, The Netherlands.
- J. VINK - Organon International B.V., Drug Metabolism R & D Laboratories, Kloosterstraat 6, Oss-4202, The Netherlands.
- R. VOGES - Sandoz AG, Biopharmaceutical Department, CH-4002 Basel, Switzerland.
- A. VON HODENBERG - Gödecke Research Institute, Biochemistry Department, 78 Freiburg, Postfach 569, Germany.
- R.Z. WALCZYNSKI - Illinois Mas.med. Center, 9856 Lauren Lane, Niles, Illinois 60648, USA.
- R. WALKER - Int.Agency for Res. Cancer, 150 Cours A.Thomas, 69008 Lyon, France.
- R.J. WARD - Division of Clinical Chemistry, MRC, Clinical Research Centre, Harrow, Middlesex HA1 3UJ, U.K.
- R.T. WILLIAMS - St.Mary's Hospital Medical School, Department of Biochemistry, London W2 1 PG, U.K.
- WOLDE-AB YISAK - Uppsala University, Biomedicum Box 579, Uppsala, Sweden.
- R.E. WOLFF - Institut de Chemie, 1 rue Blaise Pascal, B.P. 296 R/8, 67008 Strasbourg, France.
- S.E. YOUNG - S.E. Young Research Laboratories Ltd., 250 Adelaide St. West, Toronto, Ont. M5H 1X6, Canada.
- M. ZANOL - Istituto Ricerche Farmacologiche "Mario Negri", Via Eritrea 62, Milano, Italy.
- F. ZEN - Clinica Medica I, Università di Padova, Policlinico, Padova, Italy.
- L.F. ZERILLI - Laboratori Ricerche Lepetit, Via Durando 38, Milano, Italy.

INDEX